ENDURING RECORDS
THE ENVIRONMENTAL AND CULTURAL
HERITAGE OF WETLANDS

Edited by Barbara A. Purdy

Oxbow Books

Published by
Oxbow Books, Park End Place, Oxford OX1 1HN

© Oxbow Books and the individual authors, 2001

ISBN 1 84217 048 1

A CIP record for this book is available from The British Library

This book is available direct from
Oxbow Books, Park End Place, Oxford OX1 1HN
(Phone: 01865–241249; Fax: 01865–794449)

and

The David Brown Book Company
PO Box 511, Oakville, CT 06779, USA
(Phone: 860–945–9329; Fax: 860–945–9468)

and

via our website
www.oxbowbooks.com

Printed in Great Britain at
Information Press, Oxford

Dedicated to
Laurence H. Purdy

in appreciation
for his unselfish contributions of time and knowledge

Contents

Preface

Oral versions of the papers in this volume were presented at an international wetlands archaeology conference held in Gainesville, Florida, December 1–5, 1999. The theme of the conference was: The Significance of the Survival of Organic Materials from Archaeological Contexts. Individuals from seventeen countries spoke about shipwrecks, bog bodies, cenotes of sacrifice, art styles, perishable technologies, palynology, wetlands management, conservation methods, and updates on famous sites. Time periods ranged from the early Pleistocene to a few hundred years ago.

Sometimes water-saturated conditions in wetland areas entomb and preserve an entire array of organic remains in a nearly pristine state. These materials provide fresh views of ancient human groups and their behavior. They can be used to draw conclusions about activities that took place at upland sites where the organic component did not survive. And they can be compared with sites where preservation has occurred because conditions have remained continuously frozen or continuously dry. No location will contain all information of interest and that is why it is important to examine a variety of sites representing different cultural, geographic, temporal, and environmental settings.

Wetlands are often altered or destroyed before even a representative sample can be studied for cultural or environmental content, antiquity, and significance. Throughout many parts of the world, wetlands and the heritage they contain have been lost through development, shrinkage, because of agricultural use, and reduction through peat mining activities. For example, because of the huge population influx into Florida and because of Florida's extensive waterways, the state's wetlands are particulary vulnerable. Many people recognize the importance of wetlands in filtering out pollution, serving as natural brakes for flood control, reducing soil erosion, and providing critical habitat for wildlife. They do not realize that the heritage entombed in Florida wetland areas includes: (1) a 50,000-year record of Florida's environment, (2) more than 12,000 years of Florida's cultural history, (3) about 300 ancient canoes, one of which is 6000 years old, (4) prehistoric art objects

of wood and bone preserved only because they became buried in water-saturated deposits, (5) human skeletal remains that reveal age and cause of death, sex, pathologies, nutrition, and DNA in brain tissue, (6) more than 100 species of animals and more than 100 species of plants that lived in Florida thousands of years ago. At upland sites, greater than ninety-five percent of this information does not survive. Also, most people do not understand that, while artificial wetlands may replace the functions of natural ones, an artificial wetland cannot replace environmental and cultural history.

In an effort to bring the plight of wetland environments to public awareness, it might be germane to envision how people would react if, in a synchronous act of terrorism, dozens of major museums were blown up throughout Europe and the Mediterranean and hundreds of magnificent statues and other irreplaceable works of art were destroyed. Such a heinous deed would horrify the entire world and tear at the very heart of Western civilization. It would be virtually impossible to calculate the impact of the loss of these treasures. Let us hope the above scenario remains fictitious.

Yet, here is a true story. Almost daily, the environmental and cultural heritage entombed in wetlands is destroyed by ambitious dredging, draining, and development projects. It cannot be replanted, transplanted, or reproduced. Once disturbed, the record perishes forever. Did anyone take a picture? This is not rhetoric. What *is* rhetoric are statements by world leaders that the environment is the number one global priority for the twenty-first century without committing the wherewithal for its protection.

The authors of the papers in this volume furnish examples, discuss solutions, and make recommendations about the retrieval, management, preservation, and longterm curation of fragile materials that have survived only because of rare burial conditions. Hopefully, this publication will lead to a greater understanding by government, developers, and the public of the cultural and environmental heritage that is lost when wetlands are modified.

I originally hoped that topics would include sites where

materials had remained dry or frozen in addition to those that were continuously waterlogged. Obviously, the locations of sites where human remains, artifacts, and environmental information stayed dry or frozen throughout the centuries are not the same as areas with waterlogged components. Since climate and geography often dictate the kinds of items that were produced by people in the past, it should be revealing to compare the similarities and differences between the surviving cultural items from these diverse environments. Perhaps these issues will be addressed at a future conference.

Because of the great diversity of topics, time periods, and geographic areas covered, it was difficult to determine a logical order for the placement of the papers in the volume. I finally settled on the following arrangement: (1) Florida archaeology, (2) archaeology of the Americas other than Florida, (3) archaeology of the rest of the world, (4) wetland environments and management, and (5) conservation methods.

Acknowledgments
The 1999 International Wetlands Archaeology Conference was financially support through grants from: Florida Department of State, Wenner-Gren Foundation for Anthropological Research, St. Johns River Water Management District, and the University of Florida College of Liberal Arts & Sciences, Office for Research and Graduate Education, Department of Anthropology, and the Center for Wetlands. I acknowledge with gratitude the support from these sources and I am equally grateful for the time and services furnished by many individuals and organizations: Florida Museum of Natural History, Florida's Silver Springs, Hontoon Island/Blue Spring State Park, John and Bryony Coles, Wetlands Archaeological Research Project (WARP), Alachua County Official Welcome Center, Rotary Club of Crescent City, Skip Miller, Richard Hamrick, The Wilcox Family of Mount Royal, Elise V. LeCompte, Victoria Gority, and Sharyn Jones O'Day. An exhibit of artifacts from wet sites was arranged by Elise V. LeCompte for a reception held at the Florida Museum of Natural History. I wish to acknowledge the generosity of the exhibit contributors: Lorna A. Craven, Florida Atlantic University (Department of Anthropology), Florida Museum of Natural History, Florida State University (Department of Anthropology), Indian Temple Mound Museum, Museum of Science and History (MOSH), Michael E. Palmer, Sr., Rollins College (Department of Anthropology), Silver River Museum and Environmental Education Center, and Ollie Stricklin.

Barbara A. Purdy
Professor Emerita of Anthropology
University of Florida
Gainesville, Florida USA
October 2000

About the Authors

James M. Adovasio received the PhD in anthropology from the University of Utah. He is currently Professor of Anthropology and Geology, and Executive Director of the Mercyhurst Archaeological Institute of Mercyhurst College where he has been teaching since 1990. His research foci are directed to prehistoric technology, the application of interdisciplinary methods to archaeological data recovery, and Late Pleistocene-Holocene adaptations in the New World.

Kathryn Bernick has been involved in wet site archaeology on the Northwest Coast of North America since 1973, including excavation, analysis, conservation, and publication. She has also conducted research on archaeological basketry in the Near East. Bernick has an MA in anthropology from the University of Victoria, Canada. Her research interests include wetland archaeology, Northwest Coast anthropological archaeology, Coast Salish ethnology and ethnohistory, stylistic analysis of basketry, and feminist archaeology. She is a freelance archaeologist currently based in Victoria, British Columbia.

Grigory M. Burov was Scientific Worker at the Academy of Sciences of the Soviet Union (Komi Branch in Syktyvkar, Russia) from 1957 to 1968. He taught as Reader at the Pedagogic Institute of Ulyanovsk (Russia) from 1968 to 1974. From 1974 to 1988, Burov was Reader and, in 1988, became Professor at the University of Simferopol (Crimea, Ukraine). He was Candidat of History from 1964 to 1987 when he received the Doctor's degree in history. Burov's scientific works were dedicated to the prehistory (Mesolithic and other periods of the European Northeast, Middle Volga Region, Crimea), and his interests include, in the first place, wetland archaeology. Dr. Burov is currently Professor of Archaeology at the University of Tauris (the University of Simferopol has been called University of Tauris since 1999).

Scott Byram is a doctoral candidate in Anthropology at the University of Oregon. He began working in North American archaeology in 1986, and since 1994 has worked as an anthropological consultant for the Coquille Indian Tribe, other Native American tribes, and state and federal agencies. His research interests are estuarine archaeological sites, traditional technologies including lithic tool production and tidewater fishing systems, the ethnohistory of Northwest Native peoples, and the northwest fur trade.

Jorge Calvera received the Ph.D. degree in Historical Sciences at the Universidad de La Habana, Cuba in 1992 after beginning archaeological work in the Center for Anthropology of the federal Ministry of Science, Technology, and the Environment (CITMA), where he continues as Senior Investigator. From 1974 through 1995 he was also Adjunct Professor at the Universidad Pedagógica "José Marti" in Camaguey, Cuba. His archaeological investigations have focused on the Taino in the east-central region of Cuba, particularly on ceramics and pictographs. He has served since 1994 as co-director, with David Pendergast, of the Los Buchillones project.

Henry Chapman obtained his first degree at Exeter University in 1994. His initial involvement in archaeology was based in archaeological survey, working for the former Royal Commission on Historical Monuments for England (now English Heritage). His current research interests as a research fellow at Hull University focus on the application of GIS and other digital methods to landscape archaeological interpretation and wetland management, and on digital data collection, incorporating differential GPS. This has included involvement in research projects in both the UK and Egypt. As of August 1, 2000, Henry will be employed as a Research Fellow for the Wetland Archaeology and Environments Research Centre at Hull University. He is co-director of the English Heritage funded Sutton Common Project.

James L. Cheetham is studying for a PhD at the Centre for Wetland Archaeology, University of Hull, looking at in situ preservation of organic archaeological remains situated on Sutton Common, South Yorkshire. Research interests include the problems in monitoring dynamics of the burial environment by minimally intrusive means, environmental modelling using GIS, and issues relating to nature conservation and the preservation of archaeology.

Clemency Coggins. Working in the archaeology of the Maya lowland Guatemala, Clemency Coggins received a PhD in Fine Arts from Harvard University. Subsequent work involved archaeological sites and collections in Guatemala, in Honduras and in Mexico where she studied and published the wooden and other perishable artifacts excavated from the Well of Sacrifice, Chichen Itza, Yucatan. She is Professor in the departments of Archaeology and of Art History, Boston University.

John Coles was Professor of European Prehistory at the University of Cambridge, and is currently Honorary Research Professor at the University of Exeter, UK. He has a PhD from Edinburgh University, Sc.D. from Cambridge University, and Honorary doctorate from Uppsala University. He is a Fellow of the British Academy. His research interests are in wetland archaeology, Bronze Age rock carvings, experimental archaeology, and he has published 20 books on these and related subjects.

Dale R. Croes received the PhD in anthropology at Washington State University and served as Director of the Washington Archaeological Research Center at WSU from 1980 to 1987. His research focus has been on Northwest Coast wet (waterlogged) archaeological sites, and especially the analysis and comparison of prehistoric basketry and cordage artifacts from these sites. He did his PhD dissertation research on basketry and cordage artifacts from the Ozette Village wet site, and conducted post-doctoral research by directing and publishing (WSU Press) the research at the 3000-year-old Hoko River wet site. He is presently directing excavations of the Mud Bay wet site in southern Puget Sound with the Squaxin Island Tribe. Dr. Croes is currently chair of the Department of Anthropology at South Puget Sound Community College in Olympia, Washington, and is a Research Faculty member in the Department of Anthropology at Washington State University

Thomas L. Crisman is the Director of the Center for Wetlands, University of Florida and Professor of Environmental Engineering Sciences. Dr. Crisman is a broadly trained aquatic ecologist and paleolimnologist with over twenty five years of experience conducting research on the ecology, conservation and management of subtropical and tropical wetlands, lakes, estuaries and streams. His current research focus is on littoral zone ecology and ecotonal relationships with an emphasis on benthic macro-invertebrates and zooplankton, the role of exotic species in foodweb dynamics and paleolimnological reconstructions of watershed-lake interactions. He has given over 110 presentations at scientific meetings throughout the world and is the author of over 60 refereed journal articles and book chapters.

Ulrike A.M. Crisman is associated with the Center For Wetlands, the University of Florida, Gainesville, Florida.

Anne Dietrich received a Ph.D. in Archaeology/Ethnography at the Sorbonne University of Paris. Working on wooden structures and artifacts, she is employed by the Association pour les Fouilles Archeologiques Nationales (AFAN) since 1993 as a xylologist. Her research interests include wetlands archaeology, buildings and objects from the Neolithic period to the end of the Middle Ages. Dr. Dietrich is currently associated with the National Research Centre (Centre National de la Recherche Scientifique – CNRS), and is teaching seminars in various French universities as an ancient wood specialist.

Clare Ellis received a PhD in archaeomagnetism of fine-grained sediments at the University of Leicester in 1995. Her research interests include sedimentological and pedological processes operating on and off archaeological sites; environmental and landscape geoarchaeology, with particular reference to fluvial and wetland systems; the exploration of the relationship between sediments, anthropogenic activity, climatic conditions, geology and geomorphology; the application of micromorphology, soil chemistry and magnetic susceptibility to archaeologically led questions; and dating of fine-grained sediments using archaeomagnetism. Dr. Ellis is currently employed as a sedimentologist and field archaeologist by AOC Archaeology Group in Scotland, Britain.

Véronique Gallien received the Ph.D. in History in 1992 at the Sorbonne University of Paris. This work dealt with the study of middle age cemeteries and populations at Saint-Denis, France. Dr. Gallien is an anthropologist, associated with the physical anthropology laboratory of the National Research Centre (Centre National de la Recherche Scientifique – CNRS) in Sophia Antipolis. She is also an archaeological team leader. Her research now includes funeral archaeology in urban contexts and medieval populations in the northwest of France.

Naama Goren-Inbar received the PhD in prehistoric archaeology at the Hebrew University of Jerusalem, Israel and has been teaching there since 1971. Her research interests include early human behavior through the evidence of material culture preserved at archaeological sites. She has worked for many years on Early Palaeolithic sites of the Dead-Sea Rift and adjacent regions with emphasis on early human technologies. She is the director of the wet site Gesher Benot Ya'agov multi-disciplinary project.

Elizabeth Graham received the Ph.D. in archaeology at the University of Cambridge in 1983 after serving as Archaeological Commissioner of Belize from 1977 through 1979. She taught in the Department of Anthropology at York

University where she is now on leave as Associate Professor from 1988 through 1999. Her research interests center on the ancient Maya with a principal focus on Belize and include human impact on the environment, coastal adaptations, and the dynamics of Maya/Spanish interaction during the Conquest period. Dr. Graham is currently Lecturer (Mesoamerica) at the University College London; she and Juan Jardines are the vice-directors of the Los Buchillones project.

Robert van Heeringen received the PhD in archaeology at the University of Amsterdam in 1992 (The Iron Age in the Western Netherlands). His research interests lie in the field of archaeological heritage management of wetlands, in prehistory from 1800–0 BC, and in Holocene palaeo-geographical mapping. Dr. Van Heeringen is currently program leader at the National Service for Archaeological Heritage, Amersfoort, the Netherlands.

C. Andrew Hemmings received a B.A. degree in Anthropology from the University of Arizona in 1991 and an M.A degree in Anthropology from the University of Florida in 1999 where he is currently a PhD student. His research interests include Paleoindian adaptations and economy as well as lithic and bone tool technology.

Per Hoffmann received his doctorate in wood chemistry and biology at Hamburg University where he worked as a research scientist from 1974–1979. After teaching at the University of Guadalajara, Mexico he joined the Deutsches Schiffahrtsmuseum in Bremerhaven and specialized in the conservation of waterlogged wood, especially large structures like boats and ships. His main responsibility has been the conservation of the medieval "Bremen Cog," the centerpiece of the museum's collection. He has published and lectured widely on archaeological wood and its conservation, and is involved in projects for ship conservation in many countries. For many years he has been the coordinator of the ICOM Committee for Conservation Working Group on Wet Organic Archaeological Materials.

Irene Holst received a M.Sc in Botany at the Paris-London University, Salzburg, Austria in 1987. She has been working at the Smithsonian Tropical Research Institute in Panama since 1988. Since 1995, she has been actively conducting research on the identification of starch grains and phytoliths from archaeological contexts. At present, she is coordinator and curator of the phytolith-starch grain laboratory of Dr. D. Piperno at the Smithsonian Tropical Research Institute.

David C. Hyland received the PhD in anthropology from the University of Pittsburgh in 1995. He is currently an Assistant Professor of Anthropology in the Mercyhurst Archaeological Institute. His research interests include prehistoric material culture, perishable technologies, and

the archaeological records of Eastern North America and the Russian Far East.

José Iriarte received a Licenciatura degree in Anthropology with a specialization in Archaeology from the Universidad de la República Oriental del Uruguay in 1995. Since 1996 he has been pursuing a PhD in Anthropology at the University of Kentucky. His research interests include the emergence of cultural complexity in the mound-building cultures of southeastern Uruguay, human-environment interactions in wetlands, paleoethnobotany studies and lithic analysis.

Donald B. Ivy is the Cultural Resources Program Coordinator, Coquille Indian Tribe. He is enrolled member of the Coquille Tribe, actively involved in tribal government in various committee, staff, and consultant roles since the Tribe's restoration to federal recognition in 1989. Employed in current position since 1997, when the Tribe formally initiated a programmatic response to its cultural resources concerns. His experience, perspectives, and arguable competence in the field of archaeology and cultural resource management results from a decade of digging in the dirt and beating the brush alongside a great many talented and accomplished university archaeologists and anthropologists, and his close work with any number of federal and private cultural resource professionals who have consulted with the Coquille Tribe during the past decade. In particular, the collaborations and partnerships between the Tribe and the graduate students and faculties of the University of Oregon and Oregon State University Departments of Anthropology get credit for any skill or ability he has acquired that grants him a credible voice in matters concerning the archeology and prehistory of Native peoples who lived (and still do live) along the estuaries and coast of Oregon.

Juan Jardines received his Licentiate in History in 1981 from the Universidad de La Habana, Cuba. In 1984 he began work in the center for Anthropology of the federal Ministry of Science, Technology, and the Environment (CITMA) where he continues as Associate Investigator. He has also recently assumed the post of head of a CITMA east-central region research center of which archaeology forms one unit. His interests have centered on the study of Taino ceramics and art in the east-central region of Cuba; he and Elizabeth Graham are the vice-directors of the Los Buchillones project.

Dilys Johns graduated with a BA and MA in Archaeology from the University of Auckland. She then studied conservation at the International Center for the Study of the Preservation and Restoration of Cultural Property in Rome and the Canadian Conservation Institute. Her research interests include the conservation of wetland archaeological sites, waterlogged artifact conservation, and wetland archaeology. Dilys is currently employed as a Research Fellow at the University of Auckland, New Zealand.

Lars Larsson received the PhD in archaeology at the University of Lund, Sweden in 1978. He was appointed professor at the Institute of Archaeology, Lund University in 1984 and is currently professor at the same institute. His research interest has been focused upon different aspects of the Stone Age of southern Scandinavia. Interest in the Stone Age has also been the reason for international projects in Portugal and Zimbabwe. During the past few years, he has developed a new research interest in the Iron Age.

Malcolm C. Lillie received the PhD from the University of Sheffield for a study of palaeopathological analysis of diet at the Mesolithic-Neolithic transition in Ukraine. His research interests are twofold, with research in Eastern Europe focusing on early Holocene human-landscape interactions, chronology and diet in what are essentially fisher-hunter-gatherer societies. His research in wetlands, based on MSc studies at the University of Sheffield, focuses on Holocene landscape development and fluvial processes in wetlands with research interests in chronology, soils and wetland archaeology. Until recently he was employed as the Senior Palaeoenvironmentalist for the English Heritage funded Humber Wetlands Survey which was based in the Centre for Wetland Archaeology at the University of Hull, England. Dr. Lillie is currently full-time lecturer in archaeology at Hull University, and the Co-Director (along with Dr. Steve Ellis) of the Wetland Archaeology and Environments Research Centre (WAERC) at the University of Hull, England. The team at the WAERC comprises Dr. Benjamin Gearey, Henry Chapman and Helen Fenwick, all of whom were long-standing members of the English Heritage funded Humber Wetland Survey and the founding members of the CWA which WAERC replaces at Hull.

Christina Marangou received a PhD in Archaeology at the University of Paris I (Pantheon, Sorbonne) in 1989 and has been a member of the research team "Aegean Protohistory" of the CNRS. She excavated mostly in northern Greece. Her research interests include Neolithic and Early Bronze Age in Greece and neighboring areas, figurines and models, archaeology of cult, maritime and wetlands archaeology, primitive water craft and, more recently, rock art. She is currently teaching prehistoric archaeology at the University of Crete, Greece.

William Marquardt received the Ph.D. in Anthropology from Washington University, St. Louis in 1974 and is Curator in Archaeology, Florida Museum of Natural History. He has done archaeological research in New Mexico, Kentucky, South Carolina, Georgia, Florida, and Burgundy, France. Since 1985, he has directed the South-west Florida Project, focused on the ancient domain of the Calusa Indians. He has been instrumental in establishing the Randell Research Center, a research and education center at Pineland near Fort Myers. He is curator of the permanent exhibit, "People of the Estuary: Six Thousand Years in South Florida." Current research interests include the emergence of sociopolitical complexity, especially in the southeastern United States and circum-Caribbean areas; processes of archaeological site formation/deformation; and the role of archaeology and history in environmental education.

José Maria López Mazz received a PhD in Anthropology at L'Ecole des Hautes Etudes en Sciences Sociales, La Sorbonne Nouvelle (Paris III) in 1986. Since 1986 he has been coordinating archaeological excavation programs in southeastern Uruguay. His research interests include the mound-building cultures of the mid-atlantic coast of South America, the emergence of cultural complexity among maritime hunter-gatherers, lithic technology, and landscape archaeology. At present, Dr. López teaches in the Department of Archaeology at the Facultad de Humanidades y Ciencias de la Educación, Universidad de la República Oriental del Uruguay.

George P. Nicholas is Assistant Professor of Archaeology and Anthropology, and is Archaeology Program Director at the Simon Fraser University-Secwepemc Education Institute in Kamloops, British Columbia. He received the Ph.D degree at the University of Massachusetts, Amherst. His research interests include the archaeology of wetland environments, prehistoric land use, early postglacial archaeology, and archaeology and indigenous people. He has published numerous articles and book chapters on wetland archaeology and other subjects, and has edited three volumes including Holocene Human Ecology in Northeastern North America.

Robert Van de Noort is a Senior Lecturer in Archaeology at the University of Exeter, UK. From 1992 to 2000, he was the Project Manager of the English Heritage funded Humber Wetlands Project, and since 1996 also the Director of the Centre for Wetland Archaeology at the University of Hull, UK. Research interests include wetland archaeology in England and further afield, prehistoric boats and the management of the archaeological heritage.

David Pendergast received the Ph.D. degree in anthropology from the University of California at Los Angeles in 1961, and from that year until 1967 he taught at the University of Utah. From 1964 until his retirement in 1999 he was a curator and later vice president at the Royal Ontario Museum in Toronto, Canada. His principal research interest is Belize's ancient Maya with concentration on architecture, the dynamics of the Classic collapse, and especially on the Postclassic and Spanish Contact periods. He is now Curator Emeritus at the ROM and Adjunct Curator, Ancient Americas at the George R. Gardiner Museum in Toronto and is also affiliated with the Institute of Archaeology, University College, London; since 1994 he has served as co-director, with Jorge Calvera, of the Los Buchillones project in Cuba.

Leonel Cabrera Pérez received a Licenciatura degree in Anthropology with a specialization in Archaeology and a Licenciatura degree in History with a specialization in American history from the Universidad de la República Oriental del Uruguay. His research interests include ethnohistory of the River Plate Basin, historical and industrial archaeology, prehistory of southeastern Uruguay, and use wear analysis of lithic artifacts. Lic. Cabrera currently teaches at the Department of Archaeology at the Facultad de Humanidades y Ciencias de la Educación, Universidad de la República Oriental del Uruguay.

Joseph Prenger was associated with the Center For Wetlands, the University of Florida, Gainesville, Florida.

Barbara A. Purdy received a B.S. in Zoology from San Diego State University, an M.A. in Anthropology from Washington State University, and a Ph.D. in Anthropology and Geology from the University of Florida where she taught from 1970 to 1992. Her research interests include applications of physical science techniques to archaeological problems, lithic technology, wetlands archaeology, and the peopling of the Western Hemisphere. Dr. Purdy is currently Professor Emerita of Anthropology at the University of Florida and Curator Emerita in Archaeology at the Florida Museum of Natural History.

Wijnand A.B. van der Sanden studied Archaeology at the Biologisch-Archaeologisch Instituut at the Groningen University, the Netherlands. He received the PhD in Archaeology at the University of Leiden. From 1987 onwards he was head of the Archaeology Department of the Drents Museum in Assen, until he became County Archaeologist of Drenthe in 1997. His research interests focus on wetland finds from Northwest Europe with special attention to the bog bodies. Dr. Van der Sanden organized the bog body exhibition *Ansigt til ansigt mit din fortid* in Silkeborg Museum, Denmark in 1996 and wrote the accompanying book Through Nature to Eternity – the Bog Bodies of Northwest Europe.

Olga Soffer received the PhD in anthropology from the Graduate Center, C.U.N.Y. She is currently a Professor of Anthropology at the University of Illinois in Champaign-Urbana where she has been teaching since 1986. Her research interests focus on reconstructing Paleolithic lifeways in Central and Eastern Europe.

Liesbeth Theunissen finished her PhD study in 1999. Her doctoral research was aimed at societies that lived in the southern part of the Netherlands and Flanders in the Middle Bronze Age (1800 to 1050 BC). Her research interests include burial archaeology, the history of archaeological research, and the representation of the past. Currently she is working at the National Service for Archaeological Heritage on various projects. The target of the present project is an enduring conservation of Neolithic sites in a wetland area in the northwestern part of the Netherlands.

Karen J. Walker received the PhD. in anthropology in 1992 from the University of Florida. Human-environment relationships and the role of archaeological sites in building sea-level and climate histories are primary research interests. Dr. Walker is currently on the Florida Museum of Natural History's faculty and is an environmental archaeologist with the museum's Randell Research Center and Environmental Archaeology Laboratory.

S. David Webb received a B.S. in Zoology at Cornell University and a PhD in Paleontology at the University of California, Berkeley. He has served as Curator at the Florida Museum of Natural History and Professor of Zoology and Geology at the University of Florida since 1964. His research has featured extinct megafauna, intercontinental migrations, and underwater paleontology of Late Pleistocene wet sites in Florida. He edited a book on "Pleistocene Mammals of Florida" in 1974. Dr. Webb is currently Distinguished Research Curator and Professor.

Ella Werker received the PhD in botany at the Hebrew University of Jerusalem where she continues to work mainly in research. Her specialization is plant anatomy. Lately Dr. Werker works mostly with archaeologists on plant materials. She is the author of a book on "Seed Anatomy" in the series "Encyclopedia of Plant Anatomy," and co-author of the book "Wood Anatomy and Identification of Trees and Shrubs from Israel and Adjacent Regions."

Introduction

Few discoveries, other than gold-filled tombs, excite the imagination as much as archaeological enterprises that provide fresh views of ancient human groups and their activities. In rare cases, prehistoric peoples and/or their most fragile creations endure when they become accidentally or intentionally entombed in environments that have remained constantly wet, dry, or frozen. These conditions protect them from decay. They do not occur at typical terrestrial sites. The finds are particularly informative when skeletons retain flesh, internal organs, and clothes, and when they are accompanied by items of personal adornment or weaponry made of wood, cordage, or bone in addition to the more common stone and pottery objects. Preserved stomach contents sometimes reveal the ingredients of an individual's last meal. Well-known examples of this kind of survival include the bog bodies of northern Europe, the Iceman of the Alps, Egyptian and Peruvian mummies, Swiss Lake settlements, and, in North America, the Ozette Village on the Olympic Peninsula, and Key Marco on Florida's lower Gulf Coast. Since no location will contain all information of interest, it is important to examine assemblages from a variety of sites representing different cultural, temporal, and environmental settings.

These organic materials provide the only broad opportunity to discover relationships between resources available and resources utilized by prehistoric groups of people. The non-perishable objects from these sites are often so trivial that the sites would not have been discovered at all or their significance would be greatly diminished without the organic segment.

Wetlands Archaeology

A cultural heritage is preserved in many of the world's wetlands but this remarkable and interesting record, which in some cases extends back thousands of years, is neglected when other components of wetlands are studied. Archaeological wet sites are invisible since their preservation depends upon their entombment in oxygen-free deposits. As a result, they are often modified or destroyed during draining, dredging, and development projects. The plight of the wetlands heritage needs to be brought to the attention of the general public, the scientific community, developers, and government agencies in order to examine ways to reverse the situation that now exists.

Since the early 17th century, people have been fascinated by the excellent preservation of bodies in raised bogs most of which date to the late Iron Age of northern Europe around 200 to 100 B.C. (Glob 1969; van der Sanden 1996). Swiss Lake Villages became internationally and historically famous when it was discovered in the 1850s that a wide range of cultures are submerged in Swiss lakes that date from Upper Paleolithic to Medieval periods (*Helvetia Archaeologica* 1981; Stickel and Garrison 1988). In the 1890s, Frank Hamilton Cushing was so impressed with the discoveries in Switzerland that he called the famous Key Marco site the *Court of the Pile Dwellers* suggesting that the spectacular artifacts from this site on Florida's lower Gulf Coast had survived under conditions similar to the Swiss lakes (Cushing 1897; Gilliland 1975). In the 19th century, bog bodies, Lake Dwellers, and Key Marco excited the public and an international community of scholars. Very quickly, however, mention of them was reduced to a paragraph or two in archaeological textbooks.

It wasn't until the 1960s that another waterlogged site comparable to Key Marco was excavated containing wood carvings, dietary items, and environmental information. The Ozette Village Site on the Olympic Peninsula in the state of Washington was dubbed America's *Pompeii* because people and their possessions were preserved in the exact locations they occupied before the settlement was encapsulated by a mud flow (Daugherty 1988).

The sites named above were all discovered accidentally: the bog bodies and Key Marco during peat mining operations; lowered water tables at Swiss lakes; and, at Ozette, wooden artifacts were found eroding from a bank after a coastal storm. Significant but less spectacular sites and individual items were also found by chance. When scuba became available in the 1940s and 1950s, sports divers were able to explore lake bottoms, river beds, springs, and shipwrecks with an ease not known before. The sports divers, often unwittingly, began to do almost as much damage as the developer's backhoe. Even with scuba equipment, underwater sites are not easy to excavate systematically, and water-saturated sites often require the use of pumps or cofferdams.

By the 1960s, archaeologists were beginning to ask "If archaeology is not anthropology, what is it"? and they realized that the great diversity of materials recovered from sites where organic materials had survived could be used to test hypotheses about past human behavior in ways usually only possible through ethnographic interrogation.

But what had happened to the sites and the information they contained? In many parts of Europe, organic soils have been used for fuel for generations and the artifacts and environmental information contained in these peat soils have gone up in smoke. In 1980, a statewide survey of Florida's archaeological wet sites revealed that, without exception, the sites had been destroyed or modified in some major way so that the critical deposits were no longer there (Purdy 1981). Key Marco had been turned into a full scale city of homes and resort facilities (Gilliland 1975:27). And there are many more examples. Fortunately, new sites have been recorded and excavated with care, and the finds from some of these sites are available for hypotheses testing.

The whereabouts and condition of organic materials recovered from waterlogged sites prior to the mid–20th century fall into one of three categories: (1) those that survived in good condition and are on exhibit somewhere or illustrated in books, (2) those that exist in unexciting condition and are in storage somewhere awaiting analysis, and (3) those that are recorded only on notecards in some filing cabinet because the objects themselves have turned to dust.

In the 19th century, bog bodies, clothing, art objects, tools, and weapons from waterlogged sites were unprecedented objects. No one knew why they had survived in the first place or what should be done to protect them for the future. Grattan (1988) reviewed the long history of attempts to conserve waterlogged wood using various methods that usually failed. Frank Hamilton Cushing made an effort to save the fragile wooden masks, statuettes, tools, and cordage from the Key Marco site but many of them have survived only in photographs or drawings.

The invention of scuba equipment during World War II made it possible to explore deep springs containing ancient human remains, and to recover historically important shipwrecks and their artifacts. About the same time, growing populations with a coinciding increase in development projects in archaeologically sensitive areas brought to light large numbers of perishable artifacts and environmental data. These two situations, scuba and development, created a crisis. It became imperative that

adequate preservatives be found, or these materials would be lost along with all of the information they contained.

In the 1940s a number of new synthetic polymers became available. The antishrink effect on wood of polyethylene glycol (PEG) was discovered in the early 1950s. Conservators quickly realized that PEG could be used for waterlogged wood to prevent collapse and shrinkage upon drying. The method was first tried on the *Wasa*, a long lost Swedish warship uncovered in Stockholm harbor (Grattan 1988:239–240). The fragile wood carvings, tools, and basketry from the Ozette site were also treated successfully with the new preservative.

Equally important and significant are the preservative requirements for materials from dry sites and frozen sites. The question is whether or not all organic materials will submit to the same or similar procedures or will each need a very different treatment.

In the 1970s, a Working Group on Waterlogged Wood was formed within the Committee for Conservation of the International Council of Museums (ICOM). The working group, now known as Wet Organic Archaeological Materials (WOAM), includes chemists, conservators, wood anatomists, wood products specialists, foresters, historians, artists, archaeologists, and museum curators and exhibitors from around the world. The expertise of some members of this group when applied to the needs of other members has resulted in truly significant contributions. Experiments with PEG, PEG plus freeze drying, and other methods have been conducted for more than 20 years and continue to be refined (Purdy 1991).

Another group, Wetland Archaeological Research Project (WARP) has concentrated on the waterlogged sites themselves, techniques of excavation, the archaeological and environmental significance of surviving organic remains, and issues pertaining to government responsibility regarding wetlands. The organization was spearheaded in England in the late 1980s by Professors John M. and Bryony Coles who have written numerous books and articles about archaeological wetlands (e.g., Coles and Coles 1989; Coles, B. 1992). Approximately 250 individuals and 60 institutions are now members of WARP.

WOAM and WARP are extremely active. Members of both groups continuously present information about new sites, new preservation techniques, and stress how the unique categories of materials recovered from archaeological wet sites broaden the data base, increase the number of relationships available for study, and provide a more holistic view of the past.

The Dry and the Frozen

Egyptian and Peruvian mummies along with their accompanying grave offerings have been widely publicized.

Mummified burials from China and the Canary Islands are not so famous. Items as mundane as a quarter of mutton to burials of elaborately tattooed chiefs who wore woolen and leather clothes have been recovered from the kurgans of nomadic Scythian warriors on the steppes of the southern Ukraine, and from the tombs of Pazyryk in northeast Siberia (Fagan 1992:551; Davis-Kimball 1997:46) Other situations that ensure the survival of usually perishable materials are cases where sites have been covered with volcanic ash as at Herculanean (Bisel 1988:207) and Cerén (Roach 1997:74). The man in the ice, a 5300- year- old body found emerging from an alpine glacier on the Austrian – Italian border is a rare example of an ancient site with human remains and artifacts that have remained frozen through time (Spindler 1994). Other frozen remains exist, such as the mummies from St. Lawrence Island (1600 years old), Greenland (15th century) (van der Sanden 1996), and the seamen from the Franklin Expedition who were buried near Hudson Bay in the 19th century.

As mentioned, since the locations of waterlogged, dry, or frozen sites differ as to climate and geography, it should be revealing to compare and contrast the materials and the preservation requirements from each type of site.

Discussion

Prehistorians are constantly frustrated by the inadequacies of a truncated archaeological record to furnish an understanding of past human activities and behavior associated with those activities. It is a fact that more than 90 percent of all items used by people even today are composed of organic materials that decompose rapidly.

The contributions to knowledge of cultural and environmental information retrieved from sediments that have remained permanently wet, dry, or frozen are substantial. These depositional conditions are not common but one of the three types has been found at various locations throughout the world. They provide minute details about people and their activities that do not survive elsewhere and for that reason they furnish a more holistic view of the past.

The fact that wetlands contain thousands of years of environmental and cultural history apparently has not risen to the consciousness of the public, the scientific community, or to state and federal governments. One obvious reason for this situation is that this history is invisible since its very survival depends upon its entombment in oxygen–free deposits. Even large objects like canoes and totems, much less human bones and dietary items, cannot be seen prior to a backhoe cut into a black peaty matrix. When an anaerobic system is disturbed, air oxidation destroys the deposit and all of its contents within hours. When that happens, it reacts in the only way it knows how. It disappears. Of deeper significance is that this heritage has not

become part of the value system of society. Competent wetland archaeologists are rarely given the opportunity to conduct investigative surveys prior to modification of wetland areas. Even more rare are backhoe operators or pothunters who write reports about the historically significant objects they recover.

The world's wetlands have bequeathed us a virtual archive of information which is not being tapped. We should not let it wither away.

References Cited

Bisel, Sara. 1988. The Skeletons of Herculaneum, Italy. In: Wet Site Archaeology, pp. 207–218. Edited by Barbara A. Purdy. The Telford Press, Caldwell, New Jersey.

Coles, Bryony and John. 1989. People of the Wetlands: A World Survey. Thames and Hudson.

Coles, Bryony, Editor. 1992. The Wetland Revolution in Prehistory. Proceedings of a conference held by The Prehistoric Society and WARP at the University of Exeter, April 1991. Published by The Prehistoric Society and WARP.

Cushing, Frank Hamilton. 1897. Exploration of Ancient Key Dwellers' Remains on the Gulf Coast of Florida. Proceedings of the American Philosophical Society 25 (153:329–448. Philadelphia.

Daugherty, Richard D. 1988. Problems and Responsibilities in the Excavation of Wet Sites. In: Wet Site Archaeology, pp. 15–29. Edited by Barbara A. Purdy. The Telford Press, Caldwell, New Jersey.

Davis– Kimball, Jeannine. 1997. Chieftain or Warrior Priestess? Archaeology, September/October, pp. 40–41 (see also January/February 1997 issue).

Fagan, Brian M. 1992. People of the Earth. Seventh Edition. Harper Collins Publishers, NY.

Gilliland, Marion Spjut. 1975. The Material Culture of Key Marco, Florida. The University Press of Florida, Gainesville.

Glob, P.V. 1969. The Bog People: Iron-Age Man Preserved. Cornell University Press, Ithaca, NY.

Grattan, David W. 1988. Treatment of Waterlogged Wood. In: Wet Site Archaeology, pp. 237–254. Edited by Barbara A. Purdy. The Telford Press, Caldwell, New Jersey.

Helvetia Archaeologica. 1981. Zürcher Seeufersiedlungen: Von der Pfahlbau-Romantik zur modernen archäologischen Forschung, Volumes 45–48.

Purdy, Barbara A. 1981. Survey, Recovery and Treatment of wooden Artifacts in Florida, pp. 159–169. Proceedings of the ICOM Waterlogged Wood Working Group Conference, Ottawa.

Purdy, Barbara A. 1991. The Art and Archaeology of Florida's Wetlands. CRC Press, Boca Raton, Florida.

Purdy, Barbara A., Editor. 1988. Wet Site Archaeology. The Telford Press, Caldwell, N.J.

Roach, Mary 1997. New World Pompeii. Discover, February, pp. 74–80.

Spindler, Konrad. 1994. The Man in the Ice. Crown Trade Paperbacks, N.Y.

Stickel, E. Gary and Ervan G. Garrison. 1988. New Applications of Remote Sensing: Geophysical Prospection for Underwater Archaeological Sites in Switzerland. Wet Site Archaeology, pp. 69–88. Edited by Barbara A. Purdy. The Telford Press, Caldwell, New Jersey.

Van der Sanden, Wijnand. 1996. Through Nature to Eternity. The Bog Bodies of Northwest Europe. Batavian Lion International, Amsterdam.

1. Ivory and Bone Tools from Late Pleistocene Deposits in the Aucilla and Wacissa River, North-Central Florida

S. David Webb and C. Andrew Hemmings

Florida rivers produce the largest collections of late Pleistocene ivory and bone tools in the New World. In this chapter we feature the Aucilla and Wacissa Rivers because they are currently the most productive sources of late Pleistocene organic material in Florida. We give a brief account of how we explored these rivers, how we developed their archaeological and paleontological resources, and an overview of the non-lithic archaeological material that we are currently studying. Figure 1.1 suggests the natural mystery and pristine beauty of these rivers, and Figure 1.2 maps several of the prehistorically most productive rivers in the Florida peninsula.

The first indication of Florida's late Pleistocene wealth consists of the report by Jenks and Simpson (1941) of Clovis-like lithic and ivory tools from the bottom of the Ichetucknee River. Subsequent decades saw a much greater variety of Paleoindian cultural material come from submerged sites in Florida, primarily collected by SCUBA-equipped hobbyists who increasingly discovered the availability and inherent interest of such material. A few academic prehistorians also pursued this resource in those early years. Such pioneer work included that of John Goggin (1950), who worked at the confluence of the Santa Fe and Suwannee Rivers, and that of Charles Hoffman who

Figure 1.1. The rapids where a branch of the Wacissa River flows into the Half-Mile Rise segment of the Aucilla River.

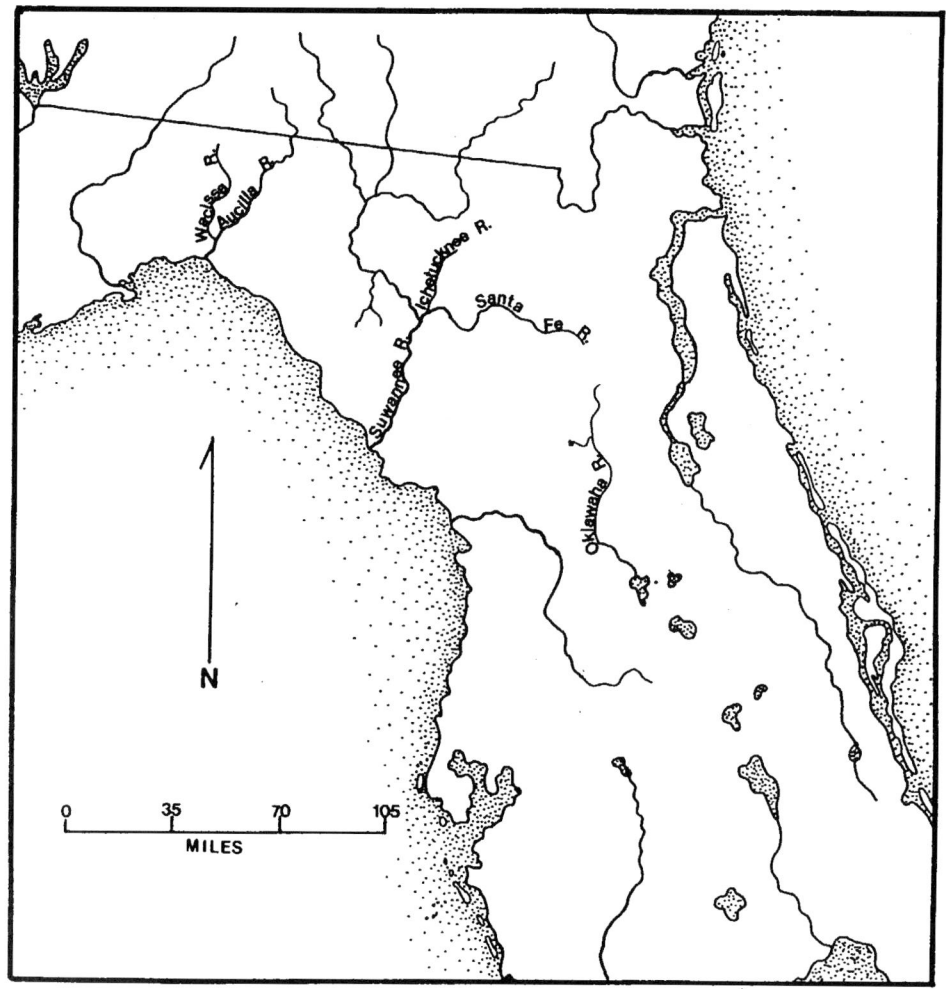

Figure 1.2. Six North Florida rivers which produce worked ivory. The Aucilla River and its tributary the Wacissa are the two featured in this chapter. The earliest worked ivory in Florida was reported from the Ichetucknee River. It flows into the Santa Fe and that in turn into the Suwannee River, the three together making up a vast rich resource of underwater paleontology and archaeology. And finally the Oklawaha River and the Silver Springs Run also produce ivory and bone that represent interactions between Late Pleistocene megafauna and Paleoindian people.

excavated the Guest Mammoth Kill Site on the Silver springs Run to the Oklawaha River (Hoffman 1983). These early results were met with considerable skepticism because they came from dark water and seemed to represent a scattered resource lacking precise stratigraphic context.

Professional paleontologists also began to explore these same rivers during the 1950's and 1960's. Their efforts seemed to be more fully rewarded, perhaps because they were less concerned about the exact age and context of their magnificently preserved Pleistocene material (Webb 1974). Among the key discoveries of underwater paleontologists in Florida rivers were outstanding specimens of *Bison antiquus* and complete skeletons of *Mammut*

americanum and *Mammuthus columbi* in the Aucilla River (Robertson 1973; Webb 1976). These paleontological sites set the stage for the inauguration of the Aucilla River Prehistory Project a decade later. Finally in 1983 came a fortuitous discovery that galvanized both archaeologists and paleontologists to greater efforts, namely the recovery of a *Bison antiquus* skull with a Paleoindian point fragment lodged in its frontal bone. This find came from the Wacissa River, a tributary of the Aucilla, and subsequent research produced other postcranial material some of which was probably the same female, and also a carbon-date of approximately 11,000 years before present (Webb *et al.* 1984). Thus the first *Bison* kill site in eastern North America stimulated a more intensive effort to investigate Florida's late Pleistocene prehistory.

Our team, known as the Aucilla River Prehistory Project (ARPP), began its work in 1984 under the leadership of James Dunbar, underwater archaeologist, and David Webb, underwater paleontologist. It sought to recover *in-situ* evidence of the interactions between the first Floridians and the late Pleistocene megafauna. We focused on the Aucilla and Wacissa rivers because these were known sources of abundant paleontological and archaeological material and specifically because they had yielded the *Bison* kill site and several pieces of worked proboscidean ivory. Through the next two decades the ARPP was fortunate to attract an extraordinarily talented and dedicated group of amateurs and professionals, who served variously as underwater excavators, equipment operators, fundraisers, conservators, data-managers, and in innumerable other jobs that have no rubrics.

This chapter presents a progress report on the ARPP's investigations of the late Pleistocene resources in the Aucilla and Wacissa Rivers. We begin by discussing the geological and sedimentary context in which archaeological and paleontological records occur, and then we review a preliminary sample of the organic cultural material which these rivers so bountifully yield.

Geological Context

In north and central Florida the rivers that trend southwestward to the Gulf of Mexico traverse extensive but low-lying exposures of Eocene and Oligocene marine formations. These Paleogene rocks dictate not only the surface features of the region, such as its sinkholes and magnificent springs, but also its deeper characteristics, notably the vast Floridan Aquifer that flows southwestward through the peninsula. Of course, the surface and deep features are connected, as exemplified by Wakulla Springs where the aquifer bursts to the surface at the rate of nearly 300 cubic feet per second (Rosenau *et al.* 1977).

Our surveys and excavations in the Aucilla and Wacissa rivers have emphasized the essential role of this karst physiography in preserving a rich organic record of late Pleistocene prehistory. In its first two seasons ARPP discovered that the bottom of the Aucilla River consists of a string of semicircular sinkholes connected by short stretches of shallow limestone, as illustrated in Figure 1.3. The sinkholes contain poorly consolidated sediments and therefore are generally more deeply eroded than the connecting strips of relatively firm limestone. Avid collectors had long known that the deeper holes in Florida's rivers were the productive areas. But the reason for this was not clear. Some suggested that deep holes would trap more material as it washed down river; and others believed simply

that the deep holes in dark water were less accessible to competing collectors. Now, for reasons set forth below, we know that the sinkhole sediments themselves, in the Aucilla River at least, are the sources of paleontologically and archaeologically productive sediments.

The ARPP's recognition that productive sites in the Aucilla River consist of sinkhole sediments exposed by subsequent erosion affected not only underwater reconnaissance methods but also subsequent excavation techniques. We learned to approach the river banks with the possiblity that they represented critical ages and environments in pristine condition. This hypothesis instructed the ARPP's early exploration strategy. It was soon rewarded by discovery of intact columns of fine-grained sediments spanning the late Pleistocene/early Holocene boundary. A further practical discovery was that these sediments retained their stratigraphic integrity from season to season (Dunbar *et al.* 1989).

The correspondence between major sites and ancient sinkhole sediments in the Aucilla River leads naturally to considerations of the Quaternary geology of the region, known as the Woodville Karst Plain. In such a lowland karst region the most potent environmental influence must surely have been changing sealevel. From the time of the last full glacial about 18,000 years before present (b.p.), sealevel may have risen some 150 meters during the late Pleistocene and into the Holocene (Curay 1960). Such a major eustatic rise surely must have raised water tables dramatically throughout the peninsular Florida's karst lowlands. In the Woodville Karst Plain the effect on the piezometric surface was particularly powerful because of the very large aquifer that feeds this area from the north. Wakulla Springs, and several other first-order springs in this region, reflect the immense scope of this aquifer (Rosenau *et al.* 1977). Presumably the rate at which the piezometric surface rose a few miles inland was rather closely correlated with the eustatic sealevel rise in the Gulf of Mexico. Likewise the sediment accumulation rate in any given sinkhole in the karst plain would be expected to fill most rapidly during the earliest and steepest phases of sealevel rise. A curve tracing the approximate sealevel rise in the Gulf of Mexico during the last deglaciation is presented in Figure 1.4, coupled with a model indicating when coastal plain sedimentation might reach a maximum. The most complete stratigraphic columns developed in any Florida river are ARPP excavations at the Page-Ladson site complex in the Middle Aucilla River. The Test C section, summarized in Table 1.1, consists of about five meters of fine-grained organic sediments and a score of carbon dates. When the dates are interpolated with the sediment record, it becomes evident that the history of these sediments corresponds rather well with the sealevel rise of the last deglacial hemicycle, as suggested by Figure 1.4. This

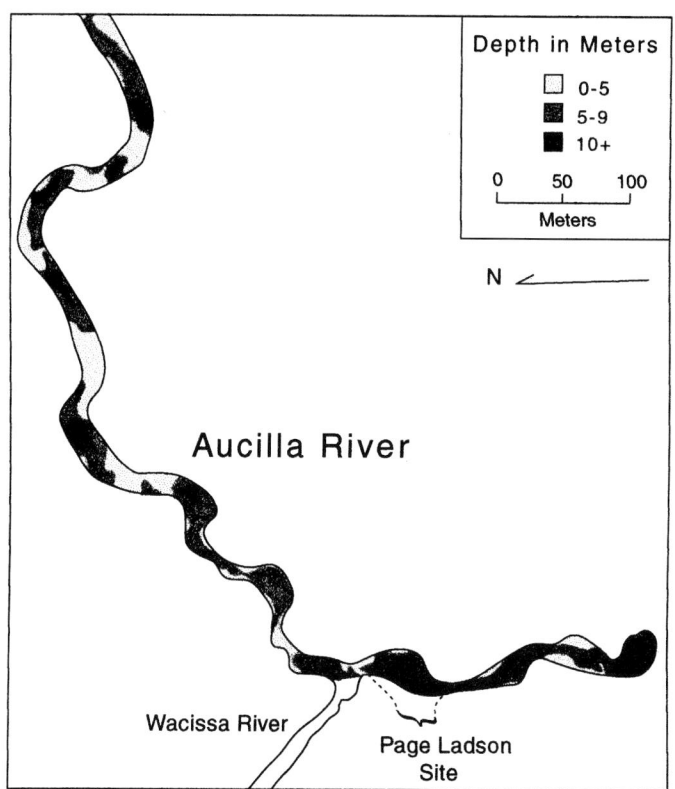

Figure 1.3. Bathymetric contour map of most of Half-Mile Rise segment of Aucilla River. Darker contours are below 15 meters deep and represent eroding sinkholes. The Wacissa River enters the lower third from the West and just south of that is the largest deep hole. Developed by the ARPP as the Page/Ladson Site complex. This segment of the river disappears into the southernmost deep sinkhole.

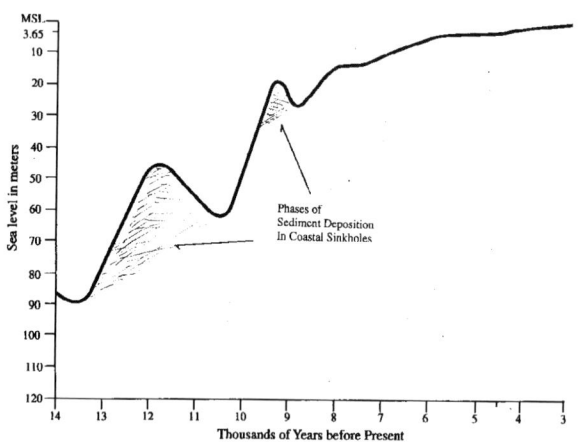

Figure 1.4. Sea level curve for latest Pleistocene deglaciation, based on Curay (1960) for northwestern Gulf of Mexico. The early phase of most rapid rise, between about 14,000 and about 10,000 years before present, is interpreted as the time when sedimentation in the karst lowlands of Florida would be most intense. This is labelled as the "filling cycle".

stratigraphic section tracks the rise of sealevel from about 13,000 to about 9,000 years b.p. at an average sedimentation rate of approximately one millimeter per year.

After the sinkhole pond represented by the Page-Ladson site had filled with sediments, it presumably began to spill over during wet seasons and flood stages. As early Holocene sealevels continued to rise, albeit at a slower rate, the spillover from sinkholes and springs throughout the karst lowlands began to integrate into short segments of rivers such as the Aucilla and Wacissa. Carbon dates associated with excavation of Test B at the Page-Ladson site indicate that the collapse of a large limestone bridge there dated younger than about 5,000 years b.p. (Dunbar et al. 1989). Even today these rivers appear physiographically immature. For example, a quarter mile below the Page-Ladson site the Half-Mile Rise segment of the Aucilla River disappears underground through a deeply eroded sinkhole, as shown at the lower end of Figure 1.3. The next segment of the Aucilla River wells up nearly a mile farther seaward by way of another eroded sinkhole conduit. The resistant ledge of silicified limestone by which one mouth of the Wacissa

Table 1.1. Stratigraphic Section with Radiocarbon Dates (Page/Ladson Site, Test C).

Strati. Unit.	Thickness (in meters)	Age/1 (C14)	Lithology
Unit 7	0.7	8,900–9,900	Recent river bottom sediments, loose organic peats and fine sands
Unit 6	0.9	9,950–10,000	Gray sandy silt with abundant molluscs, upper surface eroded, shallow eastward dip
Unit 5	0.3	10,000–10,300	Dark-brown clayey silt with wood, charcoal, abundant ostracods and cultural material
Unit 4	2.7	10,600–12,400	Gray sandy silts with wood
Unit 3	1.0	11,600–12,600	Sandy silt with abundant peat-like digesta
Unit 2	0.2	13,100–14,200	Reddish, compact peat
Unit 1	Not Known	14,300–15,400	Organic sandy silts (contacted in adjacent excavations)

1. Uncalibrated (uncorrected) carbon dates rounded to nearest century. The entire radiocarbon data set for this and adjacent cores and sections is being published more fully elsewhere.

River spills energetically into the Half-Mile Rise of the Aucilla, shown in Figure 1.1, is also an immature geomorphological feature. In geologic perspective these are still very young karst rivers.

The sediments that accumulated in karst sinkholes during the last rise of sealevel were derived by erosion of adjacent exposures of the Suwannee Limestone and clays and silts of the St. Marks and Hawthorne Formations. These locally-derived fine clastic sediments include high percentages of peat and other organic material which provided an excellent reducing medium in quiet water settings. Thus diverse organic materials, including wood, seeds, pollen, bone, ivory, mollusks, and ostracods, are well-preserved in a nearly continuous sequence of sediments. The plant remains, both pollen and macrophytic remains, clearly reflect the changing local environments from the late Pleistocene into the early Holocene. The vertebrate remains are predominantly aquatic forms, but several coarser strata include samples of terrestrial vertebrates as well. Several of the peaty sediments consist predominantly of *Mammut americanum* digesta which provide a fascinating window on the diets and seasonal patterns of these proboscideans (Webb *et al.* 1992).

The large vertebrates from late Pleistocene strata in the Aucilla River are listed in Table 1.2. These are recorded in the Page-Ladson Site from the sandy, peaty silts of Unit 3, spanning the period from about 12,600 to 11,600 radiocarbon years b.p. As shown below, these megafaunal elements provided an important part of the resources available to the first Floridians. Koch, Hoppe and Webb (1998) studied the stable carbon isotopes of many of the large extinct animals from Aucilla River late Pleistocene underwater sediments, and this provides an objective first approximation of their roles in the local ecosystem. Those data provide the basis of our "feeding ecology" assignments

of large herbivores in Table 1.2. Of further interest is the indication from strontium isotopes that *Mammut americanum* at the Page-Ladson Site were migratory and left the coastal plain for granitic terrain, the nearest of which is in what is now central Georgia (Hoppe *et al.* 1999). As suggested below, this probably influenced early Paleoindian strategies for hunting and protein provisioning.

Organic Cultural Material

Archaeological remains spanning the Pleistocene/Holocene boundary abound in the Aucilla River. *In situ* inundated components roughly predate 10,000 uncalibrated radiocarbon years before present. The circa 10,000 year old Bolen surface of Page-Ladson (8JE591) is the youngest intact human occupation known in the Aucilla (Dunbar *et al.* 1989). Older, Middle-Paleoindian, archaeological cultures are represented by Suwannee and Simpson Points and associated unifacial tools (Purdy 1981). The Ryan-Harley Site (8JE1004), a single component Suwannee site roughly 2 km northwest of Page-Ladson, is the best preserved Middle-Paleoindian site found to date (Dunbar *et al.* 1999).

Surface finds of Bolen, Suwannee, Simpson and Clovis diagnostic stone tools are being incorporated in regional technological studies, particularly those focusing on lithic reduction strategies and tool function (Gerrell *et al.* 1991).

The most rewarding aspect of the underwater research is of course the recovery of organic cultural materials not typically preserved in terrestrial sediments. Surface finds of over 100 deer bone tool forms are recorded in the collections of the Florida Museum of Natural History (FLMNH). Nearly 3000 individual bone tools have been recovered from Sloth Hole (8JE121) alone. The vast majority of these bone tools are made of unidentifiable or extant faunal elements. The density and diversity of these

Table 1.2. Megafaunal Species in Late Pleistocene Aucilla River Sites.

Genus and Specoes	Feeding Mode (carbon 13 values)[1]	Human Impacts
Mammut americanum	browser (-10.9)	Ivory shafts and otherbone tools; also cuts
Mammuthus columbi	grazer (0.1)	Beamer from spine; cuts
Equus sp	grazer (-5.6)	Awl handle from tibia; daggers from metatarsals
Tapirus veroensis	browser (-11.7)	Forelimb in site context
Mylohyus nasutus	browers (-10.3)	None known
Hemiauchenia macrocephala	mixed herbiv. (-0.2)2	Metatarsal "dagger"
Palaeolama mirifica	browser (-)3	None known
Odocoileus virginianus	browser (-12.4)	Numerous tools and cuts but most are Holocene
Bison antiquus	grazer (-0.7)	Point in cranium
Canis dirus	carnivore (no C13 data)	None known
Megalonyx jeffersoni	browser (no C13 data)	Point in vertebra; cuts on ulna
Glossotherium harlani	mixed herbiv. (no C13 data)	None known
Geochelone crassiscutata	mixed herbiv. (no C13 data)	Cuts on carapace

tools is thought to reflect outstanding preservational bias rather than adaptations or behavior unique to Florida Paleoindian people. The continent-wide distribution of large bone and ivory rods uniquely in Clovis contexts seems to support this idea (Pearson 1999).

In examining extant animal tool forms, context becomes critical. Typical deer bone "pins" or "points" are recovered throughout the Aucilla River. However, in a few sites they have been found in stratified, datable contexts. For example, an antler flaker from the Bolen layer of Test C at the Page-Ladson Site was associated with a date of 9950 ± 70 b.p. (Beta – 103888; Webb ms.). Two unstained bone "pin" fragments were recovered in a stratum associated with two gastropods locally extinct by 8500 RCYBP at Fossil Hole (8JE1497) (Hemmings 1999). Fossil Hole is an inundated lithic quarry that was abandoned prior to 8500 years ago. Currently we are awaiting dates obtained directly from these bone tools.

The cultural materials from the Early Paleoindian period are the most astonishing as they broaden our view of the continent-wide Clovis culture. Diagnostic stone projectile points and unifacial tools as well as diagnostic osseous tools made from extinct Pleistocene fauna greatly expand our understanding of Paleoindian lifeways and economy. Recent discoveries of materials directly relating Paleoindian subsistence behavior and tool manufacture to various extinct Pleistocene megafauna are somewhat at odds with current models of Eastern North American Paleoindian subsistence (Meltzer 1988). We do not wish to belabor the megafaunal connection, or fall into the old trap of considering it the predominant form of subsistence. Nevertheless, our evidence demonstrates greater similarity of Paleoindian prey selection on both sides of the Mississippi River than previously thought (Dincauze 1993; Meltzer 1993; Hemmings 1998). The rich preservation of Florida's

rivers also allows us to expand the megafaunal menu considerably.

Formal tools made from identifiable elements of camel (*Hemiauchenia macrocephala*), horse (*Equus sp.*), mastodon (*Mammut americanum*), mammoth (*Mammuthus columbi*) and possibly dire wolf (*Canis dirus*) have been recovered across Florida (Dunbar and Webb 1996). The importance of wetsite preservation cannot be understated. The particular form of any given bone or ivory tool, nor the animal species involved, can never be inferred from even the most sophisticated lithic tool analysis.

The recently discovered horse and camel (unfinished?) metatarsal daggers and *Canis* mandible ornaments expand the known Clovis bone toolkit in a particularly tantalizing way. That is because these items, like several proboscidean tools reported by Dunbar and Webb (1996), are ground. The canine mandible was cut and drilled prior to grinding (Figure 1.6). However, none of the others have any evidence of ever being cut with a stone tool. This is particularly worrisome for two reasons; not only are the specific forms not inferrable from stone tool analysis but also the process used in their manufacture evidently did not involve stone tools. The more pieces of evidence for Paleoindian lifeways we are able to collect, the clearer it becomes that our views of what Clovis people were doing is very limited.

Unequivocal subsistence-related associations between extinct Pleistocene fauna and Early Paleoindians are based on site context, unambiguous cuts or modifications, and in some cases, such as the embedded projectile "smoking gun" or a combination of these lines of evidence. Butchered Paleoindian food or tool stock remains include tapir (*Tapirus veroensis*), sloth (*Megalonyx jeffersonii*), bison (*Bison antiquus*), mastodon (*Mammut americanum*) and mammoth (*Mammuthus columbi*).

Dramatic evidence of Paleoindian interaction with

Figure 1.5. Sloth Hole ivory tool sample. Longest is broken just above shaft end (UF # 206412). Other two are complete forms. Shortest example is straight unlike the larger forms which are slightly curved (UF # 136493 and 136494).

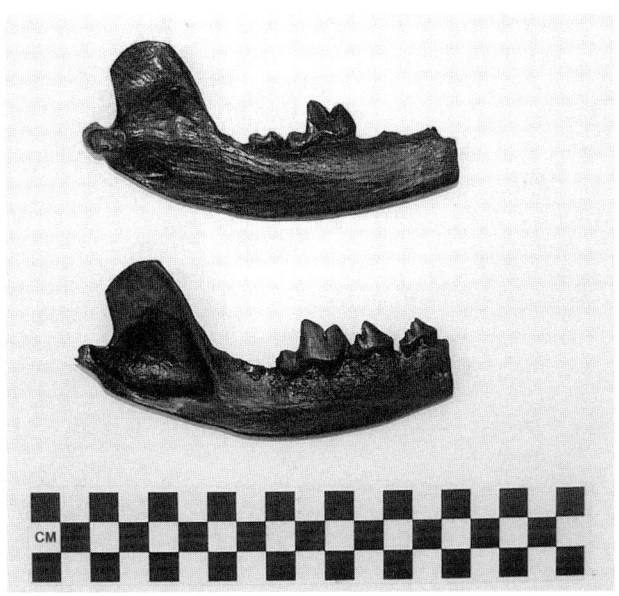

Figure 1.6. Both sides of Canis *mandible from Aucilla River. Note grinding along length of horizontal ramus and around top of coronoid process. Possibly dire wolf, extremely large for* Canis lupus *(UF # 206411 cast).*

mammoth and especially mastodon survives in the numerous forms of ivory tools found throughout Florida, the vast majority of which have been recovered in the Aucilla River. Known ivory tools can be placed into eight arbitrary categories. The first four of which are represented by unique specimens: a vestigial mastodon tusk amulet or atlatl hook, a hafted pressure flaker (?), a socketed handle fragment, a bibeveled rod square in cross section, a barb for a composite

tool, small awls or points (9–13.5 cm), medium length tools (15–25 cm) are the longest straight tools and are known in a barbed form and as straight tapered points, the most common and widely recognized forms are the long slightly curved point or lance tips (25–40 cm). The largest concentration of the long form is from Sloth Hole. The minimum ivory shaft tool number has gone to 28 after many successful refits. Figure 1.5 illustrates an ivory awl form and two of the longer shaft tools. The ivory artifact names are intended to distinguish them from each other rather than to make assertions regarding function.

The seven extinct species discussed above, with the possible addition of dire wolf, greatly expand our understanding of Paleoindian subsistence behavior. Paleoindian people were mammoth and mastodon hunters (scavengers too), but clearly they are much more than that. The unequivocal extension of our view into Paleoindian subsistence and economic behavior is what makes these finds valuable. The Clovis bone and ivory toolkit is every bit as sophisticated and widespread as the stone toolkit. As this idea spreads and the specific items recovered from Florida are added to the continent wide Clovis dataset, our collective understanding of Paleoindian lifeways should reflect reality more fully than it has in the past.

Until recently, documented interactions of Early Paleoindians and extinct Proboscideans in Eastern North American were rare enough to lead researchers to adopt subsistence models of generalized foragers (Dincauze 1993; Meltzer 1993). This also led to a perception of discontinuity with Early Paleoindian economic behavior in Western North America (Tankersley 1998). While the generalized forager model is likely to represent true behavior more accurately than near reliance on large game, as was often proposed (e.g. Martin 1969), the accumulating evidence indicates that Paleoindian and large mammal interactions were significantly broader than just reliance on proboscideans as had been proposed for Clovis sites in Western North America.

How interactions between Paleoindians and Pleistocene fauna relate to the extinction events remains unknown. Environmental change and concurrent floristic rearrangements were undoubtedly distressing to the megafauna. On the other hand, the data from the Aucilla River indicate that the activities of Paleoindians strongly impinged on the survivability of many species within the late Pleistocene ecosystem of Florida.

Acknowledgements

Special thanks go to The National Geographic Society and the Florida Department of State for grants that provided for exploration of this river system. William O. Gifford generously provided photographic expertise and image manipulation. Also, Todd Bevis, Mike Johnson, Larry Roberts and Jim Dunbar deserve thanks for access to specimens and information.

References Cited

Curay, J.P. 1960. Sediments and History of the Holocene transgression, continental shelf, northwest Gulf of Mexico, in Recent Sediments, Northwest Gulf of Mexico, edited by F.P. Shepard,, F.B. Phleger, and T.H. van Andel, pp. 221–266. Amer. Assoc. Petrol. Geol., Tulsa, Oklahoma

Dincauze, D.F. 1993. Fluted Points in the Eastern Forests. In From Kostenki to Clovis, edited by O. Soffer and N.D. Praslov, pp. 279–293. Plenum Press. New York and London.

Dunbar, James S., S. David Webb, and Dan Cring. 1989. Culturally and Naturally Modified Bones from a Paleoindian Site in the Aucilla River, North Florida. In Bone Modification edited by R. Bonnichsen and M.H. Sorg pp. 473–498. Center For The Study Of The First Americans, University of Maine, Orono, Maine.

Dunbar, J.S. and S.D. Webb. 1996. Bone and Ivory Tools from Submerged Paleoindian Sites in Florida. In The Paleoindian and Early Archaic Southeast, edited by D.G. Anderson and K.E. Sassaman, pp. 331–352. The University of Alabama Press, Tuscaloosa and London.

Dunbar, J.S., C.A. Hemmings, P. Vojnovski, W. Stanton, M. Memory, R. Means, G.H. Means, and M.C. Mihlbachler. 1999. The Ryan-Harley Site: A Suwannee Point Site in the Wacissa River, North Florida. Paper presented at Southeastern Archaeological Conference, Pensacola, Florida 1999.

Gerrell, P.R., J.F. Scarry and J.S. Dunbar. 1991. Analysis of Early Archaic Unifacial Adzes From North Florida. The Florida Anthropologist 44:3–16.

Goggin, J.M. 1950. An early Lithic Complex from Central Florida. American Antiquity 16:46–49.

Hemmings, C.A. 1998. Probable Association of Paleoindian Artifacts and Mastodon Remains from Sloth Hole, Aucilla River, North Florida. Current Research in the Pleistocene 15:16–18.

Hemmings, C.A. 1999. Fossil Hole 8JE1497: an Inundated Quarry in the Lower Aucilla River, North Florida. Paper presented at Southeastern Archaeological Conference, Pensacola, Florida 1999.

Hoffman, C. 1983. A Mammoth Kill Site in the Silver Springs Run. The Florida Anthropologist 36:83–87.

Hoppe, K.A., P.L. Koch, R.W. Carlson, and S.D. Webb. 1999. Tracking Mammoths and Mastodons: Reconstruction of Migratory Behavior Using Strontium isotope ratios. Geology 27:439–442.

Jenks, A.E. and H.H. Simpson, Sr. 1941. Beveled Artifacts in Florida of the Same Type as Artifacts Found near Clovis, New Mexico. American Antiquity 6: 314–319.

Koch, P.L., K.A. Hoppe, and S.D. Webb. 1998. The Isotope ecology of Late Pleistocene Mammals in North America Part 1. Florida. Chemical Geology 152: 119–138.

Martin, Paul S. 1969. Prehistoric Overkill. Pleistocene Extinctions: The Search for a Cause. Proceeding of the VII Congress of the International Association for Quaternary Research, 6:75–120.Yale University Press, New Haven.

Meltzer, D.J. 1988. Late Pleistocene Human Adaptation in Eastern North America. Journal of World Prehistory 2:1–52.

Meltzer, D.J. 1993. Search For the First Americans. St. Remy Press, Montreal, Smithsonian Books, Washington D.C.

Pearson, G.A. 1999. North American Paleoindian Bi-Beveled Bone and Ivory Rods: A New Interpretation. North American Archaeologist xx:81–103

Purdy, B.A. 1981. Florida's Prehistoric Stone Technology. University Presses of Florida, Gainesville.

Robertson, J.S. 1974. Fossil Bison of Florida. In Pleistocene Mammals of Florida, edited by S.D. Webb, pp. 214–246. The University Presses of Florida, Gainesville.

Rosenau, J.C. Faulkner, G.I., Hendry, C.W. and Hull, R.W. 1977. Springs of Florida. Florida Bureau of Geology Bulletin 31 (revised), pp. 1–461. Florida Bureau of Geology, Tallahassee.

Tankersley, Kenneth B. 1998. Variation in the Early Paleoindian Economies of Late Pleistocene Eastern North America. American Antiquity 63:7–20.

Webb, S.D. 1974. Pleistocene Mammals of Florida. The University Presses of Florida, Gainesville.

Webb, S.D. 1976. Underwater Paleontology of Florida's Rivers. National Geographic Society Reports 1968 Projects: 479–481. Washington, D.C.

Webb, S.D., J.T. Milanich, R. Alexon, and J.S. Dunbar. 1984. A Bison antiquus Kill Site, Wacissa River, Jefferson County, Florida. American Antiquity 49(2):384–392.

Webb, S.D., J.S. Dunbar, and L.A. Newsom. 1992. Mastodon Digesta from North Florida. Current Research in the Pleistocene 9:114–116.

2. The View from Windover: 15 Years after Excavation

Glen H. Doran

Archaeological wet sites in Florida essentially span the human occupation of the New World. Like wet sites around the world the range of materials recovered from these sites provides striking contrasts to many terrestrial sites (Purdy 1988, 1991). Wet sites in general have the capability to dramatically alter our interpretation of the past (Coles 1984; Coles and Lawson 1987). The site being considered here is the Windover site (8BR246) located on the midpeninsular Atlantic coastal margin of Florida, very near the well known Cape Canaveral/Kennedy space installation and home to one of the world's most popular vacation destinations, Disney World in Orlando. As is often the case of wet sites Windover was found accidentally during road construction (Doran and Dickel 1988a). The developers, EKS (Eckerd, Kirshenbaum, and Swan), Inc., specifically, Jack Eckerd, Malcolm Kirshenbaum, and Jim Swann, not only worked with the archaeological team from Florida State University, but were essential to the funding and field operations of the project. In recognition of his support Jack Eckerd was awarded the United States Department of Interior's, Public Service Award (the highest honor the DOI secretary can give a private citizen). Jim Swann was awarded one of the Florida Archaeological Council's first Stewards of Heritage award (the FAC is the professional archaeological organization in the state of Florida). Without their assistance and support the Windover project never would have even begun, much less accomplished what we did. EKS was the project's supporter in the state legislature. Over a four year period the Florida legislature provided $870,000 for field work and initial laboratory analysis.

Florida's wetsite archaeologists, as elsewhere, are heavily reliant upon accidental discovery. Predictive models of where wet sites will be found, are largely undeveloped. They, from the Florida perspective at least, are so primitive, that the most realistic statement is that wet sites will be found associated with water or saturated sediments. Public involvement, and the interest of non-archaeologists, is of the utmost importance in site discovery and often in their excavation. At Windover we benefited not only from the discovery process but from the thousands of hours local volunteers contributed to the project.

Windover, is a burial, or mortuary pond containing intentional burials of a people occupying the east coast of Florida during what the archaeologists know as the Early Archaic (Figure 2.1). For unknown reasons, Florida's south central populations, at least in some cases, chose to inter the dead in small, shallow ponds. The basic pattern was to excavate shallow graves in the pond's margins. Because of the near anaerobic and chemically neutral environment, preservation of a wide array of organic materials, seldom recovered in terrestrial sites, survived, some in extraordinary condition. Over 20 radiocarbon dates clearly place the Windover burials between a maximum interval of 8,120 14C years before present to 6,900 14C years BP with a very tight clustering around 7,400 14C years BP. Earlier dates (10,400 years BP) obtained on basal peat deposits provide detailed chronological control of a variety of environmental indicators spanning the close of the Pleistocene to the modern epoch (Doran and Dickel 1988b). Multidisciplinary investigations involving a devoted group of researchers, literally from coast to coast, and in several countries, made the investigations particularly informative. There are few such charnel ponds known in North America and all are restricted to central and southern Florida (Bense 1994).

The unusual challenge of removing in controlled manner burials from the 7th millennium BP burials (168 partial and complete burials), many with associated organic artifacts, was accomplished by the installation of a well point system. The well point system, ultimately, encircled most of the

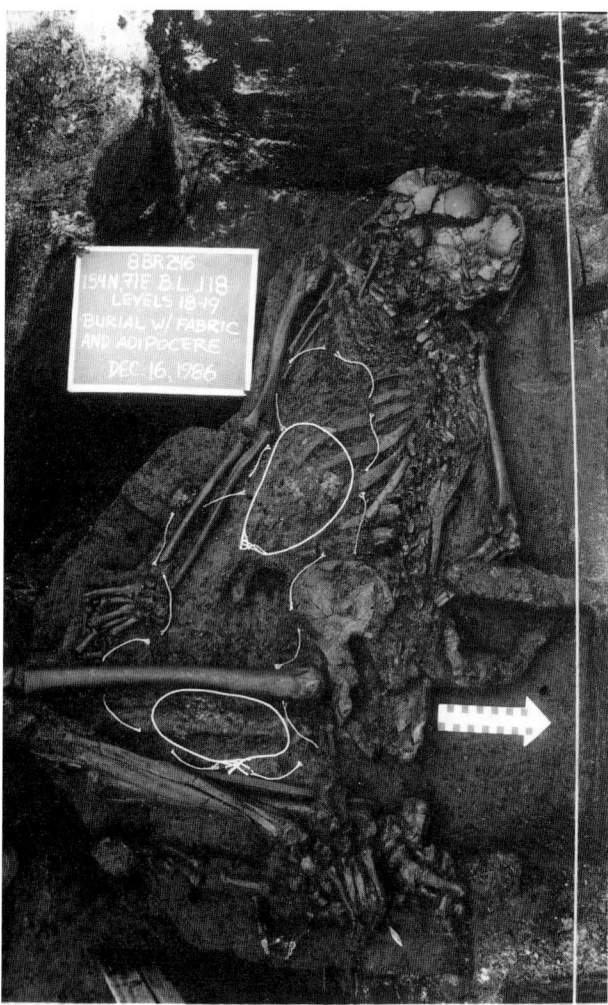

Figure 2.1 (photo 1004). Most burials were flexed and many were oriented with the heads to the west. This burial also exhibited both fabric materials and 'adipocere' ultimately best biochemically identified as the residue of tanned deer hide (this volume).

small pond and allowed excavations as much as 4 meters below the modern water level into a continuously deposited series of peat layers providing very specific information about changes in depositional context, water levels and environmental fluctuations. The well point system was capable of removing thousands of gallons of water an hour, allowing maximum excavation control matching the most carefully excavated terrestrial sites. The burials are approximately midway down in a nearly 4 meter peat deposit and roughly follow the curvilinear lenticular peat deposit forming the modern pond's bottom. Burials are not uniformly distributed but are clustered along earlier now buried pond margins. In some cases burials were held in place by intentionally cut stakes. Some smaller stakes penetrated

fabric wrappings. Larger stakes, sometimes grading into piles of natural wood, apparently helped hold burials in place. Some of the larger stakes whose tops protruded above the prehistoric pond surface, may have served as markers for socially distinct groups, possibly reflecting family, kin, or other social groupings. These and other aspects of biometrics and DNA characterization are still being examined.

Wood was clearly a critical element in the material inventory and a wide range of wooden artifacts include a double ended pestle, mortar, composite spear, and snare apparatus. The most common burial items include bone and antler tools, primarily but not exclusively manufactured from white tail deer. Other species were also used in tool manufacture and include manatee, Canidae, Felidae, shark, opossum, and turtle (Figure 2.2). A bottle gourd (Lagenaria siceraria) was recovered with a burial and provides evidence of this semidomesticate north of Mexico in an early, clearly nonagricultural context (Doran, Newsom and Dickel 1990; Newsom and Scarry 1998). What was of even greater interest, and provide unparalleled evidence of the sophistication of these early Floridians was the recovery of remarkably well preserved hand made, nonheadle loom manufactured fabrics, textiles, and cordage (Figure 2.3). The items were derived from processed plant materials of a restricted species, most likely sabal palm or saw palmetto. Adovasio *et al.* (this volume) provide additional detailed information on both these materials and on preservation issues associated with their conservation. These kinds of items, no doubt, were ubiquitous in other eastern United States contexts but they simply don't survive in most terrestrial sites (Bense 1994; Smith 1986). In contrast to these organic materials, and speaking to the diversity and complexity of the prehistoric material culture of these early Floridian's was a small, nay almost trivial, unexceptional series of only four lithic bifaces, and one biface tip, and a small set of modified and unmodified conch shells. Most of the specimens were found in questionable contexts (remember the initial backhoe discovery) and were not clearly associated with the Archaic burials. It is possible, they date to more recent depositional events. When no organic materials survive the passage of time, archaeologists by necessity focus on the durable elements of a society's inventory, and often dwell on lithics, or in later time intervals, ceramics, as the hallmarks of cultural markers, population sophistication, and subsistence strategies. Frequently we must make these cultural and chronological distinctions in the absence of the infinitely more informative and diverse organic inventories. Wet sites, if nothing else, are reminders, of what terrestrial sites lack. They provide a more accurate picture of the technological diversity and sophistication of an infinitely more critical technological corpus of material culture. Each and every

Figure 2.2 (photo 762). A wide variety of bone, antler, wood and dentary tools were recovered. White tailed deer, bobcat, canid, pelicanid and other species provided the raw material both for food and tool manufacture. Stable isotope data clearly indicates a terrestrial-inland marsh orientation.

Figure 2.3 (photo 1089). One of the best illustrated pieces of fabric clearly shows the amazing detail surviving after 8,000 years. Apparently made of saw palm or palmetto fibers the materials require both careful analysis and painstaking conservation (see this volume as well).

site, regardless of preservation conditions, can potentially provide important insights to prehistoric cultures, and it is up to the archaeologists, to capitalize on what materials are recovered, regardless of the inventory. No single site provides all the answers, but it is clear wet sites worldwide, can, and have, provided an astonishingly broader array of

materials. The greater expense and care required in wet site excavation, is the trade off.

Conservation becomes a paramount issue and must be central to any wet site investigation. None of the organics can survive beyond initial exposure. Without a strong, and often flexible program, to conserve each and every item, in a manner dictated by material composition and individual artifact category features, wet site materials will not survive. Some wet materials, at Windover and other sites, provide challenges that can tax the patience, knowledge, budget, and talents of even the most devoted. New techniques and strategies are both needed and constantly evolving. The Windover wood, for example, did not respond to traditional strategies. We performed many experiments on unmodified wood items, branches, deadfall wood, etc. These efforts graphically illustrated older wood, does not respond well to traditional techniques. Various immersions coatings and consolidation with sucrose, ethol-dammar resin, Rhoplex, PEG, etc. did not prove suitable to the Windover wooden artifacts, and certainly, not to the Windover fabrics. They did however, respond far more favorably to more involved techniques. Ultimately a combination of freeze drying and paralyne conformal coating (see Adovasio *et al.* in Doran nd.). Other soft tissue materials, specifically preserved brain masses from 91 individuals (Figure 2.4) of both sexes and all ages, while providing gross anatomical detail, have a greater potential for biomolecular analysis, and required

Figure 2.4. A few individuals during the first year produced samples of preserved brain tissue. In the northern section of the burial pond most burials contained preserved brain tissue. DNA from both bone or brain tissue show similar degrees of deterioration. Material from tooth pulp cavities appears to provide the best preserved and least contaminated DNA.

freezing at low temperatures (Doran *et al.* 1986). A residual "smear" of material identified in several burials, and analyzed by Adovasio's team was apparently the degraded remains of tanned deer hide inextricably enmeshed with a fabric wrapper. Rhoplex AC33, an acrylic emulsion which can be applied directly or by emersion to bone and antler provides superior conservation features rendering these harder, more durable bone and keratin materials indistinguishable on a gross level, from freshly recovered materials. No degradation has taken place in these items and it is clearly the best choice (Stone *et al.* 1990). The only caveat is that Rhoplex complexes with the material, and while it can be removed for DNA analysis, it makes stable isotope analysis difficult if not impossible. Freezing samples of materials possibly important in chemical analysis pertinent to dietary and even DNA analysis, should, we argue from experience, be an essential element of any wet site excavation serious about long term curation and continued expansion of analytical strategies, currently unanticipated, or now technically unforseen or impossible. Additionally, some items should be cast as a supplement to conservation. Such cast materials are superior for museum and educational displays and have the advantage that they are infinitely replaceable, durable. In most gross visible morphological aspects they are indistinguishable from the original specimens.

A primary focus of the investigations has been the bio-archaeological investigations of these individuals. As a part of this effort a variety of comparative data has been collated with some interesting results. Based on a continuing inventory of skeletal material from North America there are fewer than 80 specimens older than Windover. Windover appears to be the largest single sample of this antiquity, in the entire New World. The majority, around 88% of the North American skeletal samples providing the fundamental corpus of bio-archaeological interpretation come from the last 2,000 years. There was a dramatic increase in sample size and sample frequency after 2,000 years B.P. Basically the majority of our interpretations of New World health, demography, and morphology, is of necessity, based on these more recent material

Increasingly, and from morphological, paleopathological and genetic standpoints, these pre-7,000 samples, though few in number, and frequently poorly studied, appear consistently different from later New World populations. Explanatory hypotheses of these differences include dramatically different colonization scenarios, multiple migrations, in-place evolutionary processes with emphasis on population extinction, differential survival, and differential population contribution. Recently, an "Iberian" hypothesis, has even been discussed (Straus 2000). Some of these possibilities are more likely than others but clearly the traditional paradigms of colonization and population evolution are undergoing substantial re-evaluation in light of significant changes in our interpretation of the archaeological and osteological record. With respect to Windover's biological affiliation, DNA analysis, to date, has failed to identify any descendant living populations among modern Native Americans. Specifically an early haplotype, referred to as the "x-haplotype," has yet to be identified in living groups for which we have comparable data (Brown *et al.* 1998; Smith *et al.* 1999 and Smith *et al.* 2000).

These and other analyses, both ongoing and presented in other contexts, indicate that in some areas, particularly in areas like the central east coast of Florida, with densely packed rich inland marsh, riverine and lacustrine resources substantial departures from a simplistic model of small mobile populations at this early time interval can be observed. Resource richness, demographic success, indicate, compared to many populations, the east coast of Florida was a beneficent environment to which local populations like Windover, had effectively adapted. While bio-archaeological analysis is ongoing, and the details may change, some of the features indicate an effective early adaptation to freshwater river, marsh and ponds of the region with the exploitation of the obvious and more traditionally accepted terrestrial fauna and flora. While a series of artifacts indicate an awareness of marine resources, primarily in the form of shark's teeth, a few marine shell items frequently showing at best minimal alteration (and of uncertain provenience), the primary focus is clearly not on marine resources. The artifacts and stable isotope analysis support a terrestrial and freshwater subsistence orientation (Tuross *et al.* 1994).

From a demographic standpoint some individuals were living into the seventh decade. Sub-adult mortality was lower than in many later more complex societies. Maternal death rates may have been higher but were offset by possibly higher fertility rates than observed in many hunter-gatherer-fisher groups. To put these features in context, if given a choice of places and time to live, life was good at Windover. It was however, not without its challenges. Several individuals died from traumatic injury, only interpretable as intentional interpersonal conflict, and infections were not uncommon. These were, however, a robust people well adapted to their world. The details of their lives are still being obtained through ongoing analysis.

When first discovered, the brain tissue presented, what we thought at the time, a unique opportunity to obtain potentially better preserved DNA materials from archaeological samples. After a flurry of research activity worldwide, several things seem clear with respect to DNA analysis of archaeological materials. First, damage, and to a lesser extent, contamination are very real problems. Regardless of whether the focus is on the numerically less abundant nuclear DNA or on the more numerically abundant mitochondrial DNA, archaeological material is consistently damaged. Researchers are still unclear about what promotes damage (depurination, denaturation and fragmentation) but it is consistent. All materials can potentially produce DNA but in some cases the remnants are more badly damaged and the vagaries of preservation are so poorly understood it is best thought of, at least in my mind, as quasirandom. Until you run the samples you don't know whether useable DNA will have survived or not. As it turns out, Windover bone is just as useful, with the same preservation limitations, as the brain tissue. One material is clearly not superior to the other. We are engaged in a new protocol using dental pulp extractions which hypothetically are more productive but it is unclear what the impact will be on our interpretive capability. From a broad, and admittedly, not too technical reading of the archaeological DNA literature, several issues seem clear. My suspicion, is that most of the damage probably occurs in the first 60 years, perhaps even in a shorter duration than that. Once past this initial deterioration, it seems that the DNA is relatively "chronostable." There are numerous samples 1000 years old that are in no worse condition than those from Windover and others that are only 500 years old with no preserved DNA. You have to run the samples and hope. There is not an effective predictive strategy for recovery or quality of material recovered. The potential, however, is clear (Kolman and Tuross 2000). With adequate recovery and a larger number of samples from each population, the possibilities are exciting. It potentially can provide information on sex in sub-adults, which is notoriously difficult to assess from osteological material. This would allow more

careful assessments of gender differences in early life expectancy, mortality, disease incidence, etc. (Forster *et al.* 1996). Hypothetically, improved genetic information would also be valuable in assessments of interpretation of population colonization, gene flow. Such approaches will also allow a new way of assessing within population differences and certainly will supplement (maybe even supplant) studies of differences between populations and how we interpret them from a cultural historical standpoint. These are all critical issues, in a fuller more holistic, understanding of human prehistory.

This expanding interest in such issues also leads to other considerations. Our understanding of living populations is not as fully developed as it should be. Our comparisons between prehistoric and modern materials highlight this difficulty. There are many Native American populations for which we have little quality genetic information and there are others for which we have data, but only from a few individuals. Archaeology is the only mechanism by which we can get a better picture of what must have been a dramatic evolutionary bottleneck with the rise of western diseases and the tremendous mortality experienced after colonization. These issues can only be addressed by increased sampling efforts. This observation is certainly not unique to me. Brown and Brown noted several years ago 'Archaeologists can not hope to obtain from ancient DNA information that molecular biologists have not extracted from modern DNA' (1991:12).

There are clearly other Windovers in Florida. We know of several already. Bay West, Republic Groves and the peat materials in the slough at Little Salt Springs all slightly postdate Windover but are reflective of the same Archaic burial strategy. From the excavations at Windover we know that burials are not uniformly spaced around the pond. They occur in clusters with large intervening sections with essentially no burials. It is quite likely however, that each of these sites could have contained 300–400 individuals. We had hoped the wide publicity during excavations (1984-Jan. 1987) would have sparked public curiosity and other sites would have been reported to us. This has not been the case. Several years ago, however, a site like Windover was discovered in similar circumstances but with very different results. A large multiacre peat deposit associated with Ryder Pond (8LL1850) in south Florida was discovered by construction and no archaeologists were called in until most of the peat had been transported (Figure 2.5). The scattered skeletal material is visually very reminiscent of Windover but several things were different. Destruction/construction was sufficiently complete that excavation was limited to the point of being irrelevant and collection seemed the only appropriate strategy. Additionally, there was a loud and vocal outcry from representatives of the Native American community sufficient to attract the local press. True excav-

Figure 2.5. Ryder Pond (8LL1850) archaeologists were notified well after significant development (destruction) of the peat deposit and burials had take place. Dates indicate a Middle Archaic cemetery and from many perspectives represents an incalculable loss to the scientific community (photos and reported dates courtesy of Robert Carr).

ation was impossible even if the funds had been available. Robert Carr (a critical figure in the investigations of Bay West) and David Dickel (co-PI of the Windover project) were precluded from any serious investigative effort at Ryder Pond. Carr (personal communication 2000) obtained three radiocarbon dates on Ryder Pond (6000 BP, 7780 BP and 5870 BP) with the 5870 date being on a wooden burial stake or pole. As a response to the public outcry the materials were repatriated after, at best, a very modest analytical effort. Was affiliation really demonstrated? It seems the details of NAGPRA were ignored and one of the possible responses – reburial, was implemented, though there are other options that scientifically should have been considered. It is important for archaeologists and scientists outside the US to realize that scientific importance is no longer a sufficient criterion for investigation (excavation), collection, analysis or publication. This raises a very simple question.

NAGPRA, the Native American Graves Protection and Repatriation Act, is a 1990 federal statute (http://ww.cast.uark.edu/other/nps/nagpra/DOCS/lgm003.html). The simplest summary of NAGPRA is that it provides substantial protection for burials on federal property and requires all federally supported institutions (and that is virtually all United States universities and museums) inventory their collections of artifacts, burials and other items of potential cultural patrimony. It additionally, provides mechanisms by which federally recognized tribes (United States and Hawaii) can claim any or all of these materials – skeletal material, wampum pouches, art, ceramics, beads, etc. After following appropriate procedures, the materials may be repatriated to the appropriate tribal entities. The ultimate disposition of repatriated materials is up to the tribes. Skeletal material is frequently reburied, often on reservations or other protected locales sometimes close to the point of origination. Cultural items may be used or disposed of in any manner deemed appropriate by the tribes. While reburial is not automatic, communication is required of the holders of any such materials, and the mechanisms for claiming any and all such materials are provided for. Where there are disputes between the "holders" of the material and a legitimate claimant, a committee will decide the ultimate fate of the materials in question. NAGPRA does not mandate reburial but it is one outcome and a common one with respect to human skeletal material.

If another Windover was found during the early phase of a construction project and if there was ample funding for its investigation, could archaeologists and physical anthropologists actually undertake such an investigation? It seems unlikely. Many feel the expected outcry from even a small number of Native American's and/or their advocates concerned about disturbing prehistoric graves would be the death knell for any further investigations. It may well be that there will never be another opportunity to investigate a site like Windover. These excavations may not only have been a once in a lifetime opportunity for us as excavators, but may in fact have been a singular opportunity in New World archaeology. Clearly, the climate for excavations of Native America cemeteries has changed the practice of archaeology on a continental basis. Our understanding of the biological aspects of the lives of the people responsible for our species' greatest colonization event of the last 20,000 years is, for practical purposes, being largely limited to existing data sets. Was this the intent of NAGPRA?

When NAGPRA was first being discussed and debated on a national basis several things seemed clear. First and foremost, the operational assumption, both prior to passage and after passage, was that it basically dealt with materials of the last several hundred years. No one that I have ever talked to in the professional community, envisioned that it would functionally reach into the depths of antiquity. The Kennewick Man, Spirit Cave, Buhl, and others being cases in point (Chatters 2000). Operationally, and increasingly so from the perspective of land, forest and collection managers, there is a willingness to "Just Say No" to excavation and analysis, and presume the expectation of reburial. What is happening is that most claims are being honored regardless of the antiquity of the material and regardless of the demonstration of a clear connection. Those previously "unaffiliated" materials are suddenly within the reach of just about anyone wanting to claim them. Again, was this the intent of NAGPRA?

Over the last several years I have been involved in federally funded conferences or workshops on NAGPRA, primarily with southeastern federally recognized tribes. Initially, there was little coherence on the tribal position on unaffiliated materials (generally anything older than 500 years, and certainly most materials older than 1,500 years). In the past year this position has coalesced dramatically and at the last National Park Service/Florida State University sponsored NAGPRA conference, attended by representatives of most of the southeastern federally recognized tribes, a position paper was presented clearly stating that all NAGPRA related material was to be repatriated and there was no recognition of any validity in the concept of "unaffiliated" materials. This is a dramatic change and one that, again raises the question of, "Was this the intent of NAGPRA?"

The passage of NAGPRA is, in some opinions, one of the 20th century's most profound federal legislative mandates affecting North American archaeology. It has the ability to dramatically change the course of a great deal of research in North America. This is the law of the land and one we are all seeking to abide by. It clearly has several

elements which are laudable. In the most simplistic of terms it has dramatically increased dialogue between tribal peoples and the archaeological and academic community. This is positive. It has compelled the bio-archaeological community to make an effort to establish a more codified set of data collection protocols and requirements (Buikstra and Ubelaker 1994). This is positive. The NAGPRA inventories provide the basis for construction of a database that potentially will allow a better picture of sample distributions even if it has inherent weaknesses and limitations. This is positive as well. Will the next generation of bioarchaeologists of necessity turn their back on research in North America? Will our knowledge base become static? Whether the positives will outweigh the negatives will be answered by the future generations of researchers. My fear is that this truly is the death knell for the expansion of our understanding of many unique elements of the New World prehistoric record.

These changes also highlight problems with physical anthropology on several different levels. Right or wrong, the majority of bio-archaeological anthropologists of this and earlier generations, with few exceptions, obtained their training using Native American skeletal collections. Most instructors had only a handful of non-Native American materials (the occasional forensic specimen and perhaps a small series of prepared specimens from India and more recently China). Use of Native American materials in these contexts is increasingly problematic and potentially subjects the instructor to harsh criticism from ethical, legal and other grounds. As a discipline we have not adequately addressed this issue. Where can quality samples of human skeletal material be obtained for pedagogical uses but also for research questions? Is the modern "unclaimed forensic specimen" going to ultimately be the sole provider for such materials. Here too, there are ethical issues and issues of chain of custody that remain problems. Perhaps senior faculty should think about more personal donations. There are precedents for these kinds of donations and this is not presented lightly. Even if widely accepted it would however result in a continuing issue of age bias already a recognized problem with many "modern" collections. Again, there is no simple answer to this problem but it is a very real problem.

As noted earlier, as a result of the Windover project we began collecting information on other New World skeletal samples. Initially this focused on pre-5,000 BP samples, then expanded to all samples of greater than 50 individuals (mainly those with age/sex information suitable for paleo-demographic study), then expanded to include three databases one with cranial metrics and a second focusing on long bone metrics and a third with dental metrics. The admittedly idiosyncratic nature of bio-archaeology made it clear there were no such broad databases in existence.

Individual researchers were independently accumulating datasets but in an erratic nature. Three databases which were also moving in these directions, included Howell's excellent and professionally noble data set on cranial material collected from a series of globally distributed populations. This is available for download from the University of Tennessee and has been tremendously helpful to many researchers world wide. A second effort was Jerry Rose's effort to develop a common data structure paralleling the Standards of Osteological Data Collection. At last word this had relatively few samples. The third, or at least the initial information collection, is associated with the NAGPRA inventories which contain basic archaeological information. Issues that are not well addressed, and we are trying to address in the Florida State University databases, are careful control of racial/ethnic/geographic/ population attribution, archaeological context, and chronological placement. Other disciplines have collaborated in developing discipline wide databases (geology and human genetics to mention only two) and it seems obvious that bio-archaeological databases should be given a higher priority on a discipline wide basis. Such databases will constantly be growing simply by incorporating existing published information. We at Florida State University are currently developing appropriate error checking protocols and resolving sample issues and they are currently only available to Florida State University students and faculty. The paleodemographic data-base now contains 44,000 individuals for North America and 6,200 for South America. The comparative samples span the last 15,000 years and include approximately 1,000 individuals in the dental database, 6,600 in the craniometric database and 560 in the long bone database. Collectively they contain information from over 60 countries and 400 sites. Clearly, these efforts barely scratch the surface of what exists in published and unpublished form. Ultimately, with wider community support and participation such databases have, in my opinion, an important role to play and one that deserves more attention and funding than currently is available. The need exists. Whether we as a community have the will to pursue this avenue, is another issue.

Discussion

Wet sites have a tremendous potential to open new horizons in the interpretation of human history and prehistory. Each site has a unique story to tell. Windover was, even more so than many cases, a unique convergence of opportunity, funding and luck. As indicated previously, it may well be a unique, opportunity never to be repeated. From a research perspective the things which make Windover unique and significant include the following:

It is one of the largest, if not the largest, skeletal sample of this antiquity (7,400 14C years BP, uncorrected) in the New World. It consists of a minimum of 168 individuals, roughly 50% adults and 50% subadults, many of which exhibit remarkable preservation.

A stunning variety of faunal and floral material, artifactual and nonartifactual, directly and indirectly associated with the burial activities was recovered and one burial item, a largely intact bottle gourd (*Lagenaria siceraria*) dating to 7,290 ±120 14C years BP (uncorrected) is the oldest representative of this species north of Mexico.

It has produced the largest most complex collection of textiles from this time period in New World and provides an exceptional picture of the diversity and textile practices presumably representative of most New World populations at this time period. Preserved wooden artifacts also provide a unique technological picture.

It has an excellent pollen, floral, and lithographic record of deposition and environmental change from the close of the Pleistocene to the present. This sequence provides an excellent record of both local and regional environmental changes.

It has produced some of the oldest human DNA in the world, initially from some of the 92 preserved human brains, but subsequently also from the bone itself.

It has provided a detailed reconstruction of diet from the analysis of gut contents and from the isotopic analysis of human bone samples.

At the same time, Windover highlights, some of the challenges being faced by North American archaeologist. Future generations will have to assess whether we have made, and are making, the right decisions.

References Cited

Bense, Judith A. 1994. Archaeology of the Southeastern United States. Academic Press, San Diego.

Brown, Terence A. and Keri A. Brown. 1991. Ancient DNA and the archaeologist. Antiquity 66:10–23.

Brown, M.D., S.H. Hosseini, A. Torroni, H-J, Bandelt, J.C. Allen, T.G. Schurr, R. Scozzari, F. Cruciani, and D.C. Wallace, 1998. mtDNA haplogroup X: An ancient link between Europe/Western Asia and North America? American Journal of Human Genetics 63:1852–1861.

Buikstra, Jane E. and Douglas H. Ubelaker. 1994. Standards for Data Collection From Human Skeletal Remains. Arkansas Archeological Survey Research Series No. 44. Fayetteville, Arkansas.

Chatters, James C. 2000. The recovery and first analysis of an Early Holocene human skeleton from Kennewick, Washington. American Antiquity 65:291–316.

Coles, John M. 1984. The Archaeology of Wetlands. Edinburgh University Press, Edinburgh, Scotland.

Coles, John M. and Andrew Lawson (editors). 1987. European Wetlands in Prehistory. Clarendon Press, Oxford.

Doran, Glen H. and D.N. Dickel. 1988a. Multidisciplinary investigations at the Windover Site. In Wetsite Archaeology, edited by B.A. Purdy, pp. 263–289. Telford Press, Caldwell, New Jersey.

Doran, Glen H. and D.N. Dickel. 1988b. Radiometric chronology of the archaic Windover archaeological site (8-Br-246). Florida Anthropologist 41:365–380

Doran, Glen H., David N. Dickel, William E. Ballinger, Jr., O. Frank Agee, Philip J. Laipis, and William W. Hauswirth. 1986. Anatomical, cellular and molecular analysis of 8,000-yr-old human brain tissue from the Windover archaeological site. Nature 323:803–806.

Doran, Glen H., L.A. Newsom and David N. Dickel. 1990. A 7,290 year old bottle gourd from the Windover Site, Florida. American Antiquity 55:354–360.

Forster, P., R. Harding, A. Torroni, H-J. Bandelt. 1996. Origin and evolution of Native American mtDNA variation: A reappraisal. American Journal of Human Genetics 59:935–945.

Kolman, Connie J. and Noreen Tuross. 2000. Ancient DNA analysis of human populations. American Journal of Physical Anthropology 111:5–23.

NAGPRA. 2000. http://www.cast.uark.edu/other/nps/nagpra/DOCS/lgm003.html

Newsom, Lee A. and C.M. Scarry. 1998. Homegardens and Mangrove Swamps: Pineland Archaeobotanical Research. Chapter 6 in The Archaeology of Pineland: a Coastal Southwest Florida Village Complex, A.D. 50–1600, K.J. Walker and W. Marquardt, editors. University Press of Florida, Gainesville.

Purdy, Barbara. 1988. Wet Site Archaeology (editor). Telford Press, Caldwell, New Jersey.

Purdy, Barbara. 1991. The Art and Archaeology of Florida's Wetlands. CRC Press, Boca Raton, Florida.

Smith, B.D. 1986. The Archaeology of the Southeastern United States: From Dalton to de Soto, 10,500–500 B.P. In Advances in World Archaeology, Vol.5, pp. 1–92, edited by Fred Wendorf and Angela E. Close, Academic Press, Orlando.

Smith, David G., Joseph Lorenz, Becky K. Rolfs, Robert L. Bettinger, Brian Green, Jason Eshleman, Beth Schultz, and Ripan Malhi. 2000. Implications of the distribution of albumin Naskapi and albumin Mexico for New World Prehistory. American Journal of Physical Anthropology 111:557–572.

Smith, David. G., Ripan S. Malhi, Jason Eshleman, Joseph G. Lorenz and Frederika A. Kaestle 1999. Distribution of mtDNA haplogroup X among Native North Americans. American Journal of Physical Anthropology 110:271–284.

Stone, Tammy, David N. Dickel and Glen H. Doran. 1990. The preservation and conservation of waterlogged bone from the Windover Site Florida: a comparison of methods. Journal of Field Archaeology 17(2):177–186.

Straus, Lawrence Guy. 2000. Solutrean settlement of North America? A review of reality. American Antiquity 65:219–226.

Tuross, Noreen, Marilyn L. Fogel, Lee Newsom and Glen H. Doran. 1994. Subsistence in the Florida Archaic: the stable-isotope and archaeobotanical evidence from the Windover site. American Antiquity 59:288–303.

3. Textiles and Cordage from Windover Bog (8BR246)

R. L. Andrews, J. M. Adovasio, D. C. Hyland, D. G. Harding and J. S. Illingworth

Introduction

During the 1986 and 1987 excavations at the Windover Site (8BR246), 87 perishable fiber artifacts representing 67 separate and once complete items, and an associated suite of modified wood products were recovered, temporarily preserved, and ultimately shipped to the Perishables Analysis Facility, University of Pittsburgh, for conservation, analysis, and interpretation. Upon the transfer of that research entity to Mercyhurst College in 1990, and the completion of suitable laboratories and environmentally controlled facilities, the Windover perishable artifact collection was moved to the R.L. Andrews Center for Perishables Analysis for final study and conservation.

Preliminary results of the Windover perishable artifact analyses include Andrews and Adovasio (1988), Andrews and Adovasio (1996), Andrews, Adovasio, and Harding (1988) and the final reports on the textile and related perishable remains, as well as the wood artifacts, are currently in press (Adovasio *et al.* 2000; Andrews *et al.* 2000). The observations which follow are derived from those sources.

Analytical Procedures

Methodology of Analysis

Prior to conservation and stabilization (Andrews *et al.* 2000; Gardner 1988), each individual sample was analyzed and photographed in the wet state. Perishable class, subclass, and type were ascertained with a hand lens, Olympus stereoscopic scanner, and where warranted, a Wild microscope. Due to the desiccatory effect of high-intensity microscope lighting, measurements were obtained with a Helios needle-nose dial caliper rather than with an optical micrometer. Raw material analysis involved both light and scanning electron microscopy as well as comparative pyrolysis mass spectrometry. Scanning electron microscopy, energy dispersive X-ray analysis, nuclear magnetic resonance imaging, diffuse reflectance infrared spectroscopy, light microscopy with stains, amino acid analysis, and immunologically based testing all were employed to determine the composition of a red-brown veneer of debris that encrusts the surface of a few textile samples. Osteological and pathological analyses of the human skeletal remains associated with three Windover fabric samples were facilitated via X-rays and a Helios needle-nosed dial caliper.

All data were recorded on standardized analysis forms. A sketch of each item with notation of its salient attributes (i.e., number and orientation of folds and layers; orientation of weft rows, plaiting elements, and/or grass bundles; and horizontal position of floral and faunal associations) and its overall dimensions also were prepared. The degree of flexibility, the degree of attrition wear, possible form and function, as well as raw material and method of its preparation were recorded, when possible. Subclass attributes also were recorded and are described below (see Criteria of Classification).

All metric and nominal twining attributes were coded and computerized utilizing Condor 3, a relational database management system. SIMS was employed to calculate unstandardized and standardized "dissimilarity coefficients between cases" (Drennan, personal communication 1986) on six metric (ratio) variables by and among types. The subsequent matrices were scaled via a SYSTAT multidimensional scaling (MDS) program in one through five dimensions. Plots of the dimensions with the appropriate "elbow" and acceptable Kruskal stress level (Kruskal and Wish 1976), with concomitant low-level exploratory and confirmatory statistics, were examined for the following factors:

1. The effects of air desiccation (prior to shipment to the University of Pittsburgh) on different samples from the same burial lot.
2. The degree of standardization of manufacture by and across types and within and between burial lots.
3. The number of forms and originally complete items represented per burial lot.
4. Horizontal and vertical correlation of form and type.
5. The relationship between structural type, ratio data, age, sex, orientation, associated grave goods, and burial lots.
6. The existence of "populations" of specimens which may reflect discrete groups of weavers.
7. The range of individual or idiosyncratic variation in weaving.
8. The Euclidean "distance" between residual untypeable twining samples and the bona fide types in order to propose potential classifications for the residual specimens

As previously noted, all of the Windover perishables were photographed in a 35 mm format with high-resolution black-and-white copy film prior to the initiation of any conservation/stabilization measures. Both blue and green filters were employed to enhance the structural detail of the items. Close-up photographs were also taken of unique phenomena and/or structural features. Plant material and tissue samples were photomicrographed with a Wild microscope and a scanning electron microscope. Additionally, computer enhanced images of several specimens were also generated.

After the completion of conservation/stabilization, all specimens were re-examined in the dry state to ascertain if any attributes or features were missed during the wet-state observation. The results of the reanalyses were consistent with those obtained during wet-state study with several exceptions that are noted below.

Criteria of Classification

Basketry, as discussed here, includes several distinct kinds of items, including rigid and semi-rigid containers, or baskets proper, matting and bags. Matting includes items which are essentially two-dimensional or flat, whereas baskets are more clearly three-dimensional. Bags may be viewed as intermediate forms because they are more or less two-dimensional when empty and three-dimensional when filled. As Driver (1961:159) points out, these artifacts can be analyzed in a similar fashion because the overall technique of manufacture is the same in all instances. Specifically, all forms of basketry are manually woven without frame or loom. Since all basketry is woven, it is technically a class or variety of textile. In the present context, the term textile is restricted to fully or infinitely

flexible materials, such as cloth or fabric, produced with the aid of a frame or loom.

There are three major subclasses of basket weaves that are usually viewed as mutually exclusive: twining, coiling, and plaiting. Twining denotes a subclass of basket weaves manufactured by passing moving (active) horizontal elements called wefts around stationary (passive) vertical elements called warps. Twining techniques may be employed to produce containers, mats, and bags as well as fish traps, cradles, hats, clothing, and other "atypical" basketry forms. Coiling denotes a subclass of basket weaves manufactured by sewing stationary horizontal elements (the foundation) with moving vertical elements (stitches). Coiling techniques are used almost exclusively in the production of containers, hats, and, rarely, bags. Mats and other forms are seldom, if ever, produced by coiling. Plaiting denotes a subclass of basket weaves in which all elements pass over and under each other without engagement. For this reason, plaited basketry is technically described as unsewn. Plaiting may be used to make containers, bags, and mats, as well as a wide range of other non-standard forms.

Cordage herein denotes a class of elongate fiber constructions whose components are generally subsumed under the English terms string and rope. At Windover, these items are either spun and twisted or braided.

As noted above, 87 perishable artifacts (excluding wood and bone), probably representing 67 once-complete items, were recovered during the 1986–1987 excavations at Windover. The basketry includes four samples of twining. The textiles are represented by 49 pieces of twining and by a single example of plaiting or plain weave. Of the cordage samples, 10 consist of spun and twisted fiber, and two of braided elements.

Twined specimens were allocated to four bona fide structural types, as well as to three residual twining categories that cannot be completely classified as to type based on the number and sequence of warps engaged at each weft crossing and/or on the spacing of the weft rows. These items also were analyzed for selvage, method of starting, method of insertion of new warp and weft elements, method of preparation of warps and wefts, form, wear patterns, function, decorative patterns and mechanics, type and mechanics of mending, and raw materials. Width of individual warp and warp unit, number of warps per centimeter, width of individual weft ply and weft unit, weft gap, and number of wefts per centimeter also were recorded for each twined specimen.

The single plaited textile specimen was allocated to a structural type based on interval of element engagement. This specimen was analyzed for selvage treatment, shifts, method of preparation of elements, form, wear patterns, function, type and mechanics of mending, and decorative patterns and mechanics. Width, orientation, and angle of plaited element crossing also were recorded.

The 10 pieces of spun and twisted cordage were assigned to a single structural type based upon the number of plies and the direction of spin and twist. Length, overall diameter and individual ply diameter, angle of twist, and number of twists per centimeter were recorded for the cordage.

The two braided specimens were allocated to a single structural type based on the number of plaited elements. Length, overall diameter, and diameter of individual plies were noted for each item.

The standard descriptive terms utilized in the examination of the Windover perishables are those employed by Emery (1966), Adovasio (1977), and Shaw (1972). Cordage formulae follow Hurley (1979) and are shown in Table 3.1.

The Windover Perishable Artifact Industry

Twining

Five formal structural types and two residual untyped twining categories account for 98.1% of the Windover basketry/textile assemblage. They are described in detail in Andrews *et al.* (2000). The 73 samples in this category represent a minimum of 53 originally complete forms, and it is upon these forms that the following observations are based. Four of the five twining types represented at the Windover site are close twined, one is open twined. The untyped residual categories are both close twined. Table 3.2 shows the associations, distribution, and attributes of specimens ascribing to the five formal twining types and the single formal plaiting type in the Windover assemblage.

Warp Treatment

All of the close-twined specimens (i.e., all those containing typeable warp material) exhibit twisted and/or spun and twisted cordage warps. (It should be noted that the small diameter, wet state, and friability of the warps themselves sometimes hindered the exact determination of ply formulae). The identifiable variants include S twisted; Z twisted; two-ply, Z spun, S twisted; and two-ply, S spun, Z twisted cordage warps. It is possible, however, that the single-ply variants were originally part of multiple-ply constructions. For the purpose of this discussion, all warps are described only on the basis of their final twist.

Of all the close-twined specimens with typeable warps, S twist and Z twist warps each compose 47.6% of the total. Curiously enough, two samples (which represent the remaining 4.8% of the total) of Type III exhibit both S twisted and Z twisted warp material in unequal proportions within the same pieces of cloth. Type I comprises 7.5% of the total twining sample. Unlike the three other twining types identified in the assemblage, Type I specimens

Table 3.1. Description and ply formulae of cordage recovered from the Windover Bog site (8BR146).

Cordage Type	Description	Ply Formula
Type I	Two-Ply, Z Spun, S Twist	S^{Z}_{Z}
Type II	Three Strand Braid, S Spun	$\#^{S}_{S\,S}$
Type III	Two-Ply, S Spun, Z Twist	Z^{S}_{S}

contain only Z twisted warps. Due to the low numerical frequency of this type in the assemblage, its correlation with warp type is not statistically significant.

Type II, Type III, and Type IV account for 3.8%, 49.1%, and 26.4% of the total twining assemblage, respectively. Specimens within these types contain either Z or S twisted cordage warps. Further, the Z and S twisted warp variations are represented almost equally within each type. Specifically, the S twisted warp variants make up 50%, 57%, and 55% of Types II, III, and IV, respectively.

The single example of Type VI represents only 1.9% of the total twining assemblage. The warps of this specimen consist of unspun bundles of grass.

Weft Treatment

All of the Windover close twining specimens contain S twisted wefts. Further, the individual weft ply is inevitably composed of loosely Z twisted plant material. The weft itself is composed of either two or three of these loosely twisted plies. Paired and trebled weft varieties compose some 67.3% and 30.8% of the close-twined sample, respectively. The number of weft plies cannot be ascertained on one specimen (1.9% of the close-twined sample). In direct contrast to the close twining, the single open twined item exhibits a predominance of S slant, or twist, weft rows. However, *at least one* row is Z twist. While this reversal of twining slant is unusual and unparalleled at Windover, it is by no means unique in the archaeological record of North America (see Adovasio 1977).

Centers

Unfortunately, none of the centers of the Type III bags are intact; thus, the method by which twining was initiated cannot be determined. Based on specimen configuration, however, it is likely that radial twining on crossed warps was employed. This technique is often seen on Basketmaker II and III bags from the Southwest (Guernsey and Kidder

1921; Kent 1983) as well as similar forms from the Great Basin (Cressman 1942).

Selvages

Only five of the Windover twining samples exhibit intact portions of side selvages. All selvages are of the continuous weft variety and are found on examples of Type III (n=3) and Type IV (n=2). In one of these Type IV selvages, the weft courses are possibly reinforced with four rows of simple looping spaced ca. 8 mm apart. The sewn "thread" consists of Two Ply, Z spun, S Twist Cordage, and the work direction is left-to-right.

Only one of the close-twined specimens from Windover preserves an end selvage. This sample, a Type IV globular bag fragment, exhibits an elaboration of a 180° self end selvage. Each Two Ply, Z Spun, S Twist cordage warp "end" was given three counterclockwise twists as it was drawn over a Two Ply, S Spun, Z Twist cord which probably served to close the bag. The warp end is twisted in a counterclockwise direction twice about itself and is stabilized with a single binding course of close simple twining with paired wefts.

The single example of Type VI recovered from the Windover site does not retain side or end selvages.

Splices

Among the Windover perishables, the "laid in" splice is the most frequent technique for inserting new weft elements into close twining. In this method, additional weft material is simply inserted beneath or next to the exhausted weft ply. Examples of this splice type are evident in Type I (n=3); Type II (n=3); Type III (n=15); Type IV (n=9); and one of the residual specimens of close twining with paired, S twist wefts. Additionally, in three examples, "laid in" splices occur in combination with a technique in which new material is looped around a warp. In another Type III specimen, standard "laid in" splices are coupled with a method in which the additional weft material is folded, and then "laid in." Finally, a third variation is illustrated by a Type IV example, where "laid in" splices are combined with a technique where new material is looped around the exhausted weft ply.

Warp splices employed by the Windover weavers in close twining include the simple insertion of new warp material into a pre-existing weft crossing (n=1) and the "V" splice, in which additional warp material is folded into a "V" and inserted into pre-existing weft crossings. Windover examples of the "V" technique include single "V" (Type II [n=2]) and double "V" (Type III [n=2]) variations. "V" splices are common techniques to control the shape and size of bags and certain types of garments (e.g., hoods) and hence are often indicative of form.

The single open twining specimen recovered from the Windover site does not exhibit splices.

Mending, Charring, and Decoration

One of the Windover close-twined specimens (Type III) apparently was mended in antiquity. A group of five cords consisting of loose warps, dangling wefts, and perhaps lengths of Type I cord are joined with a square knot. Another repair occurs ca. 3.5 cm from this mend and consists of knotted wefts, though the knot itself is unidentifiable. The open-twined specimen is unmended. Only one example of Type III twining from the Windover site is charred.

Unless the stitching adjacent to the Type III close-twined 180° end selvage represents an attempt at structural decoration (i.e., wrapped twining), none of the Windover perishables are decorated, at least not by any technique that has been preserved or that has been detected to date in this study.

Attrition Wear

Excluding the grass warps contained in the single open-twined specimen, 54.9% of the Windover twining specimens show wear patterns typical of pre-depositional attrition/abrasion. The vast majority (88.24%) of these items are lightly worn. The frequency of wear on these lightly worn specimens (indicated in parenthesis by type) is variable: Type II (100.0%), Type VI (100%), Type IV (44.4%), Type III (41.2%), and Type I (25.0%). Some 6.5% of the twined specimens that show wear are moderately worn and include examples of Type I (25%) and Type III (5.9%). The specimens without attrition wear account for 45.1% of the twined assemblage and include examples of Type III (52.9%), Type I (50.0%), and Type IV (42.0%).

Raw Materials

Three genera of raw materials were employed in the manufacture of the Windover twined perishables. All but two (8.5%) of the specimens are composed of the fibers of a member of the palm family (Arecaceae), either cabbage palmetto (*Sabal* sp.) or, less likely, saw palmetto (*Serenoa repens*) (Newsom 2000 [see below]). These exceptions include a single specimen of Type IV twining and the solitary example of open twining. In both instances, the weft material is either palm/palmetto or saw palm. However, the Type IV warps consist of highly deteriorated fibers that bear a marked resemblance to yucca (*Yucca* spp.). The warps of the open-twined specimen are grass bundles of unknown taxa.

Attempts to differentiate between the two "candidate" genera of the Arecaceae have been extensive and intensive. Initial identification techniques employed include histological studies (Andrews and Adovasio 1988; Hess,

Table 3.2. Textile/Basketry Associations, Distribution, and Attributes for Formal Types from the Windover Bog Site (8BR246).

Burial								Warp			Weft		
Lot	Sex	Age (yr)	Specimen Location[a]	Unit	Level (cm)	Specimens (N)	Whole Items (N)	Element Dia. (mm)	Unit Dia. (mm)	Per cm	Element Dia. (mm)	Unit Dia. (mm)	Per cm
Type I, Close Simple Twining, Paired S Twist Wefts													
92	male	50	head shoulders	N150 E73	24–25	5	4	0.95	1.87	5	1.81	2.47	4.5
Type II, Close Simple Twining, Trebled S Twist Wefts													
73	unknown	7	entire body	N150 E79	19	3	1	1.90	–	6	1.20	1.90	5.5
82	female	38	face	N148 E75	20	1	1	1.90	–	4	1.40	2.00	5
Type III, Close Diagonal Twining, Paired S Twist Wefts													
75	unknown	10	femur	N158 E75	20	1	1	1.00	1.70	4.5	1.30	1.90	5
86	unknown	15	under body	N150 E73	20	1	1	–	–	6	0.90	1.80	6
90	unknown	11	entire body	N148 E75	20–21	7	4	0.73	1.18	5	1.04	1.76	4.5
91	male?	46	shoulders and arms	N148 E75	21–22	2	1	1.10	1.70	5	1.10	1.80	5
95	male	23	back	N148 E73	21	1	1	0.70	1.00	4	0.90	1.50	6
102	male	29	pelvis	N148 E71	18	1	1	0.40	–	9	0.50	0.80	12
103	female	37	back	N154 E71	18	2	2	0.55	1.70	6	1.00	1.75	5.25
113	unknown	9	pelvis	N152 E73	21	1	1	0.60	1.80	7	1.10	1.60	5.5
114	male	39	feet and legs	N154 E71	18–19	2	1	0.40	1.40	8	1.25	1.85	5
116	unknown	11	pelvis	N152 E71	19	1	1	–	–	–	0.90	1.50	6
117	female	36	ribs	N152 E71	18	2	2	0.65	1.45	7.5	0.90	1.55	5.75
118	male	37	torso	N154 E71	18	2	2	–	–	–	1.30	2.15	5.75
131	male	65	neck and jaw	N148 E69	27	1	1	0.80	2.10	4	1.80	2.50	4
137	unknown	0.5	jaw	N158 E69	26	1	1	0.50	1.10	9	0.80	1.40	7
139	unknown	2	back shoulder	N148 E69	28	1	1	0.60	1.90	8	1.20	2.00	4.5
150	unknown	13	entire body	N148 E69	25	4	2	0.78	1.83	8	1.20	1.78	5
156	unknown	10	humerus	N148 E69	26	1	1	0.90	2.10	8	1.30	2.00	4
157	female	46	uncertain	N150 E69	24	2	2	0.75	2.10	7.5	1.45	2.10	4
Type IV, Close Diagonal Twining, Trebled S Twist Wefts													
82	female	38	torso, arms	N148 E75	20	3	2	1.79	3.79	4	1.42	2.32	3.5
92	male	50	pelvis	N150 E75	24	2	1	0.72	1.66	9	0.96	2.24	4

Table 3.2. continued

Burial						Specimens (N)	Whole Items (N)	Warp			Weft		
Lot	Sex	Age (yr)	Specimen Location[a]	Unit	Level (cm)			Element Dia. (mm)	Unit Dia. (mm)	Per cm	Element Dia. (mm)	Unit Dia. (mm)	Per cm (mm)
Type IV – continued													
93	female	58	uncertain	N150 E71	21–22	4	2	0.84	2.50	7	1.01	2.17	4.25
94	male	49	entire body	N150 E71	20–22	2	1	0.79	1.61	5	1.34	2.22	4.5
104	female?	56	left side	N154 E71	18	1	1	0.48	0.97	8	1.36	2.17	4
113	unknown	9	legs	N152 E73	20–21	4	3	0.60	2.25	6	1.06	1.65	5.5
131	male	65	cranium	N148 E71	26	1	1	1.30	3.11	6	1.25	2.83	3.75
133	unknown	10	right arm	N152 E69	19	1	1	–	–	–	0.77	1.90	6
150	unknown	13	pelvis and legs	N148 E69	25	3	1	0.59	1.28	9	1.20	1.60	5.5
152	male	51	right femur	N148 E69	–	1	1	–	–	–	1.03	1.56	4.5
Type V: Simple Plaiting/Balanced Plain Weave													
90	unknown	11	pelvis	N148 E75	20–21	1	1	0.05	–	10	0.05	–	10
Type VI: Open Simple Twining, Paired S and Z Twist Wefts													
131	male	65	pelvis and thigh	N148 E69	27	1	1	1.35	1.58	4	1.46	–	–

a: Specimen location indicates location of the textile on the burial.

personal communication 1987; Newsom 2000; Newsom, personal communication 1986); phytolith identification (Piperno, personal communication 1987); and analysis of associated pollen (Barnosky and King, personal communication 1988). While all of these studies, particularly the histological examination, suggest that most of the fiber is not bast but rather consists of vascular bundle fiber strands, none are conclusive as to genus or species, principally due to the deteriorated and "peatified" state of the Windover fiber samples. Direct comparisons by Newsom (2000) and Andrews to vascular strand fiber from modern plants and archaeological specimens indicate that the principal raw material used by the Windover weavers was probably a member of the Arecaceae, either *Sabal* sp. or *Serenoa repens*.

Intensive scanning electron microscopy (SEM) scrutiny of the Windover fabrics and modern plant sources, following the protocols of Catling and Grayson (1982), confirmed, at least for these writers, that the most likely candidate for the aboriginal Windover fiber source was cabbage palmetto (*Sabal* sp.). This attribution is also favored by Newsom (2000) and is indirectly supported by the recovery of unmistakable cabbage palmetto (*Sabal* sp.) petioles in the form of a composite construction.

Raw Material Preparation

In all specimens except the open-twined example (Type VI), the plant epidermis apparently was removed from the leaf or frond before the fibers were extracted. A large majority of specimens (79.6%) retain no evidence of the original plant cortex (Residual Untyped Twining [100%]; Type IV [92.3%]; Type III [83.0%]). In the remaining twined specimens (20.4%), the fibers contain widely distributed, minute epidermal patches (Type I [100%], Type II [100%], Type [16.7%], Type IV [9.1%]).

The technique by which plant fibers were extracted from the leaf or frond is unknown, but 64.0% of the twined fibers (excluding the open-twined warp material) apparently were extracted by scraping the plant with some rigid tool (e.g., flaked stone tool, wooden scraper, shell, etc.) and/or by dental abrasion/mastication. Of these specimens, 66.6% are lightly altered (Type I [100.0%], Type II [50.0%], Type III [40.0%], Type IV [55.5%]). The remaining 36.0% exhibit a moderate degree of alteration (Type II [50.0%], Type III [30.0%], Type IV [22.2%]). No evidence of mechanical alteration or of mastication of the fibers is detected in 28.6% of the twining assemblage (Type III [30.0%], Type IV [22.2%], Residual Untyped Twining [100.0%]).

Associations

All of the Windover twining and related perishable fiber artifacts were directly associated with human skeletal remains derived from burial contexts (n=36). Aging and sexing studies (Doran, personal communication 1999) indicate that 50% of the burials which produced perishable fiber artifacts are adults (ca. 20–65 years old [n=18]) equally divided between males (n=9) and females (n=9). Subadults (birth to ca. 12 years old) account for nearly 42% (n=15) of the perishable producing interments while only 8.3% (n=3) are teenagers (ca. 13–19 years old). Interestingly, perishable fiber artifacts were recovered with five infants and toddlers (birth to ca. 2 years old). The potential significance of this and other correlations among age, sex, and perishable fiber artifact types is discussed below.

As was previously noted in the description of Type II twining and as is detailed in Andrews et al. (2000), six of the Windover twined specimens are encrusted with a veneer of yellowish red (5YR4/6) debris. This material, tentatively identified during the field excavation as red ochre (iron oxide), was subjected to a wide array of physicochemical tests in an effort to determine its exact identity. The architecture and histochemical properties of the specimen are consistent with that of collagenous connective tissue, and subsequent immunological testing concluded that this tissue most likely derives from deer (*Odocoileus virginianus*). Furthermore, it is suggested that this specimen may once have been a part of a deer-hide covering for the burial (see Hyland and Anderson 2000).

Five of the Windover twined fabric samples submitted for histological examination revealed pollen grains adhering to and caught between twisted sections of the plant material (Hess, personal communication 1987). This pollen is identified as a conifer, most probably pine (*Pinus* spp.) (Barnosky and King, personal communication 1988 [see also Newsom 2000]). A pine ascription is not unexpected given the paleodistribution of this plant within the Red Brown Peat, but the interdigitation of pine pollen and fabric fibers suggests that the raw material for at least five of the twined specimens was processed, and possibly woven, in the spring (Barnosky and King, personal communication 1988).

In addition to human skeletal remains and probable deer tissue, a variety of other items are directly associated with the perishable fiber artifacts from Windover. These notably include sharpened and unsharpened stakes (see Adovasio *et al.* 2000), bundles or wads of grass, a variety of artifacts, and various ecofactual materials, some of which are almost certainly the residue of meals from one or another decedent. Several of these associations warrant further commentary.

As described above and elsewhere (see Newsom 2000), several burials yielded stakes which directly penetrated one or another twined fabric type and entered the underlying peat. Given the location of these stakes at the margins of the fabric and associated burial, it is certain they were employed to secure the body to the bottom of the bog. Cordage (see below) may also have been employed in conjunction with the stakes.

The use of grass bundles within the burials is also notable. Whether in the form of actual matting like the Type VI specimen or in randomly arranged bunches, grass lining seems to have been placed in some graves prior to the staking down of the bodies. In the case of the Type VI open twining, it may actually have served to further enwrap the textile clad body.

Ecofactual materials recovered directly from twining another perishable fiber artifacts include non-human bone fragments, fish scales, cheno-am seeds, and gastropod tests (*Heliosoma* sp.). As noted by Newsom (2000), at least some of these materials are most parsimoniously explained as the residue of "last meals."

Of the many artifact associations, perhaps the most significant, in the context of perishable fiber artifact production, is the recovery of a "cache" of bone needles and a possible bone weft packing tool, perhaps from (or near?) the decomposed bag in Burial Lot 90.

Form and Function

Although the Windover perishables were recovered in a mortuary context, a relatively wide range of techno-morphological forms are evident. All of the data presented above suggest that the majority (n=47 [88.7%]) of the twined specimens are pieces of cloth from lightweight rectangular or square blankets, capes, toga-like garments, or shrouds. Two finer gauged fabrics may represent clothing of yet some other form, and another item adhering directly to a human cranium may be a portion of a cap, hood, or bag. This specimen exhibits "V" expansion splices that could occur in any of these forms. Two other specimens certainly represent bag fragments, probably from globular bags. As noted above, one is associated with a "cache" of bone needles that may have been contained within it (see Newsom 2000). As described above, the other bag was situated beneath or around the remains of a neonate juxtaposed between the lower limbs of a young subadult of unknown sex. The single open-twined specimen is almost certainly a mat fragment.

Plaiting

A single plaited specimen was recovered from the Windover locality and accounts for 1.5% of the perishable assemblage. This very finely woven item with minimal attrition wear was positioned directly beneath a Type III blanket, cape, toga-like garment, or shroud and is associated with a 10– to 11–year-old of unknown sex. The fine-gauge weave (10 elements per cm) and its association and position with the

burial suggest that this is an item of clothing, which although of unknown configuration, was perhaps worn next to the skin as a modern-day T-shirt or shift might be worn.

Cordage

The 10 specimens of Type I cordage account for 76.9% of the Windover cordage (Table 3.3). As noted, these specimens are highly fragmented, unknotted, and exhibit "laid in" splices. They are unmended, uncharred, and undecorated. The Windover spun and twisted cordage is always found in direct association with (and usually superimposed upon) specimens of the blanket, cape, tunic-like garment, or shroud configuration. In conjunction with the stakes, they possibly served to immobilize the body after interment.

The two specimens of Type II cordage recovered from the Windover site account for 15.3% of the cordage collection. One of the items is directly associated with a wooden paddle that probably functioned as a bent snare "spring" and that may constitute a portion of a trap (see Newsom 2000).

The single specimen of Type III cordage accounts for a meager 0.77% of the cordage assemblage and is exclusively associated with the composite construction from Burial Lot 152.

Given the complexity of the associated fabric assemblage, it is virtually certain that the range of recovered cordage is not representative of the cordage repertoire or the cordage products (e.g., netting) produced by the Windover population. It is likely, however, that the dominance of final S twist cordage reflected in the extant collection is "real" and may well represent the cultural standard for this group.

Composite Construction

As noted above, the single fiber construction from the Windover site was classified according to its attributes of manufacture. This construction was associated with an adult male burial and may, we speculate, represent a votive bundle of perishable construction material for use in the "next life".

Statistical Overview of Fiber-based Artifacts

Technology

As detailed above, the basketry and textile industry from Windover is dominated by paired and trebled weft versions of close diagonal, S twisted weft twining. These types are clearly variations on the same theme that differ only in the number of weft elements employed in the twining process.

At first glance, the apparent proclivity of the Windover

weavers for the relatively rare (see External Correlations) three-weft twining variant is enigmatic. As noted above and in External Correlations, three-weft twining is rare in any construction medium and is usually confined to selvages or decorations (Emery 1966:203). When used in the wall or body of a woven item, three-weft twining imparts additional strength. Further, though an extra expenditure of techno-manipulative energy is required, this "cost" is negated by the ca. 25–40% (depending upon raw material) of additional surface coverage per three-weft course.

The critical variable in the incidence of three-weft twining is apparently the idiosyncratic ability to master its relatively complex manipulative pattern. Limited replicative studies suggest that the three-weft twining technique is considerably more difficult to initially assimilate (and in some cases cannot be learned) than is the two-weft variant. Once the manipulation of the three wefts becomes automatic, however, the technique is no more difficult or time-consuming than its two-weft counterpart.

Unlike the Windover twined bags, whatever the production technique employed, all textiles were almost certainly produced on some type of fixed "frame". Although twined (or plaited) fabric can be made without a stationary support, the general regularity of twining execution, evidenced in the Windover sample tends to indicate otherwise. Indeed, the total lack of three-weft twined bags may suggest that for at least trebled weft twining, "frame weaving" was integral to production. A typical frame employed by the Windover aborigines probably consisted of a horizontal "staked" construction or a hanging, non-heddle, backstrap-type loom upon which decorticated, macerated, and twisted cords were woven (cf. Kent 1983:Figures 39 and 40).

In this regard, it is worth noting that contrary to normal practice, the twining of three-weft fabrics may have been accomplished with the weft rows, rather than the warp rows, parallel to the long axis of the body of the weaver. With this orientation, the three-ply wefts can be rather more easily "braided" down and up the horizontal warp rows. This modification of the basic twining process involves little more than hanging a frame "sideways" with the warps horizontal rather than vertical or approaching a staked frame from a different seated posture.

Whatever the exact mode of production, two close-twined types (Type III and Type IV) constitute the majority of the perishable assemblage. Multidimensional scaling of Euclidean-based dissimilarity matrices on these types were extensively scrutinized to isolate any trends in the data. Preliminary confirmation of these apparent correlations was then provided by Fisher's Exact Test (FET), chi-square test (X^2), and T-test results. These tests indicate that neither type correlates well with age or gender of the skeletal population. However, there is a significant correlation

between subadult (especially the very young) burials and a finer weft gauge (> 5.0 wefts per centimeter) of Type IV twining. Conversely, adults are positively correlated with a coarser (≤ 5.0 wefts per centimeter) Type IV weave (*FETp* = 5.04E-4). There is no concomitant correlation between closely spaced warps (> 5.5 warps per centimeter) with subadults or widely spaced warps (≤ 5.5 warps per centimeter) with adults. Interestingly, Type III twining recovered with remains of infants and children contains only closely spaced warps (> 6.5 warps per centimeter [*FETp* = 0.01]), whereas warp spacing in "adult" cloth is variable. No correlation exists between Type III weft gauge and age group ($X^2 = 0.30$; $p < .250$). Further, neither Type III nor Type IV warp/weft gauges are gender specific.

Z or S twisted warps are represented almost equally among the Type III and Type IV twined specimens. The Z and S twisted warp variants do not correlate with age of the burials. Nevertheless, the two specimens that contain a combination of S and Z twisted warps were in direct contact with the upper leg/pelvic region of two teenage/young adult burials, respectively. However suggestive this association appears – that this aberrant warp pattern was created by younger, less skilled "green" hands – there is no significant correlation between age and the co-occurrence of Z and S twisted cordage warps in the same piece of fabric. Whether this deviation indicates multiple makers per item, a lack of standardization in warp manufacture among the aboriginal population, or, as noted, simply the result of younger, less experienced craftsman may never be known. As with the major twining structural types themselves, the S and Z warp variants are not gender specific. However, among the Type III specimens, the S twisted warp variants exhibit less frequent attrition wear than do their Z twisted counterparts ($X^2 = 10.08$; $p < .01$; 0.4).

In Type III specimens, finely twined (≤ 5.0 wefts per centimeter) items are consistently made on Z twisted wefts, whereas coarser (< 5.0 wefts per centimeter) weaves are manufactured with S twisted weft material (*FETp* = 3.4E-6). There is no association between the closeness of weave and a particular warp variant in the Type IV twining. Similarly, warp spacing cannot be correlated with direction of warp twist in either structural type.

Though none of the Type III or Type IV twining retains centers, five specimens exhibit side selvages of the continuous weft variety. One of these is reinforced with four rows of simple looping. The single end selvage (Type IV) is of the 180° self variety. Splices are almost exclusively "laid in".

There is no correlation between the degree of processing (i.e., decortication, mastication, and maceration) of Type III and Type IV raw material and the age or sex of the associated burial. However, there is some indication that Type III ($X^2 = 7.75$; $p = 0.01$; 0.22) and Type IV (*FETp* =

Table 3.3. Associations, Distribution, and Attributes for Formal Cordage Types from the Windover Bog Site (8BR246).

Lot	Sex	Age (yr)	Unit	Level (cm)	Specimens (N)	Whole Items (N)	Cord Diameter (mm)	Ply Diameter (mm)	Angle	Per cm
Type I, Two-Ply, Z Spun, S Twist										
82	female	38	N148 E75	20	1	1	2.71	–	41.0	4
92	male	50	N150 E73	24–25	3	1	2.70	–	44.0	2
104	female?	56	N154 E71	18	1	1	2.39	–	48.0	4
131	male	65	N148 E71	26	3	1	2.85	–	48.5	3
150	unknown	13	N148 E69	25	2	1	0.53	–	–	–
Type II, Three Strand Braid										
82	female	38	N148 E75	20	1	1	2.46	1.04	–	–
90	unknown	11	N148 E75	20–21	1	1	3.06	–	–	–
Type III, Two-Ply, S Spun, Z Twist										
152	male	51	N150 E69	26	1	1	2.11	–	42.0	4

0.015) specimens constructed with more heavily processed, that is finer, material exhibit the least amount of attrition wear.

Form and Function

As noted above, given the mortuary context, a relatively broad suite of forms is represented in the Windover assemblage. A three-dimensional plot of six manufacturing variables by close twining type and form is presented in Figure 3.1. Representative basketry shapes include both flexible, open-twined mats, a possible hood, bag, or cap, as well as close-twined, probably globular bags, one of which contains a "drawstring".

One of the Windover textile forms includes frame-woven, plaited, extremely fine-gauge (up to 10 elements per centimeter) fabric that most probably functioned as "next-to-skin" clothing of unknown shape. However, the predominant textile configuration is clearly a rectangular, or possibly "squarish", cloth of undetermined function. While it is possible that all of these items are intentionally produced mortuary shrouds, quite literally used only once, the fact that many exhibit wear militates against this theory. The distribution of types by location on the interred remains does not resolve the functional issue, nor does the incidence of types by age or sex. Similarly, the incidence of wear per se is functionally uninformative. Nevertheless, attrition data in conjunction with age parameters suggest that the predominant twined Windover forms exhibit much less wear in younger burials than in adult contexts (*FETp* = 1.3E-3). This in turn, suggests that at least the textiles associated with adults were used by these individuals as clothing or blankets with some degree of frequency. The lack of wear in subadult fabrics may reflect either less frequent use or they may indeed be specially fabricated "votive" garments. This view is supported by the general fineness of the gauge on the specimens recovered from subadult contexts. Additionally, the scarcity of brain matter in the subadult burials suggests that a certain period of time elapsed before such individuals were interred. This "delay" in our view, may well reflect the time required to produce "suitable" burial accouterments.

Though far removed in time, there are ethnographic accounts from the Southeast that adults were occasionally interred in everyday garments forcefully appropriated from other people. Lawson notes:

> After the dead person has lain a day and a night in one of their hurdles of canes, commonly in some out house made for that purpose, those that officiate about the funeral go into the town and the first young men they meet withal, that have blankets or match coats on, whom they think fit for their turn, they strip them from their backs, who suffer them so to do without any resistance. [Lawson 1714:293–294]

Certain statistical trends noted among the closely twined textiles (see above) suggest that forms in addition to those previously discussed may be represented in the Windover assemblage. Specifically, the two-weft diagonal twining with S twisted warps is coarser and more worn than similarly typed specimens constructed with Z twisted warps. Additionally, both the two-weft and three-weft versions of diagonal twining made with more intensively processed (i.e., macerated/masticated) plant material exhibit very little attrition wear. To date, the configurational and functional implications of these correlations are unknown.

Similarly, the differential distribution of separate pieces of cloth within burials is not readily explainable. Two subadult interments contained five separate pieces of cloth each. One of these burials, presently unsexed, also contained hooks, awls, drilled antler, shark teeth, and a snare. The second unsexed subadult was found with a wooden bowl. A third burial, a 45-year-old male, provided four separate pieces of cloth plus an awl and bone needle. Three pieces of cloth were recovered with an adult male and a mature female who collectively possessed few grave goods. While it is tempting to suggest a hint of rank, status, or "wealth" in these cases, there is little other evidence to support this scenario.

Whatever the vagaries concerning production and use of the Windover fabrics in life, the function of these items in death is certain. These accouterments were probably specially selected (or possibly, in the case of subadults, manufactured) by family and/or friends to garb, hold the possessions of, and enwrap the deceased, and when transfixed with cordage-enlaced stakes, to secure him or her beneath the quiet surface of the Windover bog forever.

Chronology and Provenience

The burials at Windover were interred during the sixth millennium B.C. While there is considerable support for a relatively short-term "death span" at the Windover cemetery, scrutiny of two-dimensional multi-dimensional scaling (MDS) plots generated from a Type III (Close Diagonal Twining, Paired S Twist Weft) dissimilarity matrix suggests that the period of sequent interments may be artifactually divisible. Specifically, there appears to be a congruence between increasing/decreasing warp diameter and the physical location of the burials in the bog. More specifically, three clusters of perishables recovered between the 68–69 east, 70–74 east, and 74–75 east grid lines reflect a sequential and ever-decreasing trend towards smaller warp diameters. The specimens recovered within the 68–69 east lines were consistently from deeper levels (Levels 24–27) than those situated between the 70–75 east lines (levels 18–21). The westernmost, deeper cluster correlates with a large warp diameter (> 1.7 mm), and the eastern group ($\leq$ 1.7), a small ($FETp$ = 8.9E-5) cluster. Further, at the 95% level, there is a significant (t = -2.094, p = 0.05, df = 19) difference between the means of the transformed (1/warp diameter) warp diameters of each respective batch. It is unknown if these twining "groups" reflect the position of still-stands related to small-scale lake-level fluctuations during the period of aboriginal interment. Specifically, it is possible that each grave was positioned at a culturally determined, depth-dependent distance from the bog shore line. In other words, as the strand line retreated during drier periods, or advanced during wetter periods, the burial clusters reflect this phenomenon. Unfortunately, in the absence of direct ^{14}C dates on the observed clusters or paleotopographic data in the form of discernible submerged shorelines, there is little direct support for this hypothesis.

There is, however, an alternative and in some ways, more interesting, possibility.

If the observed statistical trends are "real" and penecontemporaneous, these clusters may well reflect different groups of weavers within a single population. In this scenario, variation in weaving would correspond to fabrics produced by different social units and, perhaps, interred in family specific "plots" which were somehow marked. While this interpretation may seem fanciful, it has been conclusively demonstrated that individual weavers and groups of weavers can be successfully distinguished in both prehistoric and ethnographic contexts (Adovasio and Gunn 1975, 1977, 1986). There is no significant correlation between the Windover basketry/textile types themselves and provenience (Figure 3.2). However, Type II twining, as well as Type V plaiting are restricted to the southeastern section of the excavated area. The single open-twined (Type VI) item was recovered from the extreme southwestern unit.

Type IV twining is directly associated with all other close-twined types. However, Type III twining occurs in conjunction with the single open-twined item and, in another unit, the balanced plainweave. The significance of this association, if any, is problematic.

External Correlations

The twined fabric assemblage from Windover is the oldest example of its craft in the New World. The Windover perishables are not, however, the oldest fiber products known in this hemisphere. Indeed, far older basketry and cordage are represented in eastern and western North America, Mesoamerica, and South America. In both eastern and western North America, the oldest basketry, textile, or cordage materials are assignable to the mid-twelfth millennium B.P., though very few specimens have been recovered in well-dated contexts. Indeed, North American sites with perishable fiber artifacts of any subclass or type older than ca. 11,000 B.P. number fewer than 10. A selection of these localities are summarized below (see Andrews and Adovasio [1996] for a more extensive treatment of this theme).

As of this writing, the oldest bona fide basketry (or perishable fiber artifact) of any subclass or type from eastern North America derives from middle Stratum IIa at Meadowcroft Rockshelter in southwestern Pennsylvania (Andrews and Adovasio 1996; Stile 1982). The item is a wall fragment without selvage constructed of simple plaiting with single elements in a 1/1 interval. It is bracketed by radiocarbon dates of 10,950 ± 870 B.C. (12,800 ± 870 B.P.) and 9350 ± 700 B.C. (11,300 ± 700 B.P.) and is associated with the Miller complex occupation at that site. The specimen lacks shifts, splices, and decoration. According to Stile (1982:133), while the finished form of the

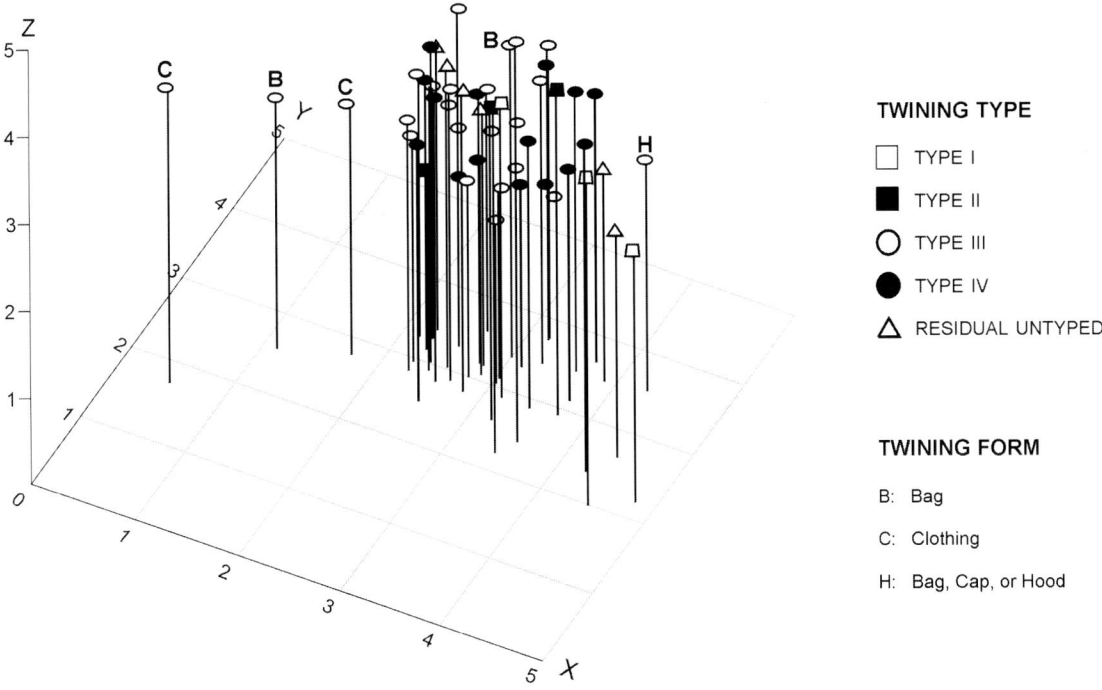

Figure 3.1. Three dimensional plot of six manufacturing variables in close twining from the Windover Bog site (8BR246). These bags may be finely made (8 wefts/cm; 10 warps/cm) or more coarsely woven (5 wefts/cm; 4 warps/cm). The Windover bags are made with more heavily processed plant material and exhibit little attrition wear.

plaiting fragment cannot be ascertained, it was manufactured of cut birch-like (cf. *Betula* sp.) bark strips. According to Andrews and Adovasio (1996:39), a far older but much more tentatively classified perishable from Meadowcroft Rockshelter derives from lowest Stratum IIa and is directly dated to 17,950 ± 2400 B.C. (19,600 ± 2400 B.P.). The specimen consists of a single element of intentionally cut birch-like (cf. *Betula* sp.) bark which is quite similar in overall morphology to the strips employed in all later Meadowcroft plaiting. If the specimen is a portion of a plaited basket and even if one sigma is subtracted from the date (i.e., 15,700 B.C. [17,650 B.P.]), it is at once the oldest basket in North or South America.

Interestingly, plaiting is also represented in an apparently ancient context some 1,500 mi (2,413.5 km) south of Meadowcroft Rockshelter at Petit Anse Island, Louisiana, on the Gulf of Mexico. Wilson (1888:674–675) reports that a single specimen of plaited matting was discovered near the surface of a salt dome, 0.6 m (2 ft) *below* the tusks and bones of a "fossil elephant" and ca. 4.3 m (14 ft) beneath the present soil horizon (Wilson 1888:674). The fragment is twill plaiting (2/2 interval) with several perhaps

intentional 2/3/2 shifts (Wilson 1888:Figure CVII). The specimen probably represents a portion of either a large burden basket or mat and does not exhibit selvage or decoration. Splicing is apparently effected by securing the new element beneath the exhausted strip close to the plaiting juncture. The fragment is made with strips of the outer bark of southern cane (*Arundinaria macrosperma* [Wilson 1888:674]). It should be noted that while Wilson (1888:675) questioned the antiquity of the Petit Anse plaiting fragment, particularly given its occurrence in what must have been a faulted salt diapir, a late Pleistocene ascription would not be out of place given the demonstrated age of the Meadowcroft plaiting. At the very least, the Petit Anse specimens would appear to confirm Goggin's (1949:166) assertion that plaited basketry has considerable antiquity in the Southeast.

In the far west, recent radiometric research has revealed that plaiting recovered in the late 1960's from Spirit Cave, Nevada, is not ca. 1,500 or 2,000 years old as was originally reported (Wheeler and Wheeler 1969), but rather, dates to 7465 ± 25 B.C. (9415 ± 25 B.P.) (Fowler *et al.* 1997). While not dating to Paleoindian times, plaited specimens of this age are unique in the Great Basin. Fowler *et al.* (1997) report that plain (i.e., simple) plaiting with paired two-ply, S spun, Z twist cordage pseudo-wefts was used to construct burial shrouds for two interments. These shrouds, which consist of a head covering for one burial and a cremation bag for the other, are made of bulrush (*Scirpus acutus*) and most likely were produced with the aid of a three-bar upright frame. The warp edges of these plaited specimens are set with two rows of close simple twining, and one of the bags is decorated. Interestingly, all of the other textiles examined from this horizon are twined and represent three different forms of this production technique.

While plaiting is represented in very early contexts at Meadowcroft Rockshelter, Pennsylvania (Andrews and Adovasio 1996; Stile 1982), and Petit Anse Island, Louisiana (Wilson 1888), as well as somewhat later at Spirit Cave, Nevada (Fowler *et al.* 1997), most early North American basketry of textiles are twined. Recently, an impression of what appears to be close diagonal twining with a Z twist weft was recovered in an alleged Clovis context at the Hiscock site in western New York State (Laub, personal communication 1996). Though not fully reported, this unique item appears to be associated with mastodon (*Mammut americanum*) remains that have direct dates ranging from 9440 ± 80 B.C. (11,390 ± 80 B.P.) to 8950 ± 80 B.C. (10,900 ± 100 B.P.), rendering it the only perishable fiber artifact potentially attributable to fluted point makers from eastern North America (Laub *et al.* 1996). By ca. 8000–6000 B.C. (9950–7950 B.P.), perishable fiber artifacts are widely represented in eastern North America (Andrews and Adovasio 1996:34–36), not only in the form of basketry and cordage but also sandals (Kuttruff *et al.* 1998:72–75).

Of relatively certain late Pleistocene/early Holocene ascription are the cordage remains from Fishbone Cave, Nevada, and Danger Cave, Utah, and basketry from several loci in the arid Great Basin. Orr (1974:47–59) reports and illustrates the same cordage type (i.e., two-ply, S spun, Z twist) both in the form of string as well as twining wefts from the base of Level 4 at Fishbone Cave, Nevada. The level and the associated specimens are putatively bracketed by dates of 9300 ± 250 B.C. (11,250 ± 250 B.P.) and 5880 ± 350 B.C. (7830 ± 350 B.P.). The older date was allegedly run on a specimen of open simple twined matting with two-ply, S spun, Z twist cordage wefts, though recently the source of the date has been questioned (see Ellis-Pinto 1994). If accurately dated, the late twelfth millennium B.P. specimen from Fishbone Cave is the oldest directly assayed fiber perishable artifact in western North America.

Stratum D1, Sand 1, at Danger Cave, Utah, has yielded the oldest definitive cordage and netting from the eastern reaches of the Great Basin (Jennings 1957:227–234). The small but informative collection includes single-ply, S twist cordage; a length of untwisted fiber; and more significantly, two segments of two-ply, Z spun, S twist cordage knotted together with a lark's head knot. Presumably, this specimens is the remnant of a section of knotted netting which, with the solitary and now unlikely exception of the possible Pendejo Cave netting, is the oldest such construction in North America. All of the Sand 1 specimens date between 9201 ± 570 B.C. (11,151 ± 570 B.P.) and 8320 ± 650 B.C. (10,270 ± 650 B.P.).

In the northern Great Basin, Cressman (1942) reports cordage of the two-ply, Z spun, S twist and single-ply, Z twist varieties from the bottom of Fort Rock Cave, Oregon. The cordage was apparently recovered with Fort Rock twined sandals and simple twined basketry with Z twist wefts. Though the age of the basal deposits at Fort Rock Cave remain controversial, these perishable specimens are at least 11,000 years old (Andrews *et al.* 1986).

Elsewhere in the northern Great Basin, specifically, and western North America, generally, the oldest basketry is invariably twined and includes open and close simple twined bags, mats, burden baskets, trays, and sandals of a variety of configurations. Though rarely directly dated, the age of these materials extends to at least 9000 B.C. (10,950 B.P.) or slightly earlier (Andrews *et al.* 1986).

Extensive studies of later prehistoric basketry, textile, and cordage manufacture in western North America, including Mexico, conclusively indicate that the production of cordage (and presumably netting) was established in the western half of the continent by at least the tenth millennium B.C. The manufacture of basketry occurred only slightly thereafter. Moreover, these industries continued to be evidenced, often in a much more elaborated state, through the period of Euro-American contact (see Adovasio 1970a, 1970b, 1971, 1974, 1975a, 1975b, 1976, 1977, 1980a, 1980b, 1980c, 1986; Adovasio and Andrews 1980, 1983, 1985, 1987; Adovasio [with Andrews] 1986; Adovasio *et al.* 1976, 1978; Adovasio and Gunn 1975, 1977, 1986; Adovasio and Hyland 1993; Adovasio and Lynch 1973; Adovasio and Maslowski 1980; Andrews and Adovasio 1980, 1989; Andrews, Adovasio, and Whitley 1988; Andrews *et al.* 1986; Cosgrove 1947; Cressman 1942; Frison, Adovasio, and Carlisle 1986; Frison *et al.* 1986; Guernsey and Kidder 1921; Heizer and Krieger 1956; Hyland 1997; Hyland and Adovasio 1997; Hyland and Adovasio with Illingworth 1995, 1999; Hyland *et al.* 1998; King 1974a, 1974b, 1979, 1986; Lindsay *et al.* 1968; Loud and Harrington 1929; MacNeish *et al.* 1967; Morris and Burgh 1941; Price 1957; Rozaire 1957, 1969, 1974; Tuohy 1970, 1974; Weltfish 1932).

By the onset of essentially modern climatic conditions, which is concomitant with the initiation of the Early Archaic period (ca. 8000 B.C.), perishables are somewhat better represented in widely separated portions of eastern North

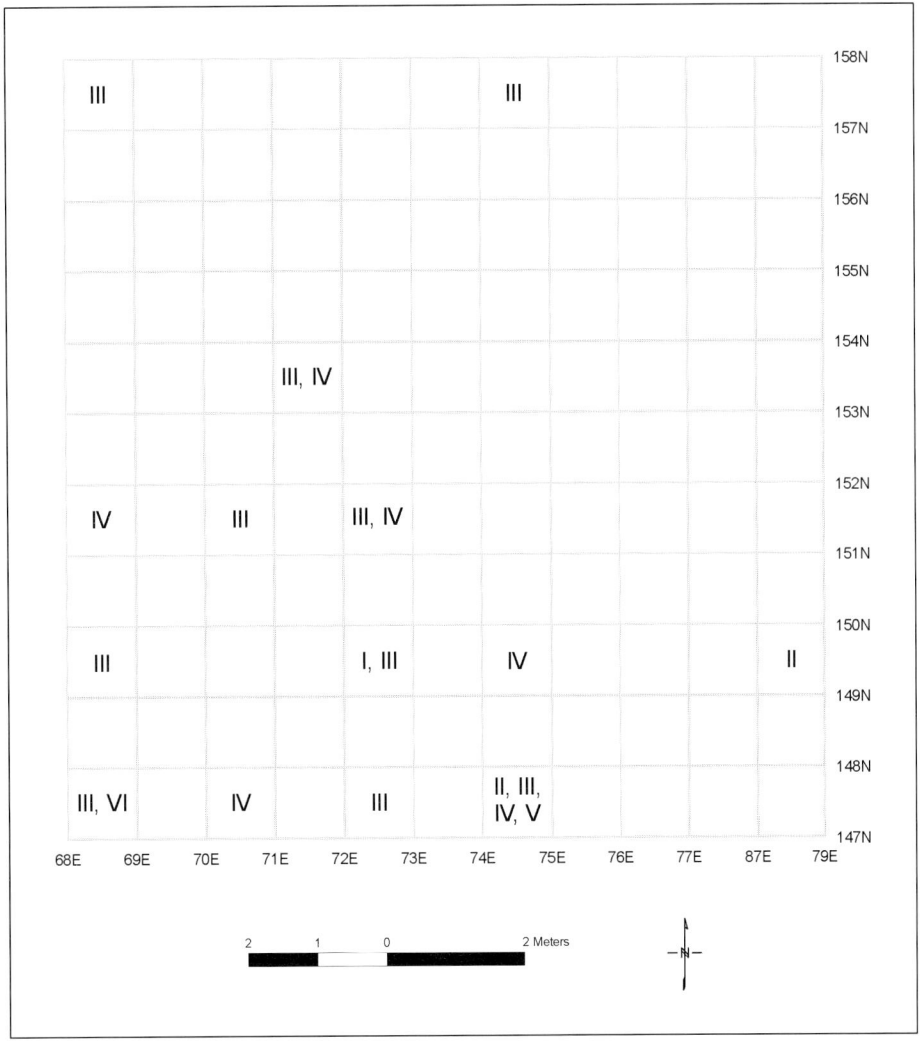

Figure 3.2. Horizontal distribution of textiles from the Windover Bog site (8BR246) by excavation unit. The Roman numerals indicate the presence of a formal textile type.

America. A single specimen from Level 6 (Zone IV) at Graham Cave in Montgomery County, Missouri, ranges from 7750 ± 500 B.C. (9700 ± 500 B.P.) to 7340 ± 300 B.C. (9290 ± 300 B.P.) in age and may be assigned to the very beginning of this period (Klippel 1971:22; Logan 1952:74). This fired clay impression, erroneously identified as "coiled" (Logan 1952:58), is actually the second-oldest evidence of twining in eastern North America. Examination of a clay positive made directly from the impression indicates that the specimen is composed of close simple twining with S twist wefts. The paired wefts appear to be single elements of loosely Z spun fibers while the composition of the warps is not discernible. If the warps were rigid, the specimen probably represents a container of some sort, and if they were flexible, it may be a bag fragment.

The condition of the impression precludes the determination of splicing techniques or any other detail of construction save to note that the specimen is not structurally decorated. Similarly, the raw material employed in construction cannot be ascertained.

Later levels (zones) at Graham Cave produced additional basketry impressions, again on fired clay (Logan 1952:58). A minimum of two types of twining were recovered (Logan 1952:Plate XXI) from Level 5 (Zone IV/III), Level 4 (Zone III), and Level 2 (Zone II). Collectively, these specimens date between 6880 ± 500 B.C. (8830 ± 500 B.P.) and 5680 ± 120 B.C. (7630 ± 120 B.P.). One of these types (Logan 1952:Plate XXIc) is an impression of open simple S twist twining over what appears to be two-ply, Z spun, S twist cordage warps, while the other (Logan 1952:Plate XXIb) is either a representative of the same type (except with Z twist warps) or it is open diagonal twining, again with Z twist warps. As the "fibrous nature" of the warps and wefts may indicate, these two wall fragments probably represent portions of flexible containers such as bags. The specimens lack selvages, splices, or decoration and are composed of indeterminate raw materials.

Southeast of Graham Cave, Icehouse Bottom in Monroe County, Tennessee, also yielded an assemblage of impressions of Early Archaic vintage (Chapman and Adovasio 1977:620). Twenty-seven of the site's 30 specimens originate from Strata M-O, the Lower Kirk horizon, and span a period of 7500–7300 B.C. (9450–9250 B.P.) (Chapman and Adovasio 1977:623). The remaining three fragments derive from Strata L and J, the Upper Kirk horizon, and may be ascribed to a 7300–6900 B.C. (9250–8850 B.P.) time interval.

Twenty-nine specimens represent impressions of open simple twining with Z twist wefts. Warps and wefts consist of two-ply, S spun, Z twist cordage. These specimens lack selvages and were probably originally flexible; thus, they probably represent impressions of the undecorated "walls" of matting or bags. Further, some of these items appear to be radially twined, although splice type and raw material cannot be determined.

Stratum O of the Lower Kirk horizon also produced the oldest netting fragment from eastern North America (Chapman and Adovasio 1977:622). It is a portion of a single element fragment built up of a series of knotted loops of single-ply, S spun cordage forming an open diamond mesh. The knot employed in the looping process is a sheet bend or weaver's knot. The specimen is unmended, undecorated, and lacks selvages and splices. The raw material is again unknown.

Of broadly contemporaneous age is the earliest sandal from Arnold Research Cave in Missouri (Kuttruff *et al.* 1998:72–75). Directly dated by two AMS assays to ca. 6375–5720 B.C. (8325–7675 B.P.), this specimen contains a simple-plaited sole dominated by longitudinal "pseudo warps" with an elaborate tie system. A slightly later specimen is also plaited with accessory rows of simple twining and some form of looping.

While the Graham Cave, Icehouse Bottom, and Arnold Research Cave perishables are clearly of Early Archaic ascription, somewhat less certain is the placement of perishable specimens from Layer G at Russell Cave, located in northeastern Alabama. This "unit" is dated between 7000 B.C. (8950 B.P.) and 5000 B.C. (6950 B.P.), and has yielded four examples of what is alleged to be over-and-under lacing – that is, simple plaiting (Griffin 1978:62). While the single published photograph does not allow exact determination of actual construction techniques or any other details of manufacture, the specimen appears to be twined.

Specifically, the illustrated item clearly seems to be functional wrapped twining with one semi-rigid fixed weft and one flexible "running weft". Though rare in the extreme, this type is represented in the archaeological records of the Pacific Northwest and the Lower Pecos in Texas. The illustrated Russell Cave specimen apparently represents a (flexible?) wall fragment (without selvage) of a container of unspecified configuration and is unspliced, unmended, and undecorated. Lamentably, the problematic identification and interpretation of the Layer G artifacts is further compounded by the fact that these materials may be intrusive from later levels. Whatever the type or age of the Russell Cave assemblage, the data from Icehouse Bottom, Graham Cave, and Arnold Research Cave clearly indicate that well-made perishables were *definitely* in use in the East at this time.

Significantly, at none of the above-noted North American sites is there any evidence of twined or plain woven-cloth predating the Windover assemblage. The same situation currently obtains in Mesoamerica (cf. Johnson 1967); however, the status of the oldest South American cloth (as opposed to basketry, bags or mats) is less easily resolved.

Ten specimens of warp-faced plain weave with single warps and wefts were recovered from Complex III at Guitarrero Cave, Peru, and are loosely "associated" with a ^{14}C determination of 5780 ± 150 B.C. (7730 ± 150 B.P.). These specimens obviously predate the Windover cloth by a few hundred radiocarbon years; however, they *may* be intrusive from younger levels, though Lynch *et al.* (1985:865) believe that the Complex III assemblage is only "minimally mixed". Elsewhere in Peru, one of two burial caves (Cave I) at Tres Ventanas, located at an elevation of ca. 3,925 m (12,874 ft) above mean sea level in the Chilca Valley, south of Lima, produced a subadult burial lying in a bed or bundle of fiber and putatively associated with a (twined?) textile or fabric of indeterminate type (Benfer, personal communication 1999). This textile as well as other grave goods were enwrapped in a fiber "mat", also undescribed, and alleged to date to the earliest occupation of the site ca. 8080 ± 170 B.C. (10,030 ± 170 B.P.) (Quilter 1989:71). Obviously, *if* the textile is a textile (i.e., a fabric) and not a bag or mat fragment, it is far and away the oldest example of this craft known in South America and ca. 2.5 millennia older then the Windover specimens. It should be noted that whatever their actual age, at least one of these Ventanas textiles is made of camelid hair, not plant fiber (Benfer, personal communication 1999).

Engel (1981) also reports fiber perishables associated with burials from Encampment 96 at Paracas. Burial IA at that site yielded a flexed burial wrapped in a rectangular open twined mat with *Juncus* sp. warps and two-ply, Z twist "reed" wefts. Side selvages are of the continuous weft or knotted (?) varieties; and selvages are clipped. Under the mat and around (?) the body is a carbonized garment or fabric made from *Tillandsia* sp. The fabric is apparently twined and structurally decorated through warp/weft manipulations though the exact production techniques are not specified. Also recovered from Burial IA are two more textiles which also allegedly wrapped the body and whose construction details are also not specified. According to Engel (1981:33), the "decorated" textile is similar to specimens recovered from La Paloma (see below).

Burial 2A at Encampment 96 also yielded an open (?) twined *Juncus* sp. warp mat with grave goods that included fiber construction material and a ball of two-ply, S spun, Z twist cordage. According to Engel (1981:32), these two burials date 7050–6050 B.C. (9000–8000 B.P.), again rendering the associated textiles substantially older than the Windover fabrics, *if* the age estimates are accurate.

Given the somewhat tenuous dating of both the Tres Ventanas and early Paracas materials, probably the best candidate for the oldest definitive South American cloth is reported from La Paloma, a Middle Preceramic site in central coastal Peru (Quilter 1977, 1989). Excavations at that site produced a corpus of perishable fiber artifacts including mats and textiles. Indeed, so numerous are the open twined, *Juncus* sp. warp mat fragments that Quilter (1989:31) states that they are the most common artifact type at the site. The mats were either wrapped around burials or occurred as floor coverings and included both open sample twining with paired wefts and, occasionally, diagonal twining. Quilter (1989:38–39) notes that mats were of two basic types based on raw material preparation and that one of these variants, Type II, made with whole *Juncus* sp. warps may have been used exclusively for burials. Also recovered from La Paloma were fine twined textiles made from *Fourcroya andina* sp. and directly associated with burials probably as clothing for the deceased. Looped fabrics and a variety of cordage were also recovered.

The La Paloma specimens are attributable to the Middle Preceramic, ca. 6050–3050 B.C. (8000–5000 B.P.) and may be at least as old if not slightly older than the Windover fabrics. Similarly, textiles, associated with burials from Arica, Chile (Allison 1985; Allison *et al.* 1984), again include open twined materials *estimated* to be as old as 5860 ± 180 B.C. (7810 ± 180 B.P.) or by some accounts much younger (Richardson, personal communication 1988). Though none of these South American assemblages appear to be as well dated or demonstrably older than the Windover collection, they do provide independent confirmations of the widespread production of twined fabric, as well as its use in mortuary contexts in the South American mid-Holocene.

Whatever the precise age of the earliest South American

twined fabric, we stress that its relatively sophisticated character, like that of the Windover assemblage, documents a technology whose ultimate origins are *far* more ancient. Indeed, given the dates on the fiber-based industry at Dolní Vestonice I and II and Pavlov I in the Czech Republic (Adovasio *et al.* 1996, 1997, 1998, 1999; Soffer, Adovasio, and Hyland 1998, 1999; Soffer, Adovasio, and Klíma 1996; Soffer, Adovasio, Hyland, and Klíma 1998; Soffer, Adovasio, Hyland, Klíma, and Svoboda 1998), even if the earliest percolations of humans across Beringia occurred more than 20,000 years ago, it is highly likely that perishable fiber technology was already a well established and a readily portable part of their techno-suite. In this deep-time perspective, the Windover materials, remarkable as they are, become much more understandable in terms of possible technological progenitors as well as their progeny.

Conclusions

1. In the mid-sixth millennium B.C., and contrary to prevailing stereotypes, the Archaic inhabitants of the Windover area produced four types of close twining, one type of open twining, and one type of plaiting in the form of fully flexible fabric, bags, and mats. In addition, they manufactured cordage and miscellaneous composite fiber or fiber and wood constructions.

2. A substantial portion of their recovered perishable suite consists of well-made, non-heddle loom, close twined or plain weave cloth. This fabric was used for clothing of indeterminate configuration(s) and/or shrouds.

3. The Windover fabrics played an important, if not critical, role in mortuary rituals for both sexes and all age groups. They served not only to enshroud the deceased but, together with stakes and accessory cordage, secure the bodies in their subaqueous graves.

4. The fabrics and most related perishables are made of highly processed cabbage palmetto (*Sabal* sp.) and represent the end products of a very labor intensive non-durable technology, the potential roles of which have been drastically underestimated in the context of the Florida Archaic, specifically, and the early Southeast, generally. As such, they have profound implications for group mobility, division and organization of labor, seasonality and subsistence practices, personal status, and a wide range of other issues virtually unapproachable through the medium of durable (i.e., lithic) technology.

5. The high degree of technological sophistication evidenced in the Windover fabric assemblage indicates that they represent not the genesis but rather the continuing evolution of a basketry and textile tradition with very ancient roots.

6. Though certainly not the oldest fiber-based

productions, the Windover twined and plain weave textiles are presently the most ancient examples of their genre in North America, if not the entire New World. As such, they collectively represent a previously undocumented, though certainly not unanticipated, level of technical virtuosity in the so-called "perishable fiber arts" of aboriginal America.

Acknowledgements
The Windover perishable fiber artifact assemblage was initially analyzed and conserved at the Perishables Analysis Facility, Cultural Resource Management Program, University of Pittsburgh. This work was supported by Florida State University and the National Park Service and was directed by R.L. Andrews and J.M. Adovasio. Final analysis and conservation was performed at the R.L. Andrews Perishables Analysis Facility, Mercyhurst Archaeological Institute, Mercyhurst College, under the direction of J.M. Adovasio. Assistance in various stages of the analysis was provided by D.C. Hyland, N. Larsen, T. Whitley, R.B. Davis, and A. Byrnes. Major assistance in the statistical examination of the Windover assemblage was graciously proffered by R. Drennan while J.B. Richardson, III, and B. Meggers kindly provided difficult-to-acquire South American textile references. We are also grateful to S.D. and J. Quilter for copies of their La Paloma studies.

Analysis of raw materials used in the production of the Windover perishables was performed and/or facilitated by a wide variety of individuals, including L. Newsom, University of Florida; F. King, Cleveland Museum of Natural History; K. Cushman, U.S. Fish and Wildlife Service, Lacey, Washington; F. Utech, The Carnegie Museum of Natural History, Pittsburgh, Pennsylvania; C. Hess, Chatham College, Pittsburgh, Pennsylvania; and S. Stout, University of Missouri, Columbia. Pollen adhering to some of the Windover fabrics was analyzed by K. Barnosky, The Carnegie Museum of Natural History, and F. King. Phytolith identification was performed by D. Piperno, Temple University, while J. Boon, Fom-Instutuut Voor Atom-en Molecuulfysica, Amsterdam, performed pyrolysis mass spectrometry. SEM scrutiny and energy dispersive analysis by X-ray (EDAX) was effected by G. Cooke, Mercyhurst Archaeological Institute. Composite enhanced images of selected textiles were provided by M.I. Siegel, Department of Anthropology, University of Pittsburgh. Osteological specimens adhering to some of the Windover fabrics were analyzed by M. Mooney, University of Pittsburgh, and D.C. Dirkmaat, Mercyhurst Archaeological Institute. Illustrations of wood and textiles were provided by W.P. Athens, Garrow and Associates, and N. Yedlowski, and S.L. Snyder, Mercyhurst Archaeological Institute. Special thanks are due to the countless laboratory assistants and students who monitored the Windover perishable artifact collection and maintained it in stasis for more than a decade; especially D.R. Pedler who has edited numerous versions of this report on short notice.

Finally, but scarcely least important, we wish to thank G. Doran and D. Dickel, Florida State University, for their unstinting cooperation in all phases of the perishables analysis as well as for initially soliciting our interest and, thereafter, securing funding for all of our efforts.

References Cited
Adovasio, J.M. 1970a. The Origin and Development of Western Archaic Textiles and Basketry. Tebiwa 13(2):1–40.

Adovasio, J.M. 1970b. Textiles. In Hogup Cave, by C.M. Aikens, pp. 133–153. University of Utah Anthropological Papers 93. University of Utah Press, Salt Lake City.

Adovasio, J.M. 1971. Some Comments on the Relationship of Great Basin Textiles to Textiles from the Southwest. University of Oregon Anthropological Papers 1:103–108.

Adovasio, J.M. 1974. Prehistoric North American Basketry. Nevada State Museum Anthropological Papers 16(5):133–153.

Adovasio, J.M. 1975a. Prehistoric Great Basin Textiles. In Irene Emery Roundtable on Museum Textiles 1974 Proceedings, edited by P.L. Fiske, pp. 141–148. The Textile Museum, Washington, D.C.

Adovasio, J.M. 1975b. Fremont Basketry. Tebiwa 17(2):67–76.

Adovasio, J.M. 1976. Basketry from Swallow Shelter (42BO268). In Swallow Shelter and Associated Sites, by G.F. Dalley, pp. 167–169. University of Utah Anthropological Papers 96. University of Utah Press, Salt Lake City.

Adovasio, J.M. 1977. Basketry Technology. Aldine Publishing Co., Chicago.

Adovasio, J.M. 1980a. Prehistoric Basketry of Western North America and Mexico. In Early Native Americans: Prehistoric Demography, Economy and Technology, edited by D.L. Browman, pp. 341–362. Mouton, The Hague.

Adovasio, J.M. 1980b. The Evolution of Basketry Manufacture in Northeastern Mexico, Lower and Trans-Pecos Texas. In Papers on the Prehistory of Northeastern Mexico and Adjacent Texas, edited by J.F. Epstein, T.R. Hester and C. Graves, pp. 93–102. The University of Texas at San Antonio Center for Archaeological Research Special Report 9. San Antonio.

Adovasio, J.M. 1980c. Fremont: An Artifactual Perspective. In Fremont Perspectives, edited by D.B. Madsen, pp. 35–40. Antiquities Section Selected Papers 16. Salt Lake City.

Adovasio, J.M. 1986. Prehistoric Basketry. In Great Basin, edited by W. D'Azevedo, pp. 194–205. Handbook of North American Indians, vol. 11, William G. Sturtevant, general editor. Smithsonian Institution, Washington, D.C.

Adovasio, J.M. 1997. Cordage and Cordage Impressions from Monte Verde. In The Archaeological Context and Interpretation, edited by T.D. Dillehay, pp. 211–228. Monte Verde, a Late Pleistocene Settlement in Chile, vol. 2. Smithsonian Series in Archaeological Inquiry, B.D. Smith and R. McC. Adams, series editors. Smithsonian Institution Press, Washington, D.C.

Adovasio, J.M., and R.L. Andrews. 1980. Basketry and Miscellaneous Perishable Artifacts from Walpi. In Textiles, Basketry and Shell Remains from Walpi, by K.P. Kent, J.M. Adovasio, R. Andrews, J.D. Nations, and J.L. Adams, pp. 1–93. Walpi Archaeological Project – Phase II (Vol. 6). A Report Submitted to the Heritage, Conservation and Recreation Service, Interagency Archeological Services, San Francisco, by the Museum of Northern Arizona.

Adovasio, J.M., and R.L. Andrews. 1983. Material Culture of Gatecliff Shelter: Basketry, Cordage and Miscellaneous Fiber Constructions. In The Archaeology of Monitor Valley 2. Gatecliff Shelter, by D.H. Thomas, pp. 279–289. Publications of the American Museum of Natural History 59(1). New York.

Adovasio, J.M., and R.L. Andrews. 1985. Basketry and Miscellaneous Perishable Artifacts from Walpi Pueblo, Arizona. Ethnology Monographs 7. Department of Anthropology, University of Pittsburgh, Pittsburgh.

Adovasio, J.M., and R.L. Andrews. 1987. Basketry and Miscellaneous Perishable Artifacts from Walpi: A Summary. In Recent Contributions to Southwestern Prehistory: A Sample, edited by J.M. Adovasio and R.L. Andrews. American Archaeology 6(3):199–213

Adovasio, J.M., with R.L. Andrews. 1986. Artifacts and Ethnicity: Basketry as an Indicator of Territoriality and Population Movements in the Prehistoric Great Basin. In Anthropology of the Desert West: Essays in Honor of Jesse D. Jennings, edited by C. Condie and D.D. Fowler, pp. 43–88. University of Utah Press, Salt Lake City.

Adovasio, J.M., R.L. Andrews, and R.C. Carlisle. 1976. The Evolution of Basketry Manufacture in the Northern Great Basin. Tebiwa 18(2):1–8.

Adovasio, J.M., R.C. Carlisle, and R.L. Andrews. 1978. An Evolution of Anasazi Basketry: A View from Antelope House. New World Archaeology 2(5):1–5.

Adovasio, J.M., and J.D. Gunn. 1975. Basketry and Basketmakers at Antelope House. The Kiva 4(1):71–80.

Adovasio, J.M., and J.D. Gunn. 1977. Style, Basketry and Basketmakers at Antelope House. In The Individual in Prehistory: Studies of Variability in Style in Prehistoric Technologies, edited by J.N. Hill and J. Gunn, pp. 137–153. Academic Press, New York.

Adovasio, J.M., and J.D. Gunn. 1986. The Antelope House Basketry Industry. In Archaeological Investigations at Antelope House, by D.P. Morris, pp. 306–397. National Park Service, Washington, D.C.

Adovasio, J.M., and D.C. Hyland. 1993. Paleo-Indian Perishables from Pendejo Cave: A Brief Summary. Paper presented at the 58th Annual Meeting of the Society for American Archaeology, St. Louis, Missouri.

Adovasio, J.M., D.C. Hyland, R.L. Andrews, J.S. Illingworth, R.B. Burgett, A.R. Berkowitz, D.E. Strong, D.A. Schmidt. 2000. Cool Wood Stuff. In Multidisciplinary Investigations at the Windover Site, edited by G. Doran. University of Florida Press, Gainesville, in press.

Adovasio, J.M., D.C. Hyland, and O. Soffer. 1999. Perishable Technology and Early Human Populations in the New World. Paper presented at the 31st Annual Chacmool Conference On Being First: Cultural Innovation and Environmental Consequences of First Peoplings. Calgary, Alberta.

Adovasio, J.M., D.C. Hyland, and O. Soffer. 1997. Textiles and Cordage: A Preliminary Assessment. In Pavlov I – Northwest, edited by J. Svoboda, pp. 403–424. Institute of Archaeology–Brno, Academy of Sciences of the Czech Republic.

Adovasio, J.M., D.C. Hyland, O. Soffer, and B. Klima. 1998. Perishable Industries and the Colonization of the East European Plain. Paper presented at the 15th Annual Visiting Scholar Conference "Fleeting Identities: Perishable Material Culture in Archaeological Research", Carbondale, Illinois.

Adovasio, J.M., and T.F. Lynch. 1973. Preceramic Textiles from Guitarrero Cave, Peru. American Antiquity 38(1): 84–90.

Adovasio, J.M., and R.F. Maslowski. 1980. Textiles and Cordage. In Guitarrero Cave, by T.F. Lynch, pp. 253–290. Academic Press, New York.

Adovasio, J.M., O. Soffer, and B. Klíma. 1996. Upper Paleolithic Fibre Technology: Interlaced Woven Finds From Pavlov I, Czech Republic, c. 26,000 Years Ago. Antiquity 70(269):526–534.

Allison, M.J. 1985. Chile's Ancient Mummies. Natural History 94(10): 74–81.

Allison, M.J., G. Focacci, B. Arriza, V. Standen, M. Rivera, and J.M. Lowenstein. 1984. Chinchorro, Momias de Preparación Complicada: Métodos de Momificacion. Chungará 13: 155–173.

Andrews, R.L., and J.M. Adovasio. 1980. Perishable Industries from Hinds Cave, Val Verde County, Texas. Ethnology Monographs 5. Department of Anthropology, University of Pittsburgh, Pittsburgh.

Andrews, R.L., and J.M. Adovasio. 1988. An Interim Statement on the Conservation and Analysis of Perishables from the Windover Archaeological Project, Florida. Department of Anthropology, University of Pittsburgh, Pittsburgh.

Andrews, R.L., and J.M. Adovasio. 1989. Knotted Cordage from Squaw Rockshelter, Aurora Run, Cuyahoga County, Ohio. *Kirtlandia* 44:59–62.

Andrews, R.L., and J.M. Adovasio. 1996. The Origins of Fiber Perishables East of the Rockies. In A Most Indispensable Art: Native Fiber Perishables from Eastern North America, edited by J.B. Petersen, pp. 30–49. University of Tennessee Press, Knoxville.

Andrews, R.L., J.M. Adovasio, and R. C. Carlisle. 1986. Perishable Industries from Dirty Shame Rockshelter, Malheur County, Oregon. University of Oregon Anthropological Papers 34, issued jointly as Ethnology Monographs 9. Department of Anthropology, University of Pittsburgh, Pittsburgh.

Andrews, R.L., J.M. Adovasio, and D.G. Harding. 1988. Textile and Related Perishable Remains from the Windover Site (8BR246). Paper presented at the 53d Annual Meeting of the Society for American Archaeology, Phoenix.

Andrews, R.L., J.M. Adovasio, B. Humphrey, D.C. Hyland, J.S. Gardner, D.G. Harding, J.S. Illingworth, and D.E. Strong. 2000. Textile and Related Perishable Remains from the Windover Site. In Multidisciplinary Investigations at the Windover Site edited by G. Doran. University of Florida Press, Gainesville, in press.

Andrews, R.L., J.M. Adovasio, and T.G. Whitley. 1988. Coiled Basketry and Cordage from Lakeside Cave (42BO385), Utah. Paper presented at the 21st Annual Great Basin Conference, Park City, Utah.

Catling, D., and J.E. Grayson. 1982. Identification of Vegetable Fibers. Chapman and Hall, New York.

Chapman, J., and J.M. Adovasio. 1977. Textile and Basketry Impressions from Icehouse Bottom, Tennessee. American Antiquity 42(4): 620–625.

Cosgrove, C.B. 1947. Caves of the Upper Gila and Hueco Areas in New Mexico and Texas. Papers of the Peabody Museum of Archaeology and Ethnography 24(2). Cambridge.

Cressman, L.D. 1942. Archaeological Researches in the Northern Great Basin. Carnegie Institution of Washington Publications No. 538. Washington, D.C.

Driver, H.E. 1961. Indians of North America. University of Chicago Press, Chicago.

Ellis-Pinto, C. 1994. Three New Dates for Basketry from the Winnemucca Lake Caves, Pershing County, Nevada. Paper presented at the 24th Meeting of the Great Basin Anthropological Conference, Elko, Nevada.

Emery, I. 1966. The Primary Structure of Fabrics: An Illustrated Classification. The Textile Museum, Washington, D.C.

Engel, F.A. 1981. Prehistoric Andean Archaeology. Humanities Press and the Department of Anthropology, Hunter College, City University of New York.

Fowler, C.S., A.J. Dansie, and E.M. Hattori. 1997. Plaited Matting from Spirit Cave, Nevada: Technical Implications. Paper presented at the 62d Annual Meeting of the Society for American Archaeology, Nashville, Tennessee.

Frison, G.C., J.M. Adovasio, and R.C. Carlisle. 1986. Coiled Basketry from Northern Wyoming. Plains Anthropologist 31: 163–167.

Frison, G.C., R.L. Andrews, J.M. Adovasio, R. Carlisle, and R. Edgar. 1986. A Late Paleoindian Animal Trapping Net from Northern Wyoming. American Antiquity 51:352–361.

Gardner, J.S. 1988. Conservation of the Windover Fabrics and Wood. Paper presented at the 53d Annual Meeting of the Society for American Archaeology, Phoenix, Arizona.

Goggin, J.M. 1949. Plaited Basketry in the New World. Southwestern Journal of Anthropology 5:165–168.

Goggin, J.M. 1951. Archeological Notes on Lower Fisheating Creek. Florida Anthropologist 4(3–4):50–66.

Griffin, J.W. 1978. Investigations in Russell Cave. Reprinted. Huntsville Chapter of the Alabama Archaeological Society. Originally published 1974, National Park Service Publications in Archaeology 13, U.S. Department of the Interior, Huntsville, Alabama.

Guernsey, S.J., and A.V. Kidder. 1921. Basketmaker Caves of Northeastern Arizona. Papers of the Peabody Museum of American Archaeology and Ethnology, No. 8. Cambridge.

Heizer, R.F., and A.D. Krieger. 1956. The Archaeology of Humboldt Cave, Churchill County, Nevada. University of California Publications in American Archaeology and Ethnology 47(1). Berkeley.

Hurley, W.M. 1979. Prehistoric Cordage: Identification of Impressions on Pottery. Aldine Publishing Company, Chicago.

Hyland, D.C. 1997. Perishable Industries from the Jornada Basin, South-Central New Mexico. Unpublished Ph.D. Dissertation, Department of Anthropology, University of Pittsburgh, Pittsburgh.

Hyland, D.C., and J.M. Adovasio. 1997. The Mexican Con-

nection: A Study of Sociotecnical Change in Perishable Manufacture and Food Production in Prehistoric New Mexico. Paper presented at the 62d Annual Meeting of the Society for American Archaeology, Nashville, Tennessee.

Hyland, D.C., and J.M. Adovasio, with J. Illingworth. 1995. Perishable Industries from Pendejo Cave, New Mexico. Paper presented at the 60th Annual Meeting of the Society for American Archaeology, Minneapolis, Minnesota.

Hyland, D.C., and J.M. Adovasio, with J. Illingworth. 1999. The Perishable Artifacts. In Excavations at Pendejo Cave, edited by J. Libby and R.S. MacNeish. University of New Mexico Press, Albuquerque, in press. Ms. 1997.

Hyland, D.C., J.M. Adovasio, and R.E. Taylor. 1998. Corn, Cucurbits, Cordage, and Colonization: An Absolute Chronology for the Appearance of Mesoamerican Domesticates and Perishables in the Jornada Basin, New Mexico. Paper presented at the 63d Annual Meeting of the Society for American Archaeology, Seattle, Washington.

Hyland, D.C., and T.R. Anderson. 2000. Biomolecular Analysis of Collagenous Tissue from the Windover Burials. In Multidisciplenary Investigations at the Windover Site edited by G. Doran. University of Florida Press, Gainesville, in press.

Jennings, J.D. 1957. Danger Cave. Anthropological Papers No. 27. Department of Anthropology, University of Utah, Salt Lake City.

Johnson, I.W. 1967. Textiles. In The Non-Ceramic Artifacts, by R.S. MacNeish, A. Nelken-Turner, and I.W. Johnson, pp. 189–226. The Prehistory of the Tehuacan Valley, vol. 2, D.S. Byers, general editor. University of Texas Press, Austin.

Kent, K.P. 1983. Prehistoric Textiles of the Southwest. School of American Research, Southwest Indian Art Series, University of New Mexico Press, Albuquerque.

King, M.E. 1974a. Medio Perishable Artifacts: Textiles and Basketry. In Casa Grandes: A Fallen Trading Center of the Gran Chichimeca, by C.C. DiPeso, J.B. Rinaldo, and G.J. Fenner, pp. 76–113. Northland Press, Flagstaff Arizona.

King, M.E. 1974b. Espanoles Period Perishable Artifacts. In Casa Grandes: A Fallen Trading Center of the Gran Chichimeca, by C.C. DiPeso, J.B. Rinaldo, and G.J. Fenner, pp. 114–125. Northland Press, Flagstaff Arizona.

King, M.E. 1979. The Prehistoric Textile Industry of Mesoamerica. In The Junius B. Bird Pre-Columbian Textile Conference, edited by A.P. Rowe, E.P. Benson, and A. Schaffer, pp. 265–278. The Textile Museum, Washington, D.C.

King, M.E. 1986. Preceramic Cordage and Basketry from Guilá Naquitz. In Guilá Naquitz: Archaic Foraging and Early Agriculture in Oxaca, Mexico, edited by K.V. Flannery, pp. 157–161. Academic Press, Orlando.

Klippel, W.E. 1971. Graham Cave Revisited: A Re-evaluation of Its Cultural Position During the Archaic Period. Missouri Archaeological Society Memoir 9. Columbia.

Kruskal, J.B., and M. Wish. 1976. Multidimensional Scaling. Sage University Paper Series on Quantitative Applications in the Social Sciences, 07–001. Sage Publications, Beverly Hills and London.

Kuttruff, J.T., S.G. Dehart, and M.J. O'Brien. 1998. 7500 Years of Prehistoric Footwear from Arnold Research Cave, Missouri. Science 281(5373):72–75.

Lawson, J. 1714. History of North Carolina. Observer Printing House, Charlotte, North Carolina.

Laub, R.S., J. Tomenchuk, and P.L. Storck. 1996. A Dated Mastodon Bone Artifact from the Late Pleistocene of New York. Archaeology of Eastern North America 24:1–17.

Lindsay, Alexander J., Jr., J. Richard Ambler, Mary Ann Stein, and Philip M. Hobler. 1968. Survey and Excavations North and East of Navajo Mountain, Utah, 1959–1962. Museum of Northern Arizona 45, Glen Canyon Series 8. Flagstaff.

Logan, W.D. 1952. Graham Cave: An Archaic Site in Montgomery County, Missouri. Missouri Archaeological Society Memoir 2. Columbia.

Loud, L.L., and M.R. Harrington. 1929. Lovelock Cave. University of California Publications in American Archaeology and Ethnology 25(1). Berkeley.

Lynch, T.F., R. Gillespie, J.A.J. Gowlett, R.E.M. Hedges. 1985. Chronology of Guitarrero Cave, Peru. Science 229 (4716): 864–867.

MacNeish, R.S., A. Nelken-Turner, and I.W. Johnson. 1967. The Non-Ceramic Artifacts. The Prehistory of the Tehuacan Valley, vol. 2, D.S. Byers, general editor. University of Texas Press, Austin.

Morris, E.H., and R.F. Burgh. 1941. Anasazi Basketry: Basket Maker II Through Pueblo II, A Study Based on Specimens from the San Juan River Country. Publication No. 533. Carnegie Institution of Washington, Washington, D.C.

Newsom, L.A. 2000. The Paleoethnobotany of Windover (8BR246): An Archaic Period Mortuary Site in Central Florida. In Multidisciplenary Investigations at the Windover Site, edited by G. Doran. University of Florida Press, Gainesville, in press.

Orr, P.C. 1974. Notes on the Archaeology of the Winnemucca Caves, 1952–1958. In Collected Papers on Aboriginal Basketry, edited by D.R. Tuohy and D.L. Rendell, pp. 47–59. Anthropological Papers No. 16. Nevada State Museum, Carson City.

Price, S. 1957. Textiles. In Danger Cave, by J.D. Jennings. University of Utah Anthropological Papers 27. Salt Lake City.

Quilter, S.M. 1977. Report on the Fiber Objects and Construction of Paloma, A Preceramic Archaeological Site (1976). Ms. on file, Department of Archaeology/Anthropology, Mercyhurst College, Erie, Pennsylvania.

Quilter, S.M. 1989. Paloma: Mortuary Practices and Social Organization of a Preceramic Peruvian Village. University of Idaho Press, Boise (in press).

Rozaire, C.E. 1957. Twined Weaving and Western North American Prehistory. Unpublished Ph.D. dissertation, Department of Anthropology, University of California, Los Angeles.

Rozaire, C.E. 1969. The Chronology of Woven Materials from Seven Caves at Falcon Hill, Washoe County, Nevada. The Nevada State Museum Anthropological Papers 14:181–186.

Rozaire, C.E. 1974. Analysis of Woven Materials from Seven Caves in the Lake Winnemucca Area, Pershing County. In Collected Papers on Aboriginal Basketry, edited by D.R. Tuohy and D.L. Rendall, pp. 60–97. Nevada. Nevada State Museum Anthropological Papers 16, Carson City.

Shaw, G.R. 1972. *Knots*. Collier Books, New York.

Soffer, O., J.M. Adovasio, and D.C. Hyland. 1998. Perishable Industries from Upper Paleolithic Moravia: New Insights into the Origin and Nature of the Gravettian. Paper presented at the Institute of Archaeology, ASCR, Masaryk University, Brno, Czech Republic.

Soffer, O., J.M. Adovasio, and D.C. Hyland. 1999. The Well-Dressed "Venus": Women's Wear ca. 27,000 B.V. (before Vogue®). Paper presented at the 64th Annual Meeting of the Society for American Archaeology, Chicago.

Soffer, O., J.M. Adovasio, D.C. Hyland, and B. Klíma. 1998. Perishable Industries from Dolní Vestonice: New Insights into the Origin of the Gravettian. Paper presented at the 63d Annual Meeting of the Society for American Archaeology, Seattle, Washington.

Soffer, O., J.M. Adovasio, and B. Klíma. 1996. Les Tissus Paléolithiques de Moravie. L'Archéologue No. 25:9–11.

Soffer, O., J.M. Adovasio, D.C. Hyland, B. Klíma, and J. Svoboda. 1998. Perishable Technologies and the Genesis of the Eastern Gravettian. In Lifestyles and Survival Strategies in Pliocene and Pleistocene Hominids, edited by Herbert Ullrich. Edition Archaea, Schwelm, Germany, in press. Ms. 1997

Stile, T.E. 1982. Perishable Artifacts from Meadowcroft Rock-shelter, Washington County, Pennsylvania. In *M*eadowcroft: Collected Papers on the Archaeology of Meadowcroft Rock-shelter and the Cross Creek Drainage, edited by R.C. Carlisle and J.M. Adovasio, pp. 130–141. Department of Anthropology, University of Pittsburgh, Pittsburgh.

Tuohy, D.R. 1970. The Aboriginal Containers of Baja California, Mexico: A Search for Origins. Tebiwa 13(2):41–51.

Tuohy, D.R. 1974. A cache of Fine Coiled, Feathered and Decorated Baskets from Western Nevada. In Collected Papers on Aboriginal Basketry, edited by D.R. Tuohy and D.L. Rendall, pp. 28–46. Nevada. Nevada State Museum Anthropological Papers 16, Carson City.

Wilson, T. 1888. Ancient Indian Matting from Petit Anse Island, Louisiana. Report of the U.S. National Museum for 1888: 673–676.

Wheeler, S.M., and G.N. Wheeler. 1969. Cave Burials Near Fallon, Churchill County, Nevada. In Miscellaneous Papers on Nevada Archaeology 1–8, edited by D.L. Rendall and D.H. Tuohy, pp. 70–78. Anthropological Paper No. 14. Nevada State Museum, Carson City.

4. Archaeological Investigations of Water-Saturated Deposits in Volusia County, Florida: Groves Orange Midden on Lake Monroe

Barbara A. Purdy

Archaeological investigations were conducted in water-saturated deposits at Groves Orange Midden (8VO2601) periodically from 1987 to 1993. The site is located on the north shore of Lake Monroe along the middle course of the St. Johns River, Volusia County, Florida. The 1993 excavations were funded by the National Geographic Society and entailed the use of cofferdams and well points. Using this method, water was evacuated from the exploratory units permitting recovery of specimens 2 m below the lake bed just as if the excavations were taking place on dry land. All environmental and cultural materials had survived in superb condition. More than one hundred species of plants and animals were identified from strata dating 4000 to 6000 years BP. The initial phases of adaptations to freshwater aquatic resources along the St. Johns River were documented (Purdy 1994a).

Florida's archaeological wetsites are numerous, including, in addition to Groves Orange Midden reported in this paper, Key Marco (Cushing 1897; Gilliland 1975), Fort Center (Sears 1982), Little Salt Spring (Clausen *et al.* 1979), Windover (Doran and Dickel 1988), Hontoon Island (Purdy 1987, 1988, 1990), and others (Purdy 1991).

The Groves Orange Midden site is important because of the vast amounts of information it contains about an environment and a way of life that existed in this area of Florida 4000 to 6000 years ago. More significant is the fact that it demonstrates how sites that have remained continuously wet, dry, or frozen contain untapped archives that furnish minute details about the past which most archaeologists never hope to recover. These results can be used to assess stability and change in cultures and environments through time and space, and to draw conclusions about materials that did not survive at upland sites.

Background

Periodically, from 1987 to 1993, I conducted archaeological investigations in water-saturated deposits at Groves Orange Midden (8VO2601), a freshwater shell midden, near the north shore of Lake Monroe, Volusia County, Florida (Figure 4.1). During the summer of 1993, the work was funded by the National Geographic Society and entailed the use of cofferdams and well points that functioned to remove lake water from the excavation units (Figure 4.2a,b). This system permitted recovery of materials 2 m below the lake bed without having to resort to scuba gear or other methods associated with underwater explorations. The excavations at Groves Orange Midden were spurred by the recovery there in 1987 of a small wooden artifact (Figure 4.3). Since ancient wood does not survive in Florida unless it has remained constantly waterlogged, the preservation of this object indicated there was a strong possibility that the site contained other culturally and environmentally significant information. The area is of interest also because Groves Orange Midden is the near to offshore component of the Old Enterprise site dug by Jeffries Wyman in the 1860s (Wyman 1875). By the time of Wyman's visit, Old Enterprise was already partially demolished by erosion and by mining the shell to fertilize orange groves and for other purposes. This terrestrial portion of the midden no longer exists.

Excavations

Over many years, artifact hunters and local residents amassed large collections of pottery, stone spearheads, and bone weapons and ornaments from the site, but no further archaeological investigations were conducted following Wyman's excavations until the late 1980s when I became interested in the offshore water-saturated deposits. The

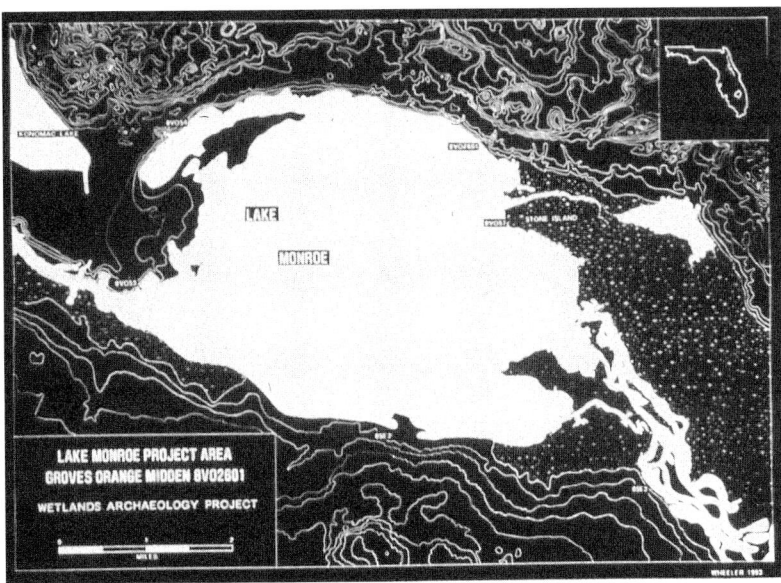

Figure 4.1. Map showing location of Groves Orange Midden (8VO2601).

 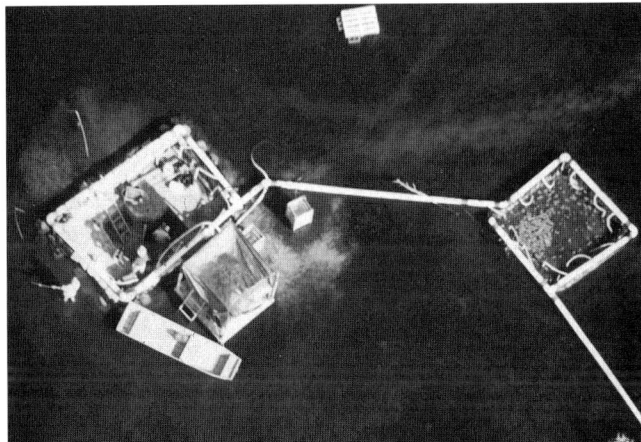

Figure 4.2. (a) Locations of cofferdams in Lake Monroe where archaeological excavations took place in water-saturated deposits at Groves Orange Midden; (b) air photo of cofferdammed areas in Lake Monroe. Individuals are working two meters below the lake bed.

deposits were examined using a bucket auger and piston corer which established that well-preserved organic materials were present.

In 1989 a single 1×2 m test pit was dug in 10 cm levels. The water table was less than 5 cm below the surface. Pumps were used to evacuate water from the unit but problems with the pumps forced the reduction of the area to a 1×1 m square halfway through the excavation. Excavated materials were water-screened using fine mesh to collect

small bones and seeds, and column samples were taken from the wall of the excavation unit in 10 cm segments.

In May 1992, additional tests were carried out at Groves Orange Midden in a slightly different area of the site. As in 1989, the 1992 project was hampered by uncooperative pumps. The primary trouble resulted when pumps malfunctioned or when they were turned off at night. The excavation units filled rapidly with water and, when the water was drawn down again, the weakened walls collapsed.

Figure 4.3. Wooden artifact recovered at Groves Orange Midden in 1987. The specimen has been successfully preserved in polyethylene glycol.

This situation was particularly discouraging because the materials recovered were ancient, abundant, and unique.

The water-saturated cultural deposits at Lake Monroe extend more than fifty meters into the lake but, because of logistics problems involved with dewatering, I did not expect to be able to conduct systematic excavations very far from shore. In 1993, through the efforts and expertise of John B. Allen and the generosity of his firm, Bumby and Stimpson, Inc. Of Orlando, specialists in dewatering operations, two locations in the lake and one location on the shoreline were cofferdammed and excavations proceeded as though the areas were on dry land. The use of a cofferdam system with well points eliminated the usual problems with wall collapse because the deposits were kept dry throughout the entire period that each unit was investigated. Extensive coring prior to the installation of the cofferdams had established the most ideal locations to place the excavation units (McGee and Wheeler 1994), i.e., where midden deposits containing preserved organic materials were thickest (Figure 4.4).

Chronology and Stratigraphy

Eighteen radiocarbon determinations were made on charcoal, wood, hickory nuts, and a fungus collected from Groves Orange Midden. In 1989, a date of 3160±65 BP was obtained from a level (Level 2) disturbed by modern activity since it contained glass and brick in addition to prehistoric objects. As far as is known, all other samples submitted for analysis came from undisturbed strata and gave results ranging from 3850±70 BP to 6210±60 BP with excellent agreement of stratigraphy and artifact styles.

The 1989 and 1992 units were excavated in 10 cm arbitrary levels. In 1993, a 1 m square was first excavated using arbitrary 10 cm levels in each of the three cofferdammed areas. This procedure revealed five major zones in the profile, and work continued in adjacent 1 m units by natural stratigraphy with each defined stratum (zone) removed in 10 cm segments (or partial segments) until the next stratum was reached. In this manner, it was possible to control chronological or cultural differences. The midden was composed primarily of freshwater shell species (Zones II and IV) with the "mystery snail" *Viviparus georgianus* dominating the assemblage. During excavations in 1993, a 10 to 20 cm stratum (Zone III) of organic material, containing numerous organic and sand lenses, was identified that intersected the otherwise massive shell midden (Figure 4.5). The thick organic zone was excavated separately to determine its contents, age, and significance. This zone represents approximately 750 years of deposition between 4930±80 BP (Beta 65860) and 4190±60 BP (Beta 65864). Zone V is the lowest excavated cultural stratum at the site. It is composed primarily of sand and underlies the shell midden.

Throughout the project, a three-tiered screen set up was used and materials from the excavation units were water-screened through 1/2, 1/4, and 1/8 inch mesh to collect small bones and seeds (Figure 4.6a). Column samples were cut from the walls of the excavation units in 10 cm segments or partial segments. The column samples assured that a sample, unbiased by collecting at the screens in the field, would be available for subsequent fine-tuned analyses in the laboratory.

Fauna and Flora

Eighty-two species of animals and plants were identified from the small test conducted in 1989 (Russo *et al.* 1992). These consisted of 20 species of shellfish, 13 species of fish, 3 species of snake, salamander, alligator, 5 species of turtle, 3 species of birds, 5 species of mammals including human, hickory, oak, cane, dogwood, persimmon, ash, sweetgum, magnolia, mulberry, palm, pine, cherry/maple, willow, elderberry, cypress, elm, grape, toothache tree, fringe tree, shelf fungus, bottle gourd, and squash/gourd. Despite the great variety of species identified in 1989, the materials from the single 1x1 m test pit and column did not constitute a very large sample.

One goal of the 1992 and 1993 excavations was to verify and expand the species list. And this was accomplished. For example, in addition to opossum, rabbit, raccoon, while-tailed deer, and human, the mammal list included dog, otter, beaver, bobcat, mouse, and squirrel (Figure 4.6b). In 1989, the role of fish in the diet appeared to be very minor accounting for only 1% of the diet. The 1992–

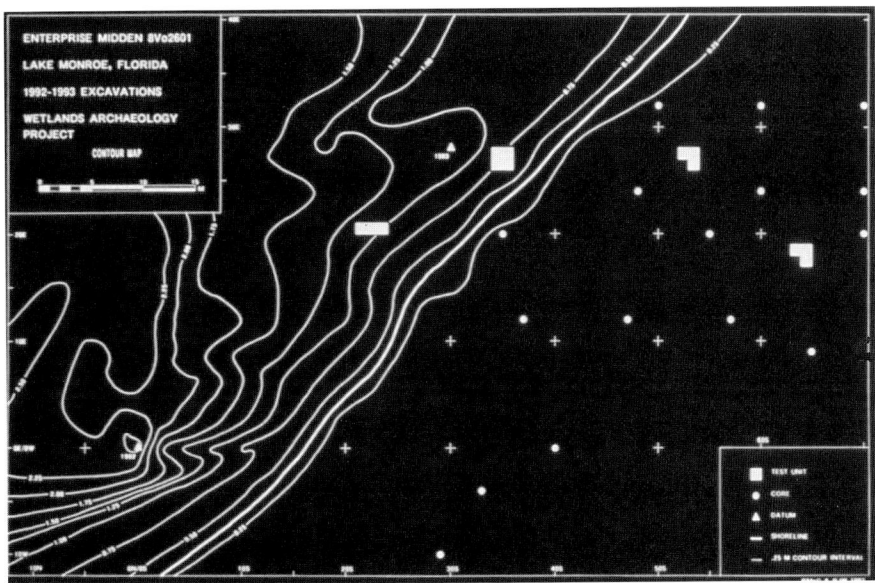

Figure 4.4. Contour map of Lake Monroe showing the on- and offshore locations of cores and excavation units in 1992 and 1993.

Figure 4.5. Stratigraphy in the water-saturated deposits at Groves Orange Midden.

1993 results demonstrated that fish made up between 11% to 63% of the diet, the variation resulting from the location in the stratigraphic profile from which the sample was taken. The largest percentage of fish occurred in the Zone III organic stratum. It seems likely that the 1992–1993 figures are more accurate because of the larger sample sizes. In addition, the 1992–1993 figures are more in line with the hypothesized prediction that aquatic resources were major food items. Why would the inhabitants have used shellfish and excluded fish in an aquatic environment?

Thirty additional archaeobotanical taxa were identified from the 1992–1993 samples, and five species identified in 1989 were not present in 1992–1993. The increase in number of species can be accounted for by the larger number of samples examined. For instance, 32 taxa were recognized from Zone V probably because three different size fractions were studied. Some of the added taxa, however, are rare and are represented by only a single specimen. Hickory nuts and acorns appear to be the most prevalent species utilized by the inhabitants. Early historic accounts mention the use of these sources for food by the Florida Indians. As with the faunal analysis, there were some interesting differences between the floral materials from the shell midden strata and the organic stratum (Zone III). Cane and grape, for example, showed a significant increase while hickory nuts declined dramatically. The

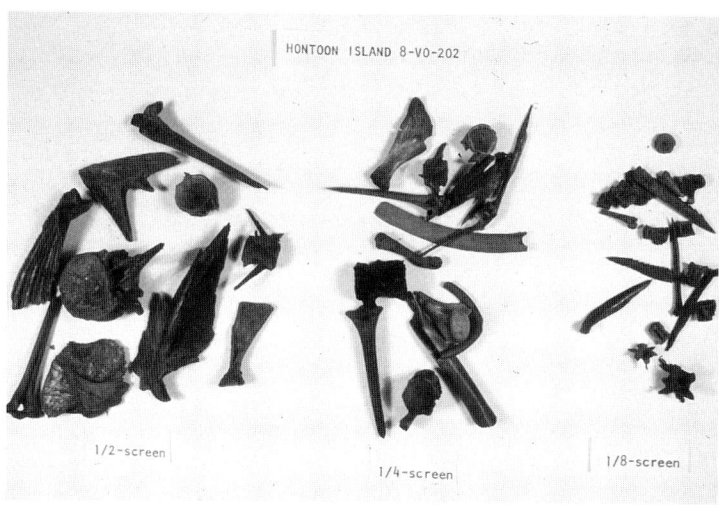

Figure 4.6(a).

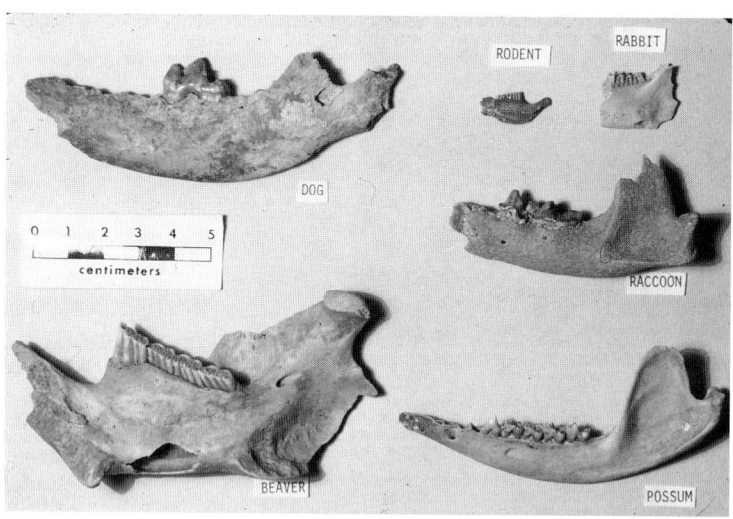

Figure 4.6(b).

Figure 4.6. (a) Small whole bones recovered in 1988 from Hontoon Island that were size-sorted through 1/2, 1/4, and 1/8 inch screens; (b) examples of mammal jaws from waterlogged deposits at Groves Orange Midden.

assemblage reveals that a diverse array of potentially edible food plants was available. Nearly all of the various nuts, fresh fruits, and small seeds have a ripening period from early summer to mid-fall. Greens, tubers, and some supplemental foods were available throughout the year. Despite the presence of two species of gourd, a plant that is usually considered cultivated, there is no clear evidence that any form of horticulture or gardening occurred in this area of Florida 4000–6000 years ago (Newsom 1994).

Many of the plants and animals identified in the excavated assemblages at Groves Orange Midden were not consumed. They may be present only as environmental indicators, or portions of them may have served as weapons, utensils, ornaments, dyes, cordage, medicines, building materials, and more. The seeds of gourds were probably eaten, but gourds function also as containers and net floats. Cane and wood can be used as building material. Cypress made up 99% of the adzed wood observed, but some of the other 13 wood species were used in various ways (Newsom 1994).

An important point to consider is that, had these excavations occurred at a terrestrial shell midden site, most of the species would not have survived or they would have been too fragmentary for identification. This statement applies to a large percentage of the fauna and virtually all of the flora. The fungus *Hexagonia hydnoides* with intact basidiocarps preserved, for example, was dated by Accelerator-AMS at 5530±80 BP (Beta 54580, CAMS-3527). Other plant identifications were made from culms, peduncles, rind, seeds, petioles, thalli, buds, thorns, nuts, woods, and charcoal. From an upland site, only charcoal or a few carbonized plant parts may have remained and not in quantity. How could anyone have guessed at the rich variety of specimens available in the environment and utilized by the human groups living there?

Just as significant as the survival of so many types of plants and animals is the fact that it is possible to determine the abundance of each species and to assess which species were the most important contributors to food and fiber. It is interesting, for example, that rattlesnake, bobcat, and otter turn up in the assemblage but their occurrence is rare. It is tempting to speculate that their presence may have been for ritualistic or prestigeous reasons. The real food staples were freshwater shellfish, fish, and nuts; there is no speculation involved in that conclusion.

The results of faunal and floral analyses can be used to compare assemblages from older and younger sites (e.g., Hontoon Island) and thus learn in what ways environments and cultures changed through time. They can be used to conclude or assume what may have been present at upland sites in the vicinity; in this case, along the middle course of the St. Johns River in Florida. Table 4.1 lists zooarchaeological remains and Table 4.2 list archaeobotanical remains recovered at Groves Orange Midden.

Artifacts

Artifacts manufactured from stone, bone, antler, shark teeth, marine shell, wood, and fired clay were recovered from Groves Orange Midden.

Stone

No chert (flint) outcrops occur in this area of Florida. Thus, the objects of this material found at the site were all imported from sources approximately one hundred miles away. Eighteen complete or broken stemmed bifaces (spearheads?) plus nearly two hundred tiny utilized and nonutilized flakes comprise the total chert industry. The biface styles are fairly typical of the Middle to Late Archaic time period in Florida and agree well with the 6000 to 4000-year BP dates. The tasks that may have been accomplished with some of the small flakes include incising,

slicing fish flesh, or cutting and scraping plant fibers in preparation for basket making. It is obvious that no heavy stone working was going on at the site (Purdy 1994b).

A nicely made steatite bead was recovered from a level dated at 4200 BP. Since steatite is a material that is not local to Florida, its presence at the site attests to the fact that the inhabitants were aware of and in contact with people in distant regions.

Bone, Antler, and Shark Teeth

The most abundant and varied artifacts recovered at Groves Orange Midden were manufactured from bone. Many are identical to bone artifacts that were made at least ten thousand years ago and to those still being made when Europeans entered the Western Hemisphere. The nearly 400 specimens ranged from nicely finished or incised "pins" (Figure 4.7a,b) and beads to utilitarian objects that probably functioned as awls, gouges, tool handles, flakers, and more (Wheeler and McGee 1994). The decorations on two of the incised bone pieces, from a stratum nearly 5000 years BP, are of interest for they resemble designs on antler objects recovered from an Archaic Period burial site at Republic Groves, Hardy County, Florida (Wharton *et al.* 1981). The specimen shown at the bottom of Figure 7b was recovered from the lowest stratum (Zone V) dated 6200 years BP. Some of the nicely made unbroken objects may have been lost from hair or clothing as people worked at offshore tasks. But, as one would expect from a trash midden, most decorative items are fragmentary, thrown away by their owners when they broke.

In addition to bone and antler, 19 modified shark teeth were recovered at Groves Orange Midden. Their presence indicates contact with coastal areas.

There was a 50% to 75% decrease in bone artifacts in the organic stratum (Zone III) compared to the shell and sand strata at the site. Also, no shark teeth were found in the organic stratum.

Marine Shell

The sophisticated marine shell industry that eventually developed in Florida appears in its incipient stage in the lowest levels at Groves Orange Midden. Some types are recognizable but the majority of the specimens are fragmentary making it difficult to determine how they were modified and how they functioned. Nine disk beads manufactured of marine shell were also recovered. The most interesting specimen, from a 6000-year level, was an extremely burned vessel made from a large Busycon contrarium (conch shell) from which the column had been removed. Other examples of this type have been found in Florida in preceramic levels suggesting that marine shell

Table 4.1. List of Zooarchaeological Species from Groves Orange Midden[1].

Scientific Name	Common Name	Scientific Name	Common Name
Land Snails		*Lepomis auritus*	Redbreast
Polygyra cereolus	Southern Flatcoil	*Lepomis macrochirus*	Bluegill
Polygyra septemvolva	Florida Flatcoil	*Lepomis* sp.	Bream
Glyphalinia indentata	Carved Glyph	*Pomoxis nigromaculatus*	Speckled Perch
Hawaiia miniscula	Minute Gem	Centrarchidae	Sunfish
Zonitoides arboreus	Quick Gloss	Osteichthyes	Unid. Fish
Euglandina rosea	Rosy Euglandid		
Gastrocopta contracta	Bottleneck Snaggletooth	Reptiles/Amphibeans	
Gastrocopta pellucida	Slim Snaggletooth	*Siren lacertina*	Great Siren
		Amphiuma means	Salamander
Freshwater Snails		*Rana* sp.	Frog
Pomacea paludosa	Applesnail		Unid. Anuran
Vivparus georgianus	Banded Mysterysnail	*Nerodia*	Water Snake
Elimia floridiensis	Rasp Elimia	Colubridae	Nonpoisonous Snake
Tryonia aequistotatus	Smoothribbed Hybrobe	Viperidae	Poisonous Snake
Littoridinops monroensis	Cockscomb Hydrobe	Serpentes	Unid. Snake
Notogillia wetherby	Alligator Siltsnail	*Chelydra serpentina*	Snapping Turtle
Hydrobiidae	Hydrobes	Kinosternidae	Mud/Musk Turtle
Planorbella duryi	Seminole Ramshorn	*Pseudemys nelsoni*	Red-bellied Turtle
Planorbella scalaris	Mesa Ramshorn	Emydidae	Pond Turtle
		Terrapene Carolina	Box Turtle
Freshwater Bivalves		*Gopherus polyphemus*	Gopher Tortoise
Unionidae	Freshwater Mussel	*Apalone* sp.	Soft-shell Turtle
		Testudines	Unid. Turtle
Marine Shell		*Alligator mississipiensis*	Alligator
Crassostrea virginica	Oyster		
Mercenaria campechensis	Quahog	Birds	
Dinocardium robustum	Hear Cockle	Anatidae	Surface-feeding Duck
	Unid. marine shell	*Meleagris gallopavo*	Turkey
		Casmerodius albus	Great Egret
Cartilagenous Fish		Passerine	Unid. Bird
Dasyatis sp.	Stingaree	Aves	Unid. Bird
Bony Fish		Mammals	
Lepisosteus sp.	Gar	*Didelphis marsupialis*	Opossum
Amia calva	Bowfin	*Sylvilagus* sp.	Rabbit
Clupidae	Herring	*Peromyscus gossypinus*	Pine Mouse
Esox niger	Chain Pickerel	*Canis familiaris*	Dog
Notemigonus chrysoleucas	Golden Shiner	*Procyon lotor*	Raccoon
Ictalurus sp.	Bullhead	*Lynx rufus*	Bobcat
Mugil sp.	Mullet	*Odocoileus virginianus*	Deer
Micropterus salmoides	Largemouth Bass	*Castor Canadensis*	Beaver
Fundulus sp.	Killifish	Mammalia	Unid. small mammal
Lepomis gulosus	Warmouth	*Homo sapiens*	Human
Lepomis punctatus	Stumpknocker		
Lepomis microlophus	Shellcracker		

[1] Most of the 73 species were identified from the column samples; a few species were recovered only in the excavation units. In descending order of frequency were freshwater shellfish, fish, turtles, and mammals. Most of the bone artifacts, however, were made from mammal (from Wheeler and McGee 1994b).

functioned as cooking pots prior to the introduction of pottery making.

Wood
Artifacts of wood consist of tool handles, a broken canoe paddle, broken stirring paddles, and a few other types of uncertain function. While interesting, particularly because wood is seldom preserved, these specimens do not excite as much enthusiasm for themselves as for the overall story they tell. The handles, for instance, are virtually identical to those from Hontoon Island and Key Marco, sites that are

Table 4.2. List of Archaeobotanical Species from Groves Orange Midden[1].

Scientific Name	Common Name	Scientific Name	Common Name
Acer sp.	Maple	*Nymphaea* sp.	Waterlily
Amaranthus sp.	Amaranth, water hemp	*Nyssa sylvatica*	Black Gum
Arundinaria sp.	Native Cane	*Opuntia* sp.	Prickly Pear Cactus
Atriplex sp.	Sand Atriplex	*Palmae*	Palm family
cf *Asteraceae*	Sunflower family	*Passiflora* (*incarnata*)	Maypop
Brasenia schreberi	Water Shield	*Pinus* sp., section *diploxylon*	Hard Pine group (so. Hard)
Carpinus caroliniana	Ironwood	*Phytolacca Americana*	Pokeweed
Carya sp.	Hickory, true group	*Polygonum* spp.	Smartweed, Knotweed
Carya sp.	Water Hickory	*Polyporaceae/Polyporus*	Shelf/woodrotting fungus
cf. *Castanea* sp.	Chinquapin/chestnut	*Potamogeton* sp.	Pondweed
Cephalanthus occidentalis	Buttonbush	*Prunus caroliniana*	Carolina Laurel Cherry
Chenopodium sp.	Pigweed	*Prunus* sp.	Wild Plum
cf. *Chionanthus* sp.	Fringe Tree	*Prunus* sp.	Cherry wood
Cissus sp.	Calusa Grape	*Quercus* sp., red group	Laurel, Water Oaks, eg
Cladium jamaicense	Sawgrass	*Quercus* sp., white group	Swamp White Oak
Cornus (*florida*)	Flowering Dogwood	*Quercus virginiana*	Live Oak
Cornus (*foemina*)	Swamp Dogwood	*Rubus* sp.	Blackberry
Cucurbita pepo, subspecies *ovifera*	Cucurbita pepo gourd	*Sabal palmetto*	Cabbage Palm
Cyperaceae	Sedge family	*Salix caroliniana*	Coastal Plain Willow
Cyperus sp.	Sedge	*Sambucus Canadensis*	Elderberry
Diospyros virginiana	Native Persimmon	*Serenoa repens*	Saw Palmetto
Fraxinus sp.	Ash	*Solanaceae*	Nightshade, e.g.
cf. *Galium* sp.	Bedstraw	*Taxodium* sp.	Cypress
Lagenaria siceraria	Bottle gourd	cf. *Ulmus* sp.	Elm
cf. *Liquidambar styr.*	Sweetgum	*Vitis* sp.	Wild Grape
Magnolia grandiflora	Southern Magnolia	*Zanthoxyplum* sp.	Toothache tree, wild lime
Morus rubra	Red Mulberry	Unid. Spiny seed/fruit	Unid. Littoral/wetland
Nuphar lutea	Spatter-dock		

[1]Plant species were identified from both column samples and excavation units (from Newsom 1994).

both several thousand years younger. This situation indicates that stability in the manufacture of wooden implements had been reached and there was no need to change an object that functioned in the required manner in an environmental and social milieu that was also stable. The canoe paddle confirms that canoes were manufactured and used in this area of Florida. There is, in fact, ample evidence for the presence of canoes as early as 6000 years BP in Florida (Newsom and Purdy 1990; Purdy 1991).

The wood recovered at Groves Orange Midden is very fragile and not at all the sturdy material it was when first disposed of at the site. The lack of statuary or decorated wooden artifacts can be explained, again, by the fact that the site was a garbage dump. Artistic pieces, if they were fashioned at this time, would have been placed in cemeteries as grave offerings. There is some support for this statement from Little Salt Spring and from the Windover site. Both are waterlogged mortuary sites. It is quite probable that decorated wood was as common as decorated bone, but it simply has not survived to tell the tale.

Fired Clay

These objects include pottery sherds and "boiling balls." The sherds were primarily of a style called Orange Incised and Orange Plain, a fiber-tempered ware that is the earliest pottery in America north of Mexico. Wyman (1875) also described this as the most common type he recovered at the Enterprise site. Pottery sherds were found in all excavation units at Groves Orange Midden but it was only in 1989 that they were found in undisturbed contexts, with the earliest dating 4115±75 BP. In 1992 and 1993, the top levels of the excavation units contained sherds mixed with modern materials such as glass; the undisturbed levels predated the advent of pottery technology.

That the inhabitants of Florida knew about fired clay before they began to manufacture ceramic vessels is borne out by the presence of fired clay balls in the pre-pottery levels at Groves Orange Midden. At other sites where similar objects have been recovered, they are usually described as having functioned as hearth stones. Careful examination of the clay balls from the site by Ray M. McGee, however, suggests that they were used as boiling

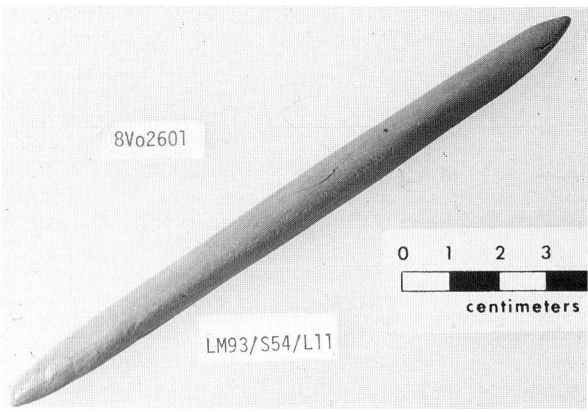

Figure 4.7a.

Figure 4.7b.

Figure 4.7. (a) A nicely made bone "pin"; and (b) incised bone objects from Groves Orange Midden.

"stones" to heat water, probably for cooking. Tightly controlled experiments by McGee demonstrated the feasibility of this practice. The presence of the fired clay objects primarily in pre-pottery levels suggests that the inhabitants needed a way to boil food (probably the freshwater snails and mussels that had recently become a food staple) before ceramic vessels became part of the cultural inventory.

Summary and Conclusions

The contributions to knowledge gained from Groves Orange Midden can be viewed in both specific and general ways. The site is located on the north shore of Lake Monroe in Volusia County, Florida. Mount Taylor is the name given to this Late Archaic culture whose people first began to exploit the aquatic resources along the middle course of the St. Johns River about six thousand years ago. Abundant freshwater species probably were not available much earlier than 6000 years ago because that approximate date marks the beginning of the widespread emergence of Florida's wetlands. Prior to that time, most of Florida's rivers did not flow, surface water was scarce, and Florida was essentially a desert (Fairbridge 1974; Gleason 1974). In shifting from nomadic hunters of terrestrial game to more sedentary lifeways, many already existing technologies were adapted to the new situation. It is logical to see continuities in technologies of long standing. Soon, technologies utilizing marine shell and fired clay were adopted.

In this paper, I described, listed, and illustrated the fauna, flora, and artifacts recovered and analyzed from Groves Orange Midden. They represent nearly an entire assemblage of cultural and environmental trash. This kind of survival is an archaeological bonanza. Only anthropologists working in libraries or in ethnographic settings with living informants are able to collect more data. From an upland site, the entire assemblage of floral remains would have disappeared, and so would the bone unless the matrix was alkaline. Bone will not survive in an acid environment. Without the plants and most of the animals to use for interpretation, the archaeologist is left with a great pile of freshwater shells, a few marine shells, and some stone. Since Groves Orange Midden was primarily a preceramic site, even the usually ubiquitous pottery would be missing. Of greater significance, however, is that stratigraphy remains intact at waterlogged sites because it has not been trampled as it would have been at an upland site, and it has not been compressed. As the large organic component degrades and eventually disappears at an upland site, strata collapse upon themselves mixing time periods.

The excellent preservation at Groves Orange Midden not only permitted an evaluation of climate and culture of 6000 years ago, but also a comparison to a 500– to 1500-year BP site with similar materials at Hontoon Island located approximately fifteen miles north. By comparing Groves Orange Midden to Hontoon Island, it was possible to conclude that the basic way of life had remained stable for more than five thousand years until dramatic changes were brought about directly and indirectly by Europeans around AD 1500. The species list is nearly identical for both sites and there is even continuity in many of the tool types. Decorative motifs on pottery and bone are the most conspicuous markers distinguishing the two sites. Changes occurred in temper, construction methods, vessel shapes, and designs in the pottery industry. Marine shell tools became more numerous, varied, and better made. Large wooden carvings of animals were recovered near Hontoon Island suggesting the development of some social complexity over time (Purdy 1987, 1991).

Of a more general nature, Groves Orange Midden points out the importance of wet, dry, or frozen sites wherever they occur throughout the world. They are not very abundant, they are invisible, and they are endangered. At these sites, there is a virtual explosion of recoverable materials compared to sites where alternate episodes of wetting, drying, or defrosting have occurred and degraded the organic component. Everything from insects to DNA has been found and analyzed, including human fecal matter (coprolites). Obviously, there are going to be logistic problems involved with excavation, preservation, and analysis. The dewatering methodology employed at Groves Orange Midden in the summer of 1993 using cofferdams and well points was ideal but economically prohibitive without funds from the National Geographic Society and the generosity of John B. Allen and his firm, Bumby and Stimpson of Orlando. Organic materials need the immediate attention of a conservator as soon as they are exposed to the air or they wil not survive. Sometimes within minutes, always within hours, most botanical specimens will degrade irreversibly. Zooarchaeologists, paleoethnobotanists, and a number of other specialists are needed to analyze and interpret the floral and faunal species recovered. The curation problems following the completion of investigations are a nightmare.

Groves Orange Midden takes its place in time with the Iceman of the Alps and some of the Swiss Lake Dwellers; it predates considerably the Iron Age bog bodies of Europe. It gives a picture in a warm climate of how people adapted to newly available aquatic resources 6000 years ago.

Acknowledgments
Many organizations and individuals were responsible for the success of investigations at Groves Orange Midden. I am grateful to Grant Groves for leading me to the site in 1987; the National Geographic Society (Grant 4946–93) for funds to conduct excavations in 1993; the St. Johns River Water Management District for financial support in 1992 and 1994; John B. Allen and his firm, Bumby and Stimpson, Inc. Of Orlando; John Russell and his family of Sanford, Florida; local residents who permitted access to their land; graduate students and colleagues; and last but not least, the many volunteers for their donations of money, food, time, labor, services, and expertise.

References Cited
Clausen, Carl J., A.D. Cohen, Cesare Emiliani, J.A. Holman, and J.J. Stipp. 1979. Little Salt Spring, Florida: A Unique Underwater Site. Science 203(4381):609–614.

Cushing, Frank Hamilton. 1897. Exploration of Ancient Key Dwellers' Remains on the Gulf Coast of Florida. Proceedings of the American Philosophical Society 25(153):329–448.

Doran, Glen H. and David N. Dickel. 1988. Multidisciplinary Investigations at the Windover Site. In Wet Site Archaeology, edited by Barbara A. Purdy, pp. 263–289. The Telford Press Inc., Caldwell, New Jersey.

Rhodes W. 1974. The Holocene Sea-Level Record in South Florida. In Environments of South Florida: Present and Past, edited by Patrick J. Gleason pp. 223–232. Miami Geological Society, Memoir No. 2.

Gilliland, Marion Spjut. 1975. The Material Culture of Key Marco, Florida. The University Press of Florida, Gainesville.

Gleason, Patrick J., Editor. 1974. Environments of South Florida: Present and Past. Miami Geological Society, Memoir No. 2.

McGee, Ray M. and Ryan J. Wheeler. 1994. Stratigraphic Excavations at Groves Orange Midden, Lake Monroe, Volusia County, Florida: Methodology and Results. Florida Anthropologist 47(4):333–349.

Newsom, Lee A. 1994. Archaeobotanical Data from Groves Orange Midden (8VO2601), Volusia County, Florida. Florida Anthropologist 47(4):404–417.

Newsom, Lee A. and Barbara A. Purdy. 1990. Florida Canoes: A Maritime Heritage from the Past. Florida Anthropologist 43(3–4):164–180.

Purdy, Barbara A. 1987. Investigations at Hontoon Island (8VO202), an Archaeological Wetsite in Volusia County, Florida: Overview and Chronology. Florida Anthropologist 40(1):4–12.

Purdy, Barbara A. 1990. Chronology of Cultivation in Peninsular Florida: Prehistoric or Historic? Southeastern Archaeology 9(1):35–42.

Purdy, Barbara A. 1991. The Art and Archaeology of Florida's Wetlands: CRC Press, Boca Raton, Florida.

Purdy, Barbara A. 1994a. Excavations in Water-Saturated Deposits at Lake Monroe, Volusia County, Florida: An Overview. Florida Anthropologist 47(4):326–332.

Purdy, Barbara A. 1994b. The Chipped Stone Tool Industry at Groves Orange Midden (8VO2601), Volusia County, Florida. Florida Anthropologist 47(4):390–392.

Purdy, Barbara A. and Lee A. Newsom. 1985. The Significance of Archaeological Wet Sites: A Florida Example. National Geographic Reserch 1(4):564–569.

Russo, Michael, Barbara A. Purdy, Lee A. Newsom, and Ray M. McGee. 1992. Groves Orange Midden Site: A Late Archaic Site on the St. Johns River. Southeastern Archaeology 11(2):95–108.

Sears, William H. 1982. Fort Center: An Archaeological Site in the Lake Okeechobee Basin. University Press of Florida, Gainesville.

Wharton, Barry R., George R. Ballo, and Mitchell E. Hope. 1981. The Republic Groves Site, Hardee County, Florida. Florida Anthropologist 34(2):59–80.

Wheeler, Ryan J. and Ray M. McGee. 1994a. Technology of Mount Taylor Period Occupation, Groves Orange Midden (8VO2601), Volusia County, Florida. Florida Anthropologist 47(4):350–379.

Wheeler, Ryan J. and Ray M. McGee. 1994b. Report of Preliminary Zooarchaeological Analysis: Groves Orange Midden. Florida Anthropologist 47(4):393–403.

Wyman, Jeffries. 1875. Fresh-Water Shell Mounds of the St. Johns River, Florida. Peabody Academy of Science, Memoirs, No. 4, Salem, Massachusetts.

5. Pineland: A Coastal Wet Site in Southwest Florida

William H. Marquardt and Karen J. Walker

Extraordinary wet-site preservation is known for several South Florida sites (e.g., Little Salt Spring, Warm Mineral Springs, Republic Groves, Bay West, Belle Glade, Fort Center, and Key Marco; see Purdy 1991 for summaries). Of these, only Key Marco is a coastal site. In this paper, we describe another coastal site with unusual preservation of organic materials – the Pineland Site Complex in Lee County, Florida (Figure 5.1) – and discuss possible reasons why such preservation exists there.

The Pineland Site Complex

Although the Pineland Site Complex was visited and noted in print as early as 1875 (Kenworthy 1875), it was Frank Hamilton Cushing who provided the first detailed description (see Figure 5.2):

> The foundations, mounds, courts, graded ways, and canals here were greater, and some of them even more regular, than any I had yet seen. ... The same sorts of channel-ways as occurred on the outer keys led up to the same sorts of terraces and great foundations, with their coronets of gigantic mounds. The inner or central courts were enormous. Nearly level with the swamps on the one hand, and with the sand flats on the other, these muck-beds were sufficiently extensive to serve ... as rich and ample gardens; and they were framed in, so to say, by quadrangles formed by great shell structures... There were no fewer than nine of these greater foundations, and within or among them no fewer that five large, more or less rectangular courts; and, beyond all, to the southward, was a series of lesser benches, courts and enclosures. ... This settlement had an average width of a quarter of a mile; ... Its high-built portions alone, including of course the five water courts, covered an area of not less than seventy-five or eighty acres. (Cushing 1897:341–342)

The site was altered in the twentieth century, notably by the removal of more than half of the Smith and Brown's Mounds, nearly all of the Old Mound, the shearing away of the northern section of the Randell Mound, and the filling of numerous low areas, including much of the ancient canal (Torrence in prep.). A great deal of this remarkable site remains, however, including not just major parts of its high mounds and plazas, but also earlier, buried waterlogged deposits, which are the focus of this paper. The Pineland Site Complex is listed in the National Register of Historic Places. Figure 5.3 is a topographic map of the Pineland Site Complex today.

We have documented deposits at Pineland dating to the following cultural periods: the latter part (A.D. 50–400) of Caloosahatchee I (500 B.C.–A.D. 500); Caloosahatchee IIA (A.D. 500–800); Caloosahatchee IIB (A.D. 800–1200); Caloosahatchee III (A.D. 1200–1350); Caloosahatchee IV (A.D. 1350–1500); and Caloosahatchee V (A.D. 1500–1750). The radiocarbon dates in this paper are discussed as generalized century date ranges based on corrected and calibrated dates with a one-sigma deviation. The details of these and other elevation details are found in Walker *et al.* (1995) and Walker and Marquardt (in prep.).

The Pineland site is centrally located within the Charlotte Harbor/Pine Island Sound estuarine system. A primary subsistence focus of Pineland's precolumbian residents throughout the centuries was the abundant fish and shellfish populations found in Pine Island Sound's mangrove-fringed, shallow-water seagrass meadows. A high diversity of mollusks, both bivalves and gastropods, and a range of fishes, dominated by smaller species, characterizes this fauna (Walker 1992). The remains of fish and shellfish

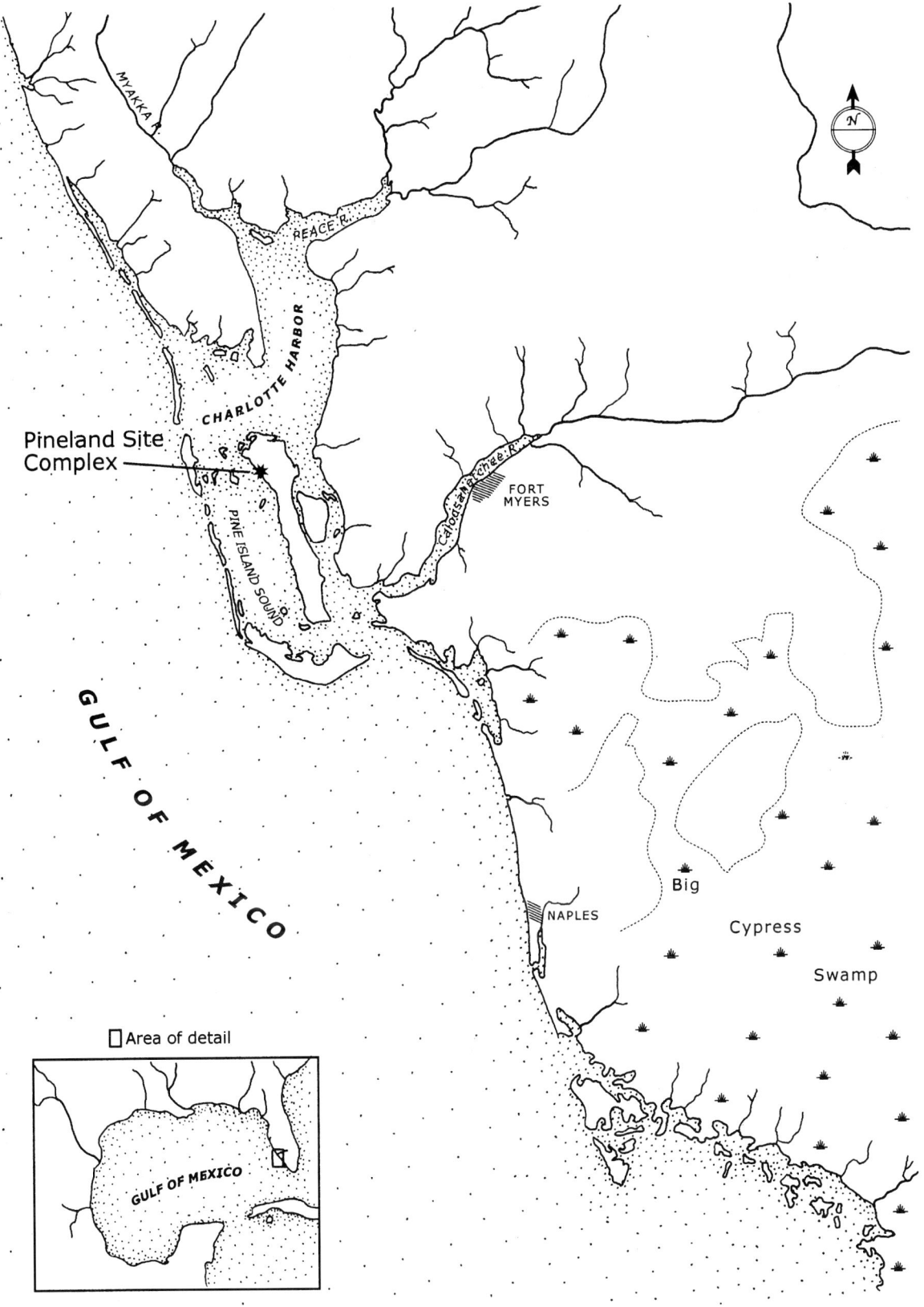

Figure 5.1. Location of the Pineland Site Complex. (Map by Corbett Torrence and Sue Ellen Hunter).

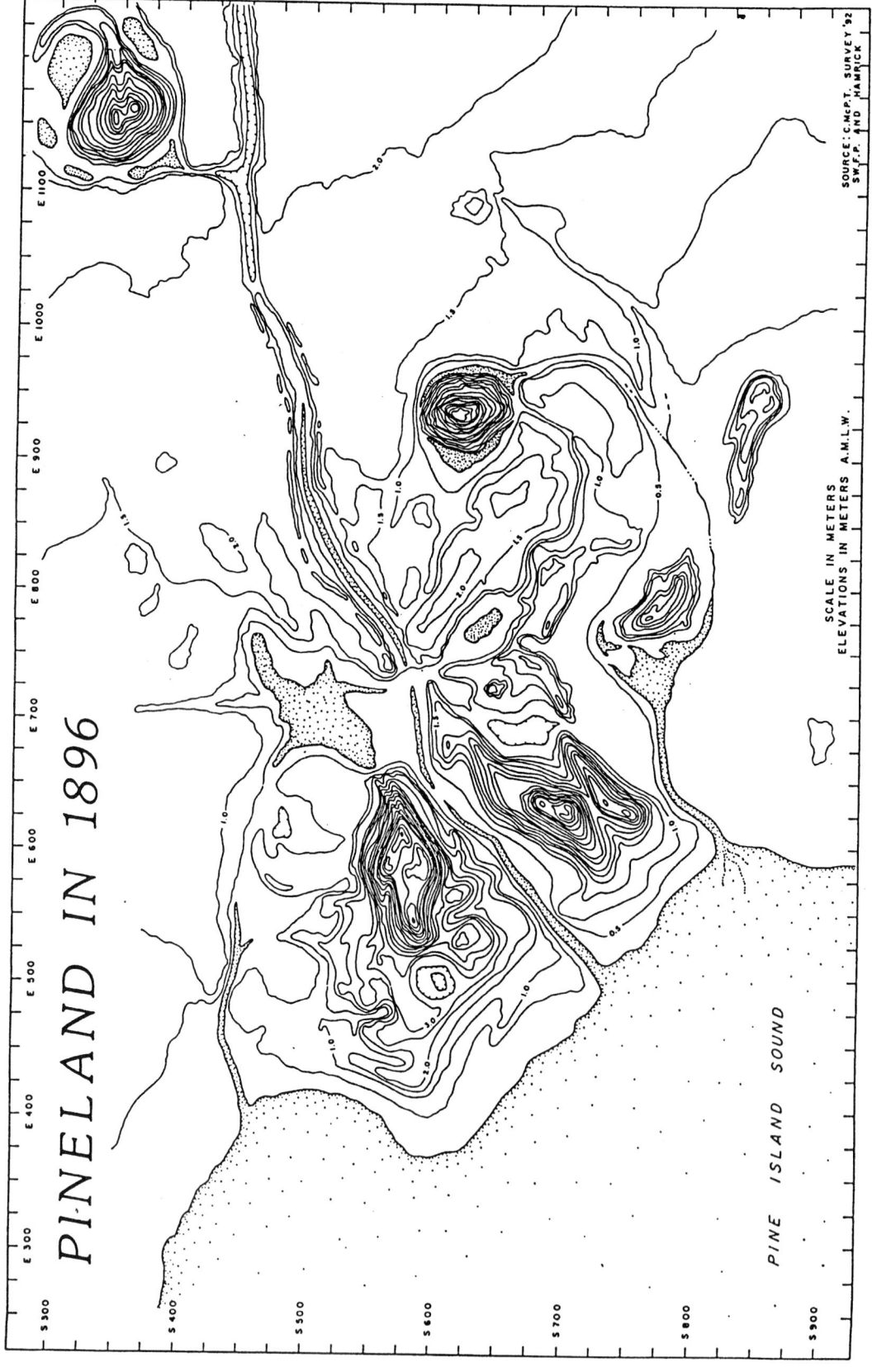

Figure 5.2. The Pineland Site Complex in 1896. (Map by Corbett Torrence, based on Cushing's written description).

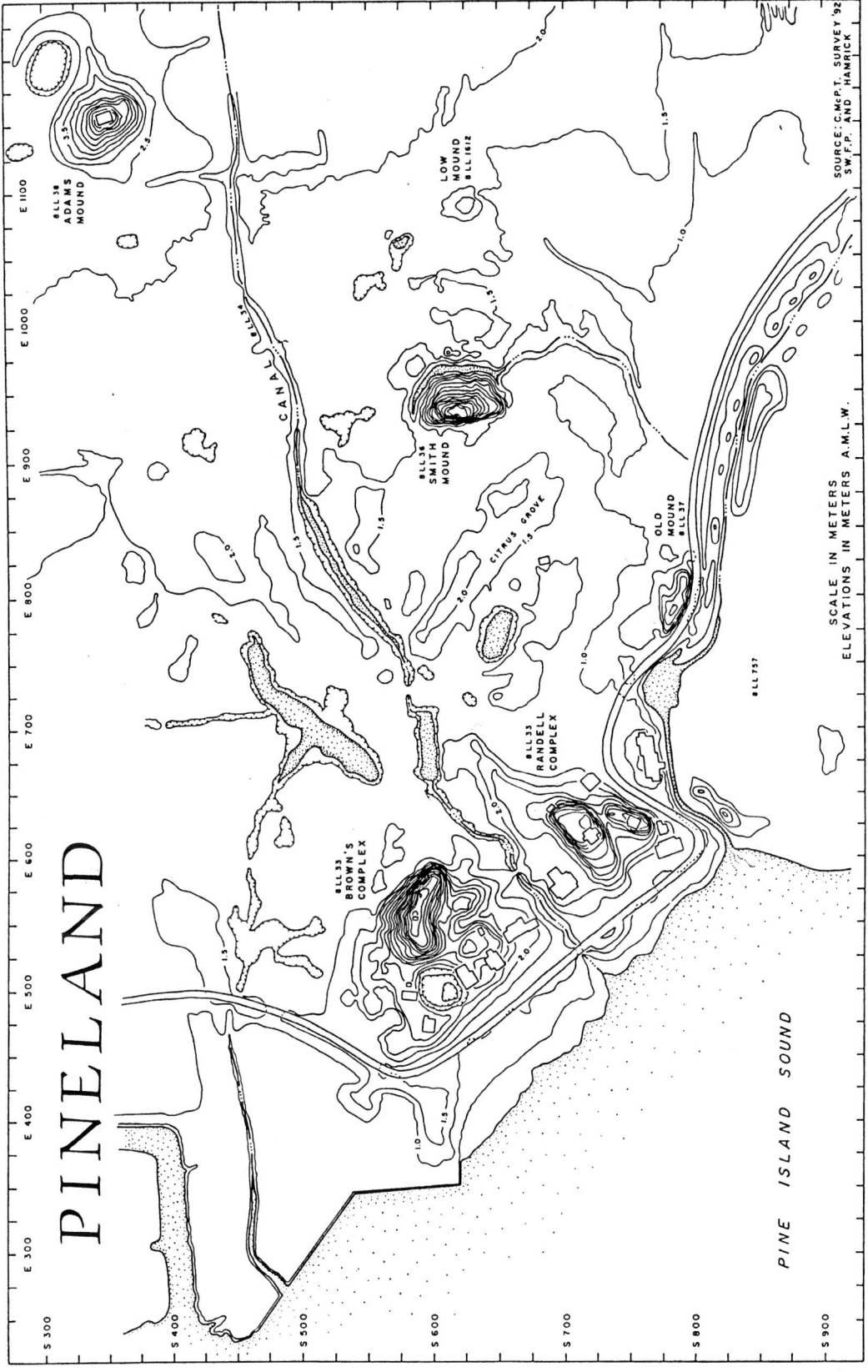

Figure 5.3. The Pineland Site Complex as it appears today. (Map by Corbett Torrence).

found in Pineland's archaeological deposits generally mirror this adjacent, rich, shallow-water environment. Thus, we infer that Pineland's residents fished and shell-fished locally for their daily needs and rarely traveled as far as the barrier islands to procure animal foods.

Pineland's Archaeological Wet Deposits

Pineland's prominent mounds are associated with the historic Calusa Indians but the site was first settled around 2,000 years ago. Distinct spatial and temporal settlement and site-formation patterns are evident at Pineland (Walker and Marquardt in prep.) and are relevant to the wet preservation of archaeological materials.

Pineland's Early Occupations (Caloosahatchee I)

Pineland's first episode of occupation began at about A.D. 50 and continued until sometime in the late fourth or early fifth century A.D. (the latter part of the Caloosahatchee I period). Habitation structures and their associated middens were at first expansive, broadly covering the shoreline and adjacent low-lying areas, with deposits averaging a meter in thickness and occurring as low as 20 cm below today's mean sea level. This area is cross-hatched in Figure 5.4. Later, during the fourth and fifth centuries, settlement shifted away from the shoreline to higher ground (to the Citrus Grove Mound and Low Mound, the darkly shaded areas in Figure 5.4).

From a taped oral interview with a previous landowner (Edic and Torrence 1990), we learned that in the early 1960s a great portion of Pineland's Old Mound was bulldozed to the north and northeast into the low-lying, black mangrove wetland and salt-flats for the purpose of creating an airstrip. We had little or no expectation of finding *in situ* deposits in those areas, so we focused our controlled excavation at the bulldozed surface where the northeastern portion of the mound had stood, hoping to document the mound's basal deposits. A 4×4 m excavation block was divided into sixteen 1×1 m units. This strategy was intended to identify activity areas and possibly structural remains. We did indeed document the basal strata of the mound, dating them to the sixth century A.D. (early Caloosahatchee IIA), but to our surprise, underlying the sixth-century base of the mound was an expansive, dense, Caloosahatchee I shell midden dating to the first through third centuries.

We changed our strategy (due to time limitations), selecting four units to take down to the base of these deposits. The lower portion (first through third century) was waterlogged, thus presenting us with the unanticipated challenge of a de-watering operation that would allow us to continue excavation. We successfully used a continuously

pumping horizontal well system (donated by Horizontal Dewatering, Inc. of Cape Coral) combined with a 3-hp Briggs and Stratton diaphragm pump (3-inch mud 'hog') (donated by Kelly Tractor of Fort Myers). Interestingly, the horizontal well system had been developed specifically for South Florida's many citrus groves and golf courses. We were able to excavate a 1×4 m area to the base of the midden. Each 1×1 m unit had an associated column sample for fine-screen archaeobotanical and zooarchaeological studies.

Extraordinarily well preserved woods, fragments of cordage, and hundreds of seeds were recovered from this waterlogged deposit. What appeared to be the end of a wooden plank located in the profile was not retrieved. The modified wood includes adzed blocks and chips; long, thin wood strips; toothpick-like splinters; a wooden stake or shaft fragment; various other roundwood; and burned wood. Of the 10,839 wooden items from the waterlogged deposit examined by Lee Newsom and Wendy Natt, the vast majority (about 95 percent) were wood debitage or chips (Newsom *et al.* in prep.). The great majority of the collection is currently being treated by a conservator. The images in Figure 5.5 show a small part of the collection that was inadvertently allowed to dry before being conserved. Evidence of modification is readily apparent.

Twenty-three fragments of cordage were analyzed by Robin Brown (Newsom *et al.* in prep.). All were made of fibers gathered in a Z-twist, then plied together by S-twist doubling. The technique of Z-twist spinning and S-twist doubling of the Pineland specimens is identical to that for specimens from Key Marco (Gilliland 1975), and the size of the Pineland two-ply cordage is equivalent to that of specimens from Key Marco.

Column samples and feature samples from the Old Mound excavations produced 247 liters for analysis, yielding about 535 gm of carbonized wood and nearly 3,000 seeds. Analysis focused on levels 100, 101, and 102 (first through third centuries, Caloosahatchee I). True anaerobic preservation begins at Level 101, with the normal present-day water table at Level 100. Plant taxa identified included fleshy fruits, ruderal taxa, wetland taxa, and woodland and hammock taxa. We will not discuss these findings in detail here (see Newsom and Scarry in prep.), but we do wish to mention that the papaya (*Carica papaya*), chile pepper (*Capsicum* sp.), and squash (*Cucurbita* sp.) seeds we recovered have presented some fascinating puzzles.

Because papaya and peppers have long been associated with neotropical homegardens (Newsom 1993), analysis focused in part on determining whether the seeds of these plants represent cultivars or wild plants. In brief, seed size, seed coat texture, relative seed size, and overall morphological variability all point to manipulation of papaya by Pineland residents. It is also possible that chile peppers were maintained in gardens, though the case is not so compelling. The squash remains are the most numerous (207 whole seeds; 355 seed fragments) and perhaps the most intriguing.

At first glance the squash seeds resemble *Cucurbita pepo*, variety *ovifera*, the summer-squash and ornamental-gourd lineage of the squash/gourd family, but closer inspection reveals three varieties: one that is smooth-surfaced; a second that has prominent marginal hair but no surface hair; and a third with both marginal hair and a wooly seed-coat surface. The third type is unprecedented, but the second group is similar to Archaic-period seeds described by Smith (cf. Smith 1997) from the Romeros and Valenzuelas caves in Mexico. Comparative work led Smith to assign one variety of the Mexican seeds to *Cucurbita pepo* and another – a variety with distinctive hairy margins – to *Cucurbita moschata*, the lineage representing butternut squash and the so-called 'Seminole pumpkin' cultivated by historic-period south Florida Native Americans.

Lee Newsom believes that the Pineland type 2 (hairy-margin) specimens are most like *Cucurbita moschata*, but are smaller, suggesting the possibility that the Pineland seeds represent an ancestral form of *Cucurbita moschata*. At any rate, the distinctive morphology and high variation represented within this sample of cucurbit seeds suggest evidence of gardening and the maintenance of at least two cucurbit cultivars, if not separate species (Newsom and Scarry in prep.).

The preservation of botanical remains in the Caloosahatchee I waterlogged strata of Old Mound is, not surprisingly, far superior to that in non-waterlogged strata. For example, whereas acorn appears to be a minor food item in the younger Brown's Mound strata in Operation I (8 percent ubiquity), it appears to be a major food in the Old Mound waterlogged deposits (80 percent ubiquity). Similarly, prickly pear and saw palmetto appear minor at 25 percent ubiquity in Brown's Mound deposits but are measured at 80 and 70 percent ubiquity in the Old Mound. The most prominent taxa from Operation I in the non-waterlogged Brown's Complex are mastic, seagrape, and cabbage palm – all of which have durable seed coats that are able to withstand burning and other post-depositional stresses. Taxa with thin shells or seed coats such as acorns and squash seeds are highly susceptible to removal from the archaeological record when not found in waterlogged contexts (Newsom and Scarry in prep.).

In the filled wetland area to the northeast of the Old Mound excavation, where we had so few expectations, we opened a narrow trench, Trench 11A (Figure 5.4), expecting to document the bulldozing event and the underlying wetland muck and then move on to more interesting work. The midden materials from the bulldozed mound indeed overlay a wetland deposit, but again, to our surprise, there was much more. Beneath the bulldozed midden is a stratum (G in Figure 5.6) of undisturbed mucky shell midden dating to the sixth century (Caloosahatchee IIA). Beneath this midden is a dark irregular layer of sand (F) with *in situ* articulated ribbed mussels (*Geukensia demissa*) radio-carbon-dated to the third and fourth centuries A.D. Beneath and at times interleaving with this layer is a dark sand deposit (D) with occasional inclusions of pottery sherds, sea turtle bone, and in another portion of the site, part of a dolphin skeleton. Also associated with this sand is a thin lens of surf-clam shells (many of which are articulated), pen shell fragments, and sea urchin remains (E). A sample of the surf clams (*Spisula solidissima*) radiocarbon-dates to the fourth and fifth centuries. Underlying these non-cultural deposits is a stratum of shell midden (C) containing pottery sherds, charcoal, and waterlogged plant remains. The stratum in part lies below today's estimated mean sea level and produced a radiocarbon date of the second to third century A.D. Below this midden is a layer of naturally deposited sand (B) and beneath this, a thin stratum (A) of the earliest midden thus far documented at Pineland, dating to the first to second centuries.

Our interpretation of the depositional sequence is as follows. Like the waterlogged deposit below nearby Old Mound, the lower two middens indicate deposition on mostly dry land (i.e., they were not deposited in standing water). Therefore, we infer a lower sea level by about 50 cm for the second and third centuries A.D. We interpret the overlying sand deposits with their washed-in articulated shells, sea-urchin spines, dolphin bones, and scattered midden material as a combination of storm-induced and sea-level rise deposition. The subsequent colonization of the upper portion of the sand by ribbed mussels points to the establishment of a black mangrove wetland. This environmental sequence of deposition occurred during the third to fifth centuries. This same area of Pineland was reoccupied later in the sixth century, as documented by the upper midden.

Trench 11B, not far to the north of Trench 11A and on slightly higher ground (Figure 5.4), exhibited a less compli-cated but no less interesting profile. A shell midden, composed of an estimated 40 percent surf-clam shells with pockets of sea urchin, appears sandwiched between two strata of sand deposits. This abundance of surf clam in a Pineland midden is not expected under present-day sea-level conditions; the habitat of the clam is high-salinity waters, usually surf conditions – the littoral zone of a barrier island, especially near an inlet. Such a locale today is too far a distance to travel on a regular (i.e., daily) basis for shellfish. Pineland's surf-clam midden radiocarbon-dates to the second and third centuries, but we believe the midden is more likely contemporaneous with the fourth to fifth century surf clam lens in Trench 11A. Whereas the surf

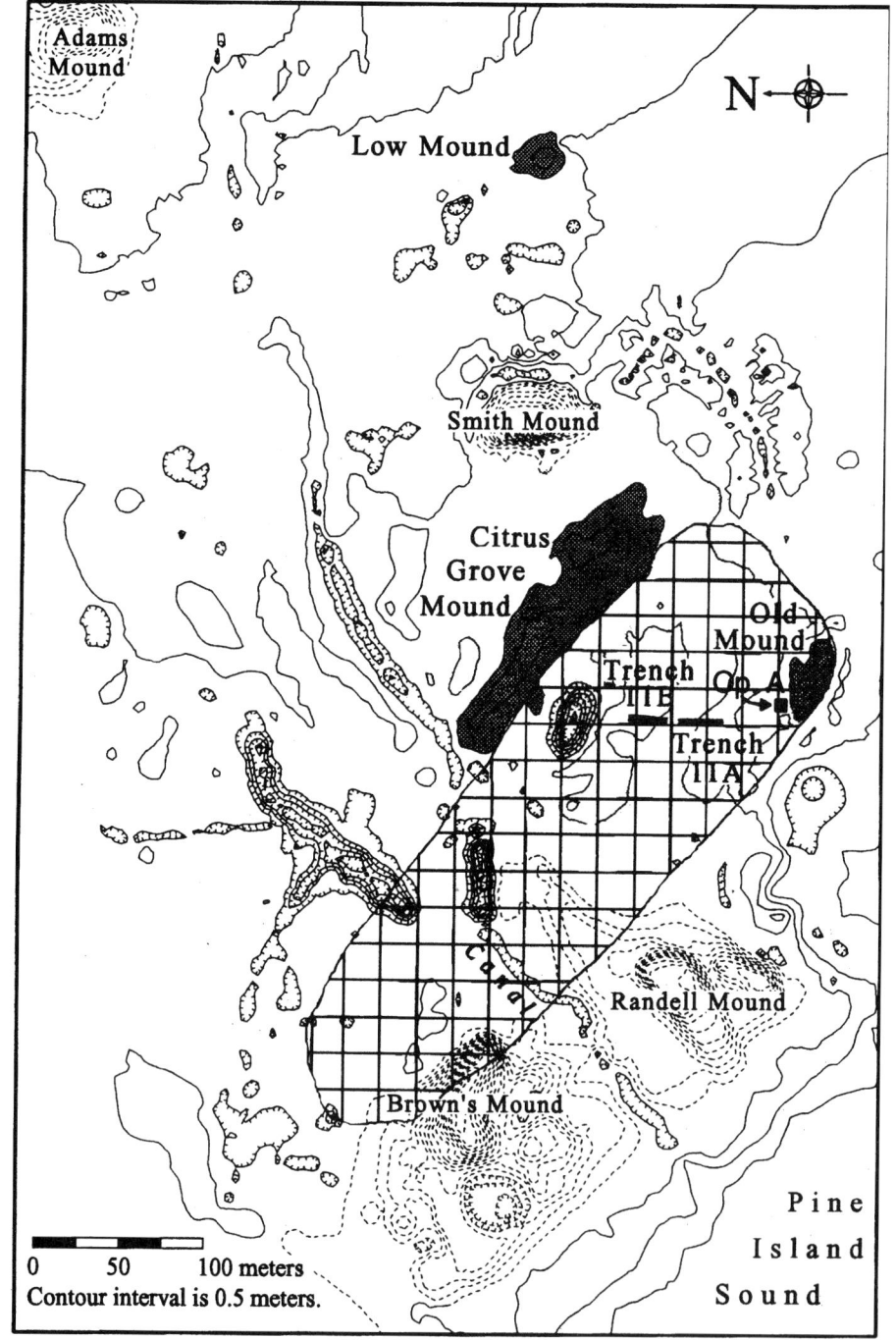

Figure 5.4. The Pineland Site Complex with its Caloosahatchee I components cross-hatched (first to fourth centuries A.D.) and shaded (fourth to fifth centuries A.D.). (Map by Corbett Torrence and Sue Ellen Hunter).

clams exhibited in Trench 11A were live wash-ins, the Trench 11B clam shells are here as a result of human collection. Both cases imply an environmental explanation. A shift in the salinity gradient and erosion of low-lying portions of barrier islands may have resulted in surf clam beds closer to Pineland (Walker *et al.* 1995).

As noted above, the base of the surf-clam midden is at a higher elevation compared with the shoreline middens. We

*Figure 5.5. Cypress (*Taxodium *sp.) wood debitage fragments from Old Mound excavations, Pineland Site Complex.*

believe that people moved even farther from the shoreline based on our tests at the Citrus Grove Mound, a raised, linear mound of sand about 2 m high situated in the middle of the site paralleling the shoreline (see Figures 5.3 and 5.4). Only sporadic deposits of shell midden were found in two small test excavations. In one, the midden was buried beneath 85 cm of sand. This midden radiocarbon-dates to the third to early fifth-century range; we believe it slightly post-dates the Trench 11A and 11B locales. Along with geologist Frank Stapor (Walker *et al.* 1995:212) and pedoarchaeologist Sylvia Scudder (in prep.), we hypo-thesize that at least part of this largely sand feature is of natural origin and its western slope was eroded by water action at some point before the sixth-century Caloosahatchee IIA reoccupation.

We believe that Pineland's residents then moved still farther inland. We infer this from a low shell midden located today in a wooded section of Pineland (Figures 5.3 and 5.4). Named the "Pineland Low Mound," it contains dense shell midden and sits conspicuously on the 5 foot (1.5 m) contour, considered the "uplands" of Pineland. The Low Mound, radiocarbon-dating to late Caloosahatchee I, may represent the final effort of Pineland's early inhabitants to tolerate a period of inundation, because it seems that the area may have been altogether abandoned by the mid-fifth century.

Re-occupation of Pineland (Caloosahatchee IIA)

Pineland experienced a dramatic re-peopling during the sixth and seventh centuries A.D., evidenced again by expansive middens along the shoreline, only this time documenting the most "sea-ward" habitation of all. The base of the most sea-ward middens, today under the later Brown's and Randell mound complexes, lay some 50 cm below today's mean sea level. Logistics prevented an adequate sampling of these deposits (i.e, we could not keep ahead of the water due in part to the density of shell and lack of any other sediment). But we extracted several small specimens of waterlogged plant material from a small exploratory test in the corner of one large excavation unit (Operation C). We dated an associated quahog clam shell (*Mercenaria campechiensis*) with a sixth-century result. The midden materials indicate that they were deposited on dry land, once again implying a lower sea level. By A.D. 700, the middens became less expansive and more mound-like. By A.D. 800, these midden-mounds were 3 to 3.5 meters thick at their centers.

Pineland's Late Occupations (Caloosahatchee IIB, III, IV, and V)

During the ninth through twelfth centuries (Caloosahatchee IIB), midden accumulation became intensive and mounded, essentially limited to the Brown's and Randell mound complexes. Mound-building using secondary deposits is

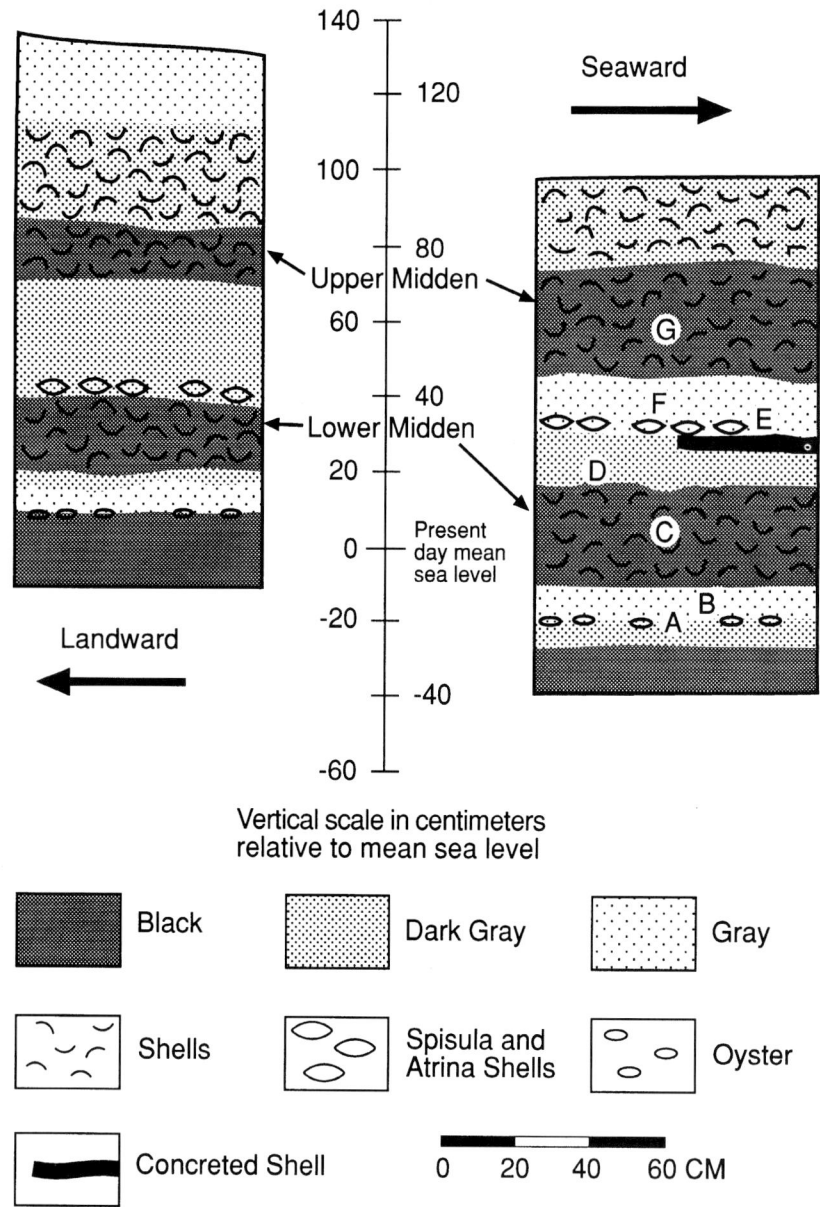

Figure 5.6. Profile of Trench 11A, Pineland Site Complex. (Graph by Sue Ellen Hunter).

clearly evident for the thirteenth century (Caloosahatchee III), but not before. Deposition continued to be limited to the Brown's and Randell mound complexes throughout the Caloosahatchee III and IV periods. Caloosahatchee V is thus far limited to the presence of several European-derived artifacts from the Smith Mound.

Because all of Pineland's post-seventh-century (Caloosahatchee IIA) deposits thus far have been limited to the major mound complexes, the location of a ninth-century

carved wooden bird head came as a surprise. The cypress-wood artifact is known as the Thomasson Figurehead, named for Phyllis Thomasson, who found it in 1971 protruding from the mangrove muck – far from the mound complexes – while walking at Pineland after a storm. The carving (Figure 5.7) is about 27 cm long, 7 cm wide, and 5 cm high. Ms. Thomasson and her husband were stunned to discover the finely carved and hollowed head and upper beak of a bird, and spent much of the afternoon trying

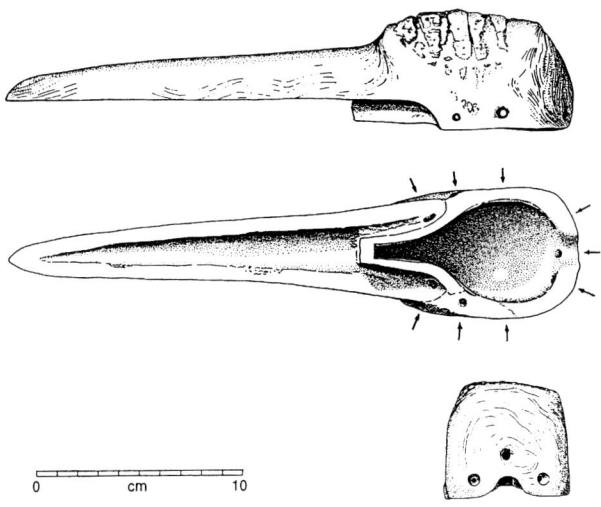

Figure 5.7. The Thomasson Figurehead. (Florida Museum of Natural History catalogue number 90–24–1; drawing by Merald Clark; source: Clark 1995:Figure 3–1).

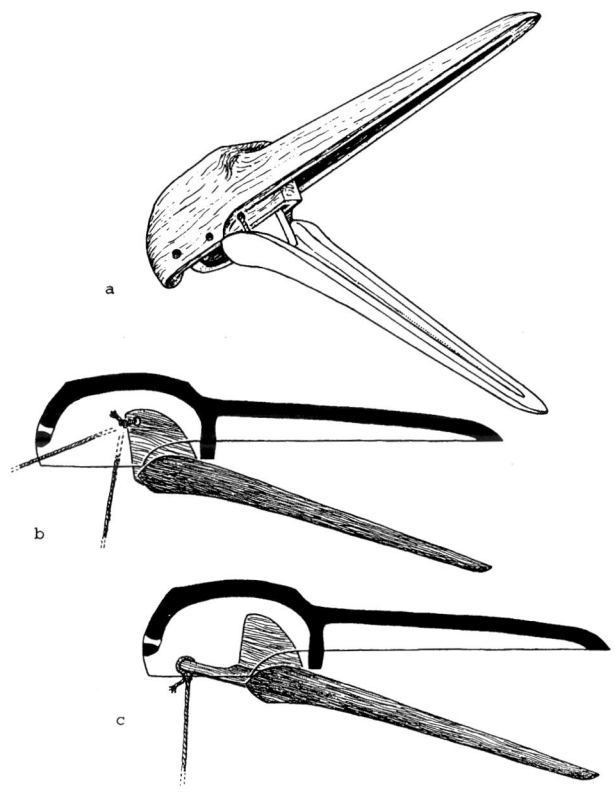

Figure 5.8. Reconstruction of the crane-head mechanism by Merald Clark. (Drawing by Merald Clark; source: Clarke 1995:Figure 3–2).

Figure 5.9. Alternate usages and methods of operation for the Thomasson Figurehead. (Drawing by Merald Clark; source: Clark 1995:Figure 3–5).

unsuccessfully to locate the lower half. They kept the artifact in good condition for two decades and, in 1990, donated it to the Florida Museum of Natural History.

The carving most closely resembles a crane. Though still in excellent shape when received, its once-waterlogged wood was now a dull gray, lightweight and fragile, and could not be further conserved. Merald Clark carved a replica, along with a lower beak to fit the upper one. He duplicated the find with care, including the nine drilled holes that may have functioned to tie the figurehead to a costume, or to manipulate the lower and upper beaks like a puppet. Based on analogies with societies of the Northwest Coast of North America, Clark (1995:24–39, in prep.) attached strings to the replica, deduced a method of opening and closing the beak by pulling the strings (Figure 5.8), and drew sketches of how the figurehead might have been used (Figure 5.9).

A tiny section of wood was removed from an

inconspicuous place inside the artifact and an AMS radio-carbon date of A.D. 865–985 (Beta-81501, corrected and calibrated) was obtained. The date surprised us because the surrounding sediments included a thick, sand-tempered plain pottery typical of the pre-A.D. 500 occupations, and, as mentioned above, because the ninth-century components of the site seem to be restricted to the mound complexes. Assuming the radiocarbon date is accurate, one plausible explanation is that the bird head was lost during a storm; another is that it was purposely buried.

Summary

Our 42 radiocarbon dates (Walker and Marquardt in prep.) demonstrate that Pineland was inhabited by about A.D. 50 and that mounds were still accumulating or being constructed as late as the sixteenth century A.D. Substantial numbers of people must have occupied Pineland, even in the earliest centuries.

Pineland's earliest human inhabitants lived in the area that we now call "Old Mound" and other low-lying areas circa A.D. 50–250. These Caloosahatchee I people did not build mounds, but instead lived extensively across the landscape of Pineland, along what was once shoreline. They exploited the surrounding estuarine/marine and pine flat-woods environments, fishing, gathering, gardening, and hunting for their material needs. Their residency there was interrupted by flooding and at least some of these people first relocated their village to higher ground (first the Citrus Grove Mound and then the Low Mound) and later may have abandoned the Pineland area altogether.

Pineland was reoccupied during the sixth and seventh centuries (Caloosahatchee IIA), and habitation moved even more sea-ward than in the previous occupation. By the beginning of the eighth century (late Caloosahatchee IIA), the middens had become less expansive, more restricted. During the ninth through twelfth centuries (Caloosahatchee IIB), midden accumulation intensified, focusing on the Brown's and Randell mound complexes. In the thirteenth century (Caloosahatchee III), at least one of the mounds was augmented using secondary deposits. During this mound building phase and later, the absence of habitation deposits in non-mound areas suggests that these low-lying areas again may have been wetlands.

The waterlogged portions of Pineland offer a rare opportunity to examine archaeological materials from a methodological perspective. The preservation of seeds and other plant remains, cordage possibly of fishing nets and lines, and wood that normally do not survive in middens allows a comparison with non-waterlogged middens and helps us to understand what is missing from most archaeological samples. The waterlogged deposits demonstrate the potential of Pineland to produce artifacts comparable to

those uncovered by Cushing's 1896 expedition to Key Marco (see Cushing 1897; Gilliland 1975).

Predicting Coastal Wet-Site Deposits

When we see a muck or peat deposit or a pond that we think may have been in the same place for a long time, we know to look for waterlogged artifacts within it. Although a coastal site, Key Marco is an example of a waterlogged muck site. Today such a site would receive the attention of any archaeologist given the opportunity to investigate it. But what of other kinds of coastal sites? The Pineland Site Complex deposits of ca. A.D. 50–250 contain perishables preserved as well as those of Key Marco, and include not only cordage and wood, but significant botanical remains that are stimulating a re-thinking of the issue of Calusa horticulture. The Thomasson cypress crane-head figure of ca. A.D. 865–985 is as sensitively carved as any of the Key Marco masks or figureheads, and its analysis has raised our awareness of the masking and drama of precolumbian people who lived more than a thousand years ago. Yet Pineland is not an obvious muck, peat, or pond site. Why, then, are its deposits so well preserved?

Our suggested explanation for the excellent preservation in the earlier two waterlogged horizons (Caloosahatchee I and IIA) at Pineland is multi-faceted. First, Pineland's waterlogged deposits are associated with the site's two expansive, shoreline habitations: its first-through-third-century occupation and its sixth-through-seventh-century occupation. In both cases, deposits come in contact with the original shoreline surface, with the sixth-century deposits being the more sea-ward of the two. In both cases, the deposits lie below today's mean sea level (i.e., both deposits were laid down during low sea-level stands). Both low stands (regionally named Sanibel I and Buck Key Low, respectively) are independently documented by Gulf coast beach-ridge seriations reported by Tanner (1991) and Stapor and his colleagues (1991) (Figure 5.10). These episodes generally correlate with cooling episodes commonly known as the Roman Minimum and Vandal Minimum documented for Europe and other areas.

Second, in both cases, the deposits were followed by relatively rapid sea-level rises with associated storminess. Sea level rose episodically but relatively rapidly at Pineland ca. A.D. 300 to 450, correlating with the latter part of a global warming episode known in Europe as the Roman Climatic Optimum and reflected locally in the Wulfert high sea-level stand (Stapor *et al.* 1991; Walker *et al.* 1995). During this warm interval, storms deposited additional materials on Pineland's shore. The water table also rose, inundating normally perishable materials near the shore. Although sea level subsequently dropped (Buck Key Low), rose again (La Costa High), dropped again (Sanibel II Low),

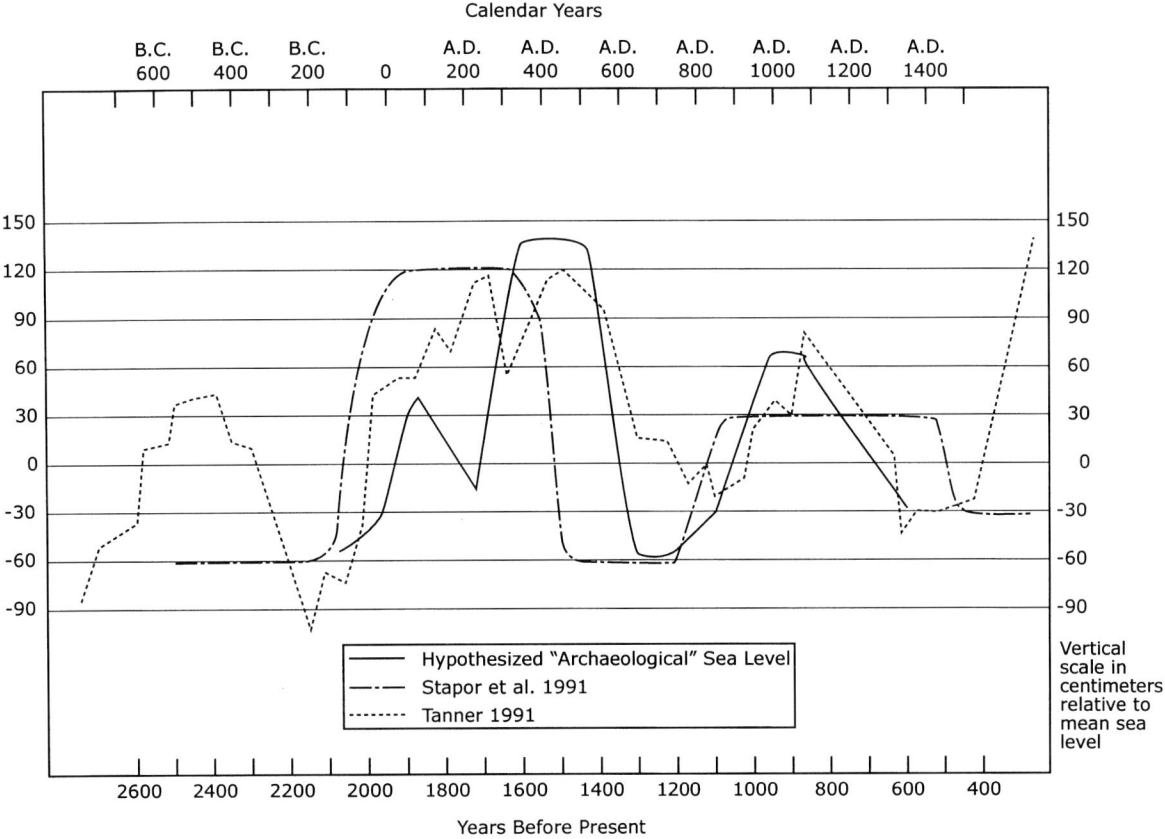

Figure 5.10. A comparison of the two regional beach-ridge-based sea-level curves with one based on archaeological evidence. (Graph by Sue Ellen Hunter).

and finally rose again to today's level (Figure 5.10), the Pineland deposits stayed moist and were never again above the water table, until our excavations brought them to the surface.

Third, we suggest that calcium carbonate leaching from mollusk shells in the overlying middens counteracted the normally acidic character of the local soils. The presence of calcium carbonate alone does not explain the recovery of perishables, because shells are present in the upper, drier strata as well. In those levels, bones and certain charred seeds are found, but wood chips, cordage, and un-charred seeds are absent. The combination of the buffering effects of the basic calcium carbonate and the constant moisture of the surrounding matrix may have favored the preservation of the perishable materials deposited ca. A.D. 50–250 and ca. A.D. 500–600.

Our explanation for the preservation of the Thomasson figurehead is quite different. We do not know the exact provenience of the figurehead, but we surmise that it was brought close to the surface by ditching for mosquito

control, then exposed by rain in 1971. A highly organic matrix may have been a contributing factor, but we cannot be sure of this in the absence of exact provenience data. An AMS radiocarbon date of A.D. 865–985 places it in the ninth or tenth century, a time characterized by global warming, a high sea level, and frequent storms (Keigwin 1996; Tanner 1991; Walker *et al.* 1995). This episode is known in Europe as the Medieval Optimum, or Medieval Warm Period, and is marked regionally by the La Costa high sea-level stand (Stapor *et al.* 1991). Walker's recent studies from nearby Useppa Island corroborate such conditions at ca. A.D. 780–945, the ninth and tenth centuries (Walker 1999). Thus, we hypothesize that in the ninth or tenth century the Thomasson figurehead was lost, re-deposited, or buried at Pineland in a low-lying area, only to become covered with additional midden or storm-wash deposits or both. A sea-level decline (Sanibel II, regionally) in the subsequent Little Ice Age apparently was not sufficient to subject the wood carving to drying conditions.

Conclusion

The Pineland Site Complex includes extraordinarily pre-served organic materials that do not occur in a peat or pond context. We hypothesize that similar preservation conditions may be discovered in other coastal contexts in which materials accumulated during low sea-level are buried relatively rapidly due to rising water table and/or storms and are covered by shells, either deposited by people or by storms or both.

If we are correct about this, both research and management implications follow. From the point of view of researchers, such sites bear witness not just to the daily lives of people who lived in the past, but also to past environmental episodes, such as sea-level fluctuations, storms, and barrier island breaches. From the point of view of cultural resource managers, sites in coastal contexts should receive attention for possible buried waterlogged deposits even if they do not contain obvious standing ponds, muck deposits, or peat. For the moment, Pineland serves as a well documented and dated example of the potential of such coastal sites for both environmental and archaeological research.

References Cited

Clark, M.R. 1995. Faces and Figureheads: The Masks of Prehistoric South Florida. M.A. thesis, Department of Anthropology, University of Florida, Gainesville.

Clark, M.R. in prep. A Mechanical Waterbird Mask from Pineland and the Calusa Masking Tradition. In The Archaeology of Pineland: A Coastal Southwest Florida Site Complex, A.D. 50–1600, edited by K.J. Walker and W.H. Marquardt. Institute of Archaeology and Paleoenvironmental Studies, Monograph 4. University of Florida, Gainesville.

Cushing, F.H. 1897. Exploration of Ancient Key Dweller Remains on the Gulf Coast of Florida. American Philosophical Society, Proceedings 35:329–448. Philadelphia.

Edic, R.F. and C. McP. Torrence. 1990. Transcript of a Taped Interview with G.W. Hyatt, March 8, 1990. On file, Florida Museum of Natural History, University of Florida, Gainesville.

Gilliland, M.S. 1975. The Material Culture of Key Marco, Florida. University Press of Florida, Gainesville.

Keigwin, L.D. 1996. The Little Ice Age and Medieval Warm period in the Sargasso Sea. Science 274 (29 November 1996):1504–1508.

Kenworthy, C.J. 1875. Ancient Canals in Florida. Forest and Stream (August 12, 1875). New York.

Newsom, L.A. 1993. Native West Indian Plant Use. Ph.D. dissertation, Department of Anthropology, University of Florida, Gainesville. University Microfilms, Ann Arbor.

Newsom, L.A. and C.M. Scarry. in prep. Homegardens and Mangrove Swamps: Pineland Archaeobotanical Research. In The Archaeology of Pineland: A Coastal Southwest Florida Site Complex, A.D. 50–1600, edited by K.J. Walker and W.H. Marquardt. Institute of Archaeology and Paleoenvironmental Studies, Monograph 4. University of Florida, Gainesville.

Newsom, L.A., R.C. Brown, and W. Natt. in prep. Pineland Cordage and Modified Wood: Material-Technological Aspects of Plant Use. In The Archaeology of Pineland: A Coastal Southwest Florida Site Complex, A.D. 50–1600, edited by K.J. Walker and W.H. Marquardt. Institute of Archaeology and Paleoenvironmental Studies, Monograph 4. University of Florida, Gainesville.

Purdy, B.A. 1991. The Art and Archaeology of Florida's Wetlands. CRC Press, Boca Raton, Florida.

Scudder, S. in prep. Soils and Landscapes: Archaeopedology at the Pineland Site. In The Archaeology of Pineland: A Coastal Southwest Florida Site Complex, A.D. 50–1600, edited by K.J. Walker and W.H. Marquardt. Institute of Archaeology and Paleoenvironmental Studies, Monograph 4. University of Florida, Gainesville.

Smith, B.D. 1997. Reconsidering the Ocampo Caves and the Era of Incipient Cultivation in Mesoamerica. Latin American Antiquity 8:342–383.

Stapor, F.W., Jr., T.D. Mathews, and F.E. Lindfors-Kearns. 1991. Barrier Island Progradation and Holocene Sea-Level History in Southwest Florida. Journal of Coastal Research 7(3):815–838.

Tanner, W.F. 1991. The "Gulf of Mexico" Late Holocene Sea Level Curve and River Delta History. Gulf Coast Association of Geological Societies, Transactions 41:583–589.

Torrence, C. McP. in prep. Topographic Reconstructions of a Calusa Village: The Pineland Site in 1896. In The Archaeology of Pineland: A Coastal Southwest Florida Site Complex, A.D. 50–1600, edited by K.J. Walker and W.H. Marquardt. Institute of Archaeology and Paleoenvironmental Studies, Monograph 4. University of Florida, Gainesville.

Walker, K.J. 1999. A Local Shift in Shellfish Availability During the Medieval Optimum Period: Environmental Archaeology on Southwest Florida's Useppa Island. Paper presented at the 1999 Annual Meeting of the Florida Academy of Sciences, Tampa, Florida.

Walker, K.J. 1992. The Zooarchaeology of Charlotte Harbor's Prehistoric Maritime Adaptation: Spatial and Temporal Perspectives. In Culture and Environment in the Domain of the Calusa, edited by W.H. Marquardt. Institute of Archaeology and Paleoenvironmental Studies, Monograph 1. University of Florida, Gainesville.

Walker, K.J., and W.H. Marquardt (editors). in prep. The Archaeology of Pineland: A Coastal Southwest Florida Site Complex, A.D. 50–1600. Institute of Archaeology and Paleoenvironmental Studies, Monograph 4. University of Florida, Gainesville.

Walker, K.J., F.W. Stapor, Jr., and W.H. Marquardt. 1995. Archaeological Evidence for a 1750–1450 BP Higher-Than-Present Sea Level Along Florida's Gulf Coast. In Holocene Cycles: Climate, Sea Levels, and Sedimentation, edited by C.W. Finkl, Jr., pp. 205–218. Journal of Coastal Research, Special Issue No. 17.

6. Subtropical Wetland Adaptations in Uruguay during the Mid-Holocene: An Archaeobotanical Perspective

José Iriarte, Irene Holst, José M. Lopez and Leonel Cabrera

Until quite recently, wetland environments have provided a critical loci for prehistoric and historic native groups. Wetlands represent one of the most environmentally diverse habitats of the world supporting a great variety of flora and fauna (Williams 1990). They provide important plant, animal, bird, and fish resources to hunter-gatherers who exploited them as part of the seasonal economic round (Janetski and Madsen 1990; Nicholas 1998). They also represent an ideal context for the adoption and intensification of agriculture (Sherrat 1980; Pohl *et al.* 1996; Siemens 1999). In addition, wetlands provide greater stability reducing risk during periods of environmental deterioration since they provide a stable source of water supply. While most articles in the present volume show how the excellent preservation of perishables in wet sites give us extraordinary insights into prehistoric peoples adapted to wetland environments, this article constitutes an example that illustrates how newly developed analytical techniques, in particular starch grain and phytolith analyses, allow us to gain a better understanding of human-environment interaction in this unique setting when preservation is not so spectacular.

The dynamic interactions between human populations and the changing environment have played a major role in the development of Early Formative societies in the Americas during the Mid-Holocene (i.e., Brown and Vierra 1983; Carr and Gibson 1997). Research shows that cultural complexity has emerged under extremely different environmental settings based on coastal (Moseley 1992; Stothert 1985, 1992; Gaspar 1998; Blasi *et al.* 1999) and inland (Roosevelt 1980; Dillehay *et al.* 1989; Heckenberger 1998; Pearsall 1999) economies, the majority of which relied on both domesticated and wild resources to different degrees (Piperno and Pearsall 1998). In this respect, wetland areas have provided one of the richest environmental settings where the earliest mound-building cultures of lowland South America developed among populations of complex hunter-gatherers during the Mid-Holocene.

In order to understand the rise of these early complex societies, it is crucial to comprehend the role that wild and domesticated plant resources played in their economies. Unfortunately, poor preservation of macrobotanical plant remains in seasonally humid environments has hampered researchers in the wetlands of southeastern Uruguay, as well as in other part of lowland South America (i.e., Piperno 1995; Pearsall 1995; Piperno and Pearsall 1998), from assessing the scope, extent, and importance that domesticated plants may have played in the emergence of these Early Formative societies. Likewise, cultural practices in the processing of food and refuse disposal can also obstruct the recovery and reconnaissance of plant remains and its subsequent interpretation. For example, plants of economic importance in lowland areas such as tubers are notorious for their failure to enter the record of carbonized remains. These plants are generally processed outside houses and heaped unburned in trash middens and therefore, are absent from the macrobotanical record (Heckenberger 1998) leaving no fossil or charred remains. In the seasonal wetland sites of southeastern Uruguay, only the "tougher" charred remains like palm kernels survive burial and recovery.

To address these shortcomings in the study of the role that domesticated plants played in the economy of the Early Formative societies of southeastern Uruguay which developed 4000 yr BP in the southern sector of the Merin Lagoon basin (Bracco, Cabrera, and Lopez 1996), we have conducted starch grain and phytolith analyses from three sites in the region. The results of this study suggest that populations of hunter-gatherers, adapting to a changing Mid-Holocene environment adopted maize, squash, beans, and possibly domesticated tubers as part of their diet.

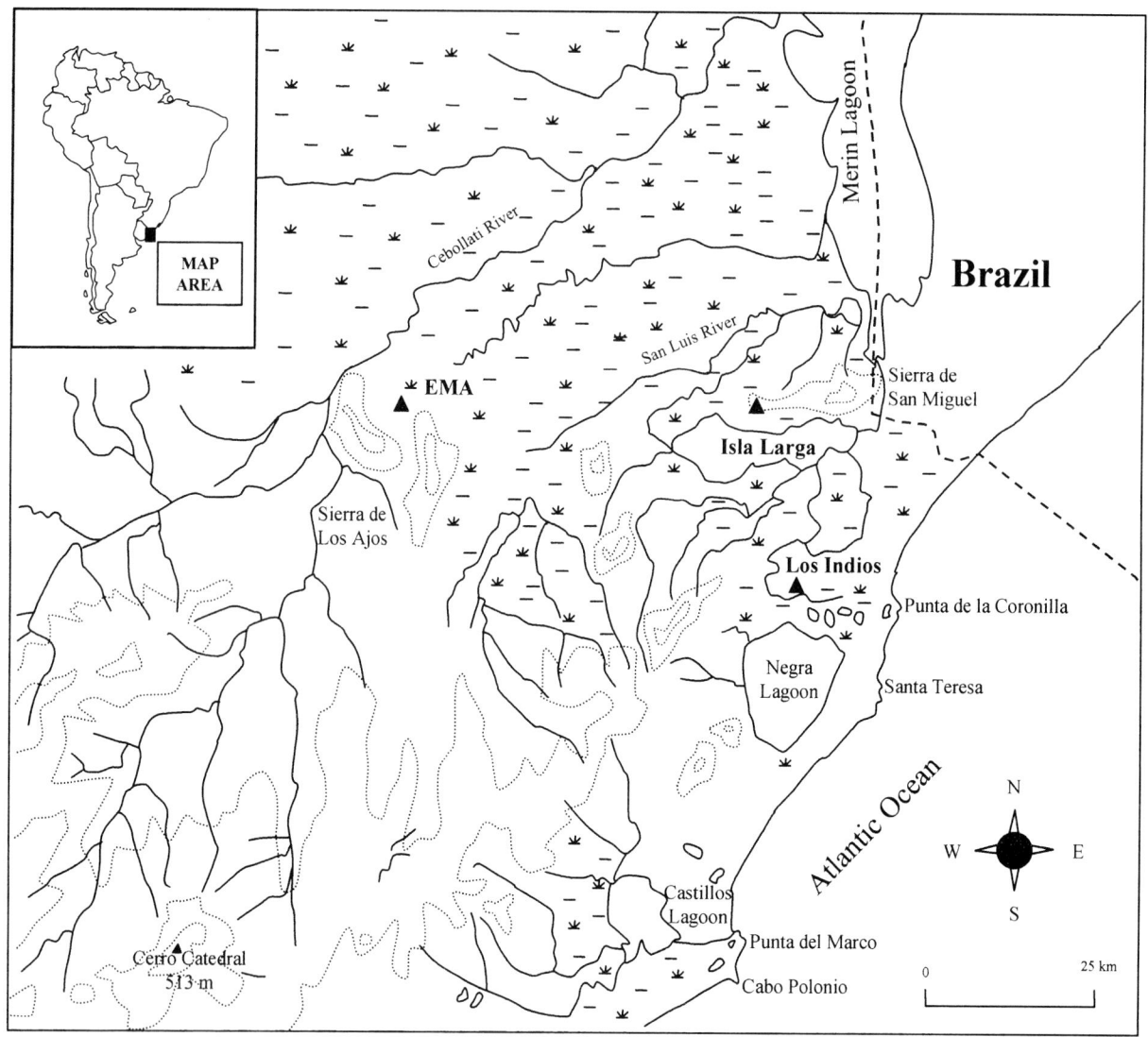

Map 6.1. Regional map of southeastern Uruguay.

The study region

The mound-building cultures known as the Vieria Tradition in the context of South American prehistory, developed ca 4000 yr BP in southeastern Uruguay (Bracco, Cabrera and Lopez 1996). The study area corresponds to the temperate sub-humid grasslands, which extend along the eastern Atlantic coast of South America from latitudes S 38° to 28° and are known as the Río de la Plata grasslands (Leon 1992) (Map 6.1). The area of southeastern Uruguay and Brazil along the Atlantic coast comprises a coastal plain (Delaney 1965) characterized by slight elevations (maximum 200 masl), generated by the Late Pleistocene and Holocene marine oscillations (Delaney 1965; Jackson 1984;

González 1989; Bracco 1992; Montaña and Bossi 1995; Tomazelli and Wilwock 1996).

The area of study, the southern sector of the Merin Lagoon basin, locally known as Bañados de Rocha, is distinguished by coastal wetlands encompassing four freshwater lagoons connected to the ocean by streams. These lagoons appear in the form of a series of microbasins connecting the ocean with the lagoons, adjacent wetlands, plains, and hills, creating a patchwork of closely packed environmental zones. Bordering these lagoons, half-million hectares of wetlands have been recognized as one of the most environmentally diverse habitats of the world (Ramsar Convention 1984), supporting a great variety of flora and fauna and including at least 120 species of water birds, 80 species of fish, and 30 species of amphibians (Probides 1997). These wetlands are characterized by low-energy meandering streams, which together with the backwater lakes and marshes impounded behind its natural levees and terraces create an extremely rich environment. The vegetation of the area is mainly characterized by prairie grasses and palm forests of *Butia capitata* in the elevated plains (10–15 masl), scrubland and hydrophytic vegetation in the wetlands, and riparian forest along the primary water courses. At present, drainage carried out by the rice-growing companies in the area has resulted in the loss of more than a quarter of the wetlands, causing the destruction of indigenous flora and fauna as well as many archaeological sites.

The archaeological record of the southern sector of the Merin Lagoon basin is characterized by the conspicuous and numerous presence of earthmounds and earthworks. Since the region encompasses a great number of sites spread over a large region and distributed over a wide span of time, the different characteristics pertaining to different sites are probably the result of temporal and regional variations. Notwithstanding, the settlement patterns are clearly associated with different features of the environment. In this regard, in the wetland floodplains, clusters of mounds are generally located over the levees and terraces of the streams showing a linear pattern. These sites are smaller in size, standing less than a meter above the water level and therefore subject to periodic flooding. In contrast, in the flattened spurs and knolls of the hills overlooking the wetlands, the sites are bigger in size reaching up to 50 ha in some cases, bearing a greater number and diversity of mounds. These sites are located 5 to 10 meters above the water level, secure from flooding, and allow for immediate access to the rich-resource wetland area (Bracco, Cabrera, and Lopez 1996; Iriarte 1999) (see EMA site, Figure 6.1).

Early archaeological work in the 1960's aimed at developing a chronological framework for the yet unstudied archaeological region of the Mid-Atlantic coast of southeastern Brazil and Uruguay by applying Ford's ceramic seriation (Ford 1962; Meggers and Evans 1969) and lithic typologies (Schmidtz 1967, 1973; Schmidtz and Brochado 1967; Prieto *et al.* 1970; Schmidtz and Baeza 1982; Brochado 1984; Cope 1992; Rodriguez 1992; Schmidtz *et al.* 1992) which divided the Early Formative societies of the area in two broad cultural traditions. The first one, the preceramic Tradition, known as Umbu Tradition antecedes the ceramic Vieira Tradition and is characterized by the presence of bifacial stemmed stone projectile points. The Umbu Tradition spreads over the states of Rio Grande do Sul, Santa Catarina, and Parana, encompassing 400 sites grouped in 17 phases in Rio Grande do Sul, Brazil (Schmitz 1987). In the State of Rio Grande do Sul, this tradition is represented by three phases: Patos Phase in Camaqua, Lagoa Phase in Rio Grande, and Chui Phase in Santa Victoria do Palmar (Schmitz *et al.* 1992, Cope 1992). Radiocarbon dates obtained for Lagoa Phase, range from 485±85 A.D. and 70±150 A.D. (Schmitz 1973). Ceramics begin to appear in the Brazilian region in the first century A.D. and correspond to Vieira Tradition (Schmitz 1963, 1976). The ceramic sequence shows a clear evolution in time. The early phase is denominated Totorama (0–200 A.D.) and is characterized by wide and shallow wares with fine walls and coarse temper. The Vieira Tradition is subdivided into Vieira I (200–900 A.D), II (900–1110/ 1300 A.D), and III (1300 A.D.). The most recent phases, Vieira II and III are characterized by deeper and bigger wares with a more uniform manufacture, bearing some decoration in the external walls, such as basket and finger impressions. Vieira III starts at 1300 A.D. and includes the arrival of the Tupi-Guarani groups in the region until the total disappearance of the indigenous groups. On-going research in the southern sector of the Merin Lagoon basin since 1985 by the CRALM (Archaeological Salvage Program of the Merin Lagoon basin) in Uruguay has continued to amplify and redefine the chronological sequence. In the southern sector of the Lagoon Merin basin, the preceramic Umbu component has been dated back to the second millennium B.C., and the appearance of the ceramic Vieira tradition components has also been pushed back to the first millennium A.D. (Bracco, Cabrera and Lopez 1996).

Results of research

We have isolated starch grains and phytoliths from soil sediments corresponding to features and profile walls of three adjacent sites in the southern sector of the Merin Lagoon basin: (1) Los Indios; (2) Isla Larga; and (3) Estancia Mal Abrigo (Figure 6.1).

The extraction of starch grains from sediments was done following the technique which is being developed by Piperno and Holst at the Smithsonian Tropical Research Institute. Starch grain microscopically-based morphological identification of taxa was based on the reference collection of more than 100 species of economic importance accumulated by Piperno and Holst (1998) at the Smithsonian Institution of Tropical Research in Panama. As Table 6.1 shows we have identified the presence of starch grains characteristic of maize (*Zea mays*), beans (*Phaseolus spp.*), and possibly tubers pertaining to the genera Canna and Calathea. These results must be considered preliminary and qualitative (Figure 6.1). Phytolith analysis was carried out at the Smithsonian Institution of Tropical Research using standard methods (Piperno 1988, 1993).

Based on present knowledge of maize starch grain

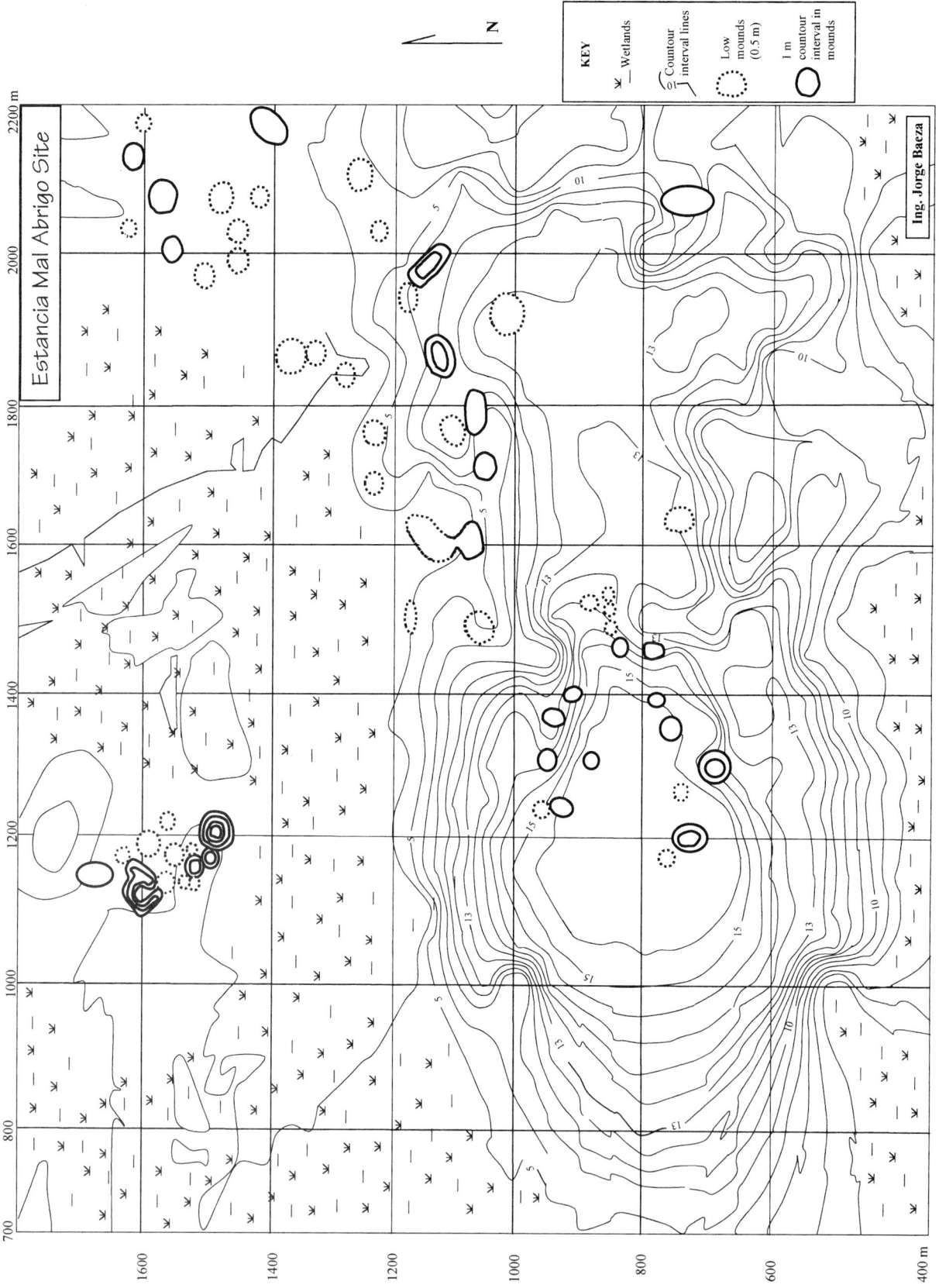

Figure 6.1. Estancia Mal Abrigo site.

Table 6.1. Results of starch grain analysis.

Site	Excavation	Level	C14 dates	Type of starch grain	N	Mean (μ)	Probable Species
EMA	II	(0.5-0.6 m)		Graminae (isolated)	12	21×18	*Zea mays*
		(0.2-0.3 m)		Graminae	1	16×16	*Zea mays*
Los Indios	I	(1.85-1.90 m)	2,800±70	Graminae (isolated)	6	14×13	*Zea mays*
	II	(1.20-1.30 cm)	1,170±60	Graminae (isolated)	42	18×16	*Zea mays*
				Graminae (aggregated)		20×17	*Zea mays*
				Leguminosae (isolated)	6	20×15	*Phaseolus spp.*
		(0.45-0.55 m)		Graminae (isolated)	18	16×13	*Zea mays*
				Graminae (aggregated)		18×18	*Zea mays*
Isla Larga	III	(0.8-0.85 m)	1,190±80	Graminae (isolated)	9	18×15	*Zea mays*
				Marantacea	1	28×16	*Calathea spp.*
		(1.90-1.95 m)	3,050±90	Graminae (isolated)	14	19×16	*Zea mays*
				Graminae (aggregated)		21×15	*Zea mays*
				Leguminoseae (isolated)	3	27×29	*Phaseolus sp.*
		(max. 2.58 m)	3,660±120	Graminae (isolated)	18	16×15	*Zea mays*
				Graminae (aggregated)		13×12	*Zea mays*
				Cannacea	2	44×26	*Canna spp.*

morphology and size, we have isolated more than 120 starch grains morphologically similar to maize from all the components of the three sites analyzed (Figure 6.2). Maize starch grains are bigger than wild grasses bearing a mean size over 14μ and presenting diagnostic shape and micro-morphological features. They can be simple or compound, spherical or may present irregular depressions and pressure facets. They often possess a distinct and continuous double border and very characteristic radiating fissures and indentations. This evidence suggests that maize was adopted at least 3,600 yr BP and continued to be cultivated until the historic period. Since the reference collection of modern maize is still small, it is premature to suggest the presence of a particular race or type of maize in this archaeological sample (Figure 6.2 and 6.3).

Single starch grains from sites Los Indios (6 grains) and Isla Larga (3 grains) compare favorably with starch grains of *Phaseoulus spp.* (Figure 6.4), although positive specific identification at the specie level cannot be made at this time. Starch of legumes, and in particular Phaseolus grains are simple, oval or kidney shape and evidently laminated. They present a large, ragged mesial fissure that extends the length of the grain. In addition, to maize and beans we have identified starch grains from plant tubers corresponding to Canna (*Cannaceae*) (Figure 6.5), and Calathea (*Maranthaceae*) (Figure 6.6). Canna starch grains occur as simple, elliptical, broad ovoid to shellshaped grains possessing prominent lamellae. The hilum is very distinct, eccentric, and slightly to the right or left of the longitudinal axis. When turned, the grains are flat. In addition to these characteristics, the grains are very big, up to 80μ. Calathea starch grains are simple, shell shaped or elongated. Grains are laminated, some lamellae more evident than others.

The hilum is small, fairly distinct and eccentric. Grains are flat when turned and frequently the proximal end is narrower than the distal end.

On-going preliminary phytolith analysis has already identified the presence of domesticated squash (*Cucurbita spp.*) phytoliths. Cucurbita phytoliths are solid, regular spheres whose surfaces are scalloped in a deep and consistent fashion (Figure 6.7). All the Cucurbita phytoliths identified in the three sites studied exceeds the range of length and thickness characteristic of wild species (Piperno and Pearsall 1998). The domesticated species of squash likely present are *C. moschata* or *C. maxima*, or possibly even both. Palm phytoliths with their characteristic spherical spinoluse morphology were also identified in all the components of all the sites (Figure 6.8). This research has also identified cross-shape type 1 maize leaf phytolith, but more research is needed to properly measure and quantify the sample for positive identification.

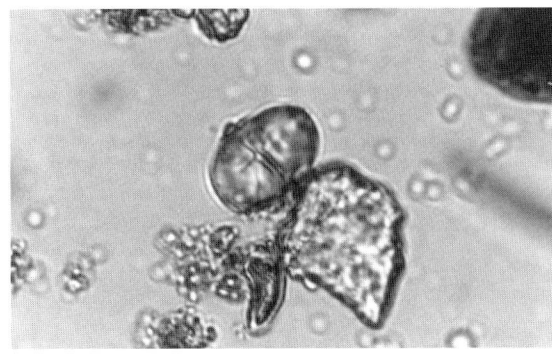

Figure 6.2. Maize starch grain from site Estancia Mal Abrigo. Exc II (50–60 cm). 360x. The grain measures 20 microns by 14 microns.

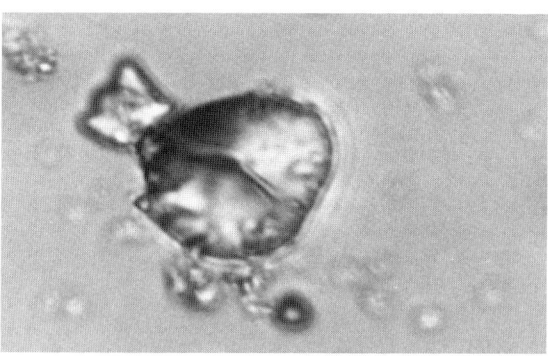

Figure 6.3. Maize starch grain from Isla Larga (max 2.85 m). 360x. The grain measures 20 microns by 20 microns.

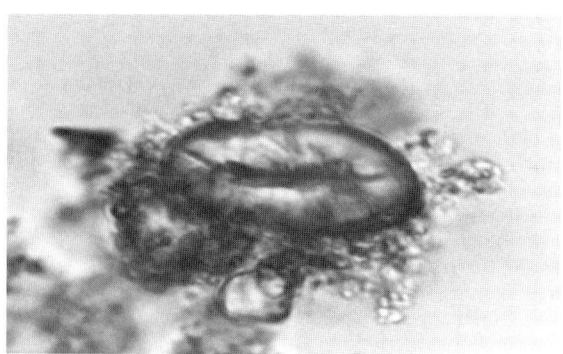

Figure 6.4. Starch grain of a Legume possible Phaseolus sp. from site Los Indios Exc III (1,20–1,30 m). 360x. The grain measures 32 microns by 20 microns.

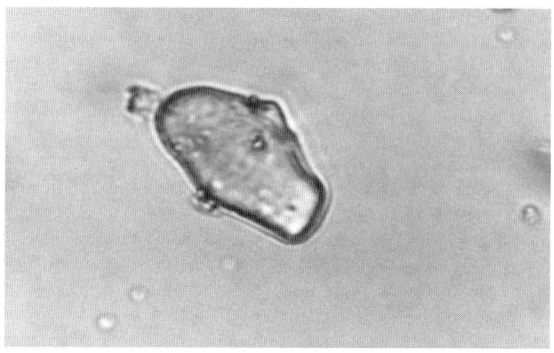

Figure 6.5. Calathea starch grain from site Isla Larga (0.80–0.85 m). 360x. The grain measures 28 microns by 16 microns.

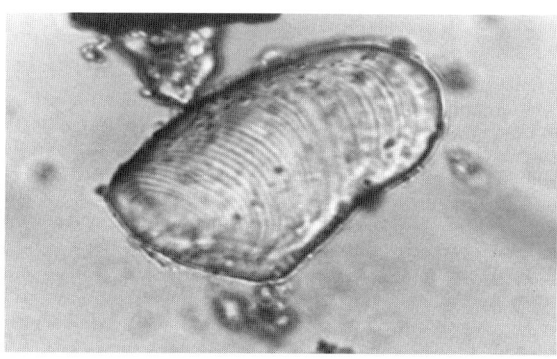

Figure 6.6. Starch grain of Canna sp. from Isla Larga (max. 2.85 m). 360x. The grain measures 58 microns by 36 microns.

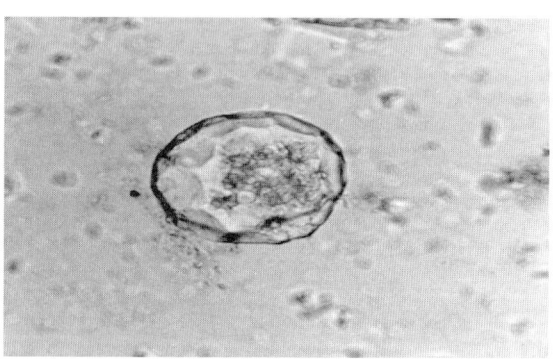

Figure 6.7. Cucurbita phytolith possibly C. moschata from site Estancia Mal Abrigo (40–50 cm). 40x. The phytolith measures 72 by 55 microns.

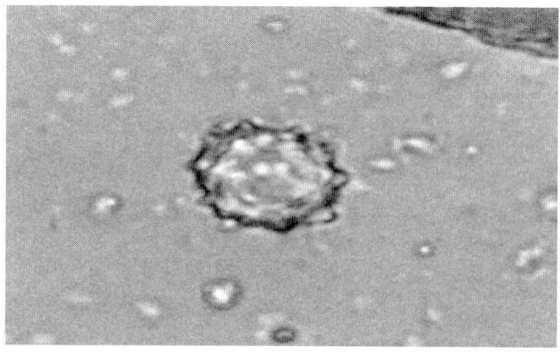

Figure 6.8. Circular spinolose palmae phytolith from site Isla Larga Exc. III (0.80–0.85 cm). x 40.

The Isla Larga site, CG14E01, is a multi-mound site located on the western extreme of the Sierra de San Miguel surrounded by the margins of the wetlands of San Miguel (Banado de San Miguel). Excavation III in the central part of the largest mound contains the longest continuous occupation of the site dating from 3600 yr BP to the historic period. This burial dome-shaped mound is roughly circular with a diameter of 40 m and 3.8 m high. It contains a wide variety of burials, including Tupi Guarani funerary urns associated with the contact period including venetian glass beads dating to the second half of the XVI century (Cabrera *et al.* 1996). Starch grain and phytolith evidence suggest that horticulture was practiced in this site since its earliest occupation. Starch grains from maize (*Zea mays*), and probably a domesticated specie of the genera Canna where recovered from sediments corresponding to the earliest component of the site (3600 yr BP) associated with a hearth containing deer bones and cutting lithic tools, which evidence the first episode of construction of the mound (Cabrera et al. 1996). Starch grains of maize and beans (*Phaseolus spp.*) were isolated from the 3000 yr BP component of the site, and starch grains of maize and Calathea were also recovered from the 1000 yr BP component. In addition, to the starch grains recovered, squash (*Cucurbita spp.*) and palm (possibly *Butia capitata*) phytolith were recovered from all these three levels analyzed of this excavation.

Los Indios site is a multi-mound, multi-component site located over a tongue-shaped spur surrounded by the margins of the wetlands of the Maravillas (Banado de las Maravillas). Its occupation dates between 5000 yr BP to the period of contact. Los Indios site comprises four mounds, two of which are connected through a ramp and facing a third one creating a central open space. The fourth is a burial mound, where so far thirteen burials have been recovered, and is located on the top of a knoll overlooking the other mounds (Lopez 1996).

Maize starch grains were recovered from the oldest occupation of Mound III (Excavation I, levels 1.85–1.90) dating to 2.800 yr BP. Starch grains of maize and beans (*Phaseolus spp.*) where isolated from the level dated to 1170 yr BP (Excavation III, level 1.20–1.30).

The EMA site is a multi-mound complex comprising 50 ha bearing more that 70 mounds located over a flat spur of the Sierra de los Ajos, which projects into the extensive wetlands of India Muerta. It represents the major concentration of mounds in the wetlands of India Muerta and like other multi-mound complexes in this area, the majority of the mounds are low flat circular mounds between 0.5–1.5 m in height, with a diameter ranging between 15 and 30 m. Starch grains and phytolith analyses were carried out from profiled walls of Excavation II on a accretional mound constituted by the vertical accumulation of midden refuse.

Starch grains of maize were recovered from the all the ceramic levels of Excavation II.

Discussion and Final Considerations

Although the interpretations we offer must be considered preliminary, the starch grain and phytolith data sets are strong enough to produce useful interpretations and suggest clear directions for future research. Previous models based on faunal and bone isotope analyses (Lopez and Bracco 1992, 1994; Bracco *et al.* 1996) conducted in various sites in the region have interpreted these societies as complex hunter-gatherers living in a rich and abundant environment in cultural complexity. Faunal analysis (Chagas 1995; Pintos and Gianotti 1995; Pintos 1996) evidenced the consumption of a wide range of terrestrial and riverine resources, including terrestrial mammals (*Blastocerus dichotomus*, *Ozotocerus bezoarticus*, *Mazama guazubira*, *Hydrochoerus hydrochaeris*, *Myocastor coypus*, among many other less represented species), marine mammals (*Arctocephalus australis*), fish, and birds (*Rhea americana*, among others), demonstrating that these groups obtained a large part of their diet from naturally available resources with an emphasis on deer (Pintos 1996). This information was further corroborated by C^{13} isotope analysis of bone collagen from skeletal remains, which indicated that the diet of these groups was predominantly based on terrestrial resources with a low incidence of maize and/or marine resources (Bracco *et al.* 1993, 1996). While these studies have significantly increased our understanding of the economy of the Early Formative societies of southeastern Uruguay, they are not without their own limitations. In the first place, bone isotope data is most useful for assessing the status of maize as a staple crop, consumed on a regular basis. In addition, the studies of C^{13} isotopes based on bone collagen fraction need to be complemented with the analysis of the bone apatite carbonate fraction to provide a more proportional assessment of their dietary regimes since 1) the bone apatite carbonate fraction provides a more accurate record of the carbon isotope composition of the diet, 2) it is less susceptible to diagenic alteration, and more importantly, 3) apatite carbonate is more sensitive for detecting and measuring maize consumption (Norr 1995). Therefore, the contribution of maize to the economy of the Early Formative societies of southeastern Uruguay is still to be determined with future research.

The results of this study complement and provide new data suggesting that the emergence of Early Formative societies in southeastern Uruguay took place within the context of a mixed economy combining hunting, fishing, and gathering with small-scale horticulture of maize, beans, squash, and possibly domesticated tubers. These groups probably engaged in wetland cultivation during the dry-season when the water table was at its lowest, employing a system similar to flood-recessional agriculture (Sherrat 1980; Pohl *et al.* 1996; Siemens 1999). Explaining the processes responsible for encouraging the adoption of these cultigens during the Mid-Holocene in the wetlands of Uruguay are undoubtedly complex and varied and require the combined articulation of paleoenvironmental and archaeological data sets. In this respect, the application and development of the methods reported here should make possible direct investigation of the plant food component of the diet of the Early Formative societies of southeastern Uruguay in sites where preserved plant remains are rarely found as macrofossils. Future studies will concentrate on expanding our reference collection of modern plants and studying close wild relatives of the major domesticated crops in the area.

Acknowledgments

This research would not have been possible without the generous support of Dr. Dolores Piperno from the Smithsonian Tropical Research Institution (STRI) in Panama, who provided Jose Iriarte with training in phytolith analysis, and the Research Challenge Fellowship of the University of Kentucky. Fieldwork in Isla Larga, Los Indios and Estancia Mal Abrigo has been supported by the Comisión Nacional de Arqueología, Ministerio de Educación y Cultura, Uruguay. We would like to thank Dr. Tom Dillehay for his continuous support and encouragement. Thanks are also extended to Digna Matias for offering valuable laboratory support at the STRI and Ing. Jorge Baeza from the Facultad de Humanidades y Ciencias de la Educación-Universidad de la República, Uruguay for his enormous help during the first stage of the Estancia Mal Abrigo Archaeology Project.

References Cited

Bracco, Roberto. 1992. Desarrollo Cultural y Evolución Ambiental en la Región Este del Uruguay. In Ediciones del Quinto Centenario, pp. 43–73. Universidad de la República. Montevideo.

Bracco, R., L. Cabrera, and J. Lopez. 1996. La prehistoria de las tierras bajas de la Cuenca de la Laguna Merin. Paper presented at the Simposio Arqueologia de las Tierras Bajas. Montevideo. Uruguay. In press.

Bracco, R. and M.I. Fregeiro. 1993. Paleo-Dietas para el "Periodo Tardio" del Sector Sur de la Cuenca de la Laguna Merin: Oligoelementos (Sr-Zn) y Carbono 13. Paper presented at the Encuentro Regional de Arqueologia. Melo. Uruguay.

Bracco, R., M.I. Fregeiro, H. Panarello, R. Odino, and B. Souto. 1996. Dietas, Modos de Producción de Alimentos y Complejidad. Paper presented at the Simposio Arqueologia de las Tierras Bajas. Montevideo. Uruguay. In press.

Brochado, J.P. 1984. Ecological model of the spread of pottery and agriculture into eastern South America. Unpublished Ph. D. dissertation. Department of Anthropology. University of Campaigne-Urbana. Illinois.

Brown, J.A. and K. Vierra. 1983. What happen in the Middle Archaic? Introduction to an Ecologic Approach to Koster Site Archaeology. In Hunter and Gatherers in the American Midwest, edited by T.D. Price and J.A. Brown, pp. 201–231. Academic Press. Orlando. Florida.

Cabrera, L., A. Duran, J. Femenias, and O. Marozzi. 1996. Investigaciones arqueologicas en el sitio CG14E01 ("Isla Larga"). Sierra de San Miguel, Dpto. Rocha. Uruguay. Paper presented at the Simposio Arqueologia de las Tierras Bajas. Montevideo. Uruguay. In press.

Carr, Phillip and Jon Gibson. 1997. A consideration of the Cultural Complexity During the Archaic in the Southeast. Paper presented at the 54th Annual meeting of the Southeastern Archaeological Conference. Baton Rouge.

Chagas, Laura C. 1995. Identificacion y analisis del material oseo de los sitios costeros del litoral Atlantico. In Arqueologia en el Uruguay, edited by M. Consens, J. Lopez Mazz and C. Curbelo, pp. 106–115. Imprenta Surcos. Montevideo.

Copé, Silvia. 1992. A Ocupacao Pre-Colonial do Sul e Sudeste do Rio Grande do Sul. In Arqueologia Prehistorica do Rio Grande do Sul, edited by A. Kern, pp. 191–221. Mercado Aberto. Porto Alegre. Brazil.

Delaney, P. 1965. Fisiografia e geografia da superficie da planicie costeira do Rio Grande do Sul. Escola de Geografia. Universidade Federal de R. S. Nro 6. Porto Alegre.

Dillehay, Tom, Netherly, Patricia and Rossen, Jack. 1989. Middle Preceramic Public and Residential Sites on the Forested Slope of the Western Andes, Northen Peru. American Antiquity 54(4):733–759.

Ford, Richard. 1962. A quantitative method for deriving cultural chronology. Technical Manual No.1. Pan American Union.

Gaspar, M.D. 1998. Considerations of the Sambaquis of the Brazilian Coast. Antiquity 72(277):592–613.

Heckenberger, Michael. 1998. Manioc agriculture and sedentism in Amazonia: the Upper Xingu example. Antiquity 72:633–48.

Hilbert, Klauss. 1991. Aspectos de la Arqueología del Uruguay. Mainz an Rheim:von Zabern. Alemania.

Iriarte, José. 1999. The Emergence of Cultural Complexity in the Mound-Building Cultures of Southeastern Uruguay. Paper presented at the 64th Annual Meeting of the Society of American Archaeology. Chicago.

Jackson, Milton. 1984. Contributions to the Geology and Hydrology of Southeastern Uruguay Based on Visual Satellite Remote Sensing Interpretation. Institut fur Geographie der Universitat Munchen. Germany.

Leon, R.J.C. 1992. Setting and Vegetation. In Natural Grasslands. Introduction and Western Hemisphere, edited by R.T. Coupland, pp. 371–378, 382–389. Elsevier.Amsterdam/London/New York/Tokyo.

Lopez Mazz, J. 1996. Construccion del paisaje y cambio cultural en las Tierras Bajas de la Laguna Merin. Paper presented at the Simposio Arqueologia de las Tierras Bajas. Montevideo. Uruguay. In press.

Lopez Mazz, J. and R. Bracco. 1992. Relaciones Hombre-Medio

Ambiente en las Poblaciones Prehistóricas del Este del Uruguay. Annals of the 46th Congreso Internacional de Americanistas. Amsterdam.

Lopez Mazz, J. and R. Bracco. 1994. Cazadores-Recolectores de la Cuenca de la Laguna Merín. In Arqueología Contemporánea 5:51–64. Buenos Aires. Argentina.

Meggers, B. and Clifford Evans. 1969. Como interpretar el lenguaje de los tiestos. Smithsonian Institution. Washington, D.C.

Montana, J. and J. Bossi. 1995. Geomorfologia de los Humedales de Humedales de la Cuenca de la Laguna Merin en el Depto. de Rocha. PROBIDES. Montevideo. Uruguay.

Moseley, Michael E. 1992. The Exploration and Explanation of Early Monumental Architecture in the Andes. In Early Ceremonial Architecture in the Andes, edited by C.B. Donnan, pp. 71–92. Dumbarton Oaks. Washington D.C.

Norr, Lynette. 1995. Interpreting dietary maize from bone stable isotopes in the American tropics: the state of the art. In Archaeology in the Lowland American Tropics, edited by Peter W. Stahl, pp. 198–223. CambridgeUniversity Press.

Pearsall, Deborah. 1995. "Doing" paleoethnobotany in the tropical lowlands: adaptation and innovation in methodology. In Archaeology in the Lowland American Tropics, edited by Peter W. Stahl, pp. 113–129. Cambridge University Press.

Pearsall, Deborah. 1999. Agricultural Evolution and the Emergence of Early Formative Societies in Ecuador. In Pacific Latin America in Prehistory. The Evolution of Archaic and Formative Cultures, edited by Michael Blake, pp. 161–170. Washington State University Press. Pullman. Washington.

Pintos, Sebastian. 1996. Economia Húmeda del Ester del Uruguay: El Manejo de Recursos Faunísticos. Paper presented at the Simposio Arqueologia de las Tierras Bajas. Montevideo. Uruguay. In press.

Pintos, S. and C. Gianotti. 1995. Arqueofauna de los constructores de cerritos: "quebra" y requiebra. In Arqueologia en el Uruguay, edited by M. Consens, J. Lopez Mazz and C. Curbelo, pp. 79–91. Imprenta Surcos. Montevideo.

Piperno, D.R. 1993. Phytolith and charcoal records from deep lake cores in the American Tropics. In Current Research in Phytolith Analysis: Applications in Archaeology and Paleoethnobotany, edited by D.M. Pearsall and D. Piperno, pp. 58–71. MASCA Research Papers in Science and Archaeology, Vol. 10. The University Museum of Archaeology and Anthropology, Philadelphia.

Piperno, D.R. 1995. Plant microfossils and their application in the New World tropics. In Archaeology in the Lowland American Tropics, edited by Peter W. Stahl, pp. 130–153. Cambridge University Press.

Piperno, D.R. 1998a. Phytolith Analysis: An Archaeological and Geological Perspective. Academic Press. San Diego

Piperno, D.R. 1998. Paleoethnobotany in the Neotropics from Microfossils: New Insights into Ancient Plant Use and Agricultural Origins in the Tropical Forest. Journal of World Prehistory 12(4):393–449.

Piperno, D.R. and I. Holst. 1998. The Presence of Starch Grains on Prehistoric Stone Tools from the Humid Neotropics: Indications of Early Tuber Use and Agriculture in Panama. Journal of Archaeological Science 25:765–776.

Piperno, D.R. and D. Pearsall. 1998. The Origins of Agriculture in the Lowland Neotropics. Academic Press. New York.

Pohl, M.D., O.P. Kevin, J.G. Jones, J.S. Jacob, D.R. Piperno, S.D. deFrance, D.L. Lentz, J.A. Gifford, M.E. Danforth, and K. Josserand. 1996. Early Agriculture in the Maya Lowlands. Latin American Antiquity 7(4):355–372.

Prieto, O., J. Alvarez, J. Arbenoiz, A. de Los Santos, P. Vesidi, P. Schmitz, and I. Basile Becker. 1970. Informe preliminar sobre las investigaciones arqueologicas en el Depto. de Treinta y Tres, R.O.U. Instituto Anchietano de Pesquisas/UNISINOS. Sao Leopoldo. Brazil.

PROBIDES. 1997. Reserva de Biosfera Banados del Ester. Avances del Plan Director. Ms. 223 pp. AECI, PNUD, GEF. Rocha, Uruguay.

RAMSAR. 1984. National Report of Uruguay for Ramsar Convention 1982. Ms. Gland. Switzerland.

Rodríguez, J. 1992. Arqueología del sudeste de Sudamérica. In Prehistoria Sudamericana. Nuevas Perspectivas, Betty Meggers, ed. pp. 177–209. Washington, D.C.

Roosevelt, Anna C. 1980. Parmana: Prehistoric Maize and Manioc Subsistence along the Amazon and Orinoco. Academic Press, New York.

Schmitz, Pedro I. 1967. Arqueologia no Rio Grande do Sul. Pesquisas, Antropologia, Sao Leopoldo, 16:1–58. Sitios de pesca lacustre em Rio Grande, RS, Brazil. Instituto Anchietano de Pesquisas, UNISINOS, Sao Leopoldo. Brazil.

Schmitz, Pedro I. 1973. Cronologia de las culturas del sudeste do Rio Grande do Sul – Brasil. In Gabinete de Arqueología. Publicacoes Nro. 4, UFRGS. Porto Alegre. Brazil.

Schmitz, Pedro I. 1978. Industrias Liticas en el Sur de Brasil. Separata de Estudos Leopoldenses. Universidade do Vale do Rio dos Sinos. Brazil.

Schmitz, Pedro I. 1987. Prehistoric hunters and gatherers of Brazil. Journal of World Prehistory 1(1):53–126.

Schmitz, P. and J. Baeza. 1982. Santa Vitoria do Palmar: una tentativa de evolución del Arroyo Chuy y su vinculación al problema de los cerritos. In. Colonia de Sacramento. Uruguay.

Schmitz, P. and J.P. Brochado. 1967. Prospecciones arqueologicas no Rio Grande do Sul. Paper presented at the 37 Congreso Internacional de Americanistas. Mar del Plata. Argentina.

Schmitz, P.I., G. Naue, and I. Becker. 1992. Os Aterros dos Campos do Sul: A Tradicao Vieiria. In Arqueologia Prehistorica do Rio Grande do Sul, edited by Arno Kern, pp. 221–251. Mercado Aberto. Porto Alegre. Brazil.

Sherrat, A. 1980. Water, Soil, and Seasonality in the Early Cereal Cultivation. World Archaeology 11:313–328.

Siemens, Alfred H. 1999. Wetlands as Resource Concentrations in Southeastern Ecuador. In. Pacific Latin America in Prehistory The Evolution of Archaic and Formative Cultures, edited by Michael Blake, pp. 137–147. Washington State University Press. Pullman. Washington.

Sombroek, G.M. 1969. Soil studies in the Merin Lagoon Basin. LM 119. CLM/PNUD/FAO. Treinta y Tres. Uruguay.

Stothert, K.E. 1985. The preceramic Las Vegas culture of coastal Ecuador. American Antiquity 50:613–637.

Stothert, K.E. 1992. Early economics of coastal Ecuador and the foundations of Andean civilization. Andean Past 3:43–54.

Tomazelli, L.J. and J.A. Villwock. 1996. Quaternary Geological Evolution of Rio Grande do Sul Coastal Plain, Southern Brazil. Annais da Academia Brasileira de Ciencias 68(3):373–381.

Williams, Michael. 1990. Wetlands. A Threatened Landscape. Blackwell. England.

7. Houses in the Sea: Excavation and Preservation at Los Buchillones, Cuba

David M. Pendergast, Elizabeth Graham, Jorge A. Calvera and M. Juan Jardines.

Introduction

The ancient Taino site of Los Buchillones, situated on Cuba's north coast at the eastern edge of the fishing village of Punta Alegre in Ciego de Avila province, saw limited dry-land excavation in 1983 and 1989 under the direction of Calvera and Jardines. Results suggested the presence of concentrations of ceramics and lithics that appeared likely to represent residential units, but the extent of the work was not sufficient to reveal any significant architectural data. No further focus on the site occurred until the activities of two Punta Alegre fishermen brought to light a unique body of perishable cultural material so extensive that it was originally seen as unlikely to be ancient (Calvera *et al.*

1996:65–66). Beginning in 1997, excavations concentrated in a large shallow lagoon where perishable objects had been collected, have provided striking evidence not only of the antiquity of the objects but also of their depositional history and of the structural contexts within which they were created.

From approximately 1989 to 1994 the fishermen, Nelson Torna and Pedro Guerra, had collected approximately 195 wooden artifacts, largely or wholly of lignum vitae (*Guaiacum officinale*) that included both utilitarian and non-utilitarian objects. The first class comprises a variety of pins; several eyed needles; at least one apparent fid; a hook of a type used elsewhere to suspend food from house

Figure 7.1. Map of Cuba, showing the location of Los Buchillones.

Figure 7.2. The site environs, seen from the west; Los Buchillones lies on the eastern curve of the embayment, in the distance beyond the village of Punta Alegre.

Figure 7.3. Small ironwood container with elaborately carved handles, recovered from the lagoon bottom prior to 1994 by Nelson Torna and Pedro Guerra.

beams; fragments of undecorated dishes of circular and other forms; a wide variety of dishes with elaborately decorated handles; several objects of unknown use; and a number of handles for petalloid axes and chisels, including two in which the stone tools remain hafted. Non-utilitarian artifacts encompass a number of complete miniature *duhos*, the stools that served as badges of rank in Taino society, as well as fragments of full-size stools; and several *zemis*,

male deity figures that played a highly important role in Taino religious practice. The fact that the collection represented a more than fourfold increase in number of wooden artifacts known from Cuban sites, and very probably exceeded the total known from all of the Antilles, contributed heavily to the view that the material could not be Precolumbian.

Based on discussions that began in 1994 (Pendergast 1994), the Royal Ontario Museum of Toronto, Canada and the Ministerio de Ciencia, Tecnología, y Medio Ambiente (CITMA) of the government of Cuba entered into an agreement that established a jointly directed and jointly funded project designed to investigate the contexts from which the wooden artifacts had come and to undertake radiocarbon dating of some of the objects (Pendergast 1996a, b). Owing to the absence of any apparent relationship between the results of the dry-land excavations and the products of the fishermen's casual collection, as well as to uncertainty regarding the significance of the lagoon in the site's history, neither the temporal nor the physical scope of the excavations was predictable before work commenced. Even had such predictions been possible, however, it is now clear that they would have been wildly short of the mark.

The Excavations

The recent erosional history at Los Buchillones led us to

believe that only a small portion remained of what was once a site of considerable size. Like all low-lying coastal terrain, the area of the site has been subject to the effects of wave action. Over the past half-century, human activity in the form of breakwater construction east of the site has intensified the action of the waves, which has exacerbated damage to the archaeological context at a dramatic pace. Examination of aerial photographs taken in the late 1940s, coupled with our on-site investigations, indicates that the force of the sea has removed approximately 50 meters of the coast throughout the site area. Preliminary investigations by Cuban and Canadian coastal geographers indicate that the lagoon has been present at the site for far longer than the period of human occupation. However, whereas at one time the lagoon was separated from the sea by a body of land, at present it is directly linked to the Caribbean through an ever-widening breach of the coastal sands and mangrove vegetation that characterize the shore. Observations by Calvera and Jardines indicate that the breach occurred as recently as a decade ago. Prior to 1997, therefore, conditions appeared less than favorable for recovery of any significant body of material *in situ*, unless from the bottom of the lagoon itself.

Initial Investigations in 1997

Prior to excavation the possibility appeared to exist that materials encountered in the lagoon had been deposited there by the Taino during ritual activity, although the presence of utilitarian wooden objects, as well as ceramic and lithic artifacts, argued against such an interpretation. In order to examine this possibility, and also because the lagoon appeared to be a more stable and more easily controllable context than the eroding shoreline, we established the 1997 excavation unit within a small northern portion of the lagoon that is separated from the remainder by a sand spit or bar. To achieve the greatest control, we decided to block the waters and work the deposit as a wet rather than an underwater site. Several factors affected our decision: water depth was only 85 to 110 cm, turbidity under strong tidal currents would have made stratigraphic precision impossible, and visual control even in the absence of tidal currents would have been difficult.

In the absence of evidence regarding the nature of artifact deposition and the range of materials that might be expectable, it would obviously have been irresponsible to employ cofferdam techniques that would have required penetration of the lagoon bottom and might have resulted in damage to cultural deposits. The hydrologic conditions, together with the limitations on technology that Cuba's present condition imposes, made use of a well-point system inadvisable. In the face of these considerations we chose a non-intrusive approach used elsewhere in flood control, which employs

dams of sandbags covered with polyethylene sheeting. Such dams effectively block the water, but of course do not prevent run-under of water driven by pressure outside the dammed area; as a result, we were forced to create a sump to allow frequent pumping, but still were faced with constant water entry that increased as we deepened the excavations. Luckily the water-control problems did not prevent full excavation of the unit established within the lagoon, which showed that the cultural material occurred in sediments that overlay a substrate of fine gray clay of a depth impossible to determine in the absence of coring. The stratigraphic situation of the artifacts demonstrated clearly that their occurrence in the lagoon was the product of secondary deposition as the result of wave action. All indications were that the sea, in the course of eroding the shoreline, had dislodged objects from its bed and carried them through the breach to be deposited in the lagoon. In times of storms, the sea may even have transported objects over the narrow spit of sand that now separates the lagoon from the sea. (Pendergast 1997a, b).

By the end of the initial investigations we had recovered 15 further wooden artifacts, including two pointed and tapered apparent lance or spear tips of forms not in the original collection, approximately half of a small plain wooden dish, and two unfinished objects that appeared to have been intended to be a tool haft and a small container with two opposing handles. Together with the formal artifacts were large amounts of worked wood, primarily small smoothed shafts of three standard-diameter sizes, broken into short sections. That the items were artifacts was beyond question, but their significance remained entirely unclear. Although the work added materially to the inventory of perishable cultural material and also yielded ceramics and a limited range of lithic artifacts, all indications were that the lagoon cultural deposit was fully reflective of the remnant nature of the site, and that artifact salvage was the primary if not sole prospect for future endeavors.

During the lagoon operation we began investigation of sea-bottom features reported to us by Torna and Guerra, which consisted of groups of post butts protruding from the sediments in a zone that extends some 20 m north of the the present shoreline. Although it was suggested initially that recent salt-production activity in the area might be the explanation for the presence of the posts, the patterns that emerged in our mapping and the characteristics of the base of a post found dislodged on the bottom quickly swept that view away. It was clear by season's end that many of the groups of posts formed rough circles or ovals, and the fact that the dislodged post bore unmistakable evidence of cutting with a stone axe argued very convincingly that the groups of posts represented ancient Taino structures. We accordingly decided to shift investigations from the lagoon to the sea in 1998.

Figure 7.4. The lagoon, with the 1997 dams in place.

The 1998 Season

Expectations at the beginning of the 1998 season rested not only on the presence of the house post groups but also on the evidence accumulated by Torna and Guerra and on the data yielded by the previous season's investigations. Recovery of wooden artifacts, ceramics, and a group of petalloid axes from the seabed, coupled with our conclusion that the lagoon-bottom cultural material had been re-deposited in recent times, suggested that examination of a post group would not be very likely to yield more than confirmatory evidence of the identity of the feature, plus perhaps some associated artifacts too heavy to have been unearthed and transported by the sea's energy.

We selected for examination a rough semi-circle of posts that stood just offshore, east of the lagoon mouth. Selection was based on logistic considerations, because the sandbag-and-polyethylene-sheet dam required to enclose the feature represented the maximum effort possible in the circum-stances, and proximity to the beach removed the concerns that would have arisen if we had attacked a post group farther out at sea. Almost as soon as the most offensive of the modern algal accumulations had been removed and the sea bottom sediments exposed, it became apparent that the quantity of structural remains present exceeded our expect-ations in very considerable measure (Pendergast 1998). Initial excavations revealed the presence of a group of long poles arranged with their butts near the post remnants and their tips converging toward a centre point just beneath the beach margin; between the long poles lay similar elements

of two shorter lengths, in greater disarray owing to their less substantial nature. Additional clearing revealed smaller transverse members in very large numbers, generally badly broken but with their original relationship to the larger members essentially preserved. The larger elements were readily identifiable as the primary, secondary, and tertiary rafters required in the construction of a conical roof, whereas the smaller transverse members were clearly the stringers on which the palm leaves of the roof thatch would have been hung. Among the stringers protruded the butts of subsidiary interior posts, some of them encircled by ver-tically oriented sticks 2 to 4 cm in diameter, which appeared to make up a partial interior circle that reflected the structure's perimeter form.

Further excavation, which sectioned a small portion of the beach margin, revealed two 7 meter-long king posts with forked tops, between which were portions of what was probably the roof's apex beam. The posts remained very close to their original positions, having fallen in opposite directions as the roof collapsed. Their size is entirely commensurate with the central support required to produce an appropriate pitch in a conical roof that spanned an interior space approximately 26 meters in diameter. Species identification of rafters and other roof elements has not yet been carried out, but the two king posts are clearly of lignum vitae, a wood of prime choice for the support function involved and also a major reason for the excellent condition of the timbers.

Additional excavation among the rafters and stringers

added data even more unexpected than the roof and support elements themselves. In several areas near the roof periphery as well as in more limited patches nearer the structure center we encountered appreciable quantities of moderately well-preserved thatch, generally in compressed pads that included both midribs and portions of the fronds. Although preservation is of surpassingly high quality throughout the structure, the survival of a material so inherently fragile and so subject to rapid decay in most environmental circumstances seemed at the time of excavation to defy all logic.

By the end of the 1998 season we had excavated and recorded only the principal elements of the northern half of the roof. In addition we encountered food remains, including seeds of hog plum (*Spondias mombin*) in a similarly excellent state of preservation, around the structure's northern perimeter. Equal quality characterized the first wooden artifact to be recovered *in situ*, a nearly complete two-handled dish found in the same area. Unfortunately continuing inflow of water both beneath our encompassing dam and through the spit at the south limit of the excavations made deeper penetration both beyond and within the mass of structural wood impossible, and indeed prevented accurate recording of stringers and other minor timbering as well. A manual examination of underlying timber near the structure's periphery showed that a minimum of 50 cm of tangled wood underlies the portion of the building that we were able to reveal, and there is no reason whatsoever to assume that this figure represents the full depth of the remains. As a result it was abundantly clear that further investigation of the structure would require a massive shift in technology, which would of necessity depend upon intrusion of caisson sheets into the seabed and across the beach spit to form a cofferdam around the entire structure. In addition, an easily alterable platform suspended above the structural mass would be needed if dissection of the remains was to be carried out without damage to the material.

In addition to the logistic challenge created by the 1998 discovery, the remains posed formidable questions regarding mechanisms of preservation, to which there were no ready answers. The quality of preservation was unquestionably attributable to emplacement of the remains in a stable anoxic environment within a relatively short time following the building's collapse, but neither the hydrologic conditions nor the possibilities regarding the form of structural collapse suggested how such emplacement could have occurred. The retention of original relationships among construction timbers and minor elements made it abundantly clear that the structure had collapsed in a comparatively gentle manner; one could therefore rule out the possibility that burial of the material in a single stroke as the result of a catastrophic event such as a hurricane had

created the conditions required for preservation. Yet at the same time the condition of the material, and especially the preservation of portions of the roof thatch, showed unequivocally that the collapsed building could not have lain exposed to sun, wind, and rain on the ground surface, or to intensive wave action beneath the sea, for even a short period. Although it was obvious that more extensive excavation was dictated by what had emerged from the initial investigation, it was also clear that further archaeology would not provide a definitive answer to the conundrum posed by the remains in the absence of sediment coring and other strategies geared to recovery of information on sea-level change and local coastal morphology through time.

Results of the 1999 Season

Owing to the fact that the technological requirements for further investigation of the 1998 structural discovery could not be met in time for the planned 1999 season, we decided to shift our attention from the house remains back to the lagoon area investigated in 1997. At the same time, we initiated intensive mapping of the lagoon and its environs, as well as studies of lagoon-bottom sediments that are expected to contribute to the reconstruction of both the human occupation and original environmental conditions. The logic behind our choice of an area for further excavation was dual: by placing a unit immediately west of our 1997 work area we would be entering a zone from which Torna and Guerra had recovered considerable numbers of wooden and other artifacts, and our earlier work provided every indication that the risk of encountering further structural remains was nil. Although we were now able to identify the smoothed shaft fragments recovered in 1997 as probable wattle from the perimeter wall of a house or other structure, not a single piece of such material occurred *in situ* in the lagoon-bottom sediments, and no other structural elements had emerged from our excavations in the area.

By the end of the second hour of excavation of the new lagoon unit we had, as any experienced archaeologist might have known we would, encountered the excellently preserved remains of another structure. The material lay along the northern margin of the lagoon and extended under both the beach sand spit and the mangrove stands and their captured sediments that separate the lagoon from the open sea. From the outset it was clear that the second structure was not a duplicate of the first, but rather appeared to have been a rectangular or square building that enclosed considerably less space than did the 1998 structure. As is expectable given its apparent form, the 1999 building had not collapsed toward its center as did the 1998 structure, but rather had gone down toward the northwest corner, where presumably the first failure of a major post occurred.

Figure 7.5. The 1999 excavation, seen from the west.

Figure 7.6. Close-up of rafters, stringers, and other elements of the 1999 structure.

The result was somewhat greater mixing of roof elements than characterized the first structure, but by and large the roof, very probably of two-slope gable form, retained its original layout in collapse almost as fully as did the circular one. Once again stringers lay perpendicular to the rafters that had supported them, and in this instance some indications were also present of what are presumably the exterior lateral strengthening members that are required in gable roofs of such construction. Beyond this, excavation among the stringers revealed far larger quantities of preserved palm-leaf thatch than we had encountered in 1998. Similar preservation of food remains was also widely in evidence within the house, and in a situation not explainable in the absence of more extensive data we recovered the nearly complete rind of a spherical fruit, species not yet identified. As before, ceramics and lithics were also present, albeit in very limited quantities.

Instead of adding to the puzzling questions posed by the first structure, the second building has provided evidence that helps to explain the seemingly miraculous state of preservation, and has also given us a picture of the site's depositional and erosional history that is different from what at least some members of the team initially proposed. The fact that a portion of the remains extends beneath the body of land (the sand spit and the mangrove stand with its captured sediments) that separated the lagoon and the sea provides a striking demonstration of the mutability of coastal environments, and is a warning to us to keep open minds concerning the rigidity of our reconstructions. Even during the time of Taino occupation, the coastline must have been ever-changing.

A catastrophic cause for the structure's collapse cannot be adduced because all elements are in or near their original relationships, and hence the 50 to 70 cm of overburden cannot be the product of a single depositional event. The stratigraphic position of the remains reflects the comparatively gentle collapse of the house into a protective context, after which the overlying sediments accumulated in conditions of minimal disturbance. One explanation that suits the circumstances is that the structure, and probably its companions, stood on pilings in shallow water in an environment protected from both intense wave action and the effects of storms. Protection from wave action could have resulted from presence in a lagoonal environment in ancient times, but without further study we cannot fully assess the possibility that storm protection was provided by an ancient vegetation canopy that was higher than the mangrove stands today. We must also consider the possibility that the houses stood at a greater distance from the sea than the present location of their remains indicates.

Although Columbus's voyage of discovery afforded the explorers numerous opportunities for extensive examination of Cuba's coast, and the natives' houses were certainly a focus of interest (de Las Casas 1951:200–251), we are unaware of any account of the island's circumnavigation that mentions buildings constructed on pilings in the sea. We are therefore left with only the archaeological evidence for the moment, and the faint hope that further archival research will reveal documentation on the house form. The ethnohistoric issue aside, the archaeological material takes on a singular new dimension when seen as the remains of piling-supported rather than land-based buildings. Two obvious concomitants of the identification are that the structures had to be provided with a floor, and that some form of perimeter platform must of necessity also have been a feature of all main buildings, and perhaps even of the related subsidiary structures of which several post groups appear to provide evidence. This perception creates a picture of remains below the lowest levels we could reach that we can expect to be far more extensive than previously recognized. It also suggests that disposal of garbage is highly likely to have involved dumping from a structure's platform into the sea, a pattern that may explain, along with the chemistry of the sediments, the excellent preservation of food remains.

Although the idea of collapse within a protected lagoonal environment may explain the retention of relationships among structural elements, it does not necessarily account fully for the survival of materials such as the thatch that are perishable in very short times on dry land. Extensive excavation of coastal sites in Belize by one of the authors (see Graham 1989) has shown that, at least in that setting, a waterlogged, protective marine environment does not in and of itself guarantee the quality of preservation of

perishable materials that we routinely encounter. We must take into account the possibility that chemical characteristics of the depositional environment may have played a part in the unique preservation, a possibility that may be strengthened by local anecdotal evidence that the coastal "muds" have therapeutic and rejuvenative value when applied to the skin. These matters obviously remain to be tested scientifically in future seasons.

Excavations in the remainder of the unit encompassed by the dams showed that although much of the area was devoid of primary cultural deposits, the zone immediately south of the structure was marked by a number of small posts and lines of very small vertical sticks, only the butts of which were preserved. If the area had been dry land one would be inclined to see the alignments as related to property limits or garden plots, but in a shallow-water setting they are much more likely to reflect fish enclosures, which the Taino are known to have used (de Las Casas 1951, Vol 2:511). None of the standard features of fish weirs, such as entry passages and entrapment pens, appears to be present, and in any event the small size of fish in such an inshore setting, protected from the full force of the sea by the country's northern reef, would hardly make fish entrapment an attractive endeavor. At present we cannot suggest a use for the alignments on any solid data basis beyond the ethnohistoric description of fish pens, though retention of trapped animals such as turtles may also be a possibility.

A further result of the 1999 excavations is that we can now see the site in its present condition not as a remnant of what was once a much larger land-based settlement but rather as a reasonable approximation of its Precolumbian form. Whereas we initially assumed that the sea had simply eroded the main body of the site, it now appears, perhaps not surprisingly, that the depositional history of the entire area is exceedingly complex. What the sea removed over the past half-century now seems quite likely, based on the evidence from the 1999 season, to have been sediments that accumulated subsequent to Taino abandonment of the site.

This reconstruction of depositional history at Los Buchillones is not without its problems, however, given the fact that a major portion of each of the two structures we have discovered lies beneath the present shoreline. If the buildings once stood on pilings surrounded by water, or if they stood in a zone that was subjected to repeated and perhaps seasonal inundation, we must posit a past coastal morphology quite different in its particulars from that of today, although not necessarily different in terms of the forces that affected coastline features. At most we can propose that the houses were built, and ultimately collapsed, in a relatively protected depositional environment, but we cannot yet be sure if the setting was lagoonal with a configuration and relatively location obviously different from those of today, or if the buildings stood off an ancient shore. The latter reconstruction seems unlikely at first glance, but because Cuba's north coast is protected by a string of islands that lie at a distance of about 25 km from the shore, wave action in general is far more subdued than would be the case if there were no land masses between the coast and the open sea.

At this point, in addition to asserting that the past coastal morphology in the area of the site had a complex configuration, we must envision considerably greater diversity in vegetational communities than is evident in modern times. The result of such diversity would have been greater variation in growth and decay cycles, which would have affected and perhaps mitigated the more destructive forces of coastal change. The Taino unquestionably knew and appreciated the past complex ecology of the area, and efficiently exploited the various niches that the coastal configuration and varied environment must have created. The depauperate modern environment, coupled with the countless changes in coastal form that have surely marked the centuries following the departure of the Taino, force us to struggle to keep in mind the difference between the present and the past we seek to reconstruct.

Dating the Occupation Span

Prior to the 1997 excavations we extracted samples from 10 of the wooden artifacts recovered by Torna and Guerra, selected on the basis of form representation because no stratigraphic or horizontal location data existed for the collection. Given the non-archaeological basis for choice of the objects, the results of AMS 14C dating are of very considerable interest because they exhibit full internal consistency. The calibrated dates range from A.D. 1220 to potentially as late as A.D. 1620 (Pendergast 1996b). Because we are on solid ground in assuming that the earliest date, in particular, does not fix a limit for the occupation span, a total period in excess of 400 years, a span significantly longer than any previously documented for Cuban sites, can be posited.

We collected an extensive suite of samples from elements of the structure encountered in 1998, in all cases from the exterior of the element. The samples come from six posts, four rafters, and the two king posts, as well as from miscellaneous pieces of wood from the assemblage. AMS determinations on 13 of the samples have produced a surprising but largely consistent group of dates. We now have calibrated dates for 12 of the 13 samples analyzed, and in all but two cases the dates produce a plausible pattern of ages for the various architectural elements. One post yielded a date of A.D. 540–690, and one rafter yielded a roughly similar date of A.D. 635–780; both of the ages are

certainly possible if one assumes careful conservation of re-usable timber by the Taino, but both seem unlikely to be correct when compared with the remainder of the dating suite. In summary, the calibrated ages of the king posts fall between A.D. 1295 and 1490, those of the posts between A.D. 1385 and 1655, and those of the rafters between A.D. 1380 and 1655. The single uncalibrated date for a miscellaneous element fits within the span, at A.D. 1420–50. The earlier dates for the king posts are credible given the size of the elements, which might well have dictated their re-use in a series of structures, and which would also have contributed to the durability of the elements. The other posts and the rafters provide identical, and surprisingly long, spans of 275–280 years, which may again indicate retention of re-usable timbers, although such action is intrinsically less likely in the case of smaller elements. In any event, the dates appear to constitute an argument that the structure was built in the first half of the 17th century. Dates for a similar suite of samples from the 1999 house are not yet available.

The Cultural Significance of the Discoveries

The post groups, of which we have now recorded almost 40, extend along the present front of the site for more than 500 m, and it appears quite likely that they will ultimately be found along much or all of the site's 1.5 km+ east-west length. The site's full extent has not yet been accurately ascertained, but even at 1 km, and with an original north-south width that may have been in the area of 100 m, Los Buchillones is by far the largest site known in Cuba. Conventional wisdom regarding Taino sites, based on ethnohistoric accounts, is that communities were occupied for periods of a few decades at most, and hence the size of Los Buchillones appears inconsistent with the recorded pattern of Taino settlement use. The strong indications of long occupation clearly raise the possibility that further excavation will reveal intrasite dynamics, which may have involved shifting settlement along the seafront. There is a very tentative suggestion in ceramic seriation data that the western portion of the site was occupied after areas farther east had fallen out of use (Bekerman 1999); the very late dates for the 1998 structure might be seen as support for this suggestion, but any judgment in the matter must be reserved until dates from the 1999 structure become available. Whatever the settlement pattern may have been, it is abundantly clear that the length of the occupation span should allow recognition of material culture change at Los Buchillones over time, which would mark a shift away from the picture of an essentially static culture that has so often been drawn for the ancient Taino (see, for example, Tabío and Rey 1966:159–217). We are certainly not in a position to assess change on the basis of the artifact assemblage currently in hand; such assessment will have to await the investigation of several structures of different forms and in widely separated locations along the seafront, from which we can hope that a horizontal stratigraphy can be developed. To the extent that we are able to judge from our investigation of the uppermost parts of two structures, the buildings conform generally in shape and construction with the very limited descriptions of Taino houses provided in ethnohistoric documentation (Colón 1984:120, 177; de Las Casas 1951, Vol.1:214; de la Cosa 1957:143). Although there are a few maddeningly limited early 16th-century observations regarding house contents (Colón 1984:120–121; de Las Casas 1951, Vol. 1:221–222), sadly no description of house framing or interior details is known to exist. As a result the significance of many internal features that future excavations can be expected to reveal may prove impossible to determine with any degree of specificity. Our discoveries prompted a local report, however, that interior posts girded by small sticks in the manner encountered in the 1998 structure occur in some buildings still in use in the easternmost part of Cuba, so we have hopes that at least one currently enigmatic feature may be explained by ethnographic analogy during the coming field season.

It is already true that the results of the excavations embrace a wide range of information on Taino construction and woodworking techniques, and with the image before us of floored structures on pilings we can expect that future investigations will add materially to our understanding of these facets of Taino life. Beyond this, the likelihood that refuse disposal around the buildings, already clearly documented for both structures, involved rapid deposition in protective sediments raises the possibility that excavations will recover a very wide range of information on Taino food habits, of particular interest as regards utilization of non-cultivated vegetal foodstuffs. Luckily, with the sharp differences in form and size that distinguish the two structures, we are quite likely to be able to exhume evidence of inter-family or inter-group differences in food consumption as well.

The differences between the two buildings surely bespeak very different types of use, and the large size of the 1998 structure just as surely points to either communal or extended-family use if it does not identify the building as the residence of the *cacique* or chief. We are not yet in a position to state that the 1998 structure exceeds all others at the site in size, but it appears fairly likely to be either unique or one of a very small number of oversize structures in the assemblage. It may therefore be reasonable to expect reflections of elevated rank or status in associated portable material culture and perhaps also in food remains; this, together with all of the other lines of evidence that remain to be pursued, argues very strongly for further investigation of the 1998 structure in coming excavation seasons.

Until now it has been true that the achievements of the Taino have perforce been understood through their portable material culture, shorn of the contexts in which the objects were made and used. Furthermore, the representation of Taino esthetic and technical abilities has been severely limited by the fact that very little of the spectrum of perishable artifacts was available for study. Although we do not have sufficient data at present to permit a real assessment of the percentage of portable material culture represented by perishable objects, it is quite likely that past assessments of the Taino have rested on less than 10 per cent of the cultural record. This figure is supported by Pendergast's study of artifacts from Antelope Cave, Arizona, which showed that only 9 per cent of the objects would have been preserved in a surface site, precisely the figure that emerges from the Water Hazard wet site in British Columbia (Bernick, this volume). In this respect, as in so many others, future work at Los Buchillones can be expected to contribute very significantly to an expanded appreciation of the heights attained by the Taino over a period that began in the 13th century if not earlier and lasted for more than 400 years.

Finally among the numerous aspects of Taino culture history touched upon by the Los Buchillones excavations comes the significance of the late dates yielded by both portable wooden artifacts and structural timber. The picture provided by the dates is given some support by the presence of one unmodified sherd of majolica from surface collection and a second, shaped into a triangle, recovered from lagoon-bottom sediments in the 1997 excavations; both show that the community survived for some time following the Spanish invasion of Cuba, but neither provides a solid basis for dating such survival. Until now it has been believed that the Taino were very largely extirpated, and otherwise were absorbed into Hispanic communities, within two to four decades of Spanish arrival in Cuba (see, for example, Dacal and Rivero 1996:51). In this view no community survived physically intact, let alone culturally, under the joint onslaughts of Spanish culture and European diseases. How is it, then, that the two groups of dates from Los Buchillones are in close agreement regarding a 17th-century date for the last currently recorded events at the site? This question was important enough when only portable objects were the focus of attention, but when radiocarbon dates indicate that house construction may have continued into the early 17th century, the picture that emerges is that of a fully functioning, if not thriving, settlement nearly a century after all such communities were thought to have ceased to exist.

If Los Buchillones did indeed maintain its existence well into the period of Spanish domination, the community's setting may have offered the concealment necessary to its survival. In the 16th and 17th centuries much what is now cleared land in the province of Ciego de Avila south of the coastal zone was covered in dense forest that could have obscured the coast from the view of those who traveled through the area by land. In addition, the range of hills that lies immediately south of the site serves as a highly effective barrier between the community and the inland area. As a result, even where bodies of water offer larger vistas than occur in forested land, an early traveler's view toward the north would not have revealed any sign of the settlement's existence. For travelers by sea, the cayes along the northern reef would probably have directed ships along their windward shores, away from the potential hazards of travel in shallow inshore waters, and in this circumstance Los Buchillones would also have gone unnoticed. Obviously we shall need much more in the way of confirmatory evidence before we can assert the survival of the community into the 17th century with full confidence, but the data available at this stage suggest that luck and location may have favored Los Buchillones for a considerable time.

Together with the posited survival of the settlement come questions regarding its abandonment. In this sphere the existence of the objects collected by Torna and Guerra may constitute evidence of pivotal importance. All data indicate that the collapse of the structures we have encountered was an essentially peaceful event such as could be expected of structures that no longer enjoyed the maintenance required to keep them proof against the forces of the sea. Had departure from the structures been unforced, however, it seems very likely that the Taino would have taken with them many of the elaborately carved wooden objects that turned up on the beach and in the lagoon bottom sediments. We now recognize that the objects are highly likely to have had their origin in remains of piling-supported structures that presently lie offshore, but that recognition offers nothing in the way of an explanation for their presence.

Among other possibilities, the profusion of high-value artifacts in the assemblage may reflect forced abandonment of the community. Combination of this possibility with the late dates for a number of objects and for the construction of the building encountered in 1998 raises the possibility that abandonment was occasioned by 17th-century or later Spanish discovery of the community. A project now in its development stage will enable us to determine if, as appears moderately likely, ethnohistoric documentation regarding the Los Buchillones area exists. At the moment we are unable to support the archaeologically based suggestion with written evidence, but there is persuasive evidence in known early records that the region in which Los Buchillones lies was not a focus of Spanish activity until perhaps as late as the early 18th century. This late date may increase the possibility that the community survived much longer than did its counterparts elsewhere in Cuba, but on the basis of archaeological evidence alone the image of the Spanish uprooting what may have been Cuba's longest-lived indigenous community is a persuasive one. As a result, the prospect that further work may yield evidence regarding Taino-Spanish interaction is intriguing both in terms of the site and in the broader context of Cuba's early colonial history.

Conservations Issues, Present and Future

It is relatively easy to chart the course for future seasons of excavation on the basis of the results thus far, and to anticipate at least some of the products of such endeavors. By far the greater challenge lies in the devising of solutions to the issues of conservation raised by the volume, character, and undeniably great importance of the assemblage of material that lies in the sediments of Los Buchillones. The present and future conservation problems raised by a collection of portable wooden artifacts that now numbers nearly 250 objects are daunting enough in themselves, though they stem only from the original surface collection and our very limited excavations. The problems raised by the future of the structural remains as we can now perceive it are more numerous still, and their solutions both vastly more complex and far more costly.

Conservation of the portable objects poses problems only insofar as selection of proper treatment method and provision of adequate treatment facilities are concerned. A number of the wooden artifacts recovered by Torna and Guerra were subjected to sucrose impregnation treatment at facilities in Havana, but it is clear from our monitoring of the collection that further treatment is required to ensure long-term preservation. Treated artifacts that were entirely stable, and effectively in almost-new condition when discovered less than a decade ago, now exhibit moderate to severe checking and cracking that argue eloquently for modification of the conservation technique. In addition, both treated and untreated objects from the surface collection require a quality of climate-controlled storage that they do not currently enjoy.

The approach to the problem of preservation that was employed by Torna and Guerra was simply to maintain the collection immersed in water. Although not an entirely feasible solution as regards preparation of artifacts for analysis and display, water immersion currently serves as our means of providing for short-term conservation of artifacts recovered in excavation. Once recorded, objects are placed in seawater-filled ziploc bags, larger heavy-gauge bags, or sealable plastic boxes, depending on artifact size. This economical and very low-tech approach will, we are convinced, permit us to retain the objects in their as-found condition, with sufficient monitoring, until it becomes possible to obtain the materials and devices required for enhanced sucrose impregnation or other suitable long-term conservation approaches.

The structural remains pose a series of logistic and technological problems immeasurably larger than those raised by the portable artifacts. The need for a cofferdam system, combined with a more effective sump system than we have managed to create thus far, is manifest in the case of each of the structures we have examined. Use of such a system will permit long-term exposure of the material, which in turn will require that we install a rehydration system to avoid desiccation damage to timber as we labor to penetrate the mass of remains. Technological considerations will very probably limit our approach to simple overhead sprayers supplied by a subsidiary pump actuated by a timing device, at least for the period of the excavations. Whatever the solution, the systems employed for protection of the material must be as low-tech as possible within acceptable limits of reliability, and yet must function round the clock without constant attention.

The challenge is made even more daunting by the remarkable quality of preservation, which has left us with virtually everything that was present when the structures collapsed. In clearing the timber encountered in 1998 and 1999 we have in each case been only at roof level, and there is no question that in order to reach lower levels of the deposits we shall be forced to remove the upper materials. Because conservation of rafters, posts, and other large structural members is not an attainable goal on-site or off, it is most likely that we shall be constrained to utilize near-shore areas of the sea, encompassed by sandbag and polyethylene sheet dams, as makeshift holding tanks for timbers that are tagged to permit ultimate re-emplacement, if this is the course chosen. Although such an approach is feasible for larger members, it appears very unlikely that it will serve for smaller elements such as stringers and wall wattle, and we have yet to devise a solution to the problem of maintenance of such material for re-emplacement or other purposes.

Beyond the requirements of excavation lies the ultimate disposition of the material. The joint team, together with representatives of other agencies of the government of Cuba, are currently considering the long-term future of the site as both an archive of information on Taino construction and an attraction for the increasing number of tourists who seek information on the country's past. It will, of course, be possible to reconstruct one or more of the buildings with modern materials, either in the sea or on dry land, once we have recovered full architectural data. It is desirable, however, that visitors to the site be able to see actual remains in addition to a reconstruction, for the interest and instructional values of authentic material are beyond denial. To address the desire to maintain structural remains in their original condition on a year-round basis we must balance conservation and public archaeology issues, a balance that will be extremely difficult to achieve and even more

difficult to maintain over time. If a cofferdam-surrounded body of material is to remain available for viewing, the simplest approach would be to flood the locale until a visit is scheduled, at which point pumping would commence. Such an approach is feasible in the short term, but frequent exposure of the wood is highly likely to generate fungal and other problems that will be extremely hard to control. Furthermore, the location of the remains within and at the margin of the sea dictates that both the energy of the Caribbean in general and the threat posed by major storms and hurricanes be taken into account in any approach to creation of a permanent display of architectural remains *in situ*. At the same time, the fragile coastal environment in which the material lies must not be altered by display techniques in such a manner as to bring about further degradation of the shoreline and associated flora.

Among solutions currently under consideration is the construction of a protective housing modeled on the approach taken at Flag Fen in England, which permits visitor examination of actual remains in an enclosed environment. It appears, however, not to offer a full solution to the problem of degradation of the remains in the long term. In the case of Los Buchillones the use of such a protective structure would entail construction of a seawall at least 25 to 30 m north of the present shoreline, to prevent the damage to the structure that wave action would inevitably bring. The effects of a seawall cannot be assessed fully, but preliminary examination of the situation by coastal specialists suggests that the approach might be feasible if a special wall design were employed.

The mix of immediate and long-term preservation issues at Los Buchillones conjures up a picture of investment of both money and time that is far beyond the scope of the present project. It is, however, clear that the excavations cannot proceed responsibly for any great distance beyond their present stage unless preservation of both the portable artifacts and the structural remains can be assured. We envision our next excavation season as a trial run for some of the procedures briefly outlined here, and we are well aware that with trial will come some measure of error. For this reason we plan to limit the compass of the work, in order to ensure that subsequent investigations rest on the best possible base of experience.

Even at their very preliminary stage the excavations at Los Buchillones have transformed our understanding of the Taino past. The work has given us not only a far greater quantity and variety of information on perishable material culture than previously existed but also the initial elements of a body of data on Taino architecture of which we have known nothing until now. At the current point in our investigations we are able to begin to see the Taino and their achievements in woodworking, as in other aspects of portable material culture, against the backdrop of the buildings in which the objects were created. It is our task to ensure that as the mass of information increases in volume, the physical evidence is preserved in such a way as to provide maximum value, both scholarly and public, now and in the years to come.

Acknowledgements

Funding for the Los Buchillones Project has been provided by the generosity of Mr. and Mrs. W.J. Simpson of Mississauga, Ontario via the Royal Ontario Museum Foundation, and by the Ministerio de Ciencia, Tecnología y Medio Ambiente of the Cuban government, which also granted us the permit for the work.

In 1999 we lost our valued colleague and beloved friend André Bekerman, whose presence had been a spark to our endeavors from the start; in recognition of his many contributions, the project has now been officially designated the *Proyecto André Bekerman* by the Cuban government contingent. In addition to André and the authors, the project's staff has consisted of the following individuals: Rafael Aguilera O., Norma Alvarez G., Antonio Barrera H., Kathryn Beeman, Anna Bekerman, Odalys Brito, Bárbaro Manuel Cabrera R., José Calvera, Antonio Luis Cancio M., Pedro Cruz R., Dr. Anthony Davis (University of Toronto), Adrián García L., Alberto García L., Agustina González L., Pedro Guerra H., Dr. José Herrera A., Julio Ibarra E., Lizet Jiménez R., Nancy López G., Rafael Angel Mangano P., Elizabeth Merino I., Anadelis Pérez S., Maribel Pérez de Calvera, Matthew Peros, Heidi Ritscher, Orbey Rivero, Nelson Rodríguez T., Roberto Rodríguez T., Norma Rojas, Alexander Sánchez G., Elgan Sánchez C., Marco Sanjil T., Yoel Sao P., Dr. David Smith (University of Toronto), Lázaro Suarez Z., Nelson Torna, Roberto Valcarcel, Francisco Velázquez, and Alicia Vera, all of whom have laboured long and hard in difficult circumstances.

References Cited

Bekerman, André. 1999. Excavation of the Los Buchillones Site, Cuba. Paper presented at the 64th annual meeting of the society for American Archaeology, Chicago.

Bernick, Kathryn. 2000. Serendipitous Discoveries: The Water Hazard Wet Site in Southwestern British Columbia. This volume.

Calvera, Jorge, Eva Serrano, Manuel Rey, Irán Perdomo, and Yudelsy Yparraguirre. 1996. El sitio arqueológico Los Buchillones (The Los Buchillones Archaeological Site). El Caribe Arqueológico 1:59–67.

Colón, Hernando. 1984. Historia del Almirante (History of the Admiral). Edited, with introductory notes, by Luis Arranz; translated by Alfonso Ulloa. Crónicas de América I. Madrid: Información y Revistas, S.A.

Dacal Moure, Ramón, and Manuel Rivero de la Calle. 1996. Art and Archaeology of Precolumbian Cuba. Translated by Daniel H. Sandweiss. Pittsburgh: University of Pittsburgh Press.

de Las Casas, Bartolomé. 1951. Historia de las Indias (History of the Indies). 3 vols. Edited by Agustín Millares Carlo. Mexico City: Fondo de la Cultura Económica.

De la Cosa, Juan. 1957. Journal de bord de Jean de la Cosa, seconde de Christophe Colomb (Ship's Log of Juan de la Cosa, Second in Command to Christopher Columbus). Edited, with notes, by Ignacio Olagué. Paris: Editions de Paris.

Graham, Elizabeth. 1989. Brief Synthesis of Coastal Site Data from Colson Point, Placencia, and Marco Gonzalez, Belize. In Coastal Maya Trade, edited by Heather McKillop and Paul F. Healy, pp. 135–154. Occasional Papers in Anthropology, No. 8. Peterborough, Ontario: Trent University.

Pendergast, David M. 1994. Up to Your Knees (Not Headfirst, Luckily) in the *Fango*: El Sitio Arqueológico Los Buchillones, Prov. de Ciego de Avila, Cuba. Royal Ontario Museum Archaeological Newsletter, Series II, No. 55.

Pendergast, David M. 1996. The Los Buchillones Site, North Coastal Cuba. NewsWARP 19:3–6.

Pendergast, David M. 1996. AMS Dates from Los Buchillones, Cuba. U 20 (1996):33.

Pendergast, David M. 1997. Up from the Shallows: A Look at the ROM's First Archaeological Excavation in Cuba. Rotunda 30, no. 2:28–35

Pendergast, David M. 1997. Los Buchillones, Cuba: 1997 Excavations. NewsWARP 22:3–15.

Pendergast, David M. 1996. The House in the Water. Rotunda 31, no 2:26–31.

Tabío, Ernesto, and Estrella Rey. 1996. Prehistoria de Cuba (Prehistory of Cuba). Havana: Academia de Ciencias de Cuba.

8. A Soft Economy: Perishable Artifacts offered to the Well of Sacrifice, Chichén Itzá

Clemency Chase Coggins

What is best known about the Cenote of Sacrifice at Chichén Itzá, Yucatán, Mexico, is that the Bishop was right. (Figure 8.1). In 1566, the Bishop Diego de Landa explained to the Spanish Crown, partly in expiation for his bad treatment of the Maya of Yucatán, that the Well of Sacrifice, or *el Cenote Sagrado,* at Chichén Itzá was a Maya pilgrimage goal like Jerusalem or Rome, where human sacrifices were made, and that if gold were to be found anywhere in Yucatán it would be in this well (Tozzer, 1941:179–184; Thompson, 1992: 4). He was right, but that is only part of the story. In addition to the human remains, and the gold and jade – obvious riches – many valuable, otherwise unknown, perishable ancient objects were also thrown into this well, where some were preserved eleven centuries in the anaerobic mud. These included: worked wooden objects, textiles, baskets, sandals, copal incense, rubber, and decorated gourds.

Yucatán is a flat, porous, limestone peninsula that lacks surface water. Ancient Maya access to fresh water was only in caves and in places where the stony surface of the earth had collapsed to reveal the water table beneath – or in sinkholes like those found in such karstic limestone formations all over the world. The Maya called such a natural well *"ts'onot,"* a word that was corrupted by the Spaniards to *"cenote."* Thus El Cenote Sagrado was a natural well that was considered sacred by the Maya; while the place name, Chichén Itzá, means "at the mouth of the well of the Itzá (people)." The third name, "the Well of Sacrifice," invokes the sixteenth century ethnographic accounts of maidens thrown into the cenote in order to bring back a prophecy, should they return alive.

In the nineteenth century these accounts, and the *Relación* of the Bishop Diego de Landa, were found in the Archivo de las Indias, Sevilla, Spain, and published. Soon, many scholars and explorers knew that at the archaeological site of Chichén Itzá, with its massive pyramid known as El Castillo, there was a sacred well that probably contained the remains of sacrificial victims and treasure of some kind. Yucatán was very far from Central Mexico, quite inaccessible by land in the nineteenth century when the peninsula was in closer contact by water with the United States and Europe – so it is not surprising that among the first explorers of Chichén Itzá were Frenchmen, Englishmen, and North Americans.

Exploration of the Cenote presented formidable problems since the round sinkhole measured 24 m down to water level, and 60 m in diameter, with vertical and overhanging walls of alternately hard and soft bedded limestone (Figure 8.2). The first man to devise a successful scheme for exploration of the Cenote was Edward H. Thompson, an American engineer, who had explored several Yucatecan sites for the Peabody Museum of Archaeology and Ethnology of Harvard University, and for the American Antiquarian Society of Worcester, Massachusetts (his hometown). He had become American Consul in Mérida, Yucatán in 1885. Thompson learned to speak Yucatec Mayan and, in 1894, bought the large cattle and wood-producing hacienda of Chichén Itzá, which included the archaeological site. Thompson decided he would dredge the Sacred Cenote. During about four winter seasons he lowered a hinged "orange-peel" type steel bucket from the northern rim to the Cenote's branch, log and rock collapse-covered, sludgy bottom, 12 m below the surface of the opaque green water. Briefly, in 1908, Thompson tried deep-sea diving, with a Greek pearl diver from Florida, and toward the end he dumped the buckets of semi-liquid mud or "muck," as he called it, onto a tethered scow, from which it was transferred to a narrow beach on the west side of the cenote, to be sifted. Thompson worked intermittently from 1904 to 1910 (Coggins 1992b).

It is impossible to quantify what was thrown in originally, what was recently taken out, or even what became of much that was found. This is because excavation, or rather

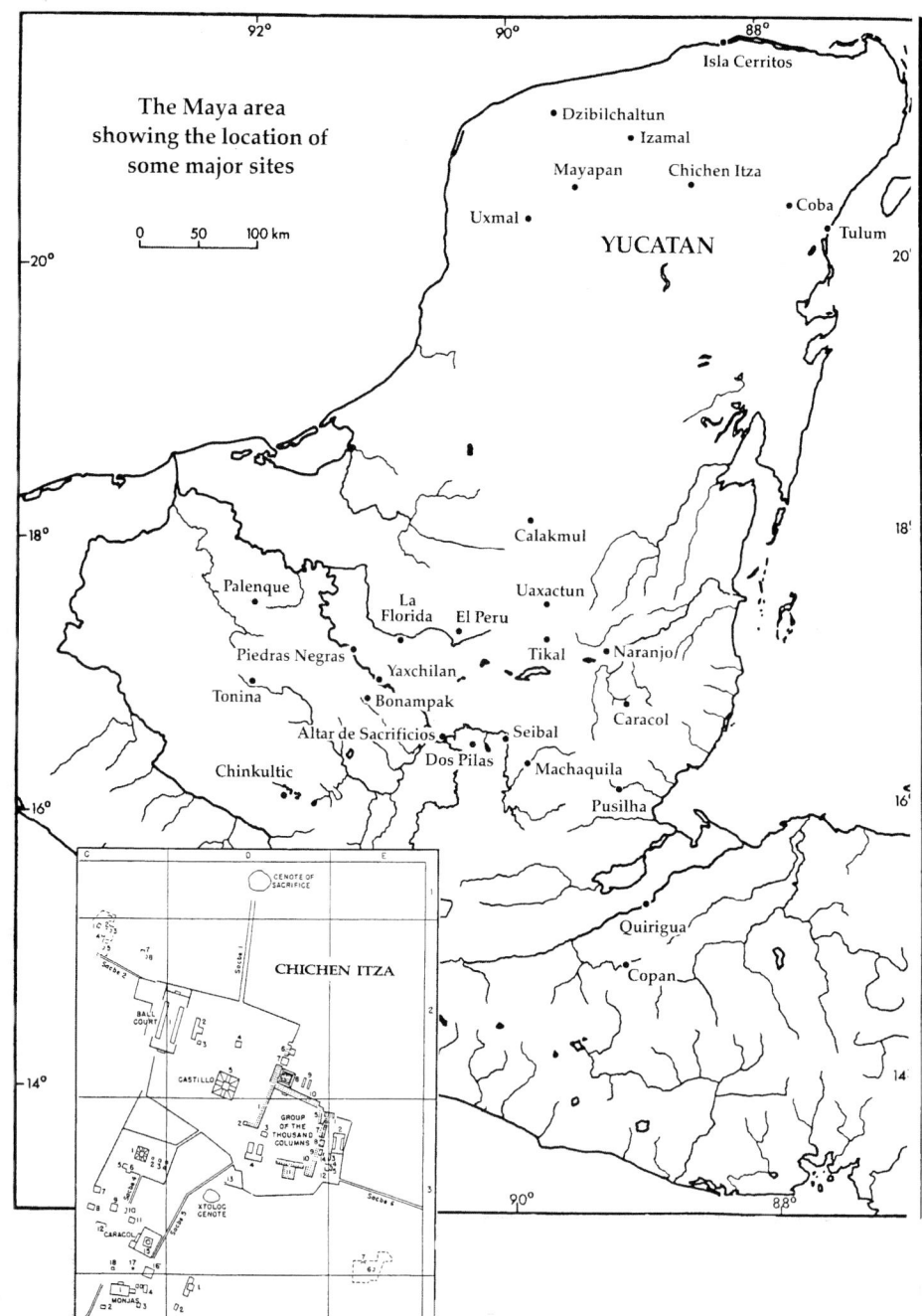

Figure 8.1. Map of Yucatán Peninsula; inset, Plan of Chichén Itzá (from Coggins, 1992a, inside cover).

extraction, techniques, were primitive early in this century. Thompson had no archaeological training, and his interests varied. Bark buckets are mentioned by later Mexican investigators, but not by Thompson; like many such artifacts they probably did not survive removal from the water due to the sudden change of environment, and to the action of the metal jaws of the dredge bucket. Other objects were not

hand-sifted from the muck until too late, and if most perishable objects were not extracted immediately, they did not survive. Thompson tried to preserve the wooden objects with a variety of ingenious methods until the Museum instructed him to pack them in water. Once they arrived at the museum, they were impregnated with paraffin.

First, the Maya workmen brought up, bones and sherds,

Figure 8.2. Edward H. Thompson at north rim of Cenote of Sacrifice, Chichén Itzá, with dredge derrick and winch, circa 1905. (Courtesy of the Peabody Museum).

Figure 8.3. Drawing of wooden manikin scepter, with "gold" mask, L. 28.8 cm. (Coggins and Ladd, 1992a. Fig. 845b).

then, after some days, balls of copal incense. From the *protium copal* tree, this gum called copal by the Aztecs, and *pom* by the Maya, is still burned in Maya ritual, and its presence confirmed Thompson's motivating belief that the Well had, indeed, been sacred. After a month, a wooden scepter was found in the shape of a small man with a golden mask covering his face (Figure 8.3). Tremendously relieved, Thompson then knew there really was gold in the Cenote, and that the Bishop de Landa had been right. The extroverted, enthusiastic, and hospitable Thompson had many friends and admirers. In 1926, one of these, T.A. Willard, wrote a book about Thompson's work at Chichén Itzá, in which the author greatly exaggerated the monetary value of the gold Thompson had found, much of which was actually made of copper-gold alloys (Willard 1926). Mexico sued for compensation for Thompson's finds, but only two decades later, after Thompson's death, did the Mexican Supreme Court rule in favor of his heirs, and could work finally begin on analysis of the collections at the Peabody

Museum, Harvard University, where they had long been stored. In 1952, after Samuel K. Lothrop's publication of the metals – gold, silver, copper, alloys of these, and lead (Lothrop 1952), the Peabody Museum gave Mexico a representative collection of 92 gold and copper objects from the Cenote, in exchange for a representative collection of ancient Mexican pottery. In 1976, two years after publication of the jade objects by Tatiana Proskouriakoff (1974) the Peabody Museum gave 246 Cenote jades to Mexico in exchange for a selection of Colonial Mexican pottery. By 1992, at the time of the publication of all the remaining artifacts (Coggins 1992a), Mexico had excellent Cenote collections of its own.

In 1961 and 1967–68 Mexican archaeologists of the Instituto Nacional de Antropología e Historia (INAH) directed work in the Cenote, using more modern techniques of retrieval, although they encountered many of the same problems as Thompson – especially lack of visibility due to silt in the water, and density of woody debris above the rocky, boulder-strewn bottom. (Folan 1968; Piña Chán 1968). Most important, in 1967–68 Román Piña Chán's expedition identified a small area with some stratigraphy near the north wall, thus establishing a sequence of offerings – something Thompson had never been able to do. While these INAH excavations discovered the same kinds of artifacts that Thompson had found, many more objects were preserved and in better condition, especially textiles and stucco-decorated gourds – thanks to their better methods.

Confirming the existence of these ancient offerings was, however, only a small part of the problem for Thompson and later archaeologists. Chichén Itzá has been the subject of scholarly controversy for more than a century, and one critical question concerns the role and significance of the Sacred Well, and how the found objects may illuminate the confusing history of the site, and its relationship to other parts of ancient Mesoamerica. They do, fortunately, throw some light on the chronology of Chichén Itzá – a particularly vexing problem. It is quite clear that no known offerings – no pottery vessels, no metal or wooden objects, no textiles, no jades or incense were offered to the Cenote before the eighth century A.D. The northern part of the Yucatán Peninsula has many cenotes, but only el Cenote Sagrado is known to have received such rich, varied, and exotic offerings. Thus, the practice of sacrificing things, and people, to a pristine source of subterranean water, may have been associated with the founding of the site near 800 A.D.

Sixth and seventh century wall paintings at the central Mexican city of Teotihuacan show priests and deities making watery offerings of jade and shell and other precious goods, and in the sixteenth century, the Aztec periodically offered jade and gold to a whirlpool in Lake Texcoco. In the 900 years between these episodes, Chichén Itzá was founded, perhaps as a southern outpost of the recently abandoned Teotihuacan. There, many of the ritual and militaristic traditions of Teotihuacan were perpetuated, as they were at Chichén Itzá's brother city of Tula, in Central Mexico.

Jade was the most treasured and precious substance for Mesoamericans and thousands of jade beads and carved pectorals were offered to the Cenote (Proskouriakoff 1974). The Spaniards and subsequent European explorers did not share this appreciation of jade. They were more interested in gold. However, in the eighth century A.D. there was no goldworking in Mesoamerica. The first gold objects, and possibly the techniques for working them, had come early in the ninth century from lower Central America, when cast gold objects were carried north to Chichén Itzá, in large coastal trading canoes. These ocean-going boats unloaded their cargo at the island port of Isla Cerritos, 60 km north

Figure 8.4. Finger grips from three broken wooden atl atls, left, L. 15.3 cm. (Coggins and Ladd, 1992a, figs. 8:20,21,22.).

of Chichén Itzá. The cast gold bells, and thousands of cast copper ones, that were thrown into the cenote, were probably part of dance and military costume (S.K. Lothrop 1952). One identifiable early phase of offerings to the cenote reflects a warrior cult. At this time (ca. A.D. 750–1150), wooden throwing sticks, known as *atlatls*, were offered with their darts and associated flint or chalcedony blades, along with a kind of wooden defensive fending stick (Figures 8.4 and 8.5). These were weapons shown carried by the many warriors depicted in paint and stone at Chichén Itzá (Figure 8.6). These men and weapons that represented a Toltec warrior elite identified with ancient Teotihuacan, and later with the Toltec capital at Tula, in Central Mexico, are the only Mesoamerican warriors known ever to have used these defensive sticks. Numerous scenes of heart sacrifice and the taking of prisoners at Chichén Itzá exemplify a conflict between Maya and Toltec, but the exact character, and the dates of these interactions are not clear. Probably foreign Maya, with strategic military connections to Central Mexico, ruled over much of the northern peninsula between about A.D. 800–1150, with Chichén Itzá as their capital.

The weapons are among the most interesting of the perishable artifacts postulated to date from the early phase, because they compose a suite of inter-related warrior offerings. Possibly also related to the military costume are carved flanged objects of unknown use that may have been ear ornaments – although pairs were not found (Figure 8.7). Another possibility is that these items were inserted through a hole in the lip or cheek – as was emblematic of later Aztec royalty. They represent men who wear the belts and "palmas" characteristic of ballplayers. Among the scores

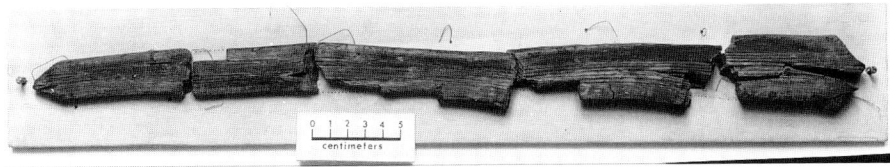

Figure 8.5. Cut wooden defensive stick, est. L. 45-47 cm. (Coggins and Ladd, 1992a, fig. 8:31).

of undistinguished cut wooden pieces found, one category may securely belong to the early period. These cut fragments were probably part of the wooden disks upon which worked metal disks were mounted and worn as pectorals (Figure 8.8). In contrast to the weapons, with some foreign associations, another extraordinary group of hand-held objects represents the local ruling Maya. These are royal scepters, later versions of the "manikin scepters" carried by Late Classic Maya rulers (Figures 8.3 and 8.9).

Contemporary with all of these were ceramic vessels made of the characteristic Yucatán slate wares, including some that had been stuccoed and painted with colorful figures and inscriptions (Ball 1992, figs. 7.22,7.23). A few fragments of stuccoed and painted gourds in the Peabody collections resemble the slate ware sherds, and may be contemporary. The Mexican archaeologists were more successful in salvaging pieces of these fragile organic vessels decorated with painted scenes. Both early and late phases included offerings made of worked bone, shell, polished stone and lithic artifacts; these were analyzed and tentatively dated in chapters devoted to them in the final volume of Cenote artifacts (Moholy-Nagy 1992; Sheets and Bathgate 1992).

In the twelfth century Chichén Itzá may have been defeated and abandoned by most of its inhabitants, but continued as a sacred place, with its sacred cenote the goal of pilgrimage. The Yucatecan capital was later moved west to Mayapán where a more purely northern Maya regional state was governed at a small facsimile of the once glorious Chichén Itzá. Evidence for this later historical period (ca. 1250–1450), and for the continued ritual use of El Cenote Sagrado, is found in a complete change in the character of the offerings. Later offerings are characteristic of middle and Late Postclassic Yucatán and of the Maya culture encountered and described by the Spaniards in the sixteenth century. At this time, numerous carved wooden figures, and ceramic tripod vessels filled with copal incense were offered to the Cenote, apparently in association with thousands of textiles, and continued human sacrifice.

The wooden figures are the rough-hewn cores of anthropomorphic "idols," as the Spanish friars called them. Originally, many of these figures had stick limbs attached to grooves on their torsos; they were coated and modeled with skin-like rubber, and with copal incense, and were painted with "Maya Blue" paint; all of these have left traces (Figure. 8.10). Late phase ritual was regularly accompanied by the use of copal incense, as it certainly had been from the beginning, but, at this time, copal, in containers, was apparently offered for its own value. The copal-filled ceramic, wooden, basketry, and cloth containers, often had jade and shell beads impressed into it. Copal was also modeled into figures, as was rubber (Figure 8.12). However, the rubber, in small balls, was most commonly used to help light the flame-resistant copal, and to produce black smoke. The second identifiable ritual practice in the late phase was the offering of textiles (J.M. Lothrop 1992). The copal, idols, and textiles are all dated to the late phase because of their repeated association with each other and with well-known late ceramics, although some might, theoretically, be much older. More than six hundred blackened cotton and vegetal fiber textile fragments survived the jaws of the dredge bucket, extraction from the muck, and transportation to Boston. At the time, they were the only known examples of ancient Mesoamerican textiles, since such organic fibers do not often survive in the wet tropical climate; these had been preserved by the anaerobic mud and alkaline waters of the Cenote. The fibers, spinning, and weaves of the textile collection were studied and classified by Joy Mahler Lothrop who also noted evidence of undetermined dyes. While simple plain weaves are commonest, there are also plain weaves with supplementary wefts or warps, and examples of tapestry, twill, crossed-warp weave, openwork with supplementary binding yarn, and borders, tassels and braids. (Figure 8.12). A few fragile fragments of sandals and basketry demonstrate that the Maya used coiling, twining, and plaiting techniques, as well as constructing complex twined sandal soles (Mefford 1992).

In the Pre-Hispanic sixteenth century cotton was a significant northern Yucatán industry, important for export and tribute, as, doubtless, it had been for centuries. A few wooden objects found in the Cenote may have been associated with textile production, like wooden battens, loom bars, and spindle whorls, of which ceramic examples were more common (Coggins and Ladd 1992a, figs. 8.100, 8.101,

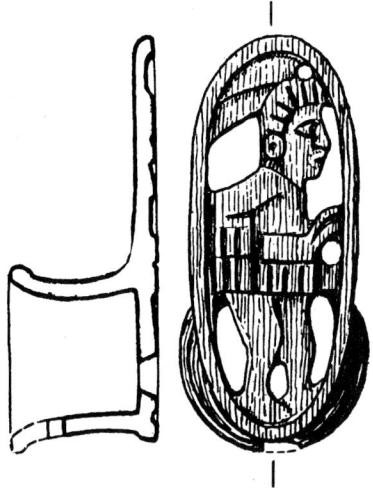

Figure 8.7. Drawing of wooden ear (?) ornament, L. 4.0 cm. (Coggins and Ladd, 1992a, fig. 8.2).

Figure 8.6. Toltec warriors, a. West wall, Lower Temple of the Jaguars, Chichén Itzá (from Tozzer, 1957, fig. 539). b. Table throne, Upper Temple of the Jaguars, (from Tozzer, 1957, fig. 574).

8.102, 8.106). These suggest that the textiles themselves may have been offerings in the late phase, as well as having been the clothing of sacrificed individuals, at every period, and possibly the wrappings for precious objects. The quantities of textiles may represent a late phase Post-Classic

cult of Ix Chel, the Maya goddess of the Moon, childbirth and weaving, who was also associated with cenotes.

A few wooden objects that survived intact are like some found by the Mexicans. These include a miniature bench, a phallus (Figure 8.8d), and jaguar tooth effigies. Much of the worked wooden objects from the Cenote are, however, pieces of cut flat wood with varying degrees of dark resinous coating. In most cases their original form, or purpose, is not identifiable, although some have perforations and might have served as pendants or their backings (Figure 8.8 a1, 2). Others, cut to a standard length of four to five cm, might indicate some specific ritual practice (Coggins and Ladd 1992a, figs. 8.130, 8.131).

Six broken sickle-like wooden tools are probably among the latest offerings to the Cenote; they resemble the common Yucatán brush-cutting tool called a *coa*, except that the hooked blade of the modern tool is steel. Some evidence of use-wear on the ancient wooden "blades," now snapped off at the end, suggests they might also have served such a function (Figure 8.13).

From the earliest jade bead, as early as the eighth century A.D, to the humble brush-cutting tool, perhaps seven centuries later, the innumerable objects found in the Cenote of Sacrifice (if not discards, or fallen in by mistake) represent ceremonial acts that embodied choice and signified cultural belief. These thousands of objects share a few traits, but differ in many more ways. All had characteristics that made them appropriate as offerings: intrinsic value, by virtue of rarity, exotic origins, or sacred character – like jade, gold, and copal; significance as emblematic of a role in society, like the weapons, textiles, or brush-cutting tools; or simply as conveyances, like ceramic vessels, and baskets. But while

Figure 8.8. Miscellaneous wood, circa 1910. Top right, L. 11.8 cm; a. Two face pendants, tapered tube, two stepped pieces; b. Three pieces cut disks, stepped piece, cut rim with perforation, c. Cut box, cut disk piece, cut length with borders, cut stepped piece, two pieces tubes, one with collar; d. Phallus effigy L. 20.2 cm. (see Coggins and Ladd 1992a; photo courtesy of the Peabody Museum).

all were offerings, they fall into two distinct offertory complexes: early ca 750–1150 A.D., and late ca 1250–1450 A.D. Virtually all of the early objects were burned or broken in some way, as part of the offering ritual. Most jade is fragmentary, cast metals are smashed and melted, thin hammered metals are torn and folded. Wooden objects were deliberately cut into pieces, seldom broken, while stone projectile points were burned. Dark resinous coatings, probably from burning, are still found on most objects.

The late phase copal and rubber were burned as part of the ceremony, and their ceramic containers were offered intact, as were the contemporary wooden idols, and it is likely the textiles were as well. The later(?) wooden tools were, however, deliberately cut and broken.

Dating of all these objects is based entirely on associations and similarities, since Thompson identified no stratigraphy in the Cenote, where the heaviest objects of jade and gold had fallen to the bottom, between the rocks, and lighter and thinner objects moved around, especially once the dredge began its work. Thus, objects were generally dated on the basis of analogous, dated artifacts from other published excavated contexts, or, occasionally, on the basis of their internal Cenote associations – as were the chalcedony point hafted in a foreshaft, the stone points with melted gold flecks, and the textiles, copal, and rubber adhering to the wooden idols.

Among the perishable artifacts, essentially none has been the subject of materials analysis as to identification, source, or dating. One hundred seventeen of the 160 copal offerings are still intact, although most once received a superficial coating of Alvar preservative. Many of the unstudied balls of copal, representing at least two kinds of gum, contain

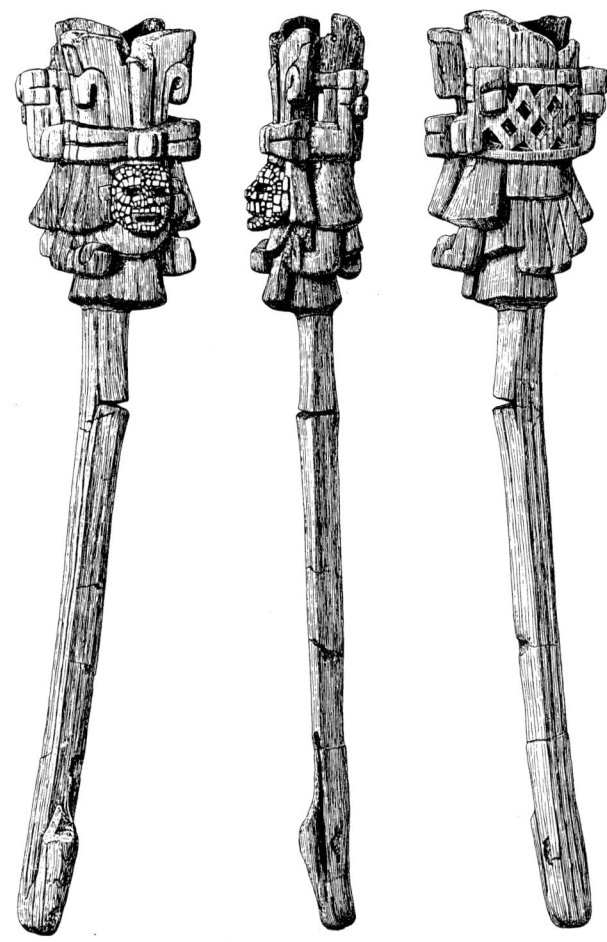

Figure 8.10.Three wooden idols with rubber and copal "skins", four rubber-copal figures, one rubber-copal ball; circa 1910. top right "idol", with textile adhering, H. 14.7 cm. (Coggins and Ladd 1992a, fig. 8.50).

Figure 8.9. Drawing of wooden manikin scepter, with mosaic-covered face. L. 39.0 cm. (Coggins and Ladd, 1992a, fig. 8.49b).

twigs and leaves, and others are known, from x-ray, to have beads inside. Most of the wooden objects were saturated with heated paraffin when they arrived in the museum, but some of the idols were not (in order to preserve their ancient coatings). This was the only treatment the collection of wood received. Little of the Cenote collection, which represents so many local and exotic proveniences, and many styles and craft traditions, has been carefully studied as to technology and workmanship. There is a lot of work to be done. New testing methods and fresh interest might rejuvenate this collection, and allow for its reevaluation in the light of ongoing Mexican excavation at Chichén Itzá, and of the attention scholars continue to focus on the abundantly documented, but still little understood, role of the Cenote of Sacrifice in Post-Classic Mesoamerica.

References Cited

Ball, Joseph W. and John M. Ladd, 1992. "Ceramics", in Coggins, 1992a, pp. 191–233.

Coggins, Clemency Chase. 1992a. Artifacts from the Cenote of Sacrifice, Chichén Itzá, Yucatán: Textiles, Basketry, Stone, Bone, Shell, Ceramics, Wood, Copan, Rubber, other Organic Materials, and Mammalian Remains. Cambridge, MA. Memoirs of the Peabody Museum of Archaeology and Ethnology, Harvard University, vol. X:3, 1992.

Coggins, Clemency Chase. 1992b. "Dredging the Cenote", in Coggins, 1992a, pp. 9–31.

Coggins, Clemency Chase and John M. Ladd. 1992a. "Wooden Artifacts", in Coggins 1992a, pp. 235–344.

Coggins, Clemency Chase and John M. Ladd. 1992b. "Copal and Rubber Offerings", in Coggins 1992a, pp. 345–357.

Folan, William J. 1968. El Cenote Sagrado de Chichén Itzá. Departamento de Monumentos Prehispanicos Informe 15. Instituto Nacional de Antropologia e Historia, Mexico.

Lothrop, J.M. 1992. "Textiles", in Coggins, 1992a, pp. 33–90.

Figure 8.13. Wooden sickle-like tools, left, L. 14.0 cm (Coggins and Ladd 1992a, fig. 8.109).

Figure 8.11. Copal-filled ceramic deer effigy vessel with rubber-skinned copal figure; five modeled copal offerings, circa 1910; lower right, H. 14.5 cm. (Coggins and Ladd 1992b, fig. 9.6).

Lothrop, Samuel K. 1952. Metals from the Cenote of Sacrifice, Chichén Itzá, Yucatán. Cambridge, MA. Memoirs of the Peabody Museum, vol. X:1.

Mefford, Jill J. 1992. "Basketry, Twined Sandal Soles, and Cordage" in Coggins, 1992a, pp. 91–97.

Moholy Nagy, Hattula and John M. Ladd. 1992. "Objects of Stone, Shell, and bone", in Coggins, 1992a, pp. 99–152.

Moholy Nagy, Hattula, Clemency Chase Coggins, and John M. Ladd. 1992. "Miscellaneous: Palm Nut Artifacts, Decorated gourds, Leather, and Stucco". In Coggins, 1992a, pp. 359–368.

Piña Chán, Román. 1968. "Exploración del Cenote Sagrado de Chichén Itzá: 1967–68", Boletin, 32:1–3, Instituto Nacional de Antropologia e Historia, Mexico.

Sheets, Payson D., John M. Ladd, and David Bathgate. 1992. "Chipped-stone artifacts", in Coggins, 1992a, pp. 153–189.

Proskouriakoff, Tatiana. 1974. Jades from the Cenote of Sacrifice, Chichén Itzá, Yucatán. Cambridge, MA. Memoirs of the Peabody Museum, Harvard University, vol. X:2.

Thompson, Edward H. 1992. "The Sacred Well of the Itzaes", in Coggins, 1992a, pp. 1–8.

Tozzer, Alfred M. 1941. Landa's Relación de las Cosas de Yucatán: A Translation. Cambridge, MA.

Tozzer, Alfred M. 1957. Papers of the Peabody Museum 18. Harvard University, Cambridge, MA

Tozzer, Alfred M. 1957. Chichén Itzá and its Cenote of Sacrifice. Memoirs of the Peabody Museum, vols. XI, XII.

Willard, T.A. The city of the Sacred Well, New York. The Century Co, 1926.

Figure 8.12. Cotton textile fragment; Inlaid supplementary weft, max. W. 11.5 cm. (J.M. Lothrop, fig. 3.13).

9. Birth to Death: Northwest Coast Wet Site Basketry and Cordage Artifacts Reflecting a Person's Life Cycle

Dale R. Croes

Introduction

On the Northwest Coast of North America we have been actively recovering perishable artifacts from wet sites over the past thirty plus years, revealing many new prehistoric cultural dimensions concerning thousands of years of cultural evolution (Figure 9.1). Early attempts to use the sensitivity of basketry and cordage had some success in identifying the individual weavers in prehistoric houses at the Ozette Village wet site (Croes and Davis 1977). Though an interesting approach to the new wood and fiber wet sites data base, a more culturally revealing tack may be to use this sensitive new artifact data (and especially basketry and cordage here) to track how these perishable artifacts might reflect a person's life-cycle, from one's birth and placement in a basketry cradle to a person's social position and occupation to one's death and shroud of basketry mats at the grave location. This represents a fairly experimental approach, so I cannot guarantee a completely successful data transformation, but this approach may reveal some new ways to view perishable artifacts becoming more abundant from along the Northwest Coast (Figure 9.1).

In this effort I will first delineate the basketry and cordage data to be presented in association with a person's life-cycle sequence as follows: (1) cradles, baskets specifically constructed to contain humans at birth, (2) hats, worn as indicators of a person's developing social status/role, (3) specific basketry and cordage items to assist in specialized occupational roles, such as whaling harpoon baskets, harpoon sheaths, harpoon ropes, and fishing tackle bags, and (4) woven mats used as shrouds following death. Though the specific individual is important in prehistoric research, they also change through their own life-cycle and this can be reflected by the basketry/cordage items associated with different parts of the life-cycle from Northwest Coast wet sites over the past several thousand years.

Cradles and other Related Infant Basketry Items

Three Northwest Coast wet sites have cradles reported: the Ozette Village wet site on the Central West Coast and the Conway and Fishtown wet sites on the Skagit Delta in central Puget Sound (Figure 9.1). As a container used to first hold an infant, these become important basketry items involved in a person's life soon after birth. According to several Northwest Coast traditions, and specifically in Northern Tlingit tradition, the cradle reflects the person's container in the return to life and, at the other end of the life-cycle, the person's funeral container is often similar to the cradle form (cf. Kan 1989). For the Northern Tsimshians, Halpin and Seguin point out the association between birth and death in considering reincarnation:

> That the same terms are used to refer both to a cradle and a grave-box (*w'o*) and that a person's baby song and mourning dirge (*li.mk'o'i*) are the same song is evidence in point (1990:279).

In another reference to the importance of cradles to the initiation of a person's life, Kew (1990:477) states that the Central and Southern Coast Salish, in reference to Spirit Dance costumes, follow these practices: "At the end of initiation the staffs and costumes are deposited in trees or caves apart from ordinary human activities and left, like cradles and baby rockers, to decay" (from Barnett 1955:281). In relating the cradle to the burial canoe, Hajda (1990:507) indicates that among the Southwestern Coast Salish the "importance of canoes is reflected in their use for burials and the shaping of cradle-boards to resemble canoes" (from Swan 1857:167–168). Swan in characterizing the cradle for Coastal Washington indicates:

> A cradle, like a bread trough, is hollowed out from a piece of cedar, and, according to the taste of the parent, is either fancifully carved, or is as simple in its artistic

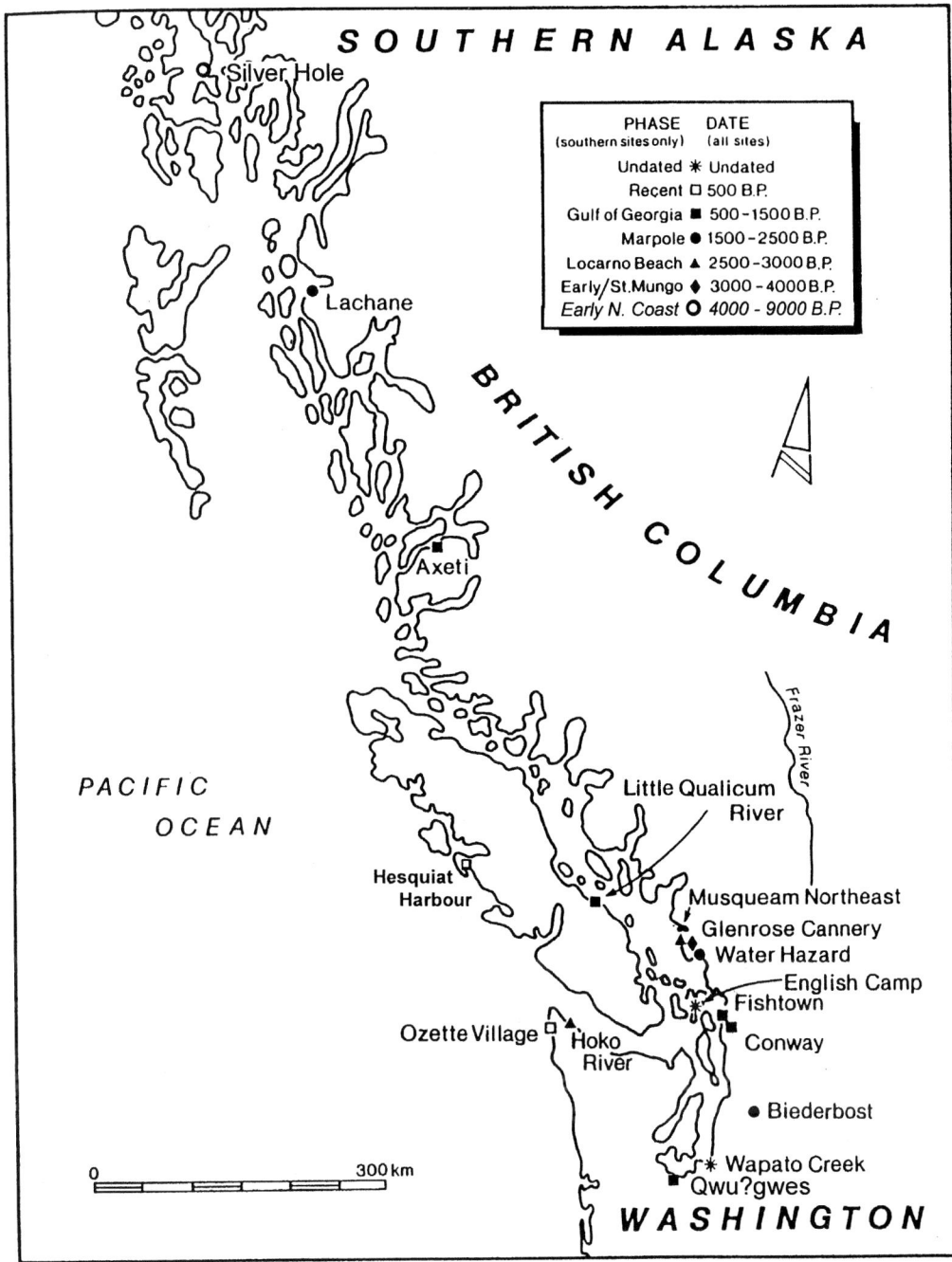

Figure 9.1. The distribution and approximate dates of investigated wet sites on the Northwest Coast of North America; also location of the Hesquait Harbour dry burial cave sites (modified from base map by Mike Rouillard).

appearance as a pig's trough. This cradle, or *canim*, or canoe, as they term it, is lined inside with the softest of cedar bark....(1857:167).

Prehistoric wet site cradles illustrate distinctive styles, with the proto-historic Ozette examples being very similar to ethnographic Makah/Nuu-chah-nulth and the central Puget Sound examples from two sites on the Skagit Delta of a very different style, though identical between the two sites.

Cradles from the Ozette Village Wet Site

Cradles from this site were constructed with wooden cedar slats interwoven into the base weave. The base was formed with very coarse gauge checker plaiting with strips averaging 3 to 4 cm wide (Figure 9.2). These were split further to constitute the body warps. The wooden slats, numbering from four to five when present, were partially smoothed before being woven into the cradle base and were tapered to produce the finished shape (see Figure 9.2). The cradle rim is finished with the hitched rim technique (Croes

1977:127). Braid loops frequently were placed along the inside edge of the cradle to aid in tying in the child (Figure 9.2; Croes 1977:111; Drucker 1951:122).

Twelve cradles of this type were reported recovered from in and around the House I area (Croes 1977:381–386). Of these, only one appears to have been in a usable condition, the others being fragmentary, discarded examples, usually from the outside refuse middens. The single usable cradle (66/VI/2) was in storage on or behind a bench platform in the northwest corner of the house. This cradle was amply padded inside with a layer of moss on the bottom covered by pads of shredded cedar bark, with worn cedar bark flat bags and thin bundles of bark on top of this. These layers of materials provided protection and comfort for the infant as described in ethnographic data.

In the 1870s James Swan describes in detail a Makah cradle and its use (I found a model cradle from his 1867 collection at the U.S. National Museum, Smithsonian Institute (Acc. #5366) and it may be the model for the following description):

> As soon as a child is born it is washed with warm urine, and then smeared with whale oil and placed in a cradle made of bark, woven basket fashion…. Into the cradle a quantity of finely separated cedar bark of the softest texture is first thrown. At the foot is a board raised at an angle of about 25 degrees, which serves to keep the child's feet elevated; or, when the cradle is raised to allow the child to nurse, to form a support for the body, or sort of a seat. This is also covered with bark, he-se'-yu. A pillow is formed of the same material, just high enough to keep the head in its natural position, with the spinal column neither elevated nor depressed. First the child is laid on its back, its legs properly extended, its arms put close to its sides, and a covering either of bark or cloth laid over it; and then, commencing at its feet, the whole body is firmly laced up so that it has no chance to move in the least. When the body is well secured a padding of he-se'-yu is placed on the child's forehead, over which is laid bark of a somewhat stiffer texture, and the head is firmly lashed down to the sides of the cradle; thus the infant remains, seldom taken out more than once a day while it is very young, and then only to wash it and dry its bedding…. The same style of cradle appears to be used whether it is intended to compress the skull or not…" (1870:18–19).

The shredded cedar bark padding Swan describes is found in the complete Ozette cradle, as were the braid loops utilized to bind in the child.

Drucker, in his Northwest Coast culture element distribution list, indicates that cedar bark cradles with "4-slat reinforced bottoms" only occurred among Nootkan groups.

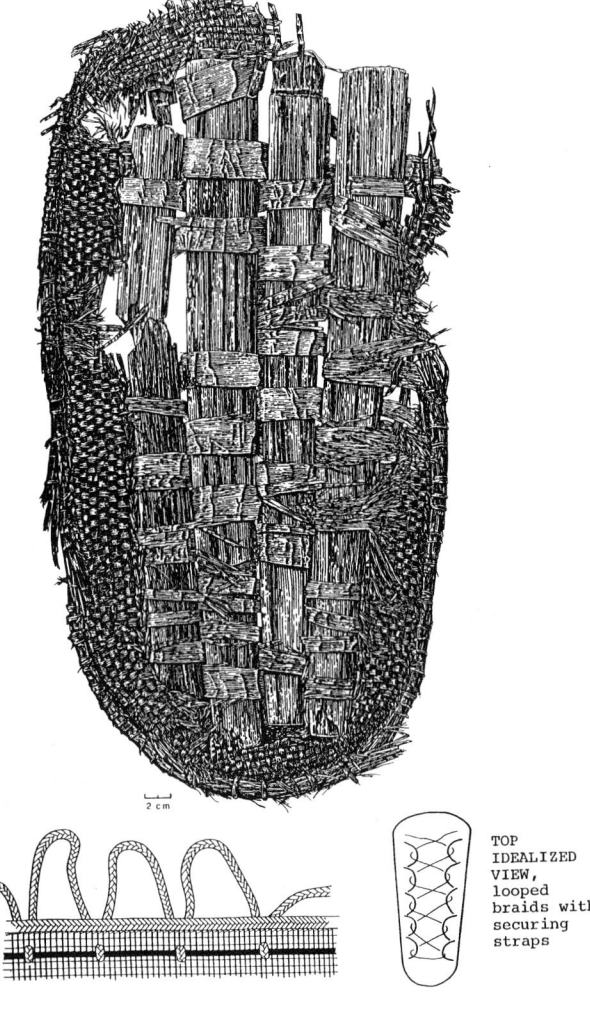

TOP IDEALIZED VIEW, looped braids with securing straps

Figure 9.2. Ozette Village wet site cedar bark cradle (163/V/ 62). Note tapered cedar wood slats interwoven to form support for the base. Missing from this example, but seen on others, is the looped braid lines along the cradle edge; used to guide additional straps across the top of the cradle to secure child inside (Croes 1977:381–386, 111; illustrations by Chris Walsh).

He described such a cradle as "an ovoid checkerwork basket, with four longitudinal splints substituted for warps in the bottom for rigidity, and reinforced rim" (1950:206, 274). This type of cradle was a Makah/Nuu-chah-nulth type at least as early as the Ozette time period.

Cradles from Conway and Fishtown Wet Sites

Both these sites are located on the Skagit Delta in central Puget Sound (Figure 9.1) and date between approximately 500–1000 years B.P. One cradle is reported from each of these wet sites, and differ considerably from the contemporary Ozette Village wet site cradles approximately 150 miles west up the Straits of Juan de Fuca. These Puget Sound cradles are essentially identical in type definition, being constructed of splint cedar limbs in a cradle trapezoid shape with low sides, and a base and body construction in a twill 2/2 weave. Both have a unique top-hitched rim construction (see Croes 1977:130 for rim definition; 223–225 for complete definitions). Therefore a very distinctive style distinguishes these prehistoric Salishan area cradles from the historic and prehistoric Makah area cradles. The Conway and Fishtown cradles were not found with padding or any indication they were in use, and may have been discarded in these site areas after the infants no longer needed their cradles (Munsell 1976b, Onat 1976).

In both the West Coast and Puget Sound areas these may have been the cradles immediately following birth and the child may have been transferred to cradle-boards or dugout cradles once it reached about 10 days to a year in age and given its first name (Codere 1990:368; Arima and Dewhirst 1990:405; Suttles 1990:465).

Plaited, Cedar Bark, Two-edged "Infant Face Covers" from the Ozette Village Wet Site

Another basketry garment associated with a person's infancy is a distinct kind of face cover (hood) ethnographically used by Nuu-chah-nulth peoples and some Salishan peoples. The only prehistoric wet site with examples of these is Ozette Village where five are reported (Croes 1977:302–308). These basketry items have two continuous opened and finished edges. They apparently were opened by spreading apart these edges, creating a conical looking shape, which was inverted over a baby's face in the cradle to keep out light, dust, smoke, and noise. One is constructed with a twill on bias weave and has one edge finished with a turned in rim (Figure 9.3); the other four reported are checker plaited and had one edge finished with a twined weave and the other with the around and back edge common to Ozette Village mats (see below). One example is extra-small in size and is probably a miniature (toy?), since it is too small to have been functional. It may

have been used with one of the miniature (toy) cradles recovered at the site. Only two of the five reported appeared to be in a usable condition and were kept with other stored items in the household. The remaining three were broken and located either within the house floor midden matrix or the outside refuse midden.

In the ethnographic literature two specific references are made to these infant face covers. Reverend Eells wrote: "A cap of cedar bark, usually of Makah make, is sometimes used by the Klallams as a cover to protect the babies from smoke" (1887:656). Significantly he characterizes these "caps" as a Makah product and identifies the function. Boas, when discussing Kwakiutl cradles describes and illustrates an infant face cover. He wrote: "A small hood made of cedar-bark matting is placed over the head of the child as a protection against light and insects" (1909:460, Fig. 135). A least two Makah Elders, Isabell Ides and Nora Barker, remembered these face covers. Nora owned a well-made miniature Makah cradle with a miniature infant face cover over the doll's face. Mrs. Ides and Mrs. Barker agree they were to keep out light, dust, and noise. Also a miniature cradle collected by James Swan in the 1860s (Smithsonian USNM #5366) mentioned above has a two-edged infant face cover tied onto the head position of the cradle. An excellent early photograph by S.G. Morse of Port Angeles (photo #48) illustrates women with her baby (identified as Lyda Colfax) in a suspended cradle next to which is an opened two-edged infant face cover (Croes 1977: Figure 47:308). The above information clearly indicates the specialized function of this unique basketry form, and I would predict that these would be a cultural style most associated with Wakashan area wet sites in the past millennium.

Hats, Worn as Indicators of a Person's Social Status/Role

After about 10 days to a year many Central Northwest Coast infants were given their first family name (Codere 1990:368; Arima and Dewhirst 1990:405; Suttles 1990:465) and began the process of gaining future names and position within the family. In some cases an infant is transferred from a basketry cradle to a dugout cradle at this time. The first naming further recognizes the individual as a true person in the hereditary ranking system of the community. A well-known indicator of personal status was in the clothing one wore and especially one's basketry hat on the Northwest Coast. Therefore looking at styles of prehistoric wet site hats reveals a person's potential status marker in the overall social system of the community.

Northwest Coast researchers often have tried to distinguish by means of the archaeological record when complex social stratification emerges on the prehistoric

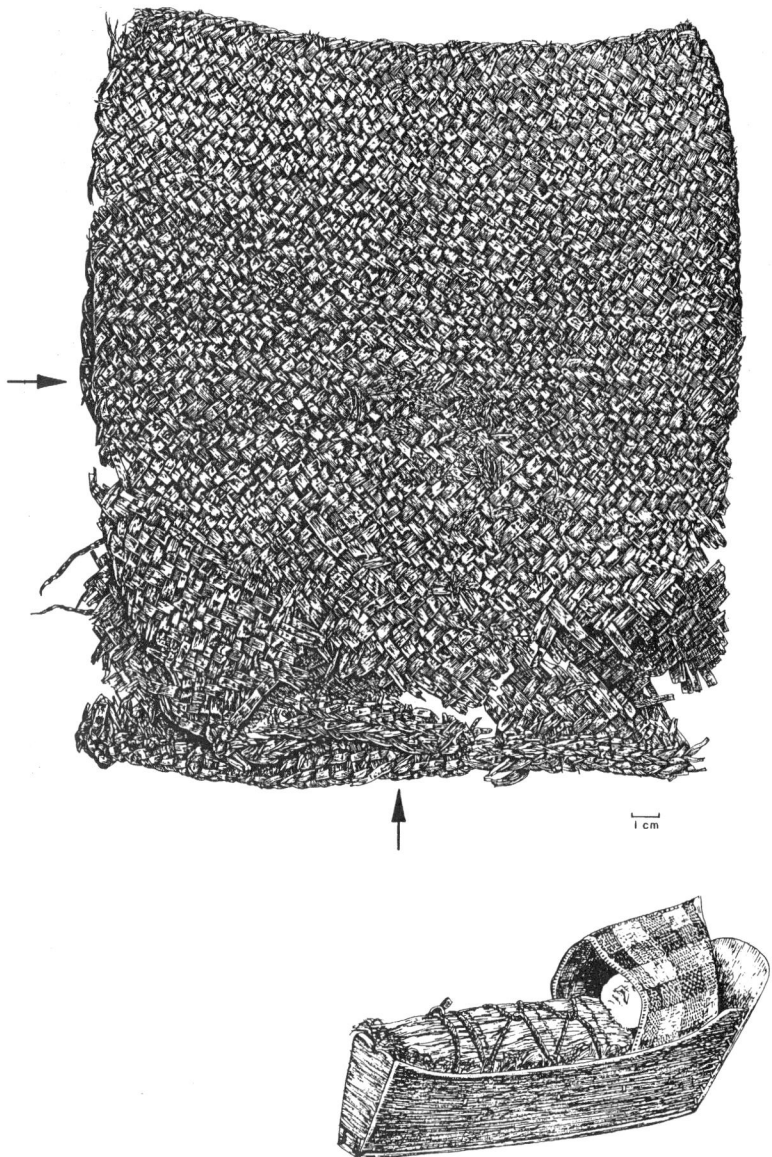

Figure 9.3. Infant-face-cover with two opened edges (open edges shown by arrows) (111/EVC/92). This example was woven with the elaborate twill on bias weave (Croes 1977:302–308; illustration by Chris Walsh). Inset illustration of infant-face-cover in use by Hilary Stewart (1984:133).

Northwest Coast. They often looked to the occurrence of head shaping or the use of labrets, as possible reflections of the emergence of differences in status. Researchers also have analyzed grave goods to see when differentials in wealth occur (Ames and Maschner 1999; Cybulski 1996). I would add that the fairly common recovery of perishable clothing, and especially hats from wet sites, may be another sensitive indicator of complex social stratification and the evolution of its emergence.

Ethnohistoric Northwest Coast social stratification included relatively well-defined (through inheritance) nobles, commoners, and slaves. Nobles appear to have emerged as a class of household and territory owners and managers.

Into the historic period, Northwest Coast nobility, as with chiefdoms elsewhere, included individuals who inherited their rank. One of the main West or Central Northwest Coast indicators of nobility was a distinctive knob-topped conical hat, worn by both noble women and

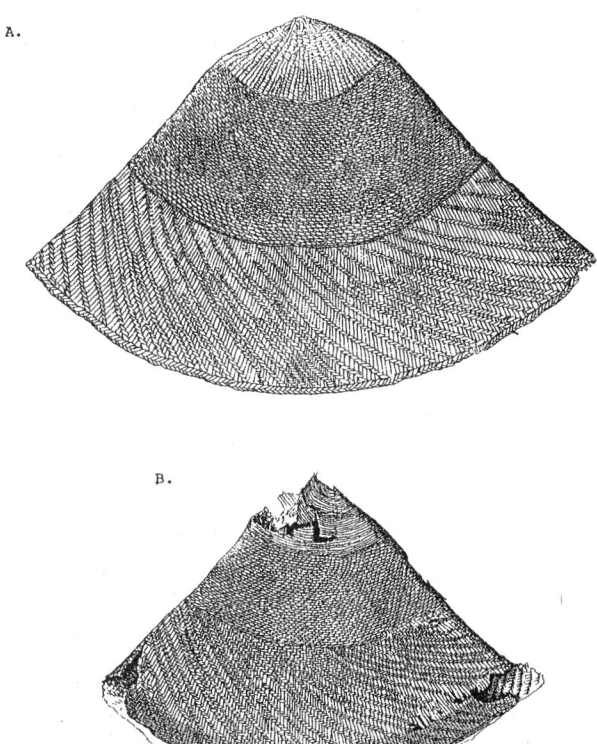

A.

B.

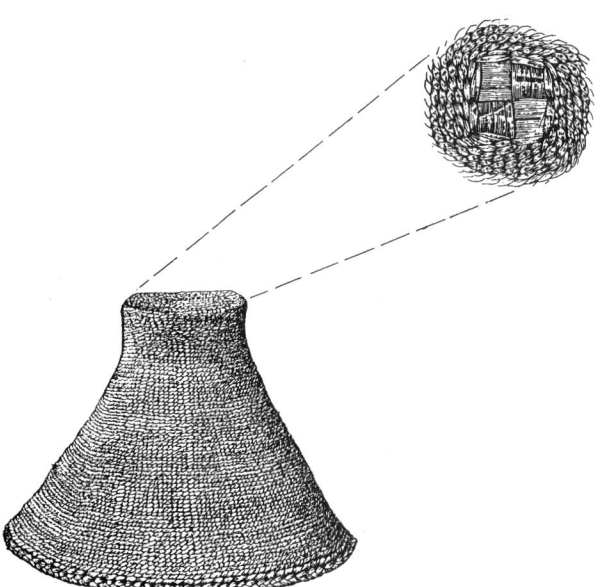

Figure 9.4. Ozette flat-top, plain twined, cedar bark hat (30/ IV/44; OH2; Croes 1977:409–417). Note checker weave initiating construction of the hat. (Illustration by Chris Walsh).

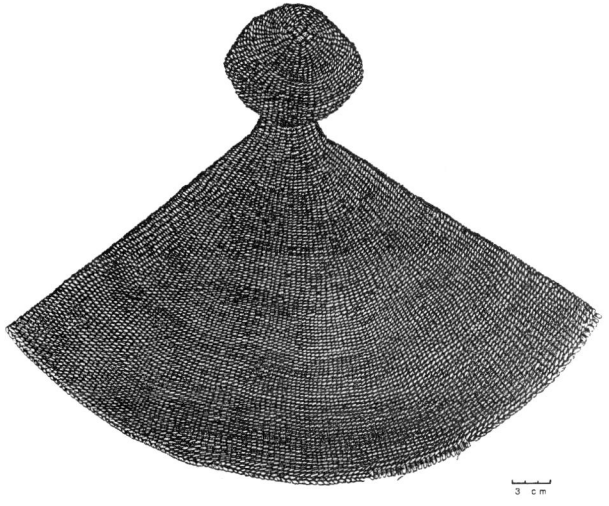

Figure 9.6. Ozette rounded-top conical hats with woven bands of, from top to bottom, plain twining, diagonal twining, and "skip-stitch" twining (see Figure 14) (OH7s; A: 160/tV/1, B: 110/IV/3; Croes 1977:427–435). Example B had been painted with a red band around the brim (stippled) and black paint on the upper body. (Illustrations by Chris Walsh).

Figure 9.5. Ozette round-knob top conical hat (FS/92; OH3; Croes 1977:417–427). Note checker weave at top, addition of new warp elements creating an expanding conical shape, and the braid-like turned in brim construction. (Illustration by Madge Gleeson).

distinct sets: (1) a plain twined, cedar bark, truncated (flat-top) conical hat with cedar bark inner layer and headband (Figure 9.4); (2) a plain twined, cedar bark, knob-top conical hat with cedar bark inner layer and headband (Figure 9.5); and (3) a complex twined, spruce root, rounded-top conical hat with an attached cedar bark headband (Figure 9.6) (Croes 1977). From ethnohistoric records, it is clear that the flat-topped hats were worn by commoners, the knob-topped hat wearers were of noble lineage, and the complex twined spruce root hats were of a northern "foreign" style (possibly Tlingit or Haida in origins) and may have also indicated the wearer's means, possibly worn mainly by noble women as illustrated ethnohistorically.

Hats from six other Northwest Coast wet sites have been recovered. These examples are not from within households, as at Ozette Village, but along with discarded or lost artifacts that had become waterlogged and settled in waterways adjacent to areas of activity. Though often

noble men (see Mozino 1970: Plate 5). Commoners wore a flat-topped conical hat (see Mozino 1970: Plate 8).

At the proto-historic Ozette Village wet site, three distinct hat types were recorded and considered functionally fragmentary, most are complete enough for reconstruction. And since they come from site dating back as early as 3,000 years ago, they become important in determining the longevity of the individual's status ranking over several millennia. I will begin with the earliest known hats from 3,000 years ago, then look at later ones throughout the coast.

Hoko River 3,000 year old basketry hat styles.

A possible indicator of early social stratification on the Northwest Coast might be the long-term existence of knob-topped conical hats as distinguished in prehistoric wet sites. At the 3,000 year old Hoko River wet site, distinct knob-top hats along with rounded-top forms were recorded (Figure 9.7; Croes 1995). These different hats probably do function as indicators of early status marking, though possibly to a different degree than in the later periods (Croes 1995). If the hat forms did reflect some degree of status differentiation, we can project that control of territories and their management by extended-family leaders may have begun to occur by 3,000 years ago. From indications derived from computer simulations of economic decision-making through time on the West Coast, the emphasis on and required management of procurement and storage activities, especially drying of fish, was established by 3000 years ago at Hoko River (Croes and Hackenberger 1988).

Undated Puget Sound Basketry Hats

A single hat was found in the wet site portion of the English Camp site (45SJ24) on the San Juan Island that has a rounded conical shape, and is twined with cedar bark (Sprague 1976:78–85). The single basket found associated with this hat is of a style typical of 2,000 year old wet sites in Puget Sound (Biederbost) and Gulf of Georgia (Water Hazard) areas (Croes 1995:131). Another hat was found near Tacoma at the Wapato Creek Fish Weir site (45PI47) which also is twined with cedar bark and constructed in a conical shape with a pointed crown (Munsell 1976:46–57). This hat may well have had a knob-top construction, where the knob is broken off. It was similar to the Ozette hats in having a double layer of weave and a head-band woven inside.

Axeti Basketry Hats from a Northerly Nuxalk (Bella Coola) wet site

Two hats were recovered from the Axeti wet site near this Nuxalk (Bella Coola) village at the mouth of the Kwatna River (Hobler 1976). Both hats are made with cedar bark with a rounded conical shape. The main difference and distinguishing feature of these northerly hats is their

complex combinations of body weaves. One hat is made with "skip-stitch" twining and a lean of the twine up-to-the-left very similar to the "foreign" spruce root hats believed to be introduced from the north to the Ozette Village. The second hat also has a unique combination of plain and diagonal twining and checker and twill 2/2 plaiting in the body weave. Both appear to reflect styles of the North Coast (Croes 1977:239–243).

Therefore wet site hats can be used to reflect both the individual's status and either their cultural area or where the hat may have been derived by these individuals through trade or contact.

Specific Basketry and Cordage Items to Assist in Specialized Occupational Roles

Basketry and cordage can also be used to reflect a person's specialized or gender related occupations. From Northwest Coast wet and some burial cave sites we have examples of these basketry items reflecting the person's occupation, and, as examples, I will particularly look at (A) whale harpoon bags, (B) harpoon sheaths, (C) harpoon line ropes, and (D) fishing tackle bags from Northwest Coast sites. Though the basketry mentioned above was probably constructed mainly by women, many of them were probably used mostly by men in the division of labor.

Whale Harpoon Bags

An indicator of a specialized occupation of the higher ranking males and owners of the household of the Makah/Nuu-chah-nulth would be the special whale harpoon bags, as good examples of basketry that reflects the occupation of high ranking Nuu-chah-nulth whalers.

Seven large whale harpoon bags are reported from the Ozette Village wet site (Croes 1977:295–302), with five badly broken and discarded in the outside refuse midden, and two in-use and lying one directly on top of the other inside the house. These bags are the largest baskets found at the Ozette Village wet site and share these technological features: cedar bark construction materials, flat trapezoid shape, base constructed with one row of plain twining, body weave of two distinct sections of plaiting: (a) the bottom one-third to one-half woven in a coarse-gauge checker weave, and (b) the upper section woven with either a medium-gauge checker weave, or a complex checker II plaid weave (see Figure 9.8). The transition between the upper and lower section is composed of one or two rows of plain twining. The rim is constructed with an open braid (the only basket with this rim at Ozette; see Croes 1977:128 for complete rim type definition).

The two bags that were in-use each contained whaling harpoon heads in harpoon sheaths and remnants of the

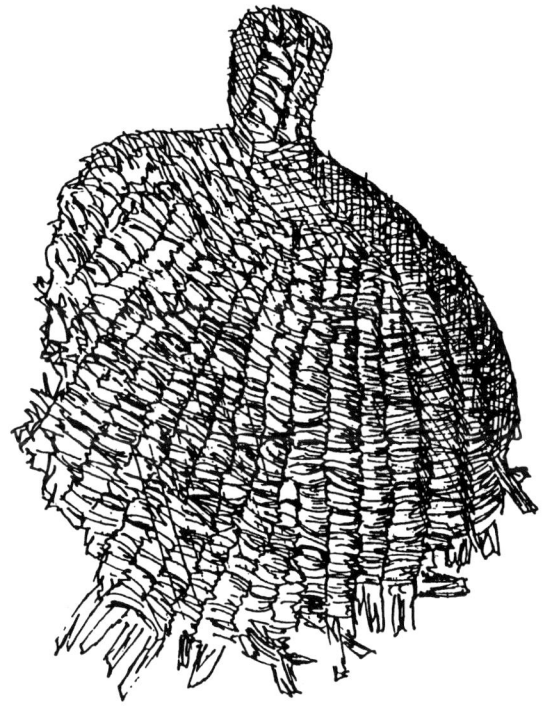

Figure 9.7. Photograph and illustration of plain-twined, knob topped hats from the 3,000 year old Hoko River site (cm. Scale; Croes 1995:134–137). (Illustration by Robin Pedersen).

harpoon lanyards. The specific contents of each bag included:

Basket 71/IV/32 (Figure 9.8): two harpoon sheaths containing harpoon heads; two separate harpoon valve halves, one plug for a seal skin float.

Basket 71/IV/33: three harpoon sheaths containing harpoons; one composite tool: a wooden shaft or rod handle with a hafted iron blade point (possibly the foreshaft for the whale-killing lance or an instrument for slitting the dead whale's jaw for tying the mouth shut).

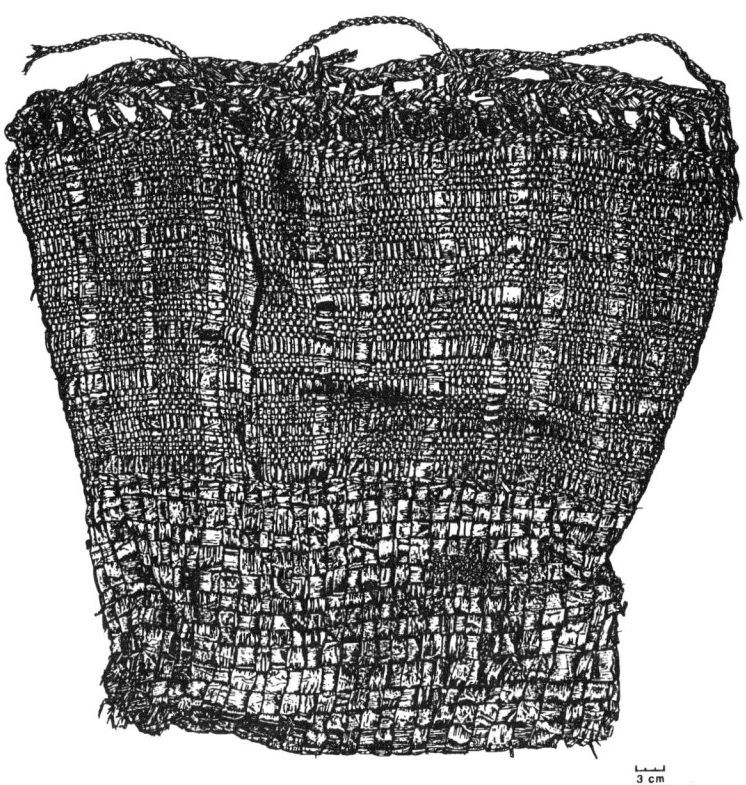

3 cm

Figure 9.8. Extra-large, plaited, whale harpoon bag with an open braid rim construction (71/IV//32). Note two sections of weave; lower coarse checker and upper checker II plaid weave. This bag was found directly on top and attached with a cordage line to another whale harpoon bag. Both contained harpoon heads in folded cedar bark sheaths (see Figures 9.11 and 9.12) (Croes 1977:295–302; illustration by Chris Walsh).

These two bags appear to have been hanging along the wall and to have fallen, scattering some of their contents. Immediately outside the mouth areas of these bags were found: four harpoon sheaths containing harpoons, one pair of bone harpoon toggle valves, killing lance heads (two unbarbed and bound wooden harpoon valves with mussel shell blades), a very large quantity of coiled cherry bark strips, once wrapping the sinew lanyards (the sinew in these harpoon lanyards has decayed leaving only the cherry bark wrapping element extending out of the harpoon bags), and a knob-top hat (71/V/18; probably outside of the bag).

These contents clearly identify the special occupational use of these baskets as the whaler's bags to store and protect the numerous whale harpoon points in their individual sheaths.

Unfortunately little has been mentioned ethnographically about whale harpoon bags and their use in the specialized occupation of whale hunters. Waterman, in *The Whaling Equipment of the Makah Indians*, does illustrate a whaling harpoon bag (Plate 5:61) and states:

Several harpoon heads are taken along on each trip, each one enclosed in a separate sheath. The collection of heads is kept in a special basket, called ha'3aL (Plate 5). This bag or basket is of checkerwork, and has a flaring top (1920:32–33).

Early photographs show these bags as part of the whaler's equipment, and all Neah Bay Elders recognize these bags as whaler's harpoon bags. Moreover several museum collections possess these bags, and usually they are described as whale harpoon baskets.

Harpoon Sheaths

Thirty-eight harpoon sheaths were reported at the Ozette Village wet site (Croes 1977:439–446). These specialized "mats" are constructed with either a single layer or a folded double layer of cedar bark (Figure 9.9). Folded in the middle, these prepared strips of bark secured, protected, and separated the whale harpoon points. The ends of the strips were split (frayed) so that they could be bound

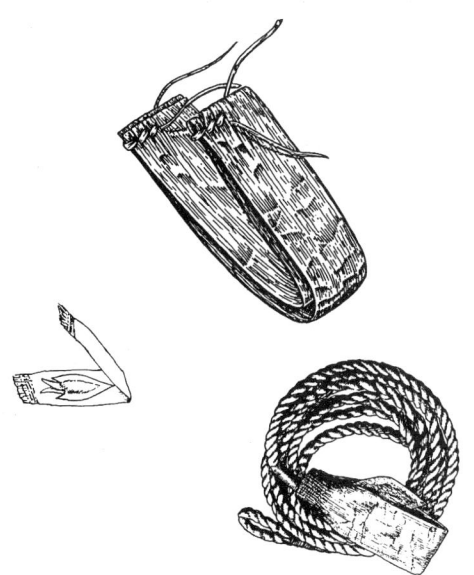

Figure 9.9. Construction of Ozette Village wet site harpoon sheaths. Note double layer of bark and twined ends of sheath (Croes 1977:439–446; Illustration by Chris Walsh). Inset illustration of harpoon sheath in use by Hilary Stewart (1984:121).

together with from two to six rows of plain twining (Figure 9.9), with one example having eight rows. This contrast with most ethnographic examples that typically have eight to sixteen plus rows of plain twining (cf. Waterman 1920, Plate 8). At Ozette there are two sizes of sheaths: a smaller and larger size (Croes 1977:442). The smaller size presumably would have contained smaller harpoons, and the larger the whaling points.

Of the thirty-eight reported harpoon sheaths, only ten appear to be in-use in the Ozette houses, with eight of those containing points. Each of these protected a single, bivalve whale harpoon point placed with the point towards the inner fold. Remnants of the harpoon lanyard – the coils of cherry bark binding – protrude from the ends of the sheaths and were once attached to the points. Two-ply cedar bark strings are bound several times around the end of the folded sheath to secure the point and lanyard. These sheaths holding points were kept together as the hunter's kit in the whale harpoon bags.

Ethnographically Waterman, discussing the whaler's equipment of the Makah, describes harpoon sheaths as follows:

> The harpoon head is kept in a sheath made of a rectangular piece of cedar bark. This is folded over across its middle. The ends are frayed out for several inches, and fine shreds of cedar-bark are worked across in plain twining (called tc3lba'tyn, cf. Tci'bat = canoe mat). Several harpoon heads are taken along on each

trip, each one enclosed in a separate sheath. The collection of heads is kept in a special basket, called ha'3aL (Plate 5) (1920:32–33).

Ethnographic evidence suggests that the Central and Northern Nuu-chah-nulth did not use these specialized whale harpoon bags or sheaths (Drucker 1951:30–31), and therefore these specialized whaler's equipment can be considered a cultural marker of the Ozette-Makah whale hunters for at least 300 years.

Harpoon Line Ropes

Harpoon line ropes usually are distinct enough in size and construction that this can be suggested as an example of cordage from a Northwest Coast wet site involved with a person's specialized occupation. Though certainly these heavy gauge ropes could and probably were used for a number of purposes, as is the case with most ropes, the need for this caliber of rope as harpoon lines for larger sea mammals, including whales, makes these larger ropes best suited for this sea mammal hunting occupation.

My main definition of heavy gauge harpoon line would be tightly twisted 3-plus-strand, cedar bough rope gauge examples (over 1.5cm in diameter) (Croes 1980:83–93, 1993). Again, while they could be used for a number of tasks, when it comes to making and using sea mammal harpoon lines, these would suit the occupation best.

Ethnographically, well-constructed ropes are often described as harpoon lines for seal and whale hunting. As a good prehistoric wet site example of a harpoon line in the Ozette Village House 1, one such rope (30/IV/43) was found in the northeast wall area attached with an overhand knot to a large wooden seal skin float plug. This rope is approximately 3.7 m long (12 feet), broken at the end of the knot, and wrap whipped at the other end of the rope. Directly associated with the end of this rope was another length of heavy gauge rope (38/IV/27) which has a small looped end, possibly for attachment to the end of the other rope (see Figure 9.10) (Croes 1980:83–93)).

Ethnographically, as mentioned before, these multi-strand ropes frequently were used as harpoon lines for seal and whale hunting. In the mid-1800s Sproat observed that the Alberni Nuu-chah-nulth whalers used a main harpoon line, "which is generally made of cedar twigs laid together as thick as a 3" rope" (1868:227–228). And Reverend Eells, around the same time period, noted that the Twana, Chemakum, and Klallam Indians made harpoon ropes from cedar withes. He wrote:

> The largest ropes I have seen made by these Indians are of cedar twigs twisted in much the same style as our hemp ropes, but they look coarser. These are very strong and lasting. The largest are made on the buoys

Figure 9.10. Three-strand, cedar bough rope with a running noose tied on its end (176/IV/10; Croes 1980:83–93; Illustration by Chris Walsh). Inset illustration of a coil of 3-strand, cedar bough rope by Hilary Stewart (1984:164).

employed in catching seal, and are ¾ of an inch in diameter (1887:636).

Other detailed accounts of cedar withe multi-strand cordage used as main harpoon lines include those by Swan (1869:21), Boas (1909:380), Curtis (1913:16; 1916:16–17), and Gunther (1927:220). These well-formed cedar with ropes were also the main lines used to tow the killed whale to shore for butchering. As Gunther points out:

> The limbs of cedar tree are stripped of their heavy leaves, soaked in water, and twisted into rope. The heavier grades are used by the whale-hunting tribes like the Quinault, Quileute, and Makah for towing home dead whales. It has remarkable strength (1945:20).

The Author had an opportunity to go with Quileute elder

Bill Penn (91 years old at the time) to gather cedar boughs used to make whale harpoon lines. He preferred "swamp" cedars that had long curved limbs near their base. We cut these ¼" to ½" diameter limbs off and he explained that, once the bark and branches were stripped off, the limbs were run into bull-kelp tubes and these were placed over highly heated stones to steam and softened. Once steamed they could be more easily twisted into withes that then were plied into 2 to 3+ strand whale harpoon ropes by strong individuals.

The well formed multi-strand cedar with cordage from the Ozette Village wet site undoubtedly was used as whaler's harpoon lines for sea mammal hunting, and to bring whales into shore.

When looking at potential sea mammal hunting with harpoon lines from Northwest Coast wet sites that have a cordage sample size of over 20 examples, those with the most emphasis on the use of multi-strand cedar bough cordage are Ozette Village, Hoko River, Musqueam Northeast, Water Hazard, Biederbost and Little Qualicum River. In fact these were the only sites with occurrences of 3- and 3+ strand, cedar bough cordage lines. The 3-strand, cedar bough cord and rope-making technology evidently was well developed at least for the last 3000 years in this Central Coast region of the Northwest Coast (see Figure 9.1). The approximately 3,000 year old Musqueam Northeast site in particular has the highest percent frequency of well formed, three-strand, cedar bough cordage (Table 9.1), followed next in frequency by the other early Gulf of Georgia/Puget Sound sites of Water Hazard and Biederbost (approximately 2,000 years old).

Table 9.1. Number of 3+ strand, cedar bough cords and ropes from Northwest Coast wet sites and their total number and percentage of combined examples of 2 and 3+ strand, cedar bough cords and ropes from these sites.

Wet Site	Number	Percentage of total number of 2 and 3+ strand ropes
1. Ozette Village (Croes 1980)	105	30
2. Hoko River (Croes 1995)	19	24
3. Musqueam Northeast (Croes 1975)	28	70
4. Water Hazard (Bernick 1989)	42	45
5. Biederbost (Croes 1980)	8	38

These well formed, 3+ strand cords and ropes at these early sites are of general interest since ethnographically this class and "grade" of cordage generally is considered a technology of Central Coast groups on the Northwest Coast (Wakashan/Salishan).

Boas, when discussing the Kwakiutl 3-strand cedar bough cordage states:

> Much better rope of the same sort is made by the Nootka and Quilleyut, who use it extensively for whaling-lines (1909:380).

The ethnographic distribution of 3+ strand, cedar bough cord and rope-making technologies is important here since the pattern emerging tends to indicate that this specific rope-making technology is attributed to Central Coast sea mammal hunting occupations in this region.

In considering this well-developed and strongly emphasized technology at the early Musqueam Northeast, Water Hazard and Biederbost Puget Sound/Gulf of Georgia sites, the prehistoric functional context of these sites needs to be further considered. The lower wet zone of the Musqueam Northeast is directly below a Locarno Beach Phase component (Borden 1976:233). Diagnostic of this phase are different forms of toggling harpoons of wapiti antler, harpoon blades, and harpoon foreshafts. Also remains of seal, sea lion and porpoise recovered from components of this phase strongly suggest the active hunting of these sea mammals with harpoons (Borden 1976:248–251). One major function of the common Musqueam Northeast rope-gauge, 3-strand lines may have been as harpoon lines (as was common at Ozette Village and historically in the Central Coast areas). As one possibility, harpooning activities at Musqueam Northeast may have been associated with the net fishing activities at the site (several nets were recovered at the site (Croes 1975, Stevenson 1989)). Seals may have been attracted to this specific location because of the net fishing – a typical problem for fishermen using nets – allowing them to be harpooned by the fishermen (Croes 1975:58; Borden 1976:248–251). Since Water Hazard and Biederbost basketry technologies appeared to be related to those from Musqueam Northeast (though 1,000 years later (Bernick 1989; Croes 1995:122–132)), and all sites had a similar stress on 3-strand cedar bough ropes, these ropes at Water Hazard and Biederbost may also be related to the occupational activity of sea mammal harpooning at these 2,000 year old sites.

The examples of the larger heavy-guage ropes of this type at Ozette Village wet site reflect the eventual development of the specialized hunting of the largest sea mammal, whales, which continues to be practiced today by the Central Coast Makah whalers.

In reference to the end of a person's life-cycle, this 3+ strand, cedar bough rope was also a common cordage type reported from the Nuu-chah-nulth Hesquiat Harbour dry cave burial site (DiSo9) (Bernick 1998a). Though 2-strand cedar bark cordage and 2-strand cedar bough cordage were more common, several examples of larger gauge 3+ strand cedar bough cordage were reported, sometimes with looped ends similar to whale harpoon lines at Ozette Village wet site. Much of the small gauge 2-strand cordage was probably used to tie together burial boxes and possibly individuals wrapped in cedar bark mat shrouds. Some of the less common examples of larger gauge 3+ strand cedar bough harpoon lines may have been left as possessions of sea mammals hunters buried in this burial cave at Hesquiat Harbour, and therefore represent their occupations at the end of their life-cycle.

Fishing Tackle Bags

This type of basket reflects the occupation of fishing, probably mostly by men, but possibly by women too. They are a plaited flat bag of cedar bark with an extra long flap extension on one side of the top edge (Croes 1977:289–295; Figure 9.11). Ethnographically these basket types are assigned to the occupation of fishing and considered a Wakashan area style of fishing bag. Drucker, in discussing the Kwakiutl states that they made a "fishhook and harpoon wallet with a long folding flap (ohLadzi)" (1950:262). With reference to the Northern and Central Nuu-chah-nulth (Nootka) he writes:

> Tackle bags, for fishhooks and other small oddments, were woven of very fine strips of bark into a form like that of a modern folding tobacco pouch: a long strip folded double and bound along the edges to make a compartment at one end and closed by folding the long flap over two or three times (1951:96).

In Jones' study of nearly 3,000 Northwest Coast museum baskets, she reported six baskets of this type and attributes them to a Nuu-chah-nulth origin (1968). This specialized fisher-person's basket type apparently is a distinctive Wakashan style of fishing occupation gear.

Two late prehistoric sites have examples of this type of occupationally indicative basketry, the Ozette Village wet site and the Hesquiat Harbour burial cave dry site.

Three of these fishing bags are reported at Ozette, and they are technologically distinguished from other plaited flat bags in that a very long flap extends from the body weave of one edge (both complete examples have flaps measuring 83 to 84 cm long, four times as long as the height of the bags (19 to 19.5 cm high) (Figure 9.11). The flap on this type of basket extends from the bag with the same cedar bark warp strips used in the body of the bag, but the strips are split to create a fine-gauge plait weave in the flap (Figure 9.11). The flap is constructed with an around and back mat selvage edge and on one of these baskets from Ozette (145/IV/38), is woven in a very complex twill

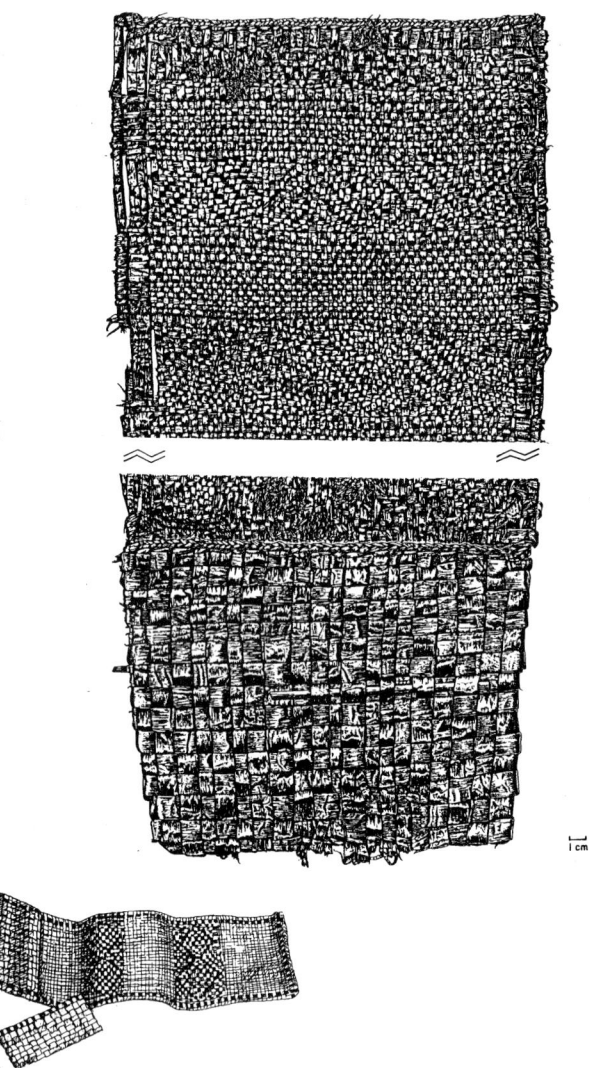

(a slightly larger box placed up-side-down over the top of the main box). This well-made box, with incised lines and remnants of red paint on its surface, also contained two woven cedar bark "pillows" full of cattail head fluff, and a well carved wooden mat creaser. The long flap on this basket was wrapped four times around the bag.

Inside were found several small fishhooks and other fishing gear. Nine complete double-barbed fishhooks were found in this fishing tackle bag (145/IV/57–65). Each hook has two bone barbs bound with cherry bark to a wooden shank. In addition to fishhooks, five cherry bark-wrapped wooden "tubes," one bundle of fiber, a fine grain sandstone whetstone (probably for sharpening bone hooks), five unused wooden fishhook shanks, and two small sharpened bone points (probably unused fishhooks barbs) were stored in this basket. Incorporated into the body of this basket (see Figure 9.11) was a long straight stick held in the weave. A similar stick was found in the other complete basket (37/IV/37). This stick may have been used as extra material for constructing additional wooden fishhook shanks. The fishing equipment found in this basket of course provides specific evidence for the function of these bags in the fishing occupation.

The only other prehistoric context for a fishing tackle bag of this type is one reported from the Nuu-chah-nulth Hesquiat Harbour burial cave dry site (DiSo 9; Bernick 1998a:42–43). In most characteristics this fishing tackle bag fits the definition of those recorded at the late Ozette Village wet site. The actual context suggests this fishing tackle bag was a grave good, probably placed with the fisher-person that owned it. At both sites this type of basket reflects the occupation of the owner both in life and at Hesquiat as an occupation of the deceased and part of the person's possessions.

Figure 9.11. Cedar bark, plaited fishing-tackle bag with long flap extension (145/IV/38). Note the complex combination of checker and twill 2/2 weave creating a chevron pattern in the flap construction (Croes 1977:289–295; illustration by Chris Walsh). Inset illustration of fishing-tackle bag by Hilary Stewart (1984:132).

Woven Cedar Bark Mats used as Shrouds Following Death

Dr. Jerome Cybulski, a physical anthropologist with the National Museum of Civilization, Ottawa, Canada, indicated to me recently that almost all Northwest Coast burial houses, canoe burials, and cave burials he has inspected over several decades have had checker weave, cedar bark matting pieces associated with the human remains (Personal Communications 1999). Dr. Cybulski analyzed the burials from a very complex dry cave site on the Central West Coast of Vancouver Island, Canada in Nuu-chah-nulth region of Hesquiat Harbour (DiSo 9; Cybulski 1978). The basketry and cordage from this Nuu-chah-nulth burial cave site recently has been thoroughly reported by Bernick (1998a). Though no attempt was made in Bernick's report to associate these items to the actual individuals buried in this site, it is clear that these basketry

2/2 and checker weave combination. The twill 2/2 is done in such a way as to produce a chevron pattern (Figure 9.11). Of the three report fishing tackle bags from Ozette Village wet site, one (FS/142) is a broken, discarded basket outside the house and within the refuse midden. The remaining two had been in use and were found inside the House 1 structure. One (37/IV/37) was found folded up and empty. This basket appears to have been stored with four raw cedar bark bundles and six halibut hooks inside a badly broken wooden box. The third reported basket (145/IV/38) was stored in a large wooden box (145/IV/37) which had a telescoping lid and cordage items were either burial goods (such as the fishing tackle bag discussed above) or mats and cordage used to cover or wrap the deceased (see ethnographic discussions of this in Arima 1983; Drucker 1950:49; Sproat 1868:178). The Hesquiat Harbour assemblage includes at least 37 mats woven in a 1/1 interval checker plaiting (Bernick 1998a:53). Many of these mats are very similar in type to those found in-use inside the proto-historic Ozette Village wet site houses, indicating that they served several purposes in the lives of Ozette/Makah/Nuu-chah-nulth peoples, as well as the matting commonly used as the a person's burial shroud.

These mats found as shrouds at the Hesquiat Harbour cave are the same types/classes of long mats as found in use in the Ozette Village wet site houses. At both sites they can be defined as intermediate to extra-large, cedar bark, checker weave mats with a constricted midline rectangular shape and around-and-back edges (Figure 9.12; Croes 1977:451–461). As the most frequent "true" long mats recorded at the Ozette Village wet site, they have been divided into three classes of mats, distinguished by different sizes in length. The Ozette mats vary from approximately 60 to 320 cm long (Hesquiat mats vary from 50–266 cm long); the width has a much smaller range, generally from 25 to 75 cm (Hesquiat: 31 to 162 cm wide), with an average of approximately 47 cm (Hesquiat: 86 cm, so normally about twice as wide) (Bernick 1998a:53–56).

The midlines of the mats at Ozette and Hesquiat typically are formed with an "anchor" row of 3-strand twining (Figure 9.12) or sometimes plain 2-strand twining (Croes 1977:451–461; Bernick 1998a:57–60). This row probably functions in anchoring the warp strands when construction of the mats is initiated. At both sites the checker weave elements on either side of the midline twining, on either end of the mat, and next to the edges are commonly of a Checker II decorative weave variation (Figure 9.12). The mat edges at both sites are formed with the around and back selvage technique (defined in Croes 1977:128; Figure 9.12). The ends of the Ozette mats typically are finished with two rows of plain twining, and the end warps cut off to leave a distinct fringe. The ends of only one Hesquiat harbour mat were fringed, with the majority having the warp ends woven back into the fabric (in a turn in and back selvage (Croes 1977:129, Bernick 1998a:69).

In the Ozette Houses these mats, particularly the large to extra-large forms, often were found covering wall boards, shelves (?), and bench platforms, or folded or rolled and stored on and under bench platforms. Of the ten large to extra-large mats recovered, two were folded several times and stored on or behind bench platforms (31/IV/81 and 71/IV/34) and two were in a worn out condition and rolled up (163/IV/9) or folded (32/IV/54) and under a bench platform with other stored objects. One 3 m long mat (31/IV/67) had

been folded in half and covered a bench platform. As this was a sleeping platform, the doubled mat probably functioned as a mattress. Directly under this was found a very worn out mat (31/IV/79) which would have provided extra padding. Two large mats (31/IV/59 and 31/IV/60) were partially folded and both covered what appears to have been either wall or shelving (?) boards. If wall boards, then these large mats may have been hung on these walls to stop air drafts in this sleeping and family area. If on shelving boards, they probably were either stored on or used to cover the shelves. Two large mats (66/IV/34 and 145/IV/155) were badly twisted and displaced by the mud-slide. Of the two intermediate sized mats apparently in use, one was folded in half along the midline and lying on a bench platform (31/IV/83; Croes 1977:Figure 111:454–460). The other (144/IV/14) also was folded in half and stored, along with several cedar bark flat bags and raw cedar bark bundles, in a large storage basket. In summary, the mats found in use in Ozette House I appear to have been used to cover walls, shelving (?), and/or floor areas.

Historically the Makah/Nuu-chah-nulth area was noted for its cedar bark mats, and they were traded to southern Washington coast and inland Puget Sound areas. Swan, who lived with the Makahs in the 1860s, noted that:

> Mats constitute one of the principle (sic) manufactures of the females during the winter months. With the Makahs, cedar bark is the only material used. Other tribes, who can obtain bulrushes and flags, make their mats of these plants, which, however, do not grow in the vicinity of Cape Flattery (1869:45).

Swan describes how cedar bark is processed into strips and:

> These are then neatly woven together, so as to form a mat six feet long by three wide. Formerly mats were used as canoe sails, but at present they are employed for wrapping up blankets, for protecting the cargoes in canoes, and for sale to the whites, who use them as lining of rooms, or as floor coverings (1869:45) (His dimensions are more consistent with the Hesquiat measurements then the narrower Ozette examples).

Swan purchased a large number of these mats (at about 40 cents each) and forwarded them to the U.S. National Museum, Smithsonian Institution. Thirty-six of these were examined by the author (mats # 54101–54135, 74789, 74704). Made in the same manner as the Ozette and Hesquiat mat types, they are generally twice as wide, averaging 225×109 cm and more similar in width to the Hesquiat mats (averaging 86 cm) in comparison with large Ozette mats averaging 233×59 cm. It should be mentioned however, that some of the mats collected by Swan (54102, 54120, 74789) are in fact as narrow and long as the Ozette mats. The wider

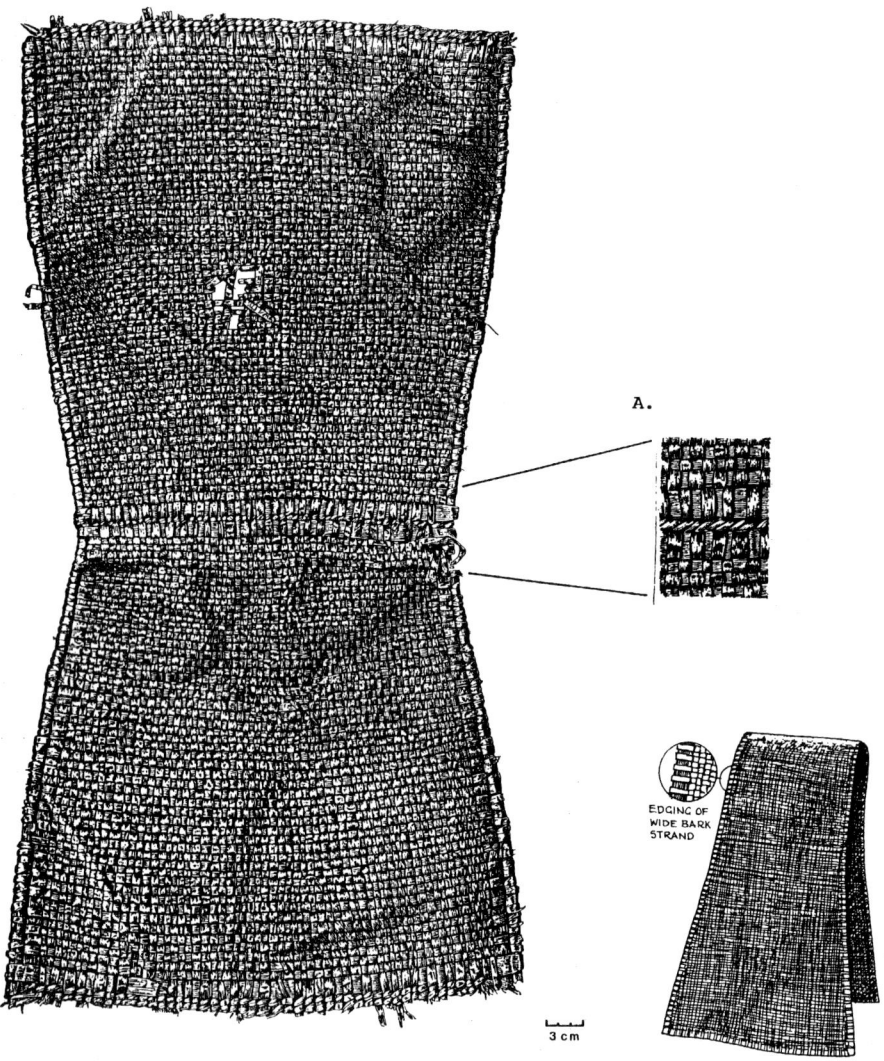

Figure 9.12. Example of Ozette Village wet site mat with constricting midline, checker weave, cedar bark mat (31/IV/83; Croes 1977:451–461; illustration by Chris Walsh). Note twined midline, checker II edging, and twined and cut off fringed ends. Insert A is a close-up of 3-strand midline twining as viewed on the reverse side of mat. Inset illustration of folded mat by Hilary Stewart (1984:139).

mats probably were preferred by Euro-Americans, especially for wall hangings and floor coverings, and the Makahs at this time probably manufactured many of them for some income through the Euro-American market.

Several ethnographers describe the function of these large cedar bark mats as feast mats to sit and eat upon, as bedding to sleep on, as makeshift sails in the contact and earlier periods, as wrappings to cover cargo, as capes, as covers to protect beached canoes from the sun's rays, as temporary shelters, as wall linings and as shrouds for the dead Sproat 1868:239; Olson 1967:111). At Ozette Village these large mats probably were used for all of these purposes. In the Ozette House I the large mats apparently

were being used on bench platforms, as wall hangings, as bedding, and also were folded and stored away under bench planks. They probably were used in numerous ways to place things on, to cover things with, and to wrap things within, including use as shrouds at the end of the person's life-cycle, as recorded in the ethnographic literature.

This common domestic mat type used in and around the Ozette Village wet site houses was the most common type of basketry found in the Hesquiat Harbour dry burial caves (at least 37 such mats reported; Bernick 1998a:53). This attests to the importance of this type of basketry as shrouds used at the end of one's life-cycle on this part of the Northwest Coast of North America.

Summary and Conclusions

I believe this study shows that basketry and cordage data from wet sites on the Northwest Coast can be associated with the full life-cycle of past peoples for thousands of years.

In investigating a person's original container, the basketry cradle, three wet sites had examples, all dating to within the last 1,000 years on the Central Coast. The two from Central Puget Sound sites of Conway and Fishtown were identical in construction and very different from the proto-historic and historic examples from the Ozette Village wet site and the Makah. Not only do we see a person's original basketry container from birth, but also see distinct styles of cradles from different ethnographic regions of the Coast Salish and Makah. Therefore cradle styles may well be good cultural markers of a person's ethnicity too.

In exploring a person's emerging status and cultural role, we found that basketry clothing and in particular hats may be the best marker of one's inherited role in one's life-cycle. Seven Northwest Coast wet sites had examples of hats, certainly indicating that this may be a common marker of a person's status from sites as early as 3,000 years old.

A person's occupation was also discernible from specialized basketry and cordage objects. I mostly stayed with baskets that were shown to be prehistorically (by contents) and historically to be used in the specific activities of whaling (harpoon bags and sheaths) and fishing (fishing tackle bags), so could link these kinds of occupations with special basketry equipment. The harpoon lines for sea mammals were more generalized and therefore somewhat more telling about harpooning activities among five Central Coast wet sites over the past 3,000 years. This large gauge, 3+ strand, cedar bough rope reflected sea mammal hunting occupations specific to Central Coast wet sites, therefore reflecting a sea mammal focus in this region versus the North Coast both prehistoric and historically.

If I could revisit this exploration of occupationally related basketry and cordage, I would probably look at less specialized basketry items as those seen at Ozette Village and, similar to exploring large gauge harpoon line cordage, would look at utility/pack baskets as found in almost all wet site contexts. Though the style of pack baskets varies considerably from site to site through time, they generally reflect the collecting, processing or carrying of resources as a division-of-labor occupation most often conducted by women, whether it is shell-fish, fish, berry crops, firewood or other resources. One characteristic of these utility/pack baskets from Northwest Coast wet sites is their significantly higher frequency at fishing stations such as Hoko River (74% of all baskets) versus major village sites such as Ozette (only 4% of all baskets) (Croes 1977:486–494). Therefore this generally female occupation of processing and packing resources at fishing stations is well reflected by the distribution of this kind of basket at these functional distinct kinds of wet sites.

The end of a person's life-cycle also appears to be associated with basketry items of one's specialized occupation (fishing tackle bags and harpoon lines found in the Hesquiat Harbour burial caves) or cedar bark, checker weave matting used as burial shrouds. All extensively excavated Northwest Coast wet sites have examples of checker weave, cedar bark matting, and these mats no doubt functioned in a wide variety of domestic roles as they did at the Ozette Village wet site. The high frequency of this kind of matting in the Hesquiat Harbour dry burial site indicates it's frequent use as one's shroud at the end of the life-cycle.

Therefore wet sites, some dating to as early as 6,000 years ago on the Northwest Coast of North America, will provide evidence of perishable cradles, clothing, occupational gear and basketry and cordage placed and used in a burial context. Typically Northwest Coast shell-midden sites might show some aspects of a person's life-cycle, but, as in other studies, wet sites provide wood and fiber artifacts representing the vast majority (95%) of the artifacts uncovered, and therefore complement greatly studies exploring the life-cycle – birth to death – through thousands of years in this region. Basketry and cordage has been shown to be very useful in identifying ethnicity of groups of people through time and space (Bernick 1988, 1989, 1998b; Croes 1977, 1989a, 1989b, 1992a, 1992b, 1995, 1997). This study begins to show how this basketry and cordage data can expand our understanding of site function, including how they provide information about a person's finite life-cycle at wet sites over the past several thousand years.

Acknowledgments

I would particularly like to thank the conference coordinator and volume editor, Barbara Purdy for her excellent job in producing the 1999 Wetlands Archaeology Research Project (W.A.R.P.) conference in Florida and compiling this volume. Barbara, besides being a wonderful conference host, provided invaluable and insightful comments for this paper and kept us all working on completing manuscripts to speed along the conference publication process. In terms of this paper, the basis of my main data came mainly from my work at the Ozette Village and Hoko River Archaeological Projects, which are co-sponsored by the Makah Tribal Nation and Washington State University. Numerous project researchers, Makah community members, field personnel, and students have contributed to data recovery, analysis, and reporting. Though this research owes its existence to these and many previous and current researchers, the summary and conclusions remain my responsibility.

References Cited

Ames, Kenneth M., and Herbert D.G. Maschner. 1999. Peoples of the Northwest Coast, Their Archaeology and Prehistory. Thames and Hudson, London.

Arima, Eugene Y. 1983. The West Coast People: The Nootka of Vancouver Island and Cape Flattery. Special Publication 6. British Columbia Provincial Museum, Victoria.

Arima, Eugene and John Dewhirst. 1990. Nootkans of Vancouver Island. In, Northwest Coast, edited by Wayne Suttles, pp. 391–411. The Handbook of North American Indians vol. 7, William C. Sturtevant, general editor, Smithsonian Institution, Washington.

Barnett, Homer G. 1955. The Coast Salish of British Columbia. University of Oregon Monographs. Studies in Anthropology 4. Eugene.

Bernick, Kathryn. 1983. A Site Catchment Analysis of the Little Qualicum River Site, DiSc 1: A Wet Site on the East Coast of Vancouver Island, B.C. National Museum of Man Mercury Series No. 118. National Museum of Man, Ottawa.

Bernick, Kathryn. 1988. The Potential of Basketry for Reconstructing Cultural Diversity on the Northwest Coast. In Ethnicity and Culture, Proceedings of the 18th Annual Chacmool Conference, edited by Reginald Auger, Margaret F. Glass, Scott MacEachern and Peter H. McCartney, pp. 251–257. Archaeological Association, University of Calgary, Calgary.

Bernick, Kathryn. 1989. Water Hazard (DgRs 30) Artifact Recovery Project Report, Permit 1988–55. Report submitted to the Archaeology and Outdoor Recreation Branch, Ministry of Municipal Affairs, Recreation and Culture, Province of British Columbia, Victoria.

Bernick, Kathryn. 1991. Wet Site Archaeology in the Lower Mainland Region of British Columbia. Report prepared for British Heritage Trust, Victoria, B.C., Department of Archaeology, Simon Fraser University, Burnaby, and Laboratory of Archaeology, Department of Anthropology and Sociology, University of British Columbia, Vancouver.

Bernick, Kathryn. 1998a. Basketry and Cordage from Hesquiat Harbour British Columbia. Royal British Columbia Museum, Victoria, British Columbia.

Bernick, Kathryn. 1998b. Stylistic Characteristics of Basketry from Coast Salish Area Wet Sites. In, Hidden Dimensions, the Cultural Significance of Wetland Archaeology. edited by Kathryn Bernick, pp. 139–156. UBC Press, University of British Columbia, Vancouver.

Boas, Franz. 1909. The Kwakiutl of Vancouver Island. American Museum of Natural History 8(2):301–522.

Borden, Charles E. 1976. A Water-Saturated Site on the Southern Mainland Coast of British Columbia. In, The Excavation of Water-Saturated Archaeological Sites (Wet Sites) on the Northwest Coast of North America, edited by Dale R. Croes, pp. 233–260. National Museum of Man Mercury Series No. 50, National Museum of Man, Ottawa.

Croes, Dale R. 1975. Musqueam Northeast Basketry and Cordage. Report appended to Charles E. Borden's Musqueam Northeast Report, University of British Columbia, Vancouver.

Croes, Dale R. 1976a. (Ed.) The Excavation of Water-Saturated Archaeological Sites (Wet Sites) on the Northwest Coast of North America, National Museum of Man Mercury Series No. 50, National Museum of Man, Ottawa.

Croes, Dale R. 1976b An Early "Wet" Site at the Mouth of the Hoko River, the Hoko River Site (45CA213). In, The Excavation of Water-Saturated Archaeological Sites (Wet Sites) on the Northwest Coast of North America, edited by Dale R. Croes, pp. 201–232. National Museum of Man Mercury Series No. 50, National Museum of Man, Ottawa.

Croes, Dale R. 1977. Basketry from the Ozette Village Archaeological Site: a Technological, Functional and Comparative Study. Ph.D. dissertation, Washington State University. University Microfilms 77–25, 762, Ann Arbor.

Croes, Dale R. 1989a. Prehistoric Ethnicity on the Northwest Coast of North America, An Evaluation of Style in Basketry and Lithics. In Research in Anthropological Archaeology, edited by Robert Whallon, pp. 101–130. Academic Press, New York.

Croes, Dale R. 1989b. Lachane Basketry and Cordage: A Technological, Functional and Comparative Study. Canadian Journal of Archaeology 13:165–205.

Croes, Dale R. 1992. An Evolving Revolution in Wet Site Research on the Northwest Coast of North America. In The Wetland Revolution in Prehistory, edited by Byrony Coles, pp. 99–111. Wetlands Archaeological Research Project (WARP) Occasional Paper No. 6. Department of History and Archaeology, University of Exeter, Exeter, England.

Croes, Dale R. 1993. Prehistoric Hoko River Cordage, A New Line on Northwest Coast Prehistory. In A Spirit of Enquiry, Essays for Ted Wright, edited by John Coles, Valerie Fenwick, and Gillian Hutchinson, pp. 32–36. Wetlands Archaeology Research Project (WARP) Occasional Paper No. 7. Department of History and Archaeology, University of Exeter, Exeter, England.

Croes, Dale R. 1995. The Hoko River Archaeological Site Complex, The Wet/Dry Site (45CA213), 3,000–1,700 B.P. WSU Press, Pullman, Washington.

Croes, Dale R. 1997. The North-Central Cultural Dichotomy on the Northwest Coast of North America: Its Evolution as Suggested by Wet-Site Basketry and Wooden Fish-Hooks. Antiquity 71:594–615.

Croes, Dale R., and Jonathan O. Davis. 1977. Computer Mapping of Idiosyncratic Basketry Manufacture Techniques in the Prehistoric Ozette House, Cape Alava, Washington. In The Individual in Prehistory: Studies of Variability in Style in Prehistoric Technologies, edited by James Hill and Joel Gunn, pp. 155–165. Academic Press, New York.

Croes, Dale R., and Steven Hackenberger. 1988. Hoko River Archaeological Complex: Modeling Prehistoric Northwest Coast Economic Evolution. In Prehistoric Economies of the Pacific Northwest Coast, edited by Barry L. Isaac, pp. 19–86. Research in Economic Anthropology, Special Supplement 3. JAI Press, Greenwich, Connecticut.

Curtis, Edward S. 1913. The North American Indian 9: Salishan Tribes of the Coast; Chimakum; Quiliute; Willapa. The University Press, Cambridge, Massachusetts.

Cybulski, Jerome S. 1978. An Earlier Population of Hesquiat Harbour, British Columbia. Cultural Recovery Paper 1. British Columbia Provincial Museum, Victoria.

Cybulski, Jerome S. 1996. Conflict and Complexity on the Northwest Coast: Skeletal and Mortuary Evidence. Paper presented in the "Northwest Coast" symposium, 96th Annual Meeting of the Society for American Archaeology, New Orleans, Louisiana.

Drucker, Philip. 1950. Culture Element Distribution: 26: Northwest Coast. Anthropological Records 9:3. University of California Press, Berkeley and Los Angeles.

Drucker, Philip. 1951. The Northern and Central Nootkan Tribes. Bureau of American Ethnology, Bulletin 144. U.S. Government Printing Office, Washington, D.C.

Eells, Reverend Myron. 1887. The Twana, Chemakum and Klallam Indians of Washington Territory. Annual Reports of the Board of Regents of the Smithsonian Institute, pp. 605–681. Washington, D.C.

Gunther, Erna. 1927. Klallam Ethnology. University of Washington Publications in Anthropology I, 5:171–314. Seattle.

Gunther, Erna. 1945. Ethnobotany of Western Washington. University of Washington Press, Seattle and London.

Hajda, Yvonne. 1990. Southwestern Coast Salish. In, Northwest Coast, edited by Wayne Suttles, pp. 503–517. The Handbook of North American Indians vol. 7, William C. Sturtevant, general editor, Smithsonian Institution, Washington.

Halpin, Marjorie M. and Margaret Seguin. 1990. Tsimshian Peoples: Southern Tsimshian, Coast Tsimshian, Nishga, and Gitksan. In, Northwest Coast, edited by Wayne Suttles, pp. 267–284. The Handbook of North American Indians vol. 7, William C. Sturtevant, general editor, Smithsonian Institution, Washington.

Hobler, Philip. 1976. Wet Site Archaeology at Kwatna. In, The Excavation of Water-Saturated Archaeological Sites (Wet Sites) on the Northwest Coast of North America, edited by Dale R. Croes, pp. 146–157. National Museum of Man Mercury Series No. 50, National Museum of Man, Ottawa.

Jones, Joan Megan. 1968. Northwest Coast Basketry and Culture Change. Research Report No. 1. The Thomas Burke Memorial Washington State Museum, Seattle.

Kan, Sergei. 1988. Symbolic Immortality, The Tlingit Potlatch of the Nineteenth Century. Smithsonian Institution Press, Washington and London.

Kew, J.E., and Michael Kew. 1990. Central and Southern Coast Salish Ceremonies Since 1900. In, Northwest Coast, edited by Wayne Suttles, pp. 476–480. The Handbook of North American Indians vol. 7, William C. Sturtevant, general editor, Smithsonian Institution, Washington.

Mozino, Jose Mariano. 1970. Noticias de Nutka. An Account of Nootka Sound in 1792. Translated and edited by Iris Higbie Wilson. University of Washington Press, Seattle.

Munsell, David A. 1976a. The Wapato Creek Fish Weir Site 45PI47, Tacoma, Washington. In, The Excavation of Water-Saturated Archaeological Sites (Wet Sites) on the Northwest Coast of North America, edited by Dale R. Croes, pp. 45–47. National Museum of Man Mercury Series No. 50, National Museum of Man, Ottawa.

Munsell, David A. 1976b. Excavation of the Conway Wet Site 45SK59b, Conway, Washington. In, The Excavation of Water-Saturated Archaeological Sites (Wet Sites) on the Northwest Coast of North America, edited by Dale R. Croes, pp. 86–121. National Museum of Man Mercury Series No. 50, National Museum of Man, Ottawa.

Olson, Ronald. 1936. The Quinault Indians. University of Washington Publications in Anthropology 6(1):1–194. University of Washington Press, Seattle.

Olson, Ronald. 1967. Social Structure and Social Life of the Tlingit in Alaska. University of California Anthropological Records 26. Berkeley.

Onat, Astrida R. Blukis. 1975. Fishtown Site, 45SK99. In, The Excavation of Water-Saturated Archaeological Sites (Wet Sites) on the Northwest Coast of North America, edited by Dale R. Croes, pp. 122–145. National Museum of Man Mercury Series No. 50, National Museum of Man, Ottawa.

Sprague, Rodrick. 1976. The Submerged Finds from the Prehistoric Component, English Camp, San Juan Island, Washington. In, The Excavation of Water-Saturated Archaeological Sites (Wet Sites) on the Northwest Coast of North America, edited by Dale R. Croes, pp. 78–85. National Museum of Man Mercury Series No. 50, National Museum of Man, Ottawa.

Sproat, Gilbert M. 1868. Scenes and Studies of Savage Life. Smith Elder, London.

Suttles, Wayne. 1988. Central Coast Salish. In, Northwest Coast, edited by Wayne Suttles, pp. 453–475. The Handbook of North American Indians vol. 7, William C. Sturtevant, general editor, Smithsonian Institution, Washington.

Stevenson, Ann. 1988. Netting and Associated Cordage. In, Water Hazard (DgRs 30) Artifact Recovery Project Report, Permit 1988–55, K. Bernick (ed.), Appendix A, Archaeology and Outdoor Recreation Branch, Ministry of Municipal Affairs, Recreation and Culture, Province of British Columbia, Victoria.

Stewart, Hilary. 1984. Cedar, Tree of Life to the Northwest Coast Indians. Douglas and McIntyre, Vancouver. 1987. The Adventures and Sufferings of John R. Jewitt, Captive of Maquinna. Annotated and illustrated by Hilary Stewart. Douglas and McIntyre, Vancouver.

Swan, James. 1857. The Northwest Coast; Or, Three Years' Residence in Washington Territory. New York: Harper.

Swan, James. 1870. The Indians of Cape Flattery. Smithsonian Contributions to Knowledge, Volume 16.

Waterman, T.T. 1920. The Whaling Equipment of the Makah Indians. University of Washington Publications in Anthropology I, 1:1–67.

10. Serendipitous Discoveries: The Water Hazard Wet Site in Southwestern British Columbia

Kathryn Bernick

Accidental discovery of archaeological sites by land-excavating machines is relatively common and in this respect the Water Hazard incident was not extraordinary. Nor is it unusual for archaeologists aided by volunteers to sort through spoil piles to recover ancient artifacts ripped from a site. But because the perishable nature of the artifacts galvanized archaeologists and local aboriginal communities (First Nations) into action and loosened government purse strings, the emergency rescue operation recovered unique archaeological evidence exceeding what is normally expected of fragmentary artifacts dissociated from their depositional context. The Water Hazard collection captivated public audiences and professional archaeologists. It has contributed significantly to knowledge of the past and stimulated research on a variety of topics. Consequently the Water Hazard site (DgRs 30) has gained a measure of distinction – even though the site itself has never been investigated.

The report of the investigations (Bernick 1989) describes the project and finds in detail, but it has not been published. In addition to communicating the information to a larger audience, this paper illustrates the potential of wetland archaeology. I highlight the Water Hazard basketry artifacts, which have been instrumental in defining my research in the past decade (Bernick 1998). In doing this, I emphasize that despite the successful transformation of a destructive incident into a positive scientific contribution, the investment of energy, time, and money could have produced much more had the artifacts been recovered in a systematic and scientific manner.

Natural and Cultural Setting

The Water Hazard site is located on the Point Roberts peninsula in south-western British Columbia, on low alluvial ground about 700 m from the Boundary Bay shoreline and 150 m from a bluff oriented parallel to the shoreline (Figure 10.1). This bluff was once an island and it became connected to the mainland by the Fraser River flood plain sometime in the past 5,000 years (Clague *et al.* 1983). Today, the site area is no more than one meter above sea level and is protected by dikes.

Regional bioclimatic environmental conditions appear to have been stable for the past 5,000 years. However, sea levels may have been one to two meters lower than today at unspecified times in the past several millennia, and significant local changes have occurred as a result of delta formation and Euro-Canadian settlement (Clague 1981:21, Clague *et al.* 1983:1325, Fladmark 1975). Eighteen hundred years ago, at the time the Water Hazard artifacts were made and used, the shoreline would have been closer to the site than it is today. Flora and fauna would have been similar to those of the early historic period before diking and the recent alteration of natural sedimentation processes, but their local availability would have differed. There would have been coniferous forests on the bluff and marshy vegetation on the flat land. Wetland fauna would have been abundant – especially waterbirds, fish, and clams – as well as seals, deer, and other land and sea mammals. The climate was much as it is today, featuring dry sunny summers and wet winters. For detailed modern environmental characteristics see Krajina *et al.* (1982) and Meidinger and Pojar (1991).

Archaeologically, the Fraser delta is one of the best known regions of the Northwest Coast culture area. Variability in the prehistoric record is manifested in a series of phases that correspond to regional culture types (Burley 1980; Matson and Coupland 1995; Mitchell 1971, 1990). Research indicates considerable similarity in material culture, subsistence activities, and settlement patterns

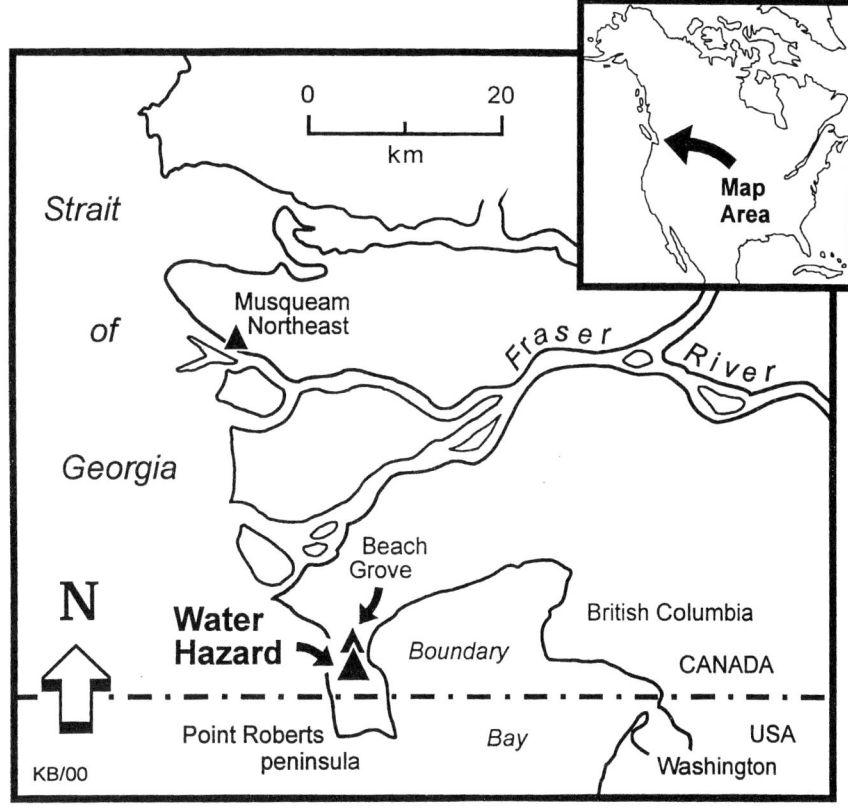

Figure 10.1. Location of the Water Hazard (DgRs 30) site. Map by K. Bernick.

across the region during the past two millennia. This enhances the credibility of ethnographic analogy for interpreting cultural remains from the Water Hazard site.

The site lies approximately in the middle of the Central Coast Salish ethnographic area, but I was unable to find any written or oral record of aboriginal occupation or use of the specific Water Hazard locality. For descriptive studies of the technology, sociology, and ideology of the Central Coast Salish see Suttles (1990) and references therein.

Archaeological investigations on the Point Roberts peninsula have identified evidence of four regional culture types: Gulf of Georgia (AD 400 – AD 1800), Marpole (400 BC – AD 400), Locarno Beach (1200 BC – 400 BC), and Charles (4500 BC – 1200 BC). Most known sites are located along the edges of the former island, though several, like Water Hazard, are located on the flood plain. In 1988 when the Water Hazard site was accidentally discovered, the only known wet-site deposits on the Point Roberts peninsula were at the Beach Grove site (DgRs 1), which is near Water Hazard. Excavations at Beach Grove in the early 1960s uncovered 1600–year-old water-saturated deposits in the bottom layers of a Marpole-phase shell-midden component,

but these were not extensively investigated. The presence of house platforms in that part of the Beach Grove site identifies it as a village (Matson *et al.* 1980, Smith 1964).

The Water Hazard Project

Construction workers enlarging an artificial pond (water hazard) on the Beach Grove golf course in Tsawwassen, British Columbia, dug up water-saturated deposits containing perishable wood and bark artifacts in late June 1988. The location where the dredging took place was not recorded as a site at the time, and there were no visible features indicating the presence of archaeological remains. A considerable but unknown portion of the culture-bearing deposits were removed by excavating machines and dumped in several locations for use in landscaping. One pile, which was kept wet by sprinklers, became the focus of an emergency salvage operation.

Initial inspection of the spoil piles confirmed the presence of stylistically sensitive perishable artifacts that could be dated by radiocarbon assay. Thus, even though they were no longer in stratigraphic context, they had considerable potential for clarifying relationships between

Figure 10.2. Recovering basketry from the spoil pile required team work. Photo by Ann Stevenson.

the regional archaeological cultures and for identifying ethnicity in the archaeological record. Proximity to the Beach Grove site and local geomorphology strongly suggested a Marpole phase assemblage. At the time, only a few perishable wood or bark artifacts dating from the Marpole era had been recovered in British Columbia.

Basketry fragments visible on the surface of the spoil pile displayed characteristics that suggested similarities to specimens from Locarno Beach phase deposits excavated in the 1970s at the Musqueam Northeast site (DhRt 4) on the north arm of the Fraser River (Archer and Bernick 1990). The prospect of obtaining a comparable collection from the Water Hazard site, especially if from the succeeding Marpole phase, defined the scope of the rescue operation. Specific goals were to recover, preserve, and analyze a sizeable assemblage of perishable wood and bark artifacts and to determine their age and relationship to regional culture history. Whereas the emergency nature of the artifact salvage generated the requisite funds and labour to do this, investigating the site was not an option at the time.

Field work consisted of retrieving perishable artifacts from the large, 120 m³, pile of DgRs 30 deposit that had been dumped in a corner of the golf course. By the time rescue work began, two weeks after the dredging, few artifacts were visible on the surface of the spoil pile. Buried items appeared to be in good condition, but they were no longer in a water-saturated environment and were deterior-

ating rapidly. Retrieving the artifacts was a matter of urgency. Since none of the artifacts was *in situ* and the stratigraphy was not intact, it was possible to accommodate a large number of volunteers with minimal professional supervision.

Excavators used shovels and trowels to search through the pile for artifacts, then exposed and removed the fragile items using water pressure supplied by garden hoses fitted with adjustable nozzles, and also plastic spray bottles. The hoses were connected to the golf club's sprinkler system and during the last week also to a municipal fire hydrant. Basketry was "floated" from the mud and placed on slabs of Coroplast, which is a light-weight rigid plastic resembling corrugated cardboard in appearance (Figure 10.2). Clumps of soil containing small, fragile items were removed *en bloc* and washed with gentle spray. After hosing off most of the mud, waterlogged artifacts were wrapped in wet cotton cloth and placed in self-sealing plastic bags. At the end of each work day they were transferred to water-filled containers in the archaeology laboratory at UBC. In addition to the wood and bark items, all artifacts of stone, bone, antler, or shell that were encountered were collected, as well as selected other remains. Although a fair portion of the pile was investigated, it was not exhausted.

The DgRs 30 wood and bark artifacts were kept in water, refrigerated, for 10 months. During that time they were cleaned of adhering mud, analyzed, and selected specimens were photographed or drawn. Some large or exceptionally fragile pieces were sealed individually in plastic bags. Most items were grouped by size and type and placed in plastic containers filled with tap water – which did not harm the artifacts as Vancouver tap water is notably soft. Fragile flat objects including all basketry were placed in folds of fibreglass screening supported on slabs of Coroplast, and if packaged in bags also wrapped with wet cotton cloth. Catalogue numbers were embossed on colored plastic Dymo tape, which was attached to or packaged with the respective artifacts. Preservation treatment was administered after completion of archaeological analysis. Conservator Jo Ann Erling treated the artifacts with polyethylene glycol (PEG 1500) followed by freeze drying and subsequently applied Rhoplex to some of the basketry (Erling 1990).

Recruiting volunteers and coordinating the participation of avocational archaeologists, members of local First Nations, and archaeology students constituted an important aspect of the project. More than one hundred people took part, in the field and in the laboratory. On-site public outreach mainly addressed golfers. A work station set up near the tee allowed them to see the latest finds and talk with the crew. News reports appeared on national and regional television, radio, newspapers, and archaeological newsletters.

Recovered Data

The salvaged wood and bark artifacts had been dredged from about two metres below ground surface, below the water table. According to BC Archaeology Branch Project Officer Steven Acheson (personal communication June 29, 1988), who briefly saw the exposed section-cut as the enlarged pond was filling up with water, the waterlogged artifacts are from a single, relatively thin stratum. Nothing is known about the horizontal extent of the deposit.

Three matrix types were discernible in the pile of earth that had been excavated from the site, but their stratigraphic relationship was not clear and the thicknesses of the respective layers is unknown. Moreover, there may be more than three strata at the site. Wood and bark artifacts were associated with two of the matrices. One consisted of sand and pebbles with some silt and lenses of crushed mussel shell, the other was a blue-gray silty clay. Both contained whole and broken shells, animal bones, and wood debris in addition to artifacts. The wood and bark artifacts appeared to be concentrated at the interface between the two matrices, often in association with non-artifactual wood and with clumps of moss. The third matrix identified in the spoil pile, a compact brownish silty clay with orange stains, was devoid of cultural material. It is typical of Fraser River alluvium and probably comprises the uppermost layer at the site.

Three samples of frayed wood-fibre cordage were submitted for radiocarbon dating. They provided uncalibrated age estimates of 1580±60 BP (SFU 585), 1670±60 BP (SFU 586), and 1980±60 BP (SFU 592). The dates correspond to the second half of the Marpole phase, which is the era represented at the nearby Beach Grove village site.

The waterlogged artifact assemblage from DgRs 30 includes basketry, cordage, and other wood and bark artifacts (Table 10.1). The basketry is all woven; there are no coiled specimens. Many pieces display ornamentation, such as decorative weaves, elaborate selvages (edge finishing), or color contrast. The cordage is made mainly by twisting. The wood and bark artifacts include complete and fragmentary woodworking wedges and bentwood fishhooks, as well as fragments of indeterminate objects and wood-chip detritus.

Other items were also recovered from the spoil pile, but their association with the wood and bark artifacts could not be confirmed due to the disturbed nature of the deposits. Among these are 37 artifacts made of stone, bone, antler, and shell (Jackson 1989). These materials and the particular types of artifacts (Table 10.1) are common in Northwest Coast shell middens. Non-artifactual materials that were collected include a few isolated human bones, as well as faunal remains comprising invertebrate shells and unmodified bones of large and small mammals, waterfowl, other birds, and fish.

Plant species determinations, conducted on a portion of the assemblage, for the most part confirm expectations that are based on other wet-site assemblages from the Coast Salish area (Archer and Bernick 1990, Bernick 1983) and the ethnographic literature (Barnett 1955). Identifications made by Jack Grey (1989) attest to reliance on locally available materials, especially wood and bark of western red cedar (*Thuja plicata*) (Table 10.1). The assemblage stands out in having a relatively large and varied sample of items made from Pacific yew (*Taxus brevifolia*). In addition to all of the woodworking wedges and one or more of their collars, elements of a basket, segments of 2-strand, 3-strand, and 4-strand twisted cordage, and an enigmatic "bound sticks" artifact were identified as Pacific yew.

The Water Hazard Basketry

The DgRs 30 assemblage features 102 basketry artifacts. Two are nearly complete baskets, the remainder are fragmentary articles. All together there are 246 pieces of basketry with a combined 21.7 m² surface area. Four weave types, defined in figure 10.3, are represented: checker plaiting (35 artifacts), twill plaiting (7 artifacts), close plain twining (5 artifacts), and open plain twining (43 artifacts). Twelve indeterminate-weave artifacts consist of severely

Table 10.1. Artifacts from the Water Hazard Site, DgRs 30.

Artifact Type

Formed wood objects: 53 (13%)	**Material[1]**	**Number**	**Comments**
Wedges	yew wood	29	20 with collars
Wedge fragments	yew wood?	15	11 tip frags., 4 shaft frags.
Bentwood fishhooks	Douglas fir wood	7	one-piece type
Comb teeth?	unspecified hardwood	1	8 tip frags. and 6 shaft frags.
Carved wood object	ocean-spray wood	1	fragment
Woven articles: 104 (27%)			
Plaited basketry	wood fiber	41	1 nearly complete basket, 4 basket bases, 36 fragmentary articles[2]
Twined basketry	wood fiber	48	1 nearly complete basket, 47 fragmentary articles[2]
Indeterminate weave basketry	wood fiber	12	fragmentary articles[2]
Textile header fragments	cedar bark	2	no intact weave
Textile side-braid fragment	cedar bark	1	6-strand braid (8 cm long)
Cordage: 135 (34%)			
Netting	cedar bark	2	479 pieces (36.7 m of cordage)
Grommets (rope rings)	wood fiber (3 cedar)	15	
Double-ring wedge collar	wood fiber	1	
Bindings	wood fiber	2	
Twisted linear cordage (2, 3, 4 strand)	cedar bark, wood fiber	94	108 pieces (25.82 m)
Single-strand knotted cordage	cedar bark, wood fiber	21	21 pieces (6.19 m)
Other modified wood and bark objects: 63 (16%)			
Creased-bark-with-handles:			
Reinforced slab	cedar bark & cedar wood	1	
Bark end-pieces	cedar bark, wood fiber	6	
Creased bark	cedar bark	10	
Fish-spreading stick	cedar wood	1	scorched at both ends
Bound sticks	yew wood	1	2 sticks bound together
Miscellaneous modified bark	cedar, cherry?	28	some may be natural
Wood chips		16	
Shaped stone, bone, and antler tools: 17 (4%)			
Chipped stone projectile point	fine-grained basalt	1	bifacially pressure flaked
Ground stone chisel	nephrite	1	broken at bit end
Ground stone knife	slate	1	fragment with beveled edge
Pecked & ground hand maul	granitic or sandstone	1	end (base?) fragment
Abraders	sandstone	6	
Bone Point	bone	1	flat, bifacially ground, no shoulder
Harpoon foreshaft	antler	1	
Haft	deer or elk tine	1	notched end
Bird bone awl	gull (Laridae) radius	1	
Drinking tube or whistle	eagle[3] ulna	1	
Drinking tube or whistle fragments	bird bone	2	
Miscellaneous stone, bone and shell objects: 20 (5%)			
Core, utilized	fine-grained basalt	1	bifacially flaked
Flakes	basalt, chert, quartz	11	
Dogfish spine	*Squalus acanthias*	1	ground at tip
Worked fragments	bone, antler	4	
Olivella shell beads	*Olivella biplicata*	2	tips of shells removed
Scallop shell fragment	*Pecten Caurinus*	1	
TOTAL (100%)		392	

Notes:

[1] The wood species identified are: yew=*Taxus brevifolia*, Douglas fir=*Pseudotsuga menziesii*, cedar=*Thuja plicata*, oceanspray=*Holodiscus discolor*.

[2] Many are fragmentary baskets

[3] *Haliaetus leucocephalus* (bald eagle)

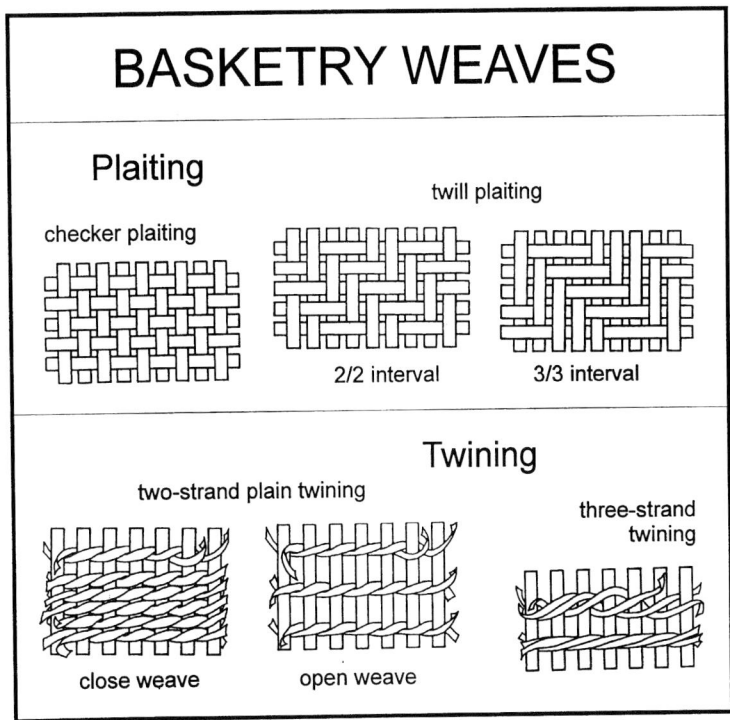

Figure 10.3. Weaving techniques documented for Water Hazard basketry. Drawing by K. Bernick.

frayed specimens and rim fragments with no associated body weave. Most of these artifacts are remnants of baskets. Other types of basketry articles may be represented, though none of the pieces displays features indicating a form other than a container. Some have twisted cordage handles. The variety of weaves and range of basket sizes indicate diverse functions, but specific uses cannot be reconstructed. None would have been watertight. Ethnographic analogy provides little insight as most information for Coast Salish baskets refers to coiled varieties.

Wood splints predominate overwhelmingly as raw material. Cedar bark elements are used only for close plain twining and as occasional decorative elements in other weaves. Species identification of selected specimens shows that nearly all are made of western red cedar withes. Warp elements in one basket were identified as Pacific yew, and the handle of another specimen is western hemlock (*Tsuga heterophylla*). Weave types show little diversity within the respective classes. Plaiting is regular, that is, warp and weft elements are of equal widths and aligned with the edges of the fabric. Twining stitches all slant up to the right (/). Open-plain-twined basketry is the most uniform weave in the assemblage, as well as the coarsest. Close plain twining is the finest, though twill-plaited baskets and some checker-plaited specimens are nearly as fine. Table 10.2 summarizes weaving gauges.

Table 10.2. Weaving Gauge of Water Hazard Basketry: Number of Warps per 10 cm.

Weave	Range	Mean	SD	N
Checker	8–30	16.0	4.9	34
Twill	14–28	21.5	7.0	4
Close Twining	14–34	26.0	8.1	5
Open Twining	10–21	13.6	3.2	42

Some checker-plaited and open-plain-twined baskets have evidence of bases, all twill-plaited and nearly all in 3/3 interval. The bases are square or rectangular and feature one or two close rows of plain twining at the perimeter. The nearly complete basket in close plain twining has a round base with radiating warps woven in the same technique as the walls.

Many specimens have reinforcement rows on the basket wall. The most prominent type of reinforcement was constructed by laying a sturdy element horizontally across the inside of the warp and wrapping it onto the warp elements with two parallel spiralling strands. This produces a row of slanted stitches on the outside of the weave. Where there are two adjacent reinforcement rows, the wrapping direction sometimes alternates resulting in a chevron motif. Two twill-plaited baskets have four adjacent reinforcement

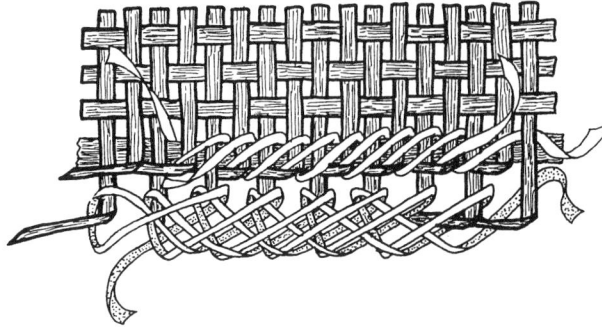

Figure 10.4. Method of applying the figure-eight-wrapped (false braid) selvage. Drawing by K. Bernick.

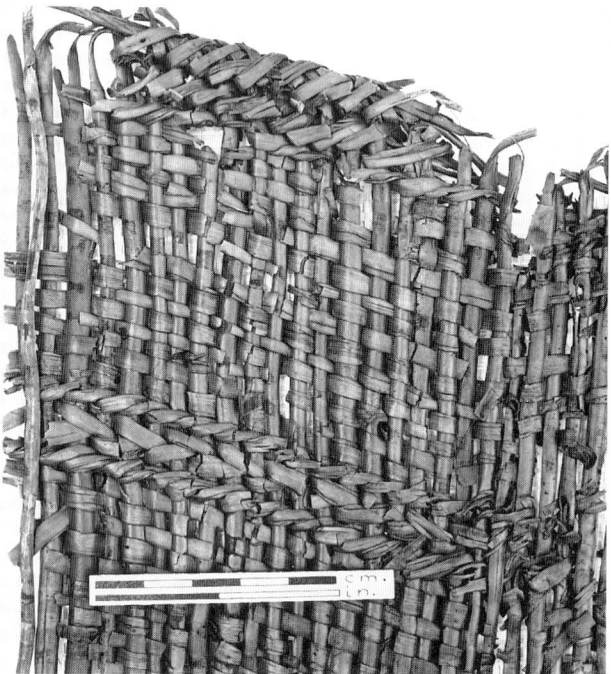

Figure 10.5. Checker-plaited basket fragment with decorative band of flat braid bounded by rows of three- strand twining (DgRs 30:49). UBC Laboratory of Archaeology photo by Michael Lay.

rows, each wrapped with three parallel strands. The reinforcements served structural and ornamental ends simultaneously. When present, they are located at the top of the basket wall just below the rim and often also several centimeters below the rim. In some cases, especially on checker-plaited baskets, two-strand wrapped reinforce-

ments form the focus for an ornamental band of weaving (Figure 10.4).

The most common selvage, which accounts for 74% of all occurrences, involves attaching an elaborate figure-eight wrapped coil bundle to the warps (Figure 10.5). This results in a "finished" appearance both inside and outside, and a chevron motif on the top edge. Other selvage types represented in the collection involve wrapping the warp ends with non-intersecting spiralling strands, which produces a finished appearance only on the outside. Twenty-three percent are wrapped with two strands, like the reinforcements. Others feature double rows of single-strand wrapping. The close-plain-twined basket has a simple composite selvage.

In addition to decorative selvages and reinforcements, many Water Hazard baskets are ornamented by means of structural variation, color contrast, or a combination of color and structure. Nearly half of the checker-plaited basketry artifacts display ornamentation. In comparison, ornamentation on open-plain-twined baskets is less frequent and simpler, but more standardized. On both these basket types structural decoration is limited to one or two narrow bands, each consisting of one to three weft rows. These bands are located at the rim or several centimeters below the rim, usually in conjunction with a reinforcement (Figures 10.4 and 10.6). In one case, a decorative band simulates a reinforcement in appearance without providing structural support.

Selective orientation of bark-covered elements (dark) versus inner split-wood surfaces (light) provide evidence of color contrast, though original colors have not been preserved. At least 25% percent of the open-plain-twined specimens exhibit color-contrast decoration consisting of vertical stripes, horizontal bands, or both stripes and bands. On checker-plaited baskets, color contrast was sometimes used to enhance horizontal bands. The twill-plaited baskets are decorated with a patterned weave that apparently covered the entire basket wall. The resulting geometric zigzag motif was enhanced by color contrast (Figure 10.7). The broken and frayed condition of the assemblage precludes reconstruction of shape and size for most specimens. A nearly complete checker-plaited basket was bowl shaped, 16 cm high and at least 101 cm in circumference at the rim. It has a twill-plaited rectangular base and numerous inserted warps indicating that the walls flared outward from the bottom becoming vertical in the upper third. It has a figure-eight-wrapped selvage and a wrapped reinforcement just below the rim. The other nearly complete basket, albeit recovered in 35 pieces, is woven in close plain twining. It was cylindrical, undecorated, and approximately 15 cm high

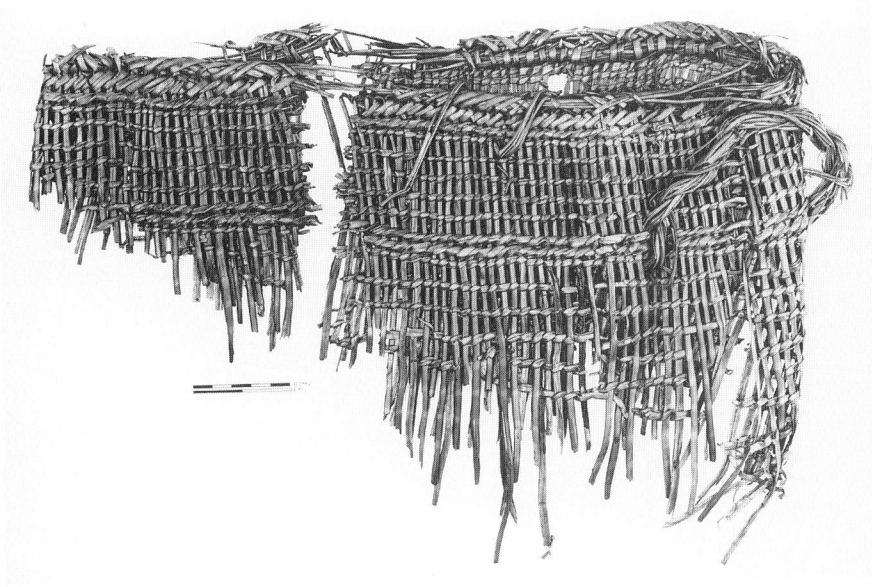

Figure 10.6. Open-plain-twined basket fragment with decorative rows of three-strand twining that also serve as reinforcements (DgRs 30:44). UBC Laboratory of Archaeology photo by Michael Lay.

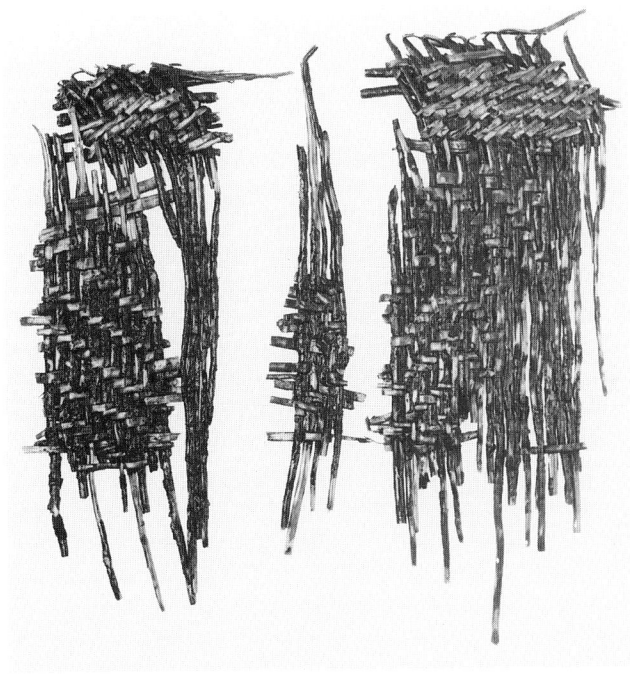

Figure 10.7. Twill-plaited fragmentary basket with zigzag decoration (DgRs 364). UBC Laboratory of Archaeology photo by Michael Lay.

and 40 cm in circumference at the rim.

Contributions, Interpretations, and Lost Information

Despite the lack of contextual data, the Water Hazard assemblage provided significant new archaeological information. This positive outcome can be credited to the organic composition of the artifacts that rendered them dateable by radiocarbon assay, the recovery of a sizeable collection, and the availability of considerable information from other sources about the Marpole culture type.

The Water Hazard artifacts provide direct evidence of woodworking techniques including steambending, carving, chopping, and splitting, as well as wood-fiber and bark construction techniques – weaving, tying, twisting, creasing, and binding. They comprise tangible examples of normally perishable items such as baskets, textiles, and cordage, and offer new perspectives on ancient technologies. Most of these activities and manufactures had been reconstructed for the Marpole era, but mainly from indirect evidence. The vegetal artifacts add details, including information about wood species and the forms and styles of objects made entirely from plant materials. In addition to informing local culture history, the Water Hazard data have stimulated research on topics such as basketry styles and fishing technology.

Figure 10.8. Stone bowl from the Marpole site (DhRs 1), approximately 9 cm high. Redrawn by K. Bernick from Smith 1923:43.

The Water Hazard discoveries provided key data for the identification of a chronological sequence of basketry styles for the Coast Salish area (Bernick 1998). Comparison of Water Hazard basketry with other assemblages from the region isolated attributes that are diagnostic of the Marpole culture type. Accordingly, characteristics of Marpole baskets are: relative techno-stylistic standardization, checker plaiting and open plain twining from wood splints as predominant weaves, elaborate false-braid (figure-eight-wrapped) selvage, structural decoration occurring as a narrow horizontal band at the rim and a second band lower on the basket wall, strikingly decorative patterned twill plaiting, and a chevron motif (Bernick 1998:152).

Recognizing diagnostic features of Marpole baskets informs other classes of archaeological data. For example, I have identified a stone bowl as a representation, not merely of a basket, but of a Marpole-style basket. The stone bowl was recovered about 100 years ago from the Eburne site (Smith 1923:42), which now is called the Marpole site (DhRs 1). It has incised decoration analogous to ornamentation on Water Hazard basketry. The chevron pattern on the lip of the bowl and the slanted unidirectional lines bounded by a straight line parallel to the rim (Figure 10.8) clearly represent a Marpole-style figure-eight-wrapped basket rim. The zigzag design on the bowl replicates the distinctive pattern on twill-plaited basketry from the Water Hazard site (Figure 10.7). Moreover, nearly identical twill-plaited basketry radiocarbon dated to the Marpole era was recovered from another location in the Fraser delta (Bernick 1998:146). The human face on the stone bowl, though not a repetition of basketry motifs, is in the same place and of similar size to handles on baskets from the Water Hazard site.

Other wood and bark artifact types recovered from the Water Hazard site also have potential to inform archaeological interpretations in important ways. Ann Stevenson's (1998) discussion of fishing technologies that were used in the Fraser River illustrates that wet-site data provide otherwise absent details that can be important for understanding the past. The Water Hazard assemblage includes fishing-related artifacts made entirely from normally perishable materials – fragments of knotted netting, bentwood fishhooks, and cordage that could have been fishing line (Stevenson 1989, 1998).

The Water Hazard project is considered a success because an informative sample of unique artifacts was serendipitously recovered and specimens from the collection have figured prominently in exhibits and in educational materials to raise awareness of wet-site archaeology. However, because the artifacts were not excavated systematically and scientifically and many were severely damaged by the excavating machines, a considerable amount of information was lost. We know virtually nothing about the site. There is no information about spatial relationships of the artifacts to one another, to other materials, or to features that might have been present. In other words, there is no contextual information to assist interpretation of artifact function. Moreover, there is no information about the nature and extent of the water-saturated deposits for use by resource managers.

That the dredged deposits were water-saturated and contained perishable wood and bark artifacts was both a devastating accident and an auspicious opportunity. Because the unearthed artifacts were rapidly disintegrating, there was little time for project planning and fundraising, nor for a prolonged field program. On the other hand, the novelty and potential significance of waterlogged wood and bark artifacts made it possible to recover, analyze, and conserve a sizeable sample. Consequently the artifacts recovered from the Water Hazard site contribute significantly to the archaeology of the Fraser delta and the Northwest Coast culture area. They provide direct evidence of technologies that would otherwise not be known and further our understanding of the Marpole culture type. In addition, the project was a catalyst for wetland archaeology in the region and promoted awareness of the importance and vulnerability of wetland sites.

Acknowledgments
The Water Hazard project, which I directed, operated under the auspices of the University of British Columbia (UBC) Laboratory of Archaeology. Most of the field work and much of the laboratory work were conducted by volunteers from UBC, the Archaeological Society of British Columbia (ASBC), and Musqueam and Tsawwassen First Nations.

Funding was provided by the British Columbia Archaeology Branch and Musqueam First Nation. The ASBC supplied field equipment, the Royal British Columbia Museum provided logistical support for conservation, and the Beach Grove Golf Club assisted with field logistics.

I sincerely thank everyone who participated or contributed to the project, particularly Professor R.G. Matson (UBC) who served as principal investigator. Heather Pratt and Ann Stevenson supervised artifact recovery in the field, Leona Sparrow organized participation of people from the Musqueam First Nation, and Loretta Williams organized the participation of the Tsawwassen First Nation. Sandra Louis (Musqueam) and Loretta Williams (Tsawwassen) worked as laboratory assistants. I am grateful to the British Columbia Archaeology Branch, especially Brian Apland and Steven Acheson, for their instrumental role in all stages of the project. I also acknowledge Leona Sparrow (Musqueam) for directing me in 1989 to publish the results of the investigation and I accept responsibility for taking so long to do it. This manuscript has benefited from comments by Sharon Keen.

References Cited

Archer, David J.W., and Kathryn Bernick. 1990. Perishable Artifacts from the Musqueam Northeast Site. Report on file, British Columbia Archaeology Branch, Victoria.

Barnett, Homer G. 1955. The Coast Salish of British Columbia. University of Oregon Press, Eugene.

Bernick, Kathryn. 1983. A Site Catchment Analysis of the Little Qualicum River Site, DiSc 1: A Wet Site on the East Coast of Vancouver Island, B.C. National Museum of Man Mercury Series, Archaeological Survey of Canada Paper 118. National Museums of Canada, Ottawa.

Bernick, Kathryn. 1989. Water Hazard (DgRs 30) Artifact Recovery Project Report. Permit 1988–55. Report on file, British Columbia Archaeology Branch, Victoria.

Bernick, Kathryn. 1998. Stylistic Characteristics of Basketry from Coast Salish Area Wet Sites. In Hidden Dimensions: The Cultural Significance of Wetland Archaeology, edited by Kathryn Bernick, pp. 139–156. UBC Press, Vancouver, B.C.

Burley, David V. 1980. Marpole: Anthropological Reconstructions of a Prehistoric Northwest Coast Culture Type. Dept. of Archaeology, Simon Fraser University, Burnaby, B.C.

Clague, John J. 1981. Late Quaternary Geology and Geochronology of British Columbia. Pt. 2: Summary and Discussion of Radiocarbon-Dated Quaternary History. Geological Survey of Canada Paper 80–35.

Clague, John J., John Luttenauer and Richard Hebda. 1983 Sedimentary Environments and Postglacial History of the Fraser Delta and Lower Fraser Valley, British Columbia.

Canadian Journal of Earth Sciences 20:1314–1326.

Erling, Jo Ann. 1990. The Conservation of Artifacts from the Beachgrove Water Hazard Site DgRs 30. Report on file, British Columbia Archaeology Branch, Victoria.

Fladmark, Knut. 1975. A Paleoecological Model for Northwest Coast Prehistory. National Museum of Man Mercury Series, Archaeological Survey of Canada Paper 43. National Museums of Canada, Ottawa.

Grey, Jack L. 1989. Wood Identifications of DgRs 30 Artifacts. In, Water Hazard (DgRs 30) Artifact Recovery Project Report, by Kathryn Bernick, appendix D. Permit 1988–55. Report on file, British Columbia Archaeology Branch, Victoria.

Jackson, Chris. 1989. Stone, Bone, Antler, and Shell Artifacts. In, Water Hazard (DgRs 30) Artifact Recovery Project Report, by Kathryn Bernick, appendix B. Permit 1988–55. Report on file, British Columbia Archaeology Branch, Victoria.

Krajina, V.J., K. Klinka and J. Worrall. 1982. Distribution and Ecological Characteristics of Trees and Shrubs of British Columbia. Faculty of Forestry, University of British Columbia, Vancouver.

Matson, R.G., and Gary Coupland. 1995. The Prehistory of the Northwest Coast. Academic Press, San Diego.

Matson, R.G., Deanna Ludowicz and William Boyd. 1980. Excavations at Beach Grove (DgRs 1) in 1980. Report on file, Laboratory of Archaeology, University of British Columbia, Vancouver.

Meidinger, Del, and Jim Pojar (editors). 1991. Ecosystems of British Columbia. British Columbia Ministry of Forests (Special Report 6), Victoria.

Mitchell, Donald H. 1971. Archaeology of the Gulf of Georgia Area, a Natural Region and its Culture Types. Syesis, Vol.4, Suppl.1. British Columbia Provincial Museum, Victoria.

Mitchell, Donald H. 1990. Prehistory of the Coasts of Southern British Columbia and Northern Washington. In, Handbook of North American Indians. Vol.7: Northwest Coast, edited by Wayne Suttles, pp. 340–358. Smithsonian Institution, Washington, D.C.

Smith, Derek G. 1964. Archaeological Excavations at the Beach Grove Site, DgRs 1, during the Summer of 1962. B.A. thesis, Department of Anthropology and Sociology, University of British Columbia. Copy on file, Laboratory of Archaeology, University of British Columbia, Vancouver.

Smith, Harlan I. 1923. An Album of Prehistoric Canadian Art. Anthropological Series 8, Bulletin 37, Canada Department of Mines, Ottawa.

Stevenson, Ann. 1989. Netting and Associated Cordage. In, Water Hazard (DgRs 30) Artifact Recovery Project Report, by Kathryn Bernick, appendix A. Permit 1988–55. Report on file, British Columbia Archaeology Branch, Victoria.

Stevenson, Ann. 1998. Wet-Site Contributions to Developmental Models of Fraser River Fishing Technology. In, Hidden Dimensions: The Cultural Significance of Wetland Archaeology, edited by Kathryn Bernick, pp. 220–238. UBC Press, Vancouver.

Suttles, Wayne. 1990. Central Coast Salish. In, Handbook of North American Indians. Vol.7: Northwest Coast, edited by Wayne Suttles, pp. 453–475. Smithsonian Institution, Washington, D.C.

11. Coquille Cultural Heritage and Wetland Archaeology

Donald Ivy and Scott Byram

Part I: Archaeology and Tribal Cultural Heritage

Native Americans and Archaeology

Native American participation is starting to bring far-reaching changes to American archaeology and cultural resource management. In the United States archaeological research has traditionally been undertaken by people of non-Native ancestry. Elsewhere in the northern hemisphere, archaeological projects are often sponsored and conducted by descendants of the people who lived and worked at the sites being studied. As such, in countries like England or Japan "prehistoric" archaeology is seen by many as an aspect of national cultural heritage. Although the situation is beginning to change, comparatively few Americans see Native American archaeological sites as a dimension of local cultural heritage worthy of preservation and study.

We hold that the value of archaeological sites can be more widely recognized when local Native communities participate in archaeological research in their traditional lands. As descendants of original site inhabitants, Native people can provide an important link between past and present, demonstrating for the public the heritage value of sites and other cultural resources. More than just laboratories for studying stone tool making or social hierarchy, sites are often places with names, and oral tradition tied to these names. They are places of celebration and tragedy, perseverance and creation. In many cases these aspects of our archaeological heritage can only be brought forth and preserved through the participation of Native American communities in cultural resource management at the local level. And due to new legislation and efforts to strengthen and diversify tribal government programs, tribes are becoming a growing force in all aspects of Native cultural heritage work.

In addition to increasing public awareness and participation, tribal involvement in cultural resources management (CRM) at the local and regional level can facilitate governmental agency communication involving site stewardship. Although a body of laws has been developed for the management of cultural resources in the United States, archaeological sites continue to be impacted at a disturbing rate by looting, vandalism, and development. In some regions local tribes have taken an active role in cultural resources protection. Their role often involves communication with individual landowners and local governments leading to site protection and access to sites for research on private as well as public lands. The Coquille Indian Tribe's efforts at archaeological site research and protection on the Coquille River illustrate the successes possible when an Indian tribe actively participates in archaeological site management at the local level, working productively with local communities, university archaeologists, agency resource managers, private landowners and others.

Archaeology and the Coquille Tribe

The Coquille Indian Tribe is a federally recognized Indian tribe headquartered in Coos County on the southern coast of Oregon in the northwestern United States (Figure 11.1). The Tribe exercises political and legal jurisdiction on more than 6500 acres of mixed use urban and forest lands, a substantial portion of these lands being located in the Coquille River drainage basin. Tribal membership exceeds 700 individuals, more than half of whom live away from the local tribal community. Tribal budgets to support governmental operations and delivery of services to the membership and the Tribal community are derived from several federal sources, and from revenues generated by Tribal enterprises. The Tribe does have a casino, and while this has not made the Coquille people rich, bills are paid on

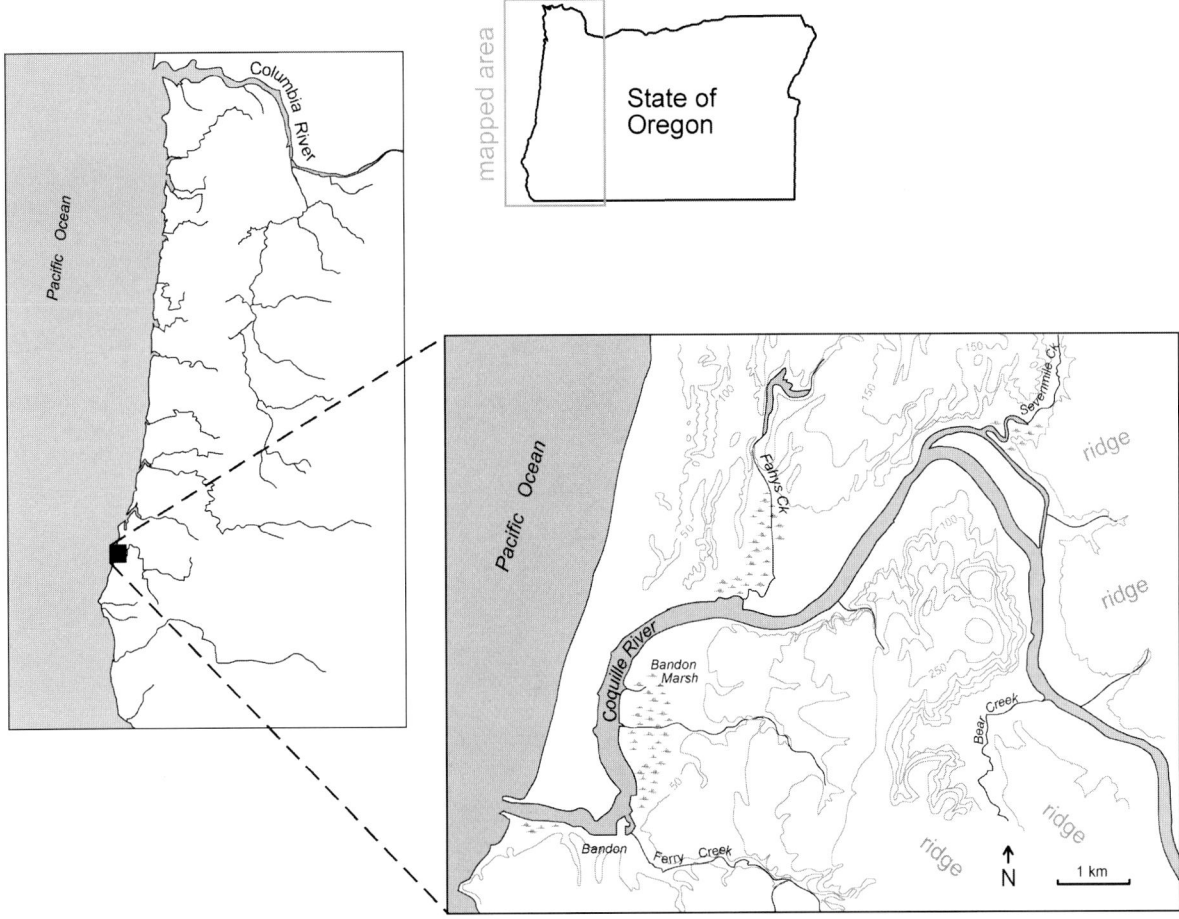

Figure 11.1. Maps showing the location of the tidally-influenced Coquille River on the coast of the state of Oregon, in Northwestern North America.

time, and banks are very happy to lend money to the Tribe.

In 1994, the Coquille Tribe began what would become an intensive five year investigation of pre-contact fishing weirs and human occupation sites along the estuary shorelines of the Coquille River (Figure 11.2). From July, 1998, to December, 1999, the Tribe's archaeological research efforts were conducted under the aegis of the Coquille River Archaeological Mapping Project, a collaborative partnership between the Coquille Indian Tribe Cultural Resources Program (CITCRP) and the University of Oregon Department of Anthropology; a project funded by Tribal dollars and a National Historic Preservation Fund grant. Much of the research highlighted in part II of this article is derived from this project's research (CITCRP 1999). Today, more than 30 archaeological sites have been documented on the shores of the 65 tidewater kilometers of the Coquille River, dating from 3500 to 150 years ago. The sites include both "wet" and "dry" sites, several with fishing weir features (Figures 11.2 and 11.3) and in some cases other perishables,

several riverbank "midden" sites containing residential debris such as lithics and faunal remains, at least six sites with weir components stratigraphically underlying midden components, and a small number of riverbank lithic scatters. Research at these sites has yielded much information about cultural activity in the Coquille wetlands, hence the CITCRP participated in information sharing at the 1999 WARP conference at the University of Florida in Gainesville.

Projects like the Coquille River Archaeological Mapping Project are undertaken by the CITCRP only after they are reviewed and endorsed by the Tribe's Culture Committee, and recommendation is passed forward to the Tribal Council. The Tribal Council, as the governing body of the Tribe, then delegates authority to the CITCRP to spend resources and represent the Tribe's cultural interests to accomplish cultural objectives. Developed through this process of committee input and governmental approvals, a primary goal of the CITCRP has been the interpretation

Figure 11.2. Coquille tribal members and University of Oregon archaeologists map and characterize wood stake fishing weirs at the Osprey Site (35CS130) on the north bank of the Coquille River near Bandon Marsh.

Figure 11.3. Rows of wooden stakes from fishing weirs at the Osprey site date between 700 and 150 years BP. Erosion of the riverbank at the site has been severe in recent years, and continues to expose new features.

and preservation of archaeological resources everywhere they exist within the ancestral territory of the Coquille Tribe. For the past five years the nexus of Coquille cultural resource work has been the Coquille river estuary and its many wetland archaeological sites.

Archaeological investigation is important to Tribal cultural preservation goals. The Coquille Tribe embraces the tools of archaeology and anthropology as critical elements of an effective cultural resource management program. Though we might dig in the dirt more often and with more enthusiasm than most other tribes, our purpose is not to just "collect" artifacts to prove our ancestors were around for a long time before white people settled here, or to "record the site" as part of some numbers game or for bragging rights on our ability to run projects or publish papers.

The reason we survey sites, dig in the dirt, – and in the case of the Coquille River, muck around in the mudflats and tidal sloughs – is that there is an inseparable relationship between people and the landscapes they live on and utilize to sustain and maintain their daily lives and way of life. We have discovered in our work at Coquille River wet sites that as the estuary landscape has changed over time, so too have the people – where they lived, the tools they used, the technologies they applied. Combined with traditional knowledge, these discoveries guide our thinking about the past, so that our CRM efforts reflect who we are and what our traditional cultural heritage is. The research also allows us to determine what is important to preserve and interpret for future generations to respect and appreciate. In this sense, research, interpretation, and preservation of archaeological sites is fundamental to the cultural legacy of the Coquille Indian Tribe.

Our research also gives the public stories which help them appreciate the richness and depth of human history on the landscape where they live, work, and recreate. For example, today's population in much of the Coquille Valley is thought to be comparable in size to what it was before American settlers moved into the area. Yet when the Coquille people were the sole residents of the valley the environment was vastly different from what it is today. In the past, thousands of people relied on the richness of the Coquille estuary, while today few people rely on these resources for their livelihood. Though heavily used, the abundance of the estuary's resources was maintained by untold generations of Native communities. Through archaeology and ethnohistoric research we are developing a better understanding of the ways people used estuary resources without diminishing ecological productivity, which in turn gives us a road map for improving our relationship with the ecosystem today.

As for more practical and programmatic reasons for the CITCRP, the Tribe has chosen to be the lead agency in managing its cultural resources and cultural interests, rather than yielding these responsibilities to the U.S. Bureau of Indian Affairs (BIA). This entails building tribal capabilities and taking responsibility for being an active participant (and frequent advocate) in the preservation of both pre-contact archaeological and contemporary cultural resource sites. The cultural resource responsibilities taken on by the Tribe today are influenced by several factors:

1) *Federal Laws* and subsequent state regulations in Oregon have directed that a process of consultation should occur between tribes and government agencies when potential threats or impacts to archaeological or cultural resources exists as a result of government actions. Two examples of that consultative process at the federal level are the Native American Graves Protection and Repatriation Act (NAGPRA) and the National Environmental Policy Act (NEPA). NAGPRA regulations provide a basis for consultation between Tribes and others in matters of cultural affiliation, cultural patrimony and the disposition of human remains found on federal lands or held in museum collections. NAGPRA also applies to archaeological materials recovered or stored by or for federal agencies at museums and other institutions. NEPA mandates that tribal cultural resource issues be considered when federally operated or funded development activities or projects are being planned and implemented within the jurisdictional authority of a tribe(s). Both NEPA and NAGPRA also address contemporary issues concerning tribal "traditional cultural properties" (plants, habitats, places) and traditional practices (ceremonies), as well as protection, interpretation and conservation of archaeological resources.

2) *Presidential Executive Orders* direct federal agencies to establish "government to government" relations with federally recognized Tribes whenever an agency action will take place within an area of tribal jurisdiction or interests. Most of these interactions between tribes and agencies have natural resource or environmental quality implications.

3) *Downsizing Federal Budgets* and subsequent reductions of federal agency budgets (particularly for agencies involved in land or natural resource management) increasingly compel federal, state, and local agencies and governments to seek "partnerships" at the local level. Where tribal governments are also operating at those local levels, they frequently are looked to as potential technical as well as funding partners.

4) *Federal Trust Responsibility* to tribes (as defined by 200 years of federal Indian law) is delegated to several federal agencies – not just the Bureau of Indian Affairs. Although the BIA maintains a lead agency role as the federal "trustee," tribes look to several agencies to perform the federal trust: the Bureau of Land Management performs cadastral surveys of Indian lands; the Bureau of Reclamation manages dams, reservoirs and water resources on Indian lands; the Army Corps of Engineers exercises permitting and regulatory authority; and the Environmental Protection Agency governs air and water quality standards and management

practices. Tribes also interact with several other federal and state agencies in management issues off tribal lands per se, but still within tribal jurisdiction authority.

5) *Other Tribes* complete the suite of agencies and governments that CITCRP interacts with in cultural resource matters. In the case of the Coquille Tribe, there are four other federally recognized tribes that also have some cultural or political interest within the Tribe's geographic territory. Cultural interests reflect the shared heritage and history of coastal tribes since the contact era of the 1850's, while political interests reflect overlapping legal jurisdictions of tribes as a result of various federal laws and agency regulations.

Although the Coquille program is still growing, we have had numerous successes in developing cultural resource management strategies and policy at the local and regional levels. These include what may be the most thorough county-level CRM policies in Oregon, wherein the Coquille Tribe and Coos County Planning Department coordinate site protection as part of the development permit process. A cultural resources GIS database is currently being developed as the framework for this planning effort. Other positive results have been generated by partnerships with the Bureau of Land Management, and focusing on wetland sites in particular, the U.S. Fish and Wildlife Service (USFWS). For example, working with the Archaeological Conservancy, the CITCRP helped to facilitate USFWS acquisition of a 500 acre parcel along the Coquille estuary, land which contains several key archaeological sites. Now part of the Bandon Marsh National Wildlife Refuge, the tidal wetlands at this location will be restored for wildlife habitat and long term site preservation (Dunbar 2000).

Partnerships have been developed with many other agencies and organizations. For several years both the University of Oregon and Oregon State University have participated in archaeological research in collaboration with the Coquille Tribe, and a handful of graduate students' theses have been based on Coquille archaeological and ethnographic work. In 1999 and 2000 the Tribe also worked with Southern Oregon University, whose field school takes place along the Coquille River and the nearby coast. Collaborative archaeological work has also been conducted with the Bureau of Land Management, including initial stages of large scale land management efforts focusing on Coquille Forest lands and surrounding federal lands. The Tribe also has frequent cultural resource interaction with the U.S. Forest Service, State Parks Department, the Division of State Lands, regional historical societies, and other western Oregon Indian tribes.

The results of this research and successful collaborative management are showcased annually at the Coquille Cultural Preservation Conference, held in May at Tribal

facilities in North Bend. Last year several geologists, archaeologists, and tribal cultural heritage specialists came together to present their research on past landscape change on the Oregon coast. Based on these presentations articles were written for a conference proceedings volume entitled *Changing Landscapes*, edited by Robert Losey (2000) of the University of Oregon. Presentations at the May, 2000 Coquille conference focused on diverse aspects of tribal history in western Oregon, emphasizing archival research initiated by collaborative efforts of the Coquille Tribe, other Oregon Tribes, and the University of Oregon Department of Anthropology (Younker 1997). Some of the highlights of Coquille wetland archaeological research through 1999 are discussed in the following section.

Part II. *Coquille Wetland Archaeology*

Among the estuaries of the Oregon Coast, the Coquille River has one of the most extensive tidewater lengths, reaching inland some 65 kilometers, with most of the estuary a comparatively narrow tidal river channel. In the lower estuary (below river mile 10) the intertidal margin ranges in size from a narrow strip of mudflat between the main channel and the steep riverbank to an expansive tidal flat and salt marsh near the mouth of the river. Although altered by drainage and diking systems for agriculture, an extensive network of tidal channels formerly meandered through the valley lowlands surrounding the main channel of the Coquille River. Historically, these tidal channels were the setting of extensive fishing weir complexes built and maintained by Indian people, many of whom lived in villages along nearby estuary shores. Typical Coquille villages were clusters of gabled wooden houses made of finely hewn split wood planks. Large cedar canoes were key to the Coquille economy, transporting people to and from fishing, hunting, and plant gathering localities within the valley and on the outer coast. Like most Oregon coast tribes, the Coquilles spent most of the year at their homes, though they frequently visited other communities for trade, family gatherings and festivals. Although the economy was centered on the estuary, seasonal visits to upriver fishing stations or upland hunting camps were undertaken by some community members (Tveskov 2000).

Since 1994 a key focus of the CITCRP and collaborating university archaeologists has been wetland fishing weir and village sites on the shores of the Coquille estuary (Byram and Erlandson 1996; Byram *et al.* 1998; Roth and Hall 1995, Tveskov 2000). Of the 30 known Coquille estuary archaeological sites, over half contain wood stake weir features, and a comparable proportion contain midden strata, including diverse residential debris such as lithics and faunal remains. Distinctive artifact classes are present in Coquille wetland sites, including projectile points, many other bifacial and unifacial tools, diverse ground stone tools including adze or celt blades, mauls, bowls, abraders, and pendants, a variety of bone and antler tools, and various worked wood and woven fiber artifacts. The perishable assemblages include artifacts primarily related to fishing apparatus such as wood stake weirs and associated split wood lattice, but other basketry, cordage, and wooden tools are also preserved in the intertidal wetland sediments. Most of these artifacts have been found in contexts dating between 1100 and 150 years ago, and a small number are much older.

Unfortunately many sites are undergoing erosion due to changes in the main Coquille River channel. We have yet to find a solution to this problem, and so far efforts have focused on documenting portions of sites being lost to erosion. By examining erosional exposures we have characterized site stratigraphy along much of the riverbank. Archaeological techniques used to investigate sites include GPS and transit mapping, cutbank profiling, feature mapping, radiocarbon dating, surface collection, and limited excavation aimed at recovering midden constituents for analysis. This suite of techniques has been proven effective over the last decade of archaeological research on the Oregon coast (Erlandson and Moss 1999).

At least six Coquille sites hold a stratigraphic sequence wherein tidal wetland sediments underlie upland riverbank levee deposits. Weirs and other perishables are preserved in the wetland sediments, and midden layers are exposed in portions of the riverbank levee stratum. Faunal remains are abundant, though predominantly fine grained, in midden layers which have been excavated. Notably, the identifiable fish bones recovered at these sites increases over 100 fold when 1 mm mesh is used instead of the more commonly used 6 mm mesh, and when water screening is used instead of dry screening (Losey 1996; Tveskov 2000). Paleobotanical macrofossils are abundant in many exposures of tidal wetland strata, including extensive buried driftwood layers, channel floor litter, buried marsh peats, and tsunami sands (Witter 1999).

Interdisciplinary collaboration has been key to the success of Coquille archaeological research at riverbank sites, providing a better understanding of estuary life history, paleoecology, fish habitat changes, and archaeological site taphonomy. The two aspects of Coquille wetland site research which have been most intensively examined at this time are the study of traditional fishing weir technology and long term wetland landscape change in the estuary.

Fishing Weir Technology at the Coquille Estuary

Interest in Coquille wetland sites on the part of tribal members and University of Oregon archaeologists began with a focus on wood stake fishing weir use on the North-

west Coast. Questions such as the antiquity of weir fishing were of concern, as weirs were thought to be a hallmark of the development of complex social structure on the Northwest Coast, founded on salmon fishing (Matson and Coupland 1995; Moss and Erlandson 1990). Archaeological weirs were often seen as indicators of large group mobilization, massive seasonal salmon harvests, and a suite of cultural activities ethnographically documented as related to weir use. However, the models of weir use which were being employed in Northwest Coast archaeology did not effectively address variation between freshwater and tidewater weir technology, nor did they allow for much diversity in weir fishing strategies.

Our research on the Coquille River, along with work elsewhere on the Oregon coast, led to new interpretations of tidal weir use as distinct from freshwater riverine weir use; archaeological and ethnohistoric data indicated that tidal weirs were used for diverse fish harvests in the estuary, emphasizing runs of fish occurring in all seasons (Byram 1998; Erlandson *et al.* 1998). The tidal weir fishing system incorporated lines of closely spaced, vertical wooden stakes (Figures 11.2, 11.3), often in pairs oriented in a V-shape across the mouths of tidal sloughs (Figure 11.4). Other broad, arc-shaped structures were built on expansive tide flats. Split wood woven lattice was used in some cases for weir panels (Figure 11.5), and elsewhere in cylindrical basketry traps or other impoundment facilities. It appears that the weirs were used in conjunction with the tides. At high tide fish would swim upstream of the weir, passing above it in the water column. As the tide went out and the water level lowered, the fish would become trapped behind the weir, often in lattice enclosures or basketry fish traps. Dip nets or open weave scooping baskets were used to remove often small fish from the impounding area, and the fish were typically placed into canoes for transport to nearby residential sites.

The new interpretation of tidal weir use on the Northwest Coast is supported by 1) the widespread distribution of archaeological weirs (over 60 known sites in Oregon estuaries) and their presence in settings such as Coquille estuary tidal sloughs and tide flats; 2) weir setting related to modern data on the habitat and abundance of numerous estuarine fishes inhabiting or spawning in these settings; 3) oral tradition recorded in early twentieth century interviews with Coquille elders and individuals from other tribes; 4) faunal data from residential sites along the Coquille estuary shore, showing a diversity of fishes were harvested and processed at these sites including herring, sardine, salmon, flounder and others; 5) the gauge (warp spacing) of archaeological fishing weir lattice (predominantly identified at Coquille estuary weir sites) which is much smaller in most recovered lattice (10 mm or less) than would be expected if salmon were the primary species targeted; 6) weir site

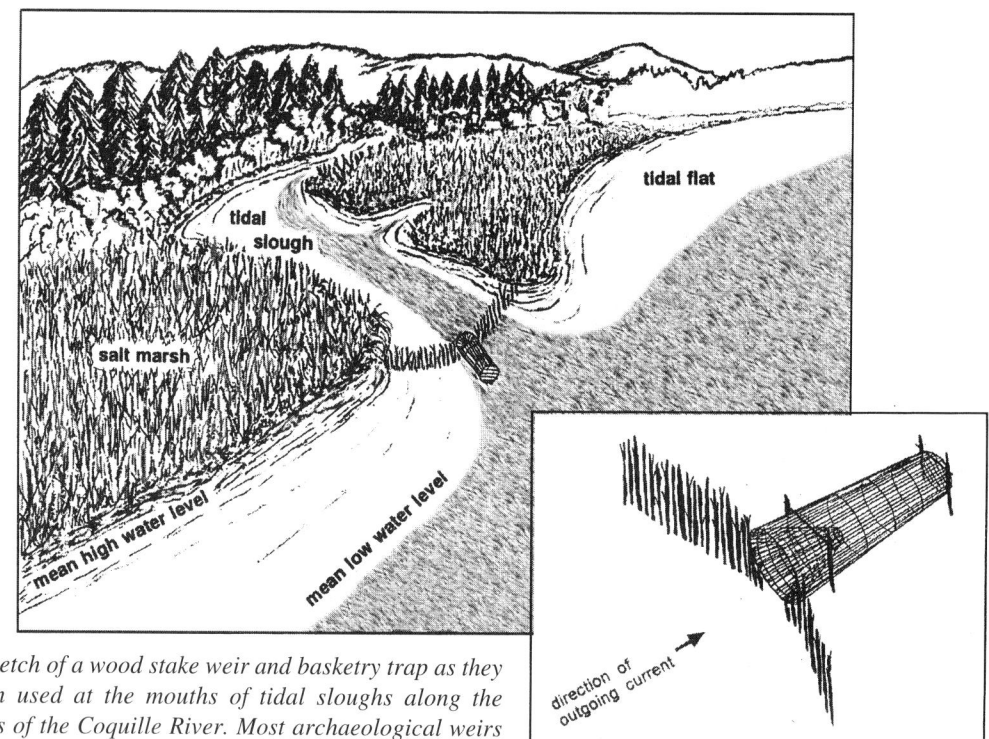

Figure 11.4. Sketch of a wood stake weir and basketry trap as they may have been used at the mouths of tidal sloughs along the tidewater banks of the Coquille River. Most archaeological weirs are known from settings such as this, and fine gauge split wood lattice from woven traps or weir panels is present at three Coquille sites.

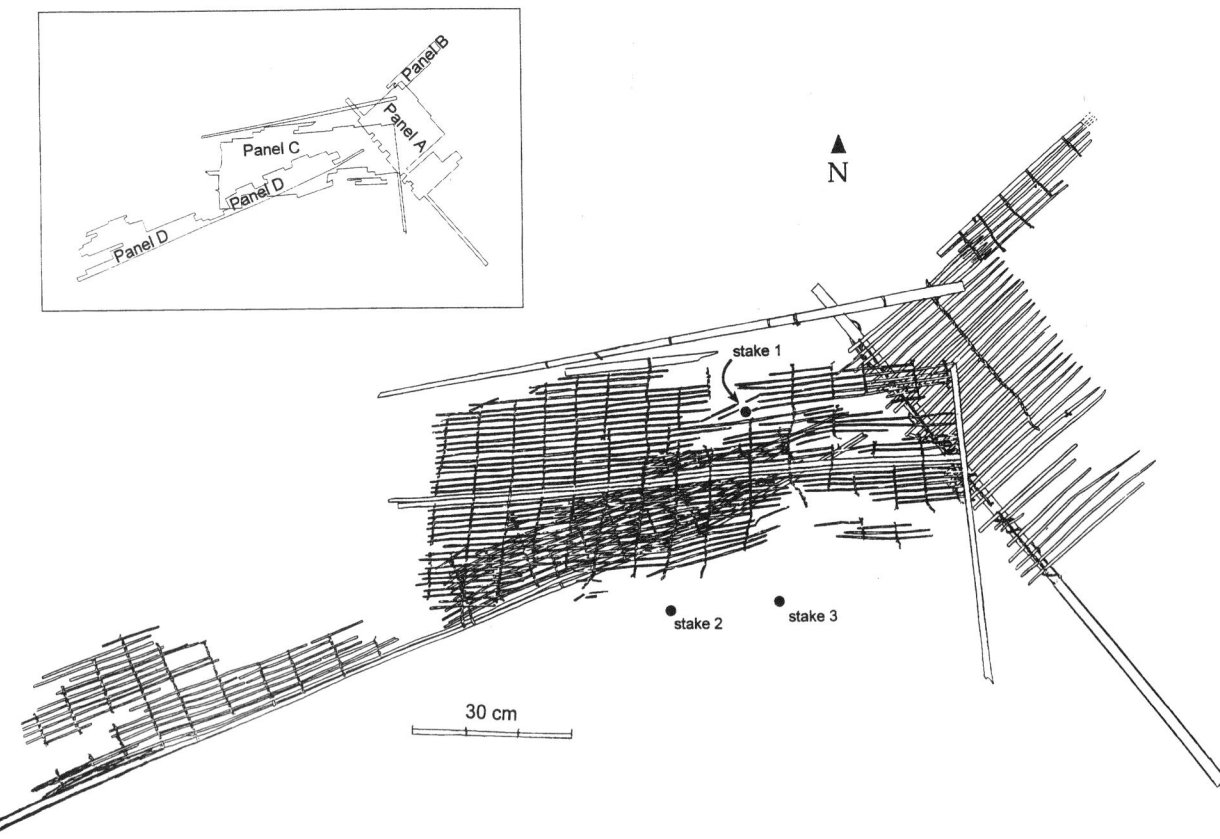

Figure 11.5. Feature consisting of four overlapping rectangular lattice panels excavated from an eroding portion of the Osprey site, dated to between 150 and 300 years ago. These panels may have collapsed during a storm or tsunami, or they may be the remains of fish traps destroyed by the U.S. Army during attacks on Coquille communities at the onset of American settlement in the region in 1851.

locations being close to major population centers on the shores of the lower estuary, indicating year round access to sites by village residents.

Overall, we came to the conclusion that weirs played a central role in the Coquille economy and elsewhere on the Oregon coast, but rather than being focused on salmon harvests, use of these weirs was a year round activity, and diverse fishes were harvested with these devices. Unpublished ethnographic accounts indicate Coquille tidewater weirs were fundamental to the local economy, and this is supported by the size of historic fish populations (Monaco *et al.* 1990–91). Oral history indicates that during spawning runs of herring, salmon, smelt, sardine and other fishes, massive numbers were caught using weirs in tidal channels or on tide flats. At other times residential fishes were caught, the fresh meat supplementing stored foods. These same ethnographic sources also indicate that anadromous fishes, including salmon and Pacific lamprey, were caught most often above tidewater. Here weirs could be built across the entire river, temporarily blocking fish passage during spawning runs. Groups of fishers from down river villages visited upriver fishing stations during runs. Although more widely documented historically, these riverine, freshwater weirs function quite differently from estuarine tidal weirs. Both systems were highly productive given appropriate conditions, and both were fundamental to the Oregon coast economy. The tidewater weirs offered the advantages of abundant catches during all seasons and deepwater canoe access, and they were more permanent due to comparatively gentle tidal current.

Fishing Weirs, Villages, and Changing Wetland Landscapes

In the course of studying fishing weir technology at Coquille wetland sites, the processes of landscape change in the Coquille estuary became a central focus of our research. The stratigraphic record at several sites showed that the wetland environment had changed drastically during periods of site use. By radiocarbon dating wooden stakes and midden organics, we began to trace patterns in the emergence of the wetlands along the Coquille estuary shore. Fortuitously, geologist Robert Witter of the University of Oregon was concurrently investigating evidence of earthquake hazards in the Coquille estuary (Witter 1999), and his willingness to collaborate greatly expanded the scope of our efforts. In terms of regional tectonics, the estuary lies in the southern portion of the Cascadia subduction zone. Previous geological research had shown that massive Cascadia earthquakes occur roughly every 500 years in this region, resulting in rapid subsidence and relative sea level rise of up to 1–2 meters. In the periods between these subsidence episodes gradual uplift occurs, in some cases

canceling out the sea-level rise caused by the subsidence.

In the wetlands of the lower Coquille Valley Witter identified a 6600 year stratigraphic sequence of repeated marsh peat development punctuated by episodes of rapid sea level rise and burial by sandy bay muds. This sequence was present in different marsh settings along the margin of the estuary, and Witter concluded that the factors which caused these changes in the Coquille estuary landscape were gradual global sea level rise, infilling by river borne sediments, and cycles of tectonic subsidence and uplift. Although global sea level rise and infilling have accounted for the greatest changes in the estuary landscape, rapid submergence and burial of wetland and riverbank strata during and after earthquakes likely accounts for the extensive, well-preserved perishable assemblages at Coquille wetland sites.

While the geological data provided a large scale record of estuary landscape change, the archaeological data allowed us to more closely examine the process of wetland emergence along the estuary shore (Byram and Witter 2000). Focusing on the margins of expansive alluvial lowlands which were once extensive tidal wetlands, we documented and radiocarbon dated repeated episodes of wood stake weir building along the riverbank. Twenty-three of the 25 radiocarbon dates obtained from Coquille wetland sites fall between ~1200 BP and ~400 BP, and therefore the processes of wetland emergence are best documented for this period. Notably, this period is roughly bracketed by two sequential subsidence episodes associated with major Cascadia earthquakes (at ~1100BP and A.D. 1700). Although many features likely remain buried under the alluvial sediments inland from the riverbank, those exposed along the modern riverbank show that the tidal wetlands expanded in a down river direction at a rate of over 200 meters per century in at least one location (Figure 11.6). As wetland accretion occurred, tidal channels became filled and new outlets to the wetland opened to the west (downstream). People adjusted to these changes by building new weirs at the new outlets to the tidal channel networks. As riverbank sediments accumulated (and interseismic uplift lowered relative sea level) several riverbank locations became suitable for seasonal or possibly permanent residence, which accounts for the extensive midden layers overlying wetland sediments along much of the riverbank (Figure 11.7). And in at least one case it appears that people may have intentionally used weirs and other debris to trap sediments and fill a tidal channel, possibly to concentrate wetland drainage at another outlet better suited to weir fishing.

There may be many weir features dating earlier than 1200 BP which remain buried beneath Coquille wetland sediments. For this period only one earlier weir has been identified, dating to ~3400 BP, at site 35-CS-167. This is

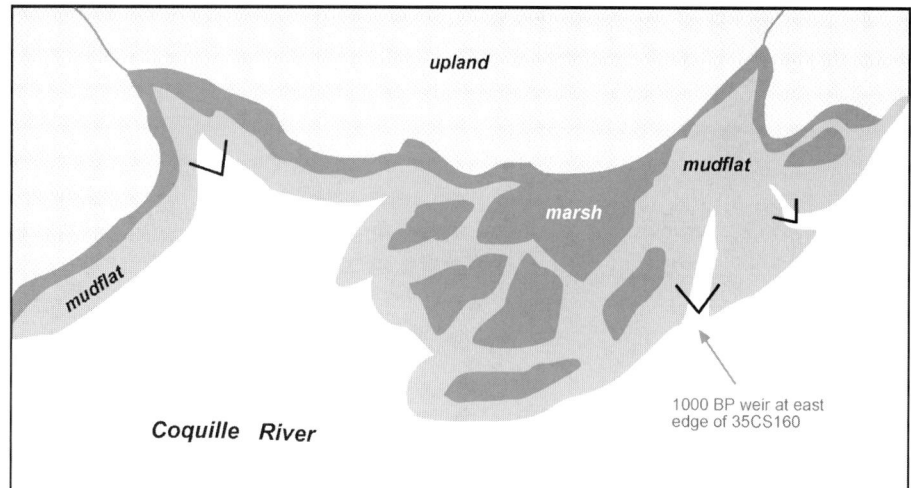

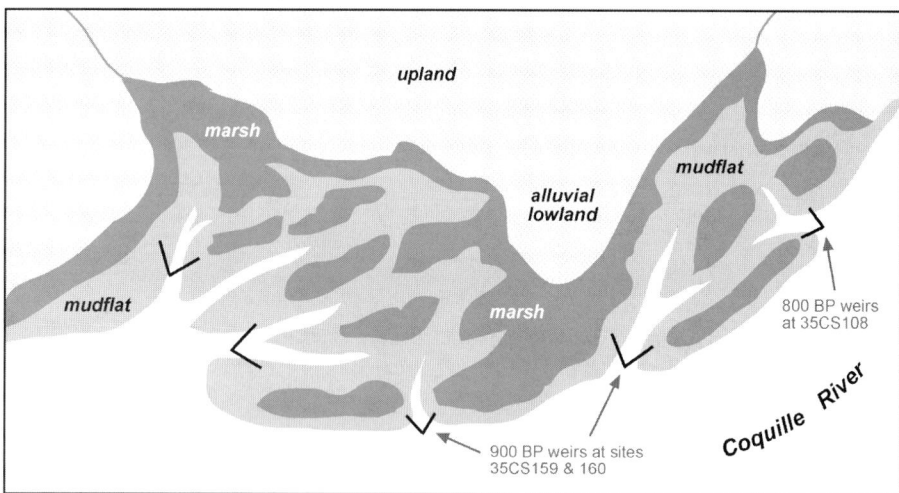

Figure 11.6. Model of wetland emergence along the north bank of the Coquille River. With relative sea-level rise decreasing after 3500 BP, infilling caused tidal marshlands and alluvial lowlands to form in the lower Coquille estuary. Archaeological and geological evidence at three riverbank sites indicate the riverbank levee accreted westward as much as 200 meters per century between 1000 and 600 BP. As older tidal channels became infilled, weirs were built at new outlets to the wetland. People may have used weirs and woody debris to trap sediments and alter wetland drainage patterns for improved weir fishing. Higher sections of the levee were used as residential sites and work areas.

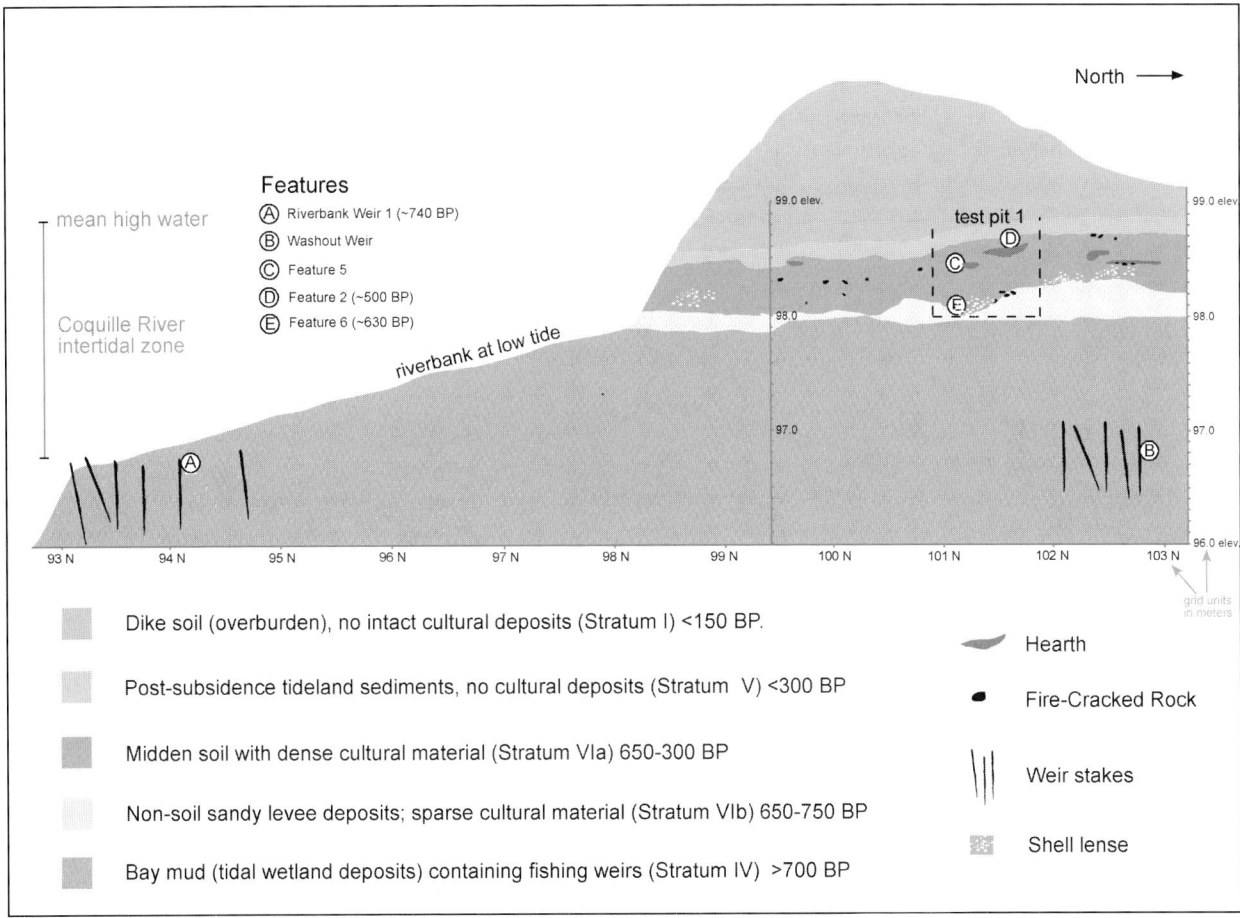

Figure 11.7. Composite profile diagram showing stratigraphy and features at the Philpott Site (35CS1) on the north bank of the Coquille River (after Byram et al. 1998). The strata here show a sequence of tidal wetland emergence, burial by riverbank levee development, and submergence due to tectonic activity. All cultural deposits are now within the intertidal zone due to sea level rise following massive subsidence in 1700 A.D. Prior to this the occupied midden layer (Str. VIa) was above tidewater. The oxygen-reduced bay mud stratum (Str. IV) underlying the midden layer contains wood stake weirs and split wood lattice panels – the remains of fishing structures used when a subsidiary tidal channel flowed through this portion of the riverbank and into the main river channel.

currently the oldest known fishing weir on the Pacific coast south of Canada. The weir is exposed at extreme low tide in a channel bank beneath over three meters of infilled sediment. It occurs in a tributary tidal channel, one of the few such channels which was not gated and drained in the past century for agricultural purposes. Located over one kilometer from the modern riverbank, the 3400 BP weir at this site was likely built in a tidal wetland at a time when the estuary was much more extensive than it is today. Since that time infilling has buried the weir and considerably narrowed this portion of the estuary. Other undated weirs have been identified in the channel where 35-CS-167 is located, and future radiocarbon dating of these will give us

the opportunity to test the model of weir building and wetland emergence in the 3500 to 1100 BP period.

Geological models indicate that between 7000 and 3500 BP the rate of sea level rise outpaced infilling in the lower Coquille estuary, and after this period the rate slowed. It is likely that tidal wetlands were less extensive before 3500 BP, though tidal channels suitable as fish habitat may have existed along the margins of the valley and in the upper portion of the Coquille valley. Notably, two radiocarbon dates from residential sites on hills above the valley floor are within a few centuries of 3500 BP. At that time, sea level rise may have constricted the valley floor, at least in the area which is now the lower estuary. Future investigations will explore the earlier years of estuary development and generate new models of weir use and residential site locations.

The archaeological research performed by the CITCRP and its collaborators has shed light on key aspects of cultural and ecological history in the wetlands of the Coquille River. In terms of Native fishing weir technology, site formation, and wetland landscape change, the Coquille estuary may be one of the most heavily studied in North America, yet many important questions have yet to be addressed by future projects: 1) What is the relationship (both in time and in character) between the sites along the estuary shoreline and nearby upland and bluff sites? Were they occupied simultaneously, or do these upland sites simply represent occasional movements to "new" estuary margins and shorelines during high water stands? 2) What additional techniques can be applied to what we know so far, that might expand interpretation of these sites and the estuary setting? Would analysis of pollen and woody debris trapped in buried soils give us more *useful* information about the past – or just more information? 3) Would excavations below the present day intertidal zone reveal new information about stratigraphy and human use of the estuary during different periods of time than what we've dated thus far? In particular, our emphasis on fishing technologies and landscape change has left residential site components comparatively less examined, yet these deposits, also threatened by erosion, hold vast evidence of diverse cultural activities which future research will emphasize.

The investigative techniques, success and discoveries of CITCRP research in tidal wetlands also pose similar studies to be done upriver – above the tide and away from the estuary lowlands. The geology of the upper river appears to be the source of raw material for lower river technologies, and geologic forces that so dramatically altered the estuary environment surely must have had similarly dramatic effects on the people and environs of the upper river. As ethnographies and Tribal oral histories demonstrated in previous research, significant relationships existed between the peoples of the upper and lower Coquille River. Did the upriver people and the landscapes they lived on – and their resources and technologies – change over time, as they did along the estuary? And what wet sites might await discovery in the gravels and valleys of the upper river? We will begin to address these questions in coming years, while continuing to monitor and evaluate eroding sites in the estuary as time permits.

Conclusions

The success of Coquille archaeology and cultural resource management can serve as a model for meaningful and productive collaboration between tribal and non-tribal interests. It demonstrates that Indian tribes can be effective and efficient lead agencies in investigating, interpreting and managing cultural sites, even when these are not on Indian lands, and it shows some of the opportunities which exist for scholars and professionals outside the tribe to gain access to Native people and traditional sources of knowledge and histories. Coquille collaborative efforts also show the significance of the inter-disciplinary/inter-agency approach to CRM issues and site investigations, particularly in understanding long-term changes to people and resources within estuary environments.

For the Coquille Tribe and its members, the research and findings of Coquille archaeological projects reinforce the significance and meaning of oral traditions and histories that, in the past, have often been overlooked, discredited, or misconstrued by science and academia. Certainly the 3400-year old fishing weir recently discovered at site 35-CS-167 brings witness to Tribal traditions that speak of "time immemorial" or "when time began," and reinforces for the public the understanding that history began long before Euro-American settlement in southern Oregon. This and other aspects of Coquille wetland archaeology contribute to a story of a magnificent relationship between a people and the landscape they have inhabited for countless generations. This story also holds a profound truth for the modern society that lives along the Coquille River today: dramatic changes will certainly come once more to this landscape, and so too will the habits and lifeways of the people who live there be changed.

Acknowledgments
Several people and agencies have contributed to CITCRP projects. We would like to thank the Coquille Indian Tribal Council and Culture Committee, the National Park Service, the U.S. Fish and Wildlife Service, the Archaeological Conservancy, the Port of Bandon, the Bussmann family, the Departments of Anthropology at the University of Oregon, Oregon State University, and Southern Oregon University, the Coos County Planning Department, and the many tribal members, students, and volunteers who participated in Coquille wetland archaeological projects.

References Cited

Byram, Scott. 1998. Fishing Weirs in Oregon Coast Estuaries. In Hidden Dimensions, the Cultural Significance of Wetland Archaeology. Kathryn Bernick, ed. University of British Columbia Laboratory of Archaeology Occasional Papers 1. Vancouver, B.C.

Byram, Scott, and Jon Erlandson. 1996. Fishing Technologies at a Coquille River Wet Site: The 1994–95 Osprey Site Archaeological Project. On file at the Oregon State Historic Preservation Office, Salem.

Byram, Scott, Mark Tveskov, Jon Erlandson, Charles Hodges,

and Robert Witter. 1998. Research in Response to Erosion at the Philpott Site, Coquille Estuary. Report prepared for the Coquille Indian Tribe Cultural Resources Program, Coos Bay, Oregon.

Byram, Scott, and Robert Witter. 2000. Wetland Landscapes and Archaeological Sites in the Coquille Estuary, Middle Holocene to Recent Times. In, Changing Landscapes: Proceedings of the 3rd Annual Coquille Indian Tribe Cultural Preservation Conference. Coquille Indian Tribe, North Bend, Oregon.

CITCRP (Coquille Indian Tribe Cultural Resource Program). 1999. Coquille River Archaeological Mapping Project Final Report, HPF-41-98-NA-4110. On file at the offices of the Coquille Indian Tribe, North Bend, Oregon.

Dunbar, Lynn. 2000. Time Well Spent. American Archaeology, Volume 4(1):31, Spring.

Erlandson, Jon M., and Madonna L. Moss. 1999. The Systematic Use of Radiocarbon Dating in Archaeological Surveys in Coastal and Other Erosional Environments. American Antiquity 64(3):431–444.

Jon M. Erlandson, Mark A. Tveskov, and R. Scott Byram. 1998. The Development of Maritime Adaptations on the Southern Northwest Coast of North America. Arctic Anthropology 35(1):6–22.

Losey, Robert. 1996. Fishing on the Lower Coquille River: An Archaeozoological Perspective. Master's Paper, Department of Anthropology, University of Oregon, Eugene.

Losey, Robert. 2000. Editor, Changing Landscapes: Proceedings of the 3rd Annual Coquille Indian Tribe Cultural Preservation Conference. Coquille Indian Tribe, North Bend, Oregon.

Maloney, Alice. 1933. Field Notes from the Coos Bay and Coquille River areas. Box 100, folder 3, Melville Jacobs Collection, University of Washington Libraries, Seattle.

Matson, R.G., and Gary Coupland. 1994. Prehistory of the Northwest Coast. Academic Press, New York.

Moss, Madonna, Jon Erlandson, and Robert Stuckenrath. 1990. Wood Stake Weirs and Salmon Fishing on the Northwest Coast: Evidence from Southeast Alaska. Canadian Journal of Archaeology 14:143–158

Monaco, M.E., R.L. Emmett, D.M. Nelson and S.A. Hinto. 1990. Distribution and Abundance of Fishes and Invertebrates in West Coast Estuaries, Volumes I and II. ELMR Report No. 4 NOAA/NOS Strategic Environmental Assessments Division, Silver Spring, MD, 232 pp.

Roth, Barbara, and Roberta Hall. 1995. Archaeological Testing at Site 35CS3. Report to the Coquille Indian Tribe. Department of Anthropology, Oregon State University, Corvallis.

Tveskov, Mark A. 2000. The Coos and Coquille Indians: A Historical Anthropology of the Northwest Coast. University of Oregon Ph.D. dissertation, University Microfilms, Ann Arbor, Michigan

Witter, Robert C. 1999. Late Holocene Paleoseismicity, Tsunamis and Relative Sea-Level Changes Along the South-Central Cascadia Subduction Zone, Southern Oregon, U.S.A. Doctoral Dissertation, Department of Geology, University of Oregon.

Younker, Jason. 1997. Revival of a Potlatch Tradition: Coquille Giveaway. Master's paper, Dept. of Anthropology, University of Oregon, Eugene.

12. From Stone Pavement to Temple – Ritual Structures from Wet Contexts in the Province of Drenthe, The Netherlands

Wijnand van der Sanden

Introduction

This article focuses on my working area – Drenthe, one of the three provinces that together constitute the northern part of the Netherlands (Figure 12.1). Over the past few centuries, the landscape of Drenthe has changed drastically. Around the end of the Middle Ages large parts of this province were still covered with bogs of varying sizes. Only very little of that landscape remains today. It was around the eleventh century that the peat of the raised bogs and fenlands began to be exploited for use as fuel. The peat-cutting activities intensified substantially in the 17th century. The greater part of the peat was cut according to plan, by specialised companies which sold the peat on the market. This gradually led to the disappearance, for example, of the large bogs near Smilde (Smildiger venen) and Hoogeveen (Echter venen) and the huge Bourtanger Moor expanding beyond the eastern border of the province of Drenthe. Numerous smaller bogs were meanwhile being exploited by local farmers, who cut the peat for their own use. The commercial production of peat peaked around the time of World War I, when annual production rose to some 65,000 days' work (1 day's work = 10,000 blocks of peat). I need not add that no quantitative information is available on the production of peat not intended for sale on the market.

Many finds came to light during the peat-cutting activities, which continued until shortly after World War II. Only a proportion of those finds have survived. Many others are mentioned in various records, but disappeared at some stage after their discovery. Then there are the finds which were never recorded on paper. How large that category is we of course don't know. Virtually no finds whatsoever are for example known from the 17th, 18th and early 19th centuries. But it's highly unlikely that peat cutters didn't come across any unusual objects in the peat in that long timespan.

There just wasn't any scientific interest in such objects at the time. We may assume that a vast number of finds disappeared forever in those days.

The range of finds that have survived is quite wide (see *e.g.,* Van der Sanden 1999). It comprises stone and flint axes, stone hammers, flint daggers and knives, bronze axes, spearheads, swords and daggers, bone tools, gold, bronze and amber ornaments, coins, clothing, hair plaits, querns, wooden ard shares, pottery, animals and parts of animals, waggons and parts of wagons, balls of wool, wooden bowls, and of course bog bodies. In comparison with the objects known from, for example, Denmark, the diversity of the Dutch finds is not very spectacular. The Dutch range includes, for example, no bronze lurs, bronze and gold vessels or bronze shields, nor any large weapon deposits. The world of the northern part of the Netherlands was clearly less complex. What is however impressive is the long timespan covered by our bog finds. The oldest objects date from the early Neolithic (*c.* 4900–3400 BC), the youngest from after the Middle Ages.

The majority of the finds recovered from the bogs of Drenthe are in my opinion votive offerings, deposited by the local population to beg supernatural powers for assistance, or thank them for favours already granted. Similar practices evidently took place in stream valleys, for those parts of the landscape have also yielded numerous intact and/or remarkable objects such as stone and bronze axes, deer antlers, querns, pottery, coins and human skeletons.

Many of the objects recovered from wet contexts in Drenthe are isolated finds. This holds for example for most of the stone and bronze axes. Others came to light in association with similar or different objects. An example of the latter category is the Roswinkel hoard. This hoard, which was found in 1924, consists of part of a bronze axe, 46 amber beads, a horn comb, a leather band and a few

Figure 12.1. Raised bogs and fenlands in the Netherlands around AD 1500. The most important raised bogs in Drenthe were: A: Bourtanger Moor; B: Echter venen; C: Smildiger venen. Dotted line: the Dutch border. Insert: the 12 provinces of the Netherlands; Dr. = Drenthe; Fr. = Friesland and Gr. = Groningen. Drawing: Groninger Instituut voor Archeologie (GIA).

pieces of textile. These objects were deposited in the peat around 1500 BC.

A small number of votive offerings were found in association with structural remains. It is the evidence available on such finds in Drenthe that I intend to discuss in this article. I shall also pay attention to structures associated with other finds indicative of rituals, and to structures near which no other finds were recovered, but which probably or most likely nevertheless played a part in rituals. But let us first take a look at what is known about such structures in the areas surrounding the Netherlands.

Ritual structures in wet contexts outside the northern part of the Netherlands

In the countries surrounding the Netherlands numerous examples are known of structures – from simple to complex – which were discovered in association with votive offerings or which can in some other way be linked with ritual acts in a wet environment. Below, I shall give an impression of the great diversity of these structures, which span the period from the Neolithic up to and including the early Middle Ages. Unfortunately only some of the finds have been extensively published.

The simplest structures are pits with rectangular or round cross-sections and depths of up to 1 m of the kind known from the bogs of northern Germany and Denmark. The majority date from the last few centuries BC and the first centuries AD (Becker 1947; *id.* 1970; Jankuhn in Diezel *et al.* 1958:193 ff.; Schwabedissen 1951:*id.* 1965). Among the finds recovered from these pits are intact vessels and potsherds, wooden and iron objects, animal bones and even human remains. Pottery seems to be the most common find. Many of the pits were also found to contain wood and/or stones. The wood may have served to line or cover the pit or to secure the objects. Various different combinations of pottery and stones have been encountered. The stones have been found inside, underneath (in the form of a pavement), next to, around and on top of the pots. One pit was found to contain pottery surrounded by rectangular blocks of peat.

Deeper pits were dug in bogs, too, as we know from a recent discovery made near Barth at Vorpommern, Germany (Kuhlmann 1998). Here, in a small peat-filled depression, excavators unearthed three 3-m-deep shafts whose fills contained pottery, deer antlers, branches and twigs. The pottery dates from the Roman period.

We also know of pits with substantial linings. Good examples from Denmark are the pit lined with wooden planks that was found in a bog near Smederup and the hollowed-out tree-trunk of Over Jersdal, both of which date from the pre-Roman Iron Age (Vebæk 1945). The fill of the 'sacrificial well' of Smederup, which was excavated in 1942, contained an alder barrel, part of a wooden spoon, the remains of 14–16 pots and some 200 stones. One year later a large hoard of bronze objects, mostly rings, was found very close to this lined pit.

The second category of structural remains from wet contexts is that of hearths. In 1886 a 10-cm-thick hearth with a diameter of 2 m came to light in a bog near Hedelisker in Denmark (Jankuhn 1967:137; Bemman and Hahne 1992:53–54). Among the charcoal the excavators found burned sheep and human bones, fragments of at least three pots and two knives. The date of the hearth is not certain, but other finds recovered from the peat, among which are weapons and a bronze sieve, can be dated to the Roman period.

Another example are the hearths that were found along bog trackway XII (IP) in the Ipweger Moor in Lower Saxony, Germany (Krämer 1992). In 1990 a platform measuring 6×3 m made of branches and brushwood was discovered on the south side of this trackway. An east-west oriented row of vertical posts came to light to the south of this platform, which the excavators referred to as a *Rastplatz*. More impressive is the 2-m-long trunk of a pine with a sharpened end which had sunk obliquely into the peat next to the latter fence. On the platform were four hearths whose ashes were mixed with large numbers of potsherds. Several layers were observable in the hearths. Noteworthy finds discovered in association with this platform are the remains of hazelnut shells, a sloe stone, parts of an axle and stones the size of a fist. The trackway was dendrochronologically dated to 713 BC.

Closely related to these hearths is the *Brandplatz* that came to light in the Aukamper Moor near Braak, in the German federal state of Schleswig-Holstein, in 1948 (Hingst 1967). Here, excavators found a horizon with a thickness of 50–70 cm consisting of layers of fire-cracked stones and layers of stone grit, charcoal and pottery fragments. The find layer had a diameter of approximately 12 m. To the north and west was a ditch with a width of 2–3 m and a maximum depth of 2 m in which a forked branch with a length of 178 cm was found. The two well-known anthropomorphic wooden figures came to light 50 to 70 m from this findspot (Capelle 1995:10–15). The pottery and recently obtained C14 dates point to the Early Iron Age.[1]

The third category of finds is that of wooden platforms. The oldest known platforms date from the Neolithic (Becker 1947:16–19, 53–58; Koch 1998:143–145; Rech 1979:53–54). They were built from roundwood, branches and twigs, often held together by vertical stakes. Some of the platforms comprise several layers. The objects that have been found on top of or next to such platforms comprise pottery, flint axes and chisels, animal bones and even human bones. The dimensions of some of these platforms are known: Salpetermosen (10×22 m), Veggerslev (50×10 m), Siggeneben Süd (40×20 m) and Alvastra (45×20 m); the first two findspots lie in Denmark, the third in northern Germany and the fourth in Sweden. The most impressive is the Swedish findspot, on the shore of lake Vättern in Ostergötlands, at the foot of the Omberg. The platform, which was constructed close to a spring, was connected to the mainland by a 75-m-long footbridge. On top of the platform, where 17 compartments could be distinguished, were more than a hundred stone hearths. The numerous finds included half-finished hammer-axes, amber beads shaped like such hammers and animal and human bones. According to the available dendrochronological evidence the platform was used for only a few decades, around 3100 BC. Alvastra, which was excavated in the early 20th century, is not the only Swedish example of such a site. In 1941 a platform consisting of roundwood, twigs and brushwood was excavated at Käringsjön (Arbman 1945). Within, on top of and next to this platform Arbman found a large quantity of stones, pottery, wooden implements and bundles of flax. The pottery dates from the 2nd to the 5th centuries AD. Another findspot has recently been discovered in Sweden:Skogsmossen near Fellingsbro. In this small fen Hallgren and his colleagues (Hallgren *et al.* 1997) found postholes which they interpreted as the remains of a platform measuring approximately 4×5 m. Near these features they found a vast quantity of pottery of the TRB (Funnel Beaker) culture, stones, lumps of daub, cereal grains, querns and axes. C14 dates show that the platform was used between *c.* 3800 and 3300 BC.

A good example of a Bronze Age timber platform is that of the site now known as Berlin-Spandau (Schwenzer 1997). In 1881, in the Stresow area, at the point where the rivers Spree and Havel flow together, rows of vertical stakes came to light in a layer of peat during digging operations for the construction of a gunpowder store. Between these palisades were beautifully patinated bronze swords, daggers and axes, but also a canoe, querns, bone implements and animal and human bones. A second excavation in 1954 brought to light horizontally arranged, closely spaced roundwood. It had long been thought that the pieces of timber were the remains of pile dwellings, but that hypothesis was abandoned after the second excavation. It is now believed that the remains represent a platform that played a part in the ritual deposition of objects in a wet environment. The platform is assumed to have been used between 1400 and 1200 BC.

Timber platforms are, incidentally, not restricted to Scandinavia and northern Germany. Similar structures are known from Poland (Makiewicz 1997) and England, too. The most complex platform is of course that of Flag Fen which Pryor excavated in the Fenlands in the 1980s (Pryor 1991:*id.* 1992).

Closely related to the timber platforms are the stone pavements that have been found at the peripheries of bogs. Their dates also go back to the Neolithic. Eva Koch (1998:145–146) mentions five in the area she studied – the Danish islands Sjælland, Møn, Lolland and Falster. No information incidentally seems to be available on the dimensions of those pavements. Pots and/or sherds were found in the immediate vicinity of the pavements, and also stone axes, querns and animal and human bones. Whether the pavements and the finds are contemporary is not absolutely certain in all of these cases.

Stone pavements are known to have been found in peat in northern Germany, too, in particular in Schleswig-Holstein. At Wees, for example, pots were found that were said to be standing on "rows of stones". More pots standing on stones came to light at Rethwischdorf (Schwabedissen 1949:68). What was described as pots set between two rows of stones were found at Royum (La Baume 1952). These assemblages date from around the beginning of our era.

My fourth category of finds concerns the cairns that have been found in several former bogs. The previously discussed *Brandplatz* near Braak actually belongs to this category too. In 1960 four find concentrations were recorded at the Danish findspot Forlev Nymølle, one of which is a cairn beneath which was a 3-m-long wooden anthropomorphic figure; on top of the cairn Andersen found sherds and two bundles of flax. The other assemblages consisted of pottery sherds, animal and human bones and a few wooden objects. They are thought to date from the pre-Roman Iron Age (Andersen 1961). The combination of a cairn and a wooden anthropomorphic figure has been encountered at other Danish findspots, too. Differently sized cairns were also found in a former bog near Rosbjerggaard, close to a concentration of cattle horns and sherds from the Roman Iron Age. On top of the highest cairn the excavators found two posts with sharpened ends, which they interpreted as the legs of a human figure. And in 1880 an ithyphallic anthropomorphic figure was found standing next to a cairn in the peat near Broddenbjerg (Capelle 1995:23–24). This wooden figure was dated to the end of the Late Bronze Age on the basis of the pottery associated with it (the C14 date obtained for the wood does not contradict such a date).[2] At a few sites a pit was found beneath a cairn. Becker (1970:147–148) describes such a find discovered at Skævinge in Denmark. The cairn had a diameter of about 2 m and the pit contained four pots. Three of the four pots were intact and each contained a stone.

I shall conclude my summary of structural remains outside the Netherlands with the enclosures. Remains of fairly small wattlework enclosures have been found at several sites (Jankuhn in Diezel *et al.* 1958:201; Bemmann and Hahne 1992:36 and 50). They were for example observed at the well-known Danish findspots Dejbjerg, Vimose and Thorsberg. There was even a separate timber causeway leading to the sacrificial site of Thorsberg. Several pots are reported to have been found standing within a wattlework enclosure in a small bog near Wees.

The largest concentration of enclosures was however found at Oberdorla in the German federal state of Thuringia. Unfortunately this findspot, which was excavated between 1957 and 1964, was never extensively published. The preliminary publications (Behm-Blancke 1957:*id.* 1960; *id.* 1989) inform us that in the last centuries BC a small lake formed at this site in a peat-filled depression. The sacrificial activities took place largely on the shores of this lake, apparently until after the year 1000, by which time the lake had filled up with peat to a great extent. All in all, Behm-Blancke recorded several dozen (about 90?) separate cult sites enclosed by palisades or wattlework fences. The enclosures vary in shape:some are round, others oval or more or less rectangular. Some of these enclosures contained altars surrounded by posts, posts arranged in a ship-shaped configuration and/or simple anthropomorphic figures. The site yielded a widely varied range of finds:animal bones (mostly of domestic animals but also of wild animals), human bones, farming implements, tools, pottery, wooden vessels, parts of wheels, bundles of flax and stones.

The different types of structures mentioned above were of course encountered in various combinations. The finest example of all is that of the Polish findspot Otąłażka. This site, situated on the banks of a meander of the Mogielanka, was excavated between 1967 and 1970 (Bender and Stupnicka 1974; Makiewicz 1988; Brzeziński 1992). On the right bank – the outside bend of the river – were three hearths in a row, surrounded by an 80 cm-thick layer of burned matter and pottery. Among the finds recovered near this layer are a spindle whorl, parts of an ard, whetstones and a quern. In the immediate vicinity of the hearths were two wooden benches and the remains of what was probably a wattle fence. Along the bank, finally, lay a 2.5–3 m-wide stone pavement, to the north of which were pottery and animal bones. The dominant element on the other bank was a round cairn with a diameter of 6 m and a height of 60 cm containing animal bones and pottery. Several timber structures were linked to this cairn, including a small bridge and an oval structure that may have been a platform. Large quantities of pottery, animal bones and wooden objects, including a felloe segment of a spoked wheel, were found on this bank, too. The datable finds derive from the late Roman period and the time of the great migrations.

The examples quoted above give an impression of the diversity of the ritual structures that have been found in bogs and stream valleys. The range of structures comprises hearths, pits and shafts, stone pavements, cairns, platforms and enclosures. The objects that have been found inside, beneath, on top of or near the structures are assumed to be votive offerings, remains of ritual meals or cult objects. The pots that were buried in the pits dug in the raised bogs, for example, are generally thought to have contained offerings of foodstuffs (*cf.* Harck 1984). We may assume that combinations of different activities took place at many of the structures. The platforms at the edges of open water, for example, will have witnessed sacrificial acts, during which offerings were deposited in the water, but also the consumption of ritual meals. A concrete example of a combination of different activities is provided by the complex cult site of Otąłażka. From the results of the analysis of the animal bones it was inferred that different activities took place on the left and right bank. It is thought that the right bank was used primarily for ritual feasts, while the left bank, where the cairn was found, was used for offering.

Stones seem to have played a leading part in all these activities. They were evidently believed to have special powers. People used them to build altars. They lit fires on top of these altars (Braak), erected figures on top of or next to them (Rosbjerggaard and Broddenbjerg) or buried figures beneath them (Forlev). A different story are the isolated stones which have frequently been found in association with bog finds. Some archaeologists are of the opinion that they, too, are to be seen as votive offerings. I would like to suggest that those stones once formed part of an altar and, as such, were believed to enhance the power of votive offerings in their new context. With this information at the back of our minds, let us now turn to the evidence available on Drenthe, to see whether the picture it presents deviates to any extent from that outlined above.

The structures found in Drenthe

A small number of structures that can be associated with rituals have come to light in Drenthe (Figure 12.2). Like the other bog finds, they are on the whole poorly documented. That is not really surprising. The finders – mostly peat cutters – had no idea what they had chanced upon, and wouldn't have been interested even if they had known. A bronze axe was worth a bit of money, but a group of timber stakes was only inconvenient. Such finds were recorded on paper only in the rare cases that their discovery was actually reported. And even those cases present plenty of problems today. For example, not one sample of the wood of the concentrations of stakes that were recorded on paper has been preserved, so we are unable to perform C14 analyses to determine the age of the structural remains.

Hearths

A few records mention the discovery of a hearth in the peat. In a survey of bog finds, the *Tegenwoordige Staat van het Landschap Drenthe* (Current State of the Landscape of Drenthe; 1792) refers to "paved hearths", which may have been stone pavements covered with a layer of charcoal. We of course know nothing about the age of these remains. Equally vague is a report from 1918 mentioning a flint dagger that was found by peat cutters working near Emmer-Compascuum, at a depth of 1.50 m, near what they described as "a layer of ashes". This layer of ashes may have been a hearth, but we don't know for sure. The dagger is of a Scandinavian type (type II according to Bloemers 1968:77) dating from the Early Bronze Age, or possibly even the Bell Beaker period. Less obscure is a record from 1924 reporting the discovery of a Scandinavian-type flint dagger in bog peat near Emmer-Compascuum. The peat cutters found the dagger at a depth of about 1 m, among stumps near a "collection of pine posts with tapered ends, which were arranged in a pattern rather like rushes in the seat of a chair". The remains of an 'extinguished, repeatedly used fire' were found 2 m from these posts. The dagger is of type III (Bloemers 1968:83) and dates from the Early

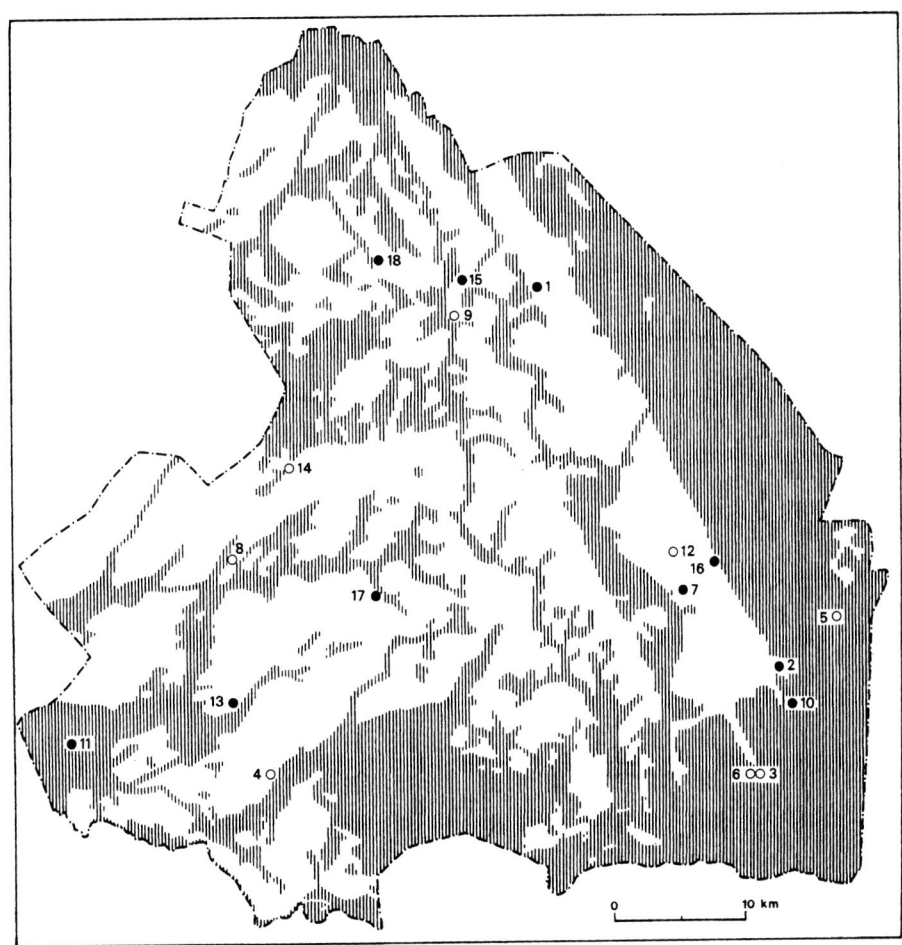

Figure 12.2. Map of Drenthe showing sites mentioned in the text. Black circle: exact location of site known; open circle: exact location of site not known. Drawing: J. Bruggink. 1. Anloo; 2. Barger-Oosterveld; 3. Barger-Westerveen; 4. Echten; 5. Emmer-Compascuum; 6. Erica; 7. Klijndijk; 8. Leggeloo; 9. Loon; 10. Nieuw-Dordrecht; 11. Nijeveen; 12. Odoorn; 13. Ruinen; 14. Smilde; 15. Taarloo; 16. Valthe; 17. Wijster; 18. Zeijen

Bronze Age, but whether the hearth is as old we can of course no longer determine.

Pits

Pits resembling those found in the bogs of Denmark and northern Germany are known in Drenthe, too. Van Giffen informs us that several such pits were found in the peat of a small bog near Zeijen called Bolleveen, which yielded many finds (Van Giffen 1950:*id.* 1952). Clason mentions pits found in a small bog near aarloo also called Bolleveen, which likewise contained many finds (Clason 1963), and Van Es recorded similar discoveries on the Looveen near Wijster (Van Es 1967:126–137, 185, 226, 239, 241 and 242). Their observations will be briefly discussed below. There are also a number of pits that have been 'reconstructed' on the basis of evidence obtained in pollen analysis. Those pits were not actually observed as such in

the field, but were inferred from, for example, hiatuses in pollen diagrams. It was such evidence, for example, that showed that the Roman-period bog body of Zweeloo (Van der Sanden 1990:96–97) and the late Neolithic disk wheel of De Eese (Van der Waals 1964:95) must have been deposited in a pit.

The Bolleveen near Taarloo yielded a wide range of finds:pottery, shoes, hair, cattle horns, a wooden bowl and a half-finished similar bowl, a half-finished nave, a complete wooden spoked wheel and a few half-finished felloes. The pottery, which includes complete vessels, dates from the period from the 1st to the 6th century AD (Van der Sanden and Taayke 1995:173–176). Some of these finds may well have been deposited in pits, but we have no records to this effect. The pits described by Clason yielded no finds whatsoever.

Figure 12.3. The excavation in the Bolleveen near Zeijen in 1927. Photo: GIA.

The two pits of Wijster were both discovered near the periphery of the Looveen. In the pits the excavators found several wooden objects and the remains of various pots. The wooden objects include bowls and a half-finished bowl, what may have been a horse collar, a loom weight, numerous half-finished felloes and a half-finished nave. The pottery dates from the Roman period.

The most interesting pits are those which Van Giffen found in the Bolleveen near Zeijen. On several occasions he performed investigations in this little bog, which repeatedly yielded interesting finds during peat-digging activities. Among the finds mentioned by Van Giffen are a wooden winnow, two bog bodies, the stomach of a bovine animal or a horse and the felloe of a spoked wheel. In 1927 Van Giffen recorded several vertical sections in which the contours of pits are clearly visible. He also excavated three pits in the southern peripheral part of the bog: a pit with a rectangular cross-section with rounded corners measuring 3×4.15 m, a pit with a cross-section shaped like a truncated oval measuring 2.3×2.7 m and a pit with a truncated-round cross-section measuring 4.3×3.7 m. Besides a few wooden objects, which Van Giffen unfortunately doesn't specify, they contained a large number of sherds, a few intact or almost complete pots and many animal bones, mainly of

cattle, but also of horse, pig, sheep and dog. In 1948 more pits were found near this bog's periphery. Unfortunately they were not recorded in drawings.[3] Waterbolk (1950 and daily reports of 1948) mentions that there was a lot of wood at the bottom of these pits, both twigs and roundwood with diameters of 10 cm. Many of the posts, which were of oak and birch, had sharpened ends. Other finds recovered from these pits are irreversibly dried lumps of peat and pottery sherds. In this last excavation campaign the excavators also found a skin, possibly of a bovine animal, and an unusually carved piece of oak with horn-shaped projections creating the impression of a cow's head (Figure 12.4). There are two rectangular holes in the bottom of the wooden object, indicating that it could be mounted on something, perhaps a wagon. Whether this object was found in a pit is not known. A C14 date places a comparable find from the Dutch province of Friesland in the Roman period. The pottery found in the Bolleveen dates from the period from the 1st to the 6th century AD (Van der Sanden and Taayke 1995:177–180).

Wells

Only one find reminiscent of the well found at Smedeby in Denmark is known in Drenthe. The find in question is

Figure 12.4a

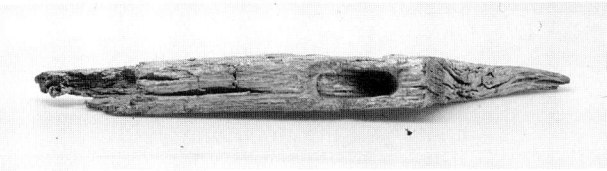

Figure 12.4b

Figure 12.4c

Figure 12.4. Carved pieces of oak with horn-shaped projections from the Bolleveen near Zeijen (top and centre) and the Burmania dwelling mound near Ferwerd, province of Friesland (bottom); heights at the centre 9.4 cm and 13.3 cm, respectively. The object from Friesland has a C14 date of 1790 ± 50 BP. Photos: Centrale Fotodienst Provincie Drenthe.

however surrounded by a lot of uncertainties (Janssen 1851). It comprises a number of oak planks lying on top of a hollowed-out oak tree-trunk which came to light in 1850 just outside Odoorn, in a peat-filled depression bordering the road to Exloo. The tree-trunk had a diameter of 1 m at the depth at which it was encountered. Further research showed that the tree-trunk became narrower further down (the 2.5 m-long trunk had been set in the ground upside down), that it was surrounded by vertically arranged planks with sharpened ends, and rested on a stone floor. This floor, which was paved with stones with diameters of 10–20 cm and was a little larger than the tree-trunk, rested on sand. Beneath the floor were numerous sherds (including rim, base and belly fragments), followed by another layer of stones and, finally, fragments of a very large pot. According to Janssen, the pottery was all hand-made, of a brown or

black fabric, undecorated and tempered with quartz grit. He described it as "Germanic". Unfortunately the remains of wood and the sherds were not kept, so we are unable to date this remarkable structure (Van der Sanden and Taayke 1995:164–165).

Stone pavements, cairns of stones and small stone chambers

Stone pavements have been unearthed at several sites in Drenthe. Unfortunately most are hard to date. This holds for example for the aforementioned paved hearth and for two stone pavements that came to light when sods were removed at the eastern edge of a pool near Anloo in 1990.[4] The southern pavement had a diameter of approximately 2 m, the other, which lay 8.5 m to the north of the first, was slightly oval and measured 4×5 m. Some of the stones were severely eroded; the largest measured 30 cm. A few tiny indeterminable sherds and fragments of flint were found lying on the northern pavement. Some 160 m to the east lies *hunebed* D11, which dates from the middle Neolithic (3400–2900 BC).

Very interesting and better datable are the pavements on top of which complete objects have been found. The oldest is the stone pavement that was found near Ruinen in 1888, or shortly before then. A small bronze flanged axe of the Oldendorf type (Butler 1995/96:207) with a length of only 8 cm was found on this pavement. The patina shows that the axe lay in a wet environment. The stones were deposited here some time in the Bronze Age, around 1700–1600 BC. A second pavement was found at Loon. Here a concave stone pavement emerged from the peat of a former small bog between 1931 and 1936. On top of this pavement lay an oak ard share (Van der Sanden 1993/94). This find most probably dates from the first half of the Iron Age, between 800 and 400 BC (see below). This was not the first time an ard share was found in association with stones. This same combination had come to light twenty-one years earlier, in 1905. On that occasion two ard shares were found in a humic layer together with what was described as "calcareous matter" and pieces of wood when a ditch was dug on the moor near Echten. These objects were covered by a large cairn, built from at least two cartloads of stones according to the newspaper that reported the find. The fact that the wood had survived implies that the findspot lay in a relatively wet environment. C14 analysis yielded a date of 2520 ± 70 BP for one of the ard shares, implying that the Echten find dates from the period 800–410 BC.

In addition to pits, stone structures have also been found in the previously mentioned Bolleveen near Zeijen, but unfortunately we have no drawings or photos of any of those finds. The first of these structures were observed by Van Giffen in 1916. He described them as 'beehive-shaped dry-stone structures built from field stones' (Van Giffen 1950:90). One was undamaged; according to Van Giffen it was one and a half times the size of a beehive and rested on the sand. In 1922 he found a few more at the centre of the Bolleveen. Inside these structures were a few sherds. The walls contained a small amount of wood, including a rectangular piece with four round holes in it. By 1927 all those stone structures had disappeared. During the last excavation campaign, in 1948, Waterbolk was able to make several important observations. He noted a few small stacks of stones, one of which had a diameter of 1.30 m. According to him, only the bottom stones were still *in situ*, arranged in a circle with the aforementioned diameter. Among the stones he found plant remains, such as twigs. He concluded that this find represented the remains of a collapsed lining of a pit built from stones and organic matter. A second stack of stones, with the same dimensions but lying further towards the centre of the bog, yielded a few sherds and an acorn cup. Here, too, organic matter was found among the stones. Waterbolk also observed a cylindrical pit that could be followed to a depth of 1 m beneath the surface of the peat. This pit contained thirteen stones with an average diameter of 30 cm. They are undoubtedly the remains of a beehive-shaped structure of the kind Van Giffen had observed in 1916.

A curious group of finds, finally, is that comprising small stone underground chambers. Janssen mentions two which were reportedly found near Leggelo, one *in* the peat and the other *beneath* the peat (Janssen 1848:35–36, 88–90). Both chambers were round. One of the two, which came to light in 1847, had a diameter of 75 cm and a height of 30 cm and was built from granite. A sandstone figure representing a reclining female figure is said to have been found in the chamber. This figure was for a long time thought to be very old, dating from the 2nd or 3rd century AD, but it is now known that it is actually much younger, possibly dating from the 18th century. So it is doubtful whether this chamber is a prehistoric find. The other underground chamber had a diameter and a height of 50 cm and was built from stones measuring 20–50 cm. Inside the chamber was a "reddish pot of a coarse fabric". Nothing is known about the age of this find.

Posts and stakes

The most obscure group of structures is that of concentrations of posts or stakes. Posts or stakes have been found at several sites, but we only very rarely have any information about their exact configuration, their number or their dimensions.

The first example has already been mentioned with reference to the hearths. This – possibly circular – arrangement of obliquely positioned posts that came to light at

Emmer-Compascuum in 1924 was associated with a Scandinavian-type flint dagger.

In 1875 two rows of stakes were found at a depth of about 1 m in the peat to the south of Erica, one of five and one of nine stakes.[5] All we know is that the stakes had been driven down into the sand underlying the peat, that they were 1.5 m long, had a diameter of about 25 cm, were spaced about 1.5 m apart and had flat tops. A simple drawing was made of this find.

Another two rows of stakes came to light on the Barger-Westerveen when the canal the Verlengde Hoogeveensche Vaart was dug here in 1878.[6] The rows were arranged somewhat in a V shape; one of the rows comprised five stakes and was 9.40 m long, the other consisted of nine stakes and was almost 15 m long. The stakes had lengths of 1.10–1.30 m and diameters of 17–30 cm.

In 1887 a large number of posts were found at a depth of 1.5 m in the peat near the canal the Oranje Kanaal near Smilde.[7] No further information is available on this find, except that a few of the posts were taken to the museum in Assen, where they were not preserved.

Far more noteworthy was the discovery made some 17 years earlier, in 1870, near Nijeveen (Leemans 1871). There, peat cutters found a canoe with a length of over 8 m whose boards had a height of only 27 cm. The boat was made from an oak tree-trunk. Fortunately C. Leemans of the National Museum of Antiquities in Leiden was able to inspect the find *in situ* and obtain information about other objects previously discovered in the immediate surroundings. Thanks to his efforts we know that stakes had been found driven into the peat in a few places to the south of the canoe. The wood of two of these stakes was analysed in Leiden. It was found to be *Taxus baccata*. One year before the discovery of the canoe, more stakes, with lengths of 2–2.5 m, had been found driven vertically into the peat in a trench approximately 15 m to the east of the canoe. And in 1869 and 1870 two bronze flanged axes of the Mägerkingen type (Butler 1995/96:220–221) had been found near the findspot. Some time before then the peat had also yielded a half-finished *Arbeitsaxt*. The bronze axes, which had acquired a fine patina in the peat, date from approximately 1700–1600 BC. The half-finished axe may well date from the same period. These implements are much older than the canoe, for which a C14 date of 2480 ± 25 BP has been obtained;[8] that corresponds to a calibrated date of approximately 760–410 BC. Whether the stakes are associable with the canoe or with the objects of the Bronze Age hoard can no longer be determined.

The last arrangement of posts to be discussed here is that which peat cutters found in a small bog near Klijndijk around 1934. The vertically arranged posts came to light when a trench was dug across the bog. Other finds that were recovered from the peat around this time are six cattle horn sheaths and the remains of six late Neolithic and Early Bronze Age vessels (Prummel and Van der Sanden 1995; 94 and 116; Van der Sanden 1997a:136–138). One of the cattle horn sheaths yielded a C14 date in the Early Bronze Age (3540 ± 70 BP, *i.e.* between *c.* 2030 and 1680 BC). So it is quite conceivable that the arrangement of posts dates from the same period, *i.e.* the late Neolithic/Early Bronze Age.

The Barger-Oosterveld temple

The small building whose remains were found in the Bourtanger Moor in 1957 may without exaggeration be termed unique (I strongly doubt the existence of a parallel described by Dieck (1987)). The structure which Waterbolk and Van Zeist (1961) reconstructed from these remains comprised eight posts which rested on broad planks. It was firmly anchored in the peat by the sturdy posts that extended far beneath these planks at the corners of the building. More posts, some of which were still covered with bark, were set between the planks (Figure 12.5). The building measured approximately 2 × 2 m and was surrounded by a ring of stones with a diameter of 4 m. Outside this ring, among twigs and wood chips, the excavators found horn-shaped pieces of wood which they interpreted as the ends of four beams from the top of the building. No other objects were found near the remains of this remarkable structure.

A C14 date obtained shortly after the find's discovery places the structure in the Middle Bronze Age:3240 ± 65 BP (approximately 1680–1400 BC). Recent dendrochronological analyses have yielded a more exact date. We now know that the timber used to build the structure came from trees felled between 1478 and 1470 BC.

The horn-shaped projections are one of the most intriguing parts of the "Barger-Oosterveld temple". What do they signify? When we look at their outlines, the individual beams could represent ships, with the projections being the stem and the stern. Ships are not an unknown phenomenon in the Bronze Age iconography of northwest Europe. Kaul has recently attempted to convincingly arrange the numerous depictions of ships known from bronze objects, mostly from Denmark, in a diagram that could represent the cosmological views of the Bronze Age inhabitants of northwest Europe (Kaul 1998). The ship motif plays a leading part in this diagram, for it transports the sun across the sky in the daytime, but also in the darkness at night. Kaul assumes that his reconstructed cosmology was widespread throughout a large part of northwest Europe. But no evidence of this in the form of depictions on bronze objects like razors or rock engravings has been found in the Netherlands. This would make the four assumed ships of the temple unique in this respect. But there is an alternative interpretation. If we focus only on the corners of the building, the projections look like cattle horns. On the basis of other finds, this interpretation seems more likely here. I will return to this below.

Bog trackways

The number of bog trackways that have been found in the northern part of the Netherlands is only very small, especially in comparison with that known from for example the adjacent German federal state of Lower Saxony. Casparie (1987) published a summary of the known trackways. He lists over 25 for the northern part of the Netherlands, rightly adding that many bog trackways will have been destroyed unobserved. The oldest trackway dates from around the middle of the 3rd millennium BC, the youngest from after the Middle Ages. The best-known trackways are that of Nieuw-Dordrecht, the northern and southern plank footpaths that were found near Bargeroosterveld and the Valtherbrug, which are known as XXI(Bou), XV(Bou), XVII(Bou) and I(Bou), respectively. I shall discuss the first and last of these structures below.

The trackway of Nieuw-Dordrecht extended from the Hondsrug – the eastern, highest part of the Drents plateau – in an easterly direction, into the expansive Bourtanger Moor, where almost a kilometre of the trackway has survived. It was built from roundwood which, in the western part at least, was placed on longitudinally arranged thin tree-trunks. The trackway's width varies from 2.5 to 3 m. The published C14 dates place the find in the late Neolithic, more specifically the time of the single-grave culture (4080 ± 55 BP, 4100 ± 55 BP and 4020 ± 35 BP).

The abrupt end of the trackway – at some point in the former bog – has raised a lot of questions. It has been suggested that it may have something to do with the sand ridge lying further to the east, the so-called Postwegrug, or with the nearby presence of a major iron-ore area. But objections can be made to both those interpretations. In the first place, as Casparie determined (1982:131), the sand ridge had disappeared beneath the expanding peat long before the trackway was constructed and, secondly, it is highly unlikely that iron ore was being exploited in the late Neolithic already. Casparie seems to assume (1982:160; *id.* 1987, 53, 64) that the trackway was found to be a failure and was therefore never completed. This is not very likely either. It is highly inconceivable that the builders would have been so poorly familiar with the terrain as to have built an entire kilometre of trackway before finding out that it was unsuitable for use, for example because it was incapable of bearing the weight of wagons as the greater part of the surface of the trackway was unsupported. The results of recent dendrochronological research and wiggle matching of new C14 dates have shown that the trackway was constructed in several phases.[9] There is a gap of about a century between the available dates. This evidence is

Figure 12.5. Excavation of the Barger-Oosterveld 'temple' in 1957 and the proposed reconstruction of the wooden structure. Photo and drawing: GIA.

inconsistent with the hypothesis that the trackway was abandoned as a failure. I would like to suggest that the trackway had a ritual function. Evidence pointing in this direction are the various objects that have been found alongside or beneath the road: a wooden disk wheel, a wooden axe handle and a mysterious wooden object shaped rather like a hockey stick. The hoard of flint objects comprising one axe and eleven long blades (Harsema 1981) that came to light at the beginning of the trackway may perhaps also be seen in this context. Only a small part of the surviving 800 m of the trackway has been excavated. Future excavation of the part which can probably not be preserved *in situ* may yield more evidence supporting this interpretation.

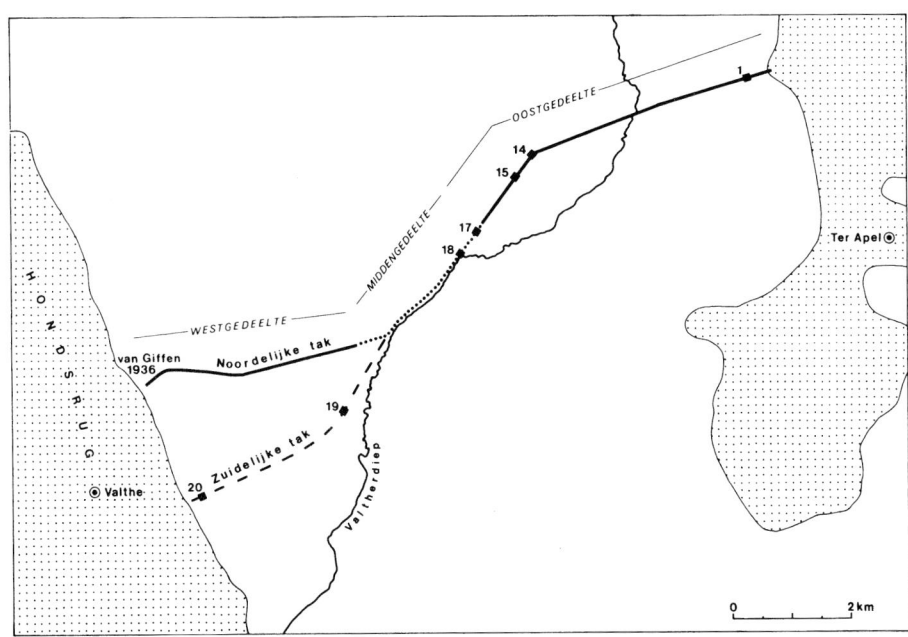

Figure 12.6. The Valtherbrug that connected the Hondsrug in the west with the Westerwolde area in the east. The numbers refer to the excavation trenches dug by Karsten in 1818. Solid line: more or less reliable reconstructed route according to Casparie; dashed line and dotted lines: unreliable reconstructed route according to Casparie. Drawing: GIA.

The other trackway, the well-known Valtherbrug, also lies in the Bourtanger Moor but is much younger. There is no trackway that has been discussed as extensively as this, which, since its discovery in the early 19th century, has always been referred to as a "Roman" timber road. The history of the archaeological research devoted to this trackway has been summarised by Van Zeist (1958), from the excavation conducted by Karsten (1818) up to and including that led by Van Giffen (1936). A reconstruction drawing made by Van Zeist shows a road with a length of 12 km linking the Hondsrug in the west to the sandy Westerwolde area in the east (Figure 12.6). There are two "access routes" in the west, a northern and a southern branch, which merge well before halfway along the trackway's overall length. C14 analysis performed thirty years ago yielded a date of 2295 ± 50 BP, which places the trackway between approximately 480 and 200 BC, in other words in the second half of the Iron Age. It is incidentally not known from which of the trackway's planks the C14 sample was taken.

Van Zeist's reconstruction has recently been contested by Casparie (1993), who is of the opinion that the remains represent more than one trackway. His arguments are the trackway's strange course, incorporating several awkward bends, the differences in the dimensions of the planks and roundwood of the western and eastern parts and, finally, the trackway's assumed course alongside the bog stream

the Valtherdiep, which he finds highly unlikely. He moreover finds Karsten's observations relating to the "southern access route" unreliable. What remains when all this has been eliminated are two separate trackways which led to the centre of the Bourtanger Moor from the west and the east. I do not agree with Casparie's interpretation. The evidence obtained in the central part does not allow us to assume that there was no trackway here. I moreover find the southern access route drawn by Karsten far from unreliable. His very precise drawings show roundwood and supports that are essentially entirely in accordance with the evidence observed in the other excavation trenches. The evidence even suggests that there was another access route, built from both roundwood and planks, between the northern and the southern branches. New dating evidence has recently been obtained:a dendrochronological date for a plank from the eastern part (Figure 12.7) and C14 dates for the annual rings of planks from the northwestern and eastern parts of the trackway. The dendrochronological date, which is incidentally based on only 60 annual rings, is after 36 ± 6 BC. The wiggle-matched C14 dates indisputably point to the 1st century AD. This means that the trackway is much younger than was assumed in the past decades. It would now seem that the trackway dates from the Roman period.

And that makes the road all the more interesting. Around that time the Westerwolde area was becoming very wet. It was indeed totally unoccupied (Groenendijk 1995). It was

Figure 12.7. The Valtherbrug during the excavations in the eastern part in 1892. Photo: courtesy of the Drents Museum.

definitely impossible to travel across it by cart. So why should people have gone to the trouble of linking an occupied area to an unoccupied area? Did the trackway perhaps form part of a much longer road leading to the river Ems? Or could it have been used for other purposes? Particularly interesting in this context are the finds that were recovered near the trackway, especially the eastern part. Several bog bodies were for example found very close to the trackway (Van der Sanden 1990; 55). Two of the four records of these bodies are extremely reliable and for one of the bodies we even have a C14 date, based on a tuft of hair from the head:2025 ± 65 BP, *i.e.* between 200 BC and AD 125. In 1904 the two Weerdinge bodies came to light a little further away from the trackway. This find has been dated to 1980 ± 70 BP (*c.* 169 BC – AD 213; Van der Sanden 1996:53 and 191).

Querns have been found near the trackway, too – complete and incomplete rotary querns made from tephrite. A newspaper report from 1854 mentions the discovery of four querns near the Oude Ossebrug, which lies fairly close to the Valtherbrug.[10] Two of the four were reportedly recovered from the peat intact. This find most probably ended up in the collection of the Drents Museum a year later. The querns are of the Westerwijtwerd type and date from the Roman period. A badly worn top stone of a rotary quern is said to have been found 'near or beneath the so-called Roman bridge' twenty-five years later. It is of the Brillerij type, which preceded the Westerwijtwerd type. This quern also made its way into the Drents Museum (Van der Sanden 1998b; 113–115).

Then there are the wagons and parts of wagons, at least seven of which are reported to have been found near the trackway, along the western, the central and the eastern parts (Hadders 1953:18; Van der Sanden 1997b:*id.* 1998a). Three or four reports speak of (a large part of) a waggon. None of these waggons have been preserved, so they are all paper finds. One of the wagons was reportedly found 150 paces from the trackway (in 1849), one had spoked wheels and was found "in the vicinity of'" the trackway (before 1897), one is said to have had four disk wheels and came to light in peat lot 60 in the Valtherveen (before 1909) and, finally, the most conspicuous find, "a cart with a seat woven from cowskin and two long cattle horn sheaths" (before 1953). The latter find lay a little further from the trackway. The wagons, wheels and bog bodies could imply that the trackway played a part in rituals, perhaps ritual processions of the kind described by Tacitus in his *Germania* (*caput* 40).

Summary

The various structures I have reviewed above – the pits, post settings, hearths, stone pavements and cairns, the timber building and the timber trackways – possibly or probably played a part in certain rituals. The probability of this having been the case is greater with some structures than with others. The association with prehistoric pottery and cattle horn sheaths makes it somewhat easier to regard the arrangement of stakes found in the small bog near Klijndijk as the remains of a fence or platform than the concentration of stakes of Smilde, in whose vicinity no finds whatsoever were recorded. Unfortunately we will never be able to eliminate the uncertainty surrounding that find.

On the whole, the structures that have come to light in Drenthe bear a close resemblance to structures found in the countries surrounding the Netherlands. Stones evidently played an important part in the northern part of the Netherlands, too. Votive offerings were deposited on a pavement or beneath a cairn. Noteworthy are the small cylindrical structures that were found in the Bolleveen near Zeijen. Whether they were used for offering organic objects can no longer be determined as they were too poorly excavated and documented. An even more remarkable structure is the Barger-Oosterveld temple, for which no parallels whatsoever are known throughout the whole of northwest Europe. We do know of other Bronze Age sanctuaries, but they are all closely associated with burial rites and were not built in bogs (see Lomborg 1957; Kaul 1985).

What rituals took place in the bogs and stream valleys of Drenthe? I think many finds suggest acts to implore and/or express gratitude for agricultural success:take for example the ard shares, the offerings of foodstuffs, the querns, *etc*. What also seems to emerge from the evidence is the importance of cattle in pre- and protohistoric man's realm of thought. We have come across cattle in the form of horn sheaths, the bones in the pits in the Bolleveen, the carved piece of wood from that same bog, the projecting parts of the Barger-Oosterveld temple and the cart that was found in association with cattle horn sheaths in the Exloërdreef. All this indicates that cattle were of considerable social and ideological significance to the local population (*cf*. also Roymans 1996:44–49). In this context the information that the pre- and protohistoric farmers of Drenthe lived with their cattle beneath the same roof will come as no surprise. The wheels and parts of wheels from the bogs near Taarloo, Zeijen and Wijster may symbolise the sun, the source of all life, and can as such be seen in the same context of agricultural success and prosperity. Perhaps the ring of stones of Barger-Oosterveld forms part of the same symbolism.

If the piece of carved wood from the Bolleveen formed part of a wagon, the cattle may be associated with wagons. Did this piece of wood perhaps adorn a wagon that was drawn by oxen and was used for ritual "processions"? May the Valtherbrug have been used for similar processions in the 1st century AD? This question brings me to the people involved in such activities. It will be clear that people on different social levels will have been involved in the actions I assume took place at these sites. The ard share of Loon will have been deposited by an individual donor, and this undoubtedly also holds for the hundreds of stone axes that

have been found in the bogs and stream valleys. It is more likely that the votive bog of Zeijen with its small cylindrical structures is to be associated with collective acts. This bog was probably used by a small community, the occupants of the various hamlets in the immediate vicinity. The Barger-Oosterveld temple, and especially the Valtherbrug, imply ritual activities of local farming communities. Prestigious objects are totally absent. There is indeed nothing to indicate that an elite was involved in the religious acts that took place at these sites.[11]

Notes
1. Written information M. Dietrich, Kiel.
2. Oral information F. Kaul, National Museum, Copenhagen.
3. Oral information Professor Dr H.T. Waterbolk, Haren.
4. See the inventories of the Drents Museum, No. 1990/Il 1.
5. Mentioned in letters dated 28 June and 2 August 1875 (in the Public Record Office of the province of Drenthe, 0028–105).
6. Mentioned in a letter dated 1 November 1886 (in the Public Record Office of the province of Drenthe, 0028–105).
7. See the reports of the Management Committee of the Provincial Museum of Antiquities in Drenthe for 1887 (p. 9) and 1888 (p. 5).
8. Oral information J.N. Lanting, Department of Archaeology, Groningen....
9. Oral information J.N. Lanting.
10. See the *Provinciale Drentsche en Asser Courant* of 15 March 1854, p. 3.
11. I would like to thank Dr J.J. Butler, Paterswolde, A. Smith, Exloo, and Professor Dr H.T. Waterbolk for the information they kindly provided. I am most grateful to J.N. Lanting for the help he gave me in various fields and for the discussions we had on the Valtherbrug. Professor Dr N. Roymans, Department of Archaeology of the Free University, Amsterdam, provided useful comments on the first version of this text. The English translation is by S.J. Mellor.

References Cited

Andersen, H. 1961. Hun er moder jord. Skalk 4:7–11.
Arbman, H. Käringsjön. 1945. Studier i halländsk järnålder. Stockholm:Kung. Vitterhets Historie och Antikvitets Akademien.
Becker, C.J. 1947a. Tørvegravning i Ældre Jernalder. Fra Nationalmuseets Arbejdsmark:92–100.
Becker, C.J. 1947b. Mosefundne lerkar fra Yngre Stenalder, studier over Tragtbægerkulturen i Danmark. Aarbøger for Nordisk Oldkyndighed og Historie:1–318.
Becker, C.J. 1970. Zur Frage der eisenzeitlichen Moorgefässe in Dänemark, in: H. Jankuhn ed. Vorgeschichtliche Heiligtümer und Opferplätze in Mittel- und Nordeuropa. Göttingen: Vandenhoeck & Ruprecht, 1970:119–166.
Behm-Blancke, G. 1957. Germanische Mooropferplätze in Thüringen. Ausgrabungen und Funde 2:129–135.

Behm-Blancke, G. 1989. Heiligtümer, Kultplätze und Religion. in:J. Herrmann ed., Archäologie in der Deutschen Demokratischen Republik. Denkmale und Funde 1. Leipzig:Konrad Theiss Verlag, 1989:166–176.

Behm-Blancke, G. 1960. Latènezeitliche Opferfunde aus dem germanischen Moor- und Seeheiligtum von Oberdorla, Kr. Mühlhausen. Ausgrabungen und Funde 5:232–235.

Bemman, J. and G. Hahne. 1992. Ältereisenzeitliche Heiligtümer im nördlichen Europa nach den archäologischen Quellen, in: H. Beck, D. Ellmers and K. Schier eds. Germanische Religionsgeschichte – Quellen und Quellenprobleme. Berlin/ New York:Walter de Gruyter, 1992:29–69.

Bender, W. and E. Stupnicka. 1974. Z badań archeologiczno-geologicznych stanowiska torfowego w miejscowości Otalazka, pow. Grójec. Archeologia Polski 19:307–366.

Bloemers, J.H.F. 1968. Flintdolche vom skandinavischen Typus in den Niederlanden. Berichten van de Rijksdienst voor het Oudheidkundig Bodemonderzoek 18:47–110.

Brzeziński, W. 1992. Recent developments in wetland archaeology in Poland. in: B. Coles ed. The wetland revolution in prehistory. Exeter:Wetland Archaeological Research Project, 1992:73–79.

Butler, J.J. 1995/96. Bronze Age metal and amber in the Netherlands (part II:1) – Catalogue of flat axes, flanged axes and stopridge axes. Palaeohistoria 37/38:159–243.

Capelle, T. 1995. Anthropomorphe Holzidole in Mittel- und Nordeuropa. Stockholm: Almqvist & Wiiksell International.

Casparie, W.A. 1982. The Neolithic wooden trackway XXI(Bou) in the raised bog at Nieuw-Dordrecht (the Netherlands). Palaeohistoria 24:115–164.

Casparie, W.A. 1987. Bog trackways in the Netherlands. Palaeohistoria 29:35–65.

Casparie, W.A. 1993. De Valtherbrug (DR. en GR.); meer dan één weg?. Paleo-Aktueel 4:95–99.

Clason, A.T. 1963. Het Bolleveen bij Taarloo; sporen van voorhistorische turfwinning in Drenthe. Nieuwe Drentse Volksalmanak 81:231–240.

Dieck, A. 1987. Das Moorheiligtum von Bentheim aus der Jüngeren Bronzezeit. Telma 17):109–127.

Diezel, P.B., W. Hage, H. Jankuhn, J. Klenk, U. Schaefer, K. Schlabow, R. Schütrumpf and H. Spatz. 1958. Zwei Moorleichenfunde aus dem Domlandsmoor, Gemarkung Windeby, Kreis Eckernförde. Prähistorische Zeitschrift 36:118–219.

Es, W.A. van. Wijster. 1967. A native village beyond the imperial frontier. Groningen:diss. Universiteit van Groningen.

Giffen, A.E. van. 1950. De nederzettingsoverblijfselen in het Bolleveen en de versterking, de zgn. 'legerplaats', aan het Witteveen op het Noordse Veld, beide bij Zeijen, gem. Vries. Nieuwe Drentse Volksalmanak 68:89–99.

Giffen, A.E. van. 1952. Het Bolleveen bij Zeijen. Gem. Vries. Naschrift. Nieuwe Drentse Volksalmanak 70:89–108.

Groenendijk, H.A. 1995. Niemandsland Ter Apel; zuidelijk Westerwolde voor de kloostervestiging, in:J.W. Boersma & C.J.A. Jörg eds. Eresaluut; opstellen voor mr. G. Overdiep. Groningen:Regio-Projekt, 1995:207–220.

Hadders, J. 1953. 100 jaren Valthermond. Emmen:N.V. Drukkerij v.h. W. ten Kate.

Hallgren, F., U. Djerw, M. af Geijerstam & M. Steineke. 1997. Skogsmossen, an Early Neolithic settlement site and sacrificial fen in the northern borderland of the Funnel-beaker Culture. Tor 29:49–111.

Harck, O. 1984. Gefässopfer der eisenzeit in nördlichen Mitteleuropa, in:Frühmittelalterliche Studien 18:102–121.

Harsema, O.H. 1981. Het neolithische vuursteendepot van Nieuw-Dordrecht, gem. Emmen en het optreden van lange klingen in de prehistorie. Nieuwe Drentse Volksalmanak 98:113–128.

Hingst, H. 1967. Ein Brandplatz der älteren Eisenzeit aus Braak, Kr. Eutin. Offa 24:108–114.

Jankuhn, H. 1967. Archaeologische Beobachtungen zu Tier- und Menschenopfern bei den Germanen in der Römischen Kaiserzeit. Nachrichten der Akademie der Wissenschaften in Göttingen, Phil.-I. Hist. Klasse 6:117–147.

Janssen, L.J.F. 1848. Drentsche oudheden. Utrecht:Kemink en zoon.

Janssen, L.J.F. 1851. Over eene merkwaardige, oud-Germaansche ontdekking te Odoorn. Nieuwe Drentse Volksalmanak 15:163–180.

Kaul, F. 1985. Sandagergård – a Late Bronze Age cultic building with rock engravings and menhirs from Northern Zealand, Denmark. Acta Archaeologica 56:31–54.

Kaul, F. 1998. Ships on bronzes – a study in Bronze Age religion and iconography. Copenhagen:Nationalmuseet.

Koch, E. 1998. Neolithic bog pots from Zealand, Møn, Lolland and Falster. Copenhagen:Det Kongelige Nordiske Oldskriftselskab.

Krämer, R. 1992. Die 'Notgrabung' am Bohlenweg XII (Ip) aus dem Jahre 713 v. Chr. im Ipweger Moor, Ldkr. Wesermarsch. Archäologische Mitteilungen aus Nordwestdeutschland 15:101–114.

Kuhlmann, N. 1998. Feuer und Wasser. Archäologie in Deutschland 14, Heft 4 (Okt.-Dez. 1998):47.

La Baume, P. 1952. Der Moorfund von Geel-Royum, Kr. Schleswig. Offa 11:42–45.

Leemans, C. 1871. Over eene oude kano, in een veen onder Nijeveen, prov. Drenthe, en over oude vaartuigen in andere landen ontdekt. Amsterdam: C.G. van der Post.

Lomborg, E. 1956. En højgruppe ved Ballermosen, jægerpris gravfund, hustomt og højryggede agre fra Ældre Bronzealder. Aarbøger for Nordisk Oldkyndighed og Historie, 144–203.

Makiewicz, T. 1988. Opfer und Opferplätze der vorrömischen und römischen Eisenzeit in Polen. Prähistorische Zeitschrift 63:81–112.

Makiewicz, T. 1997. Zapomniane bagienne miejsce ofiarne kultury Przeworskiej w południowej Wielkopolsce (Łagiewniki koło Kościana, woj. Leszczynskie). Folia Praehistorica Posnaniensia 8:121–138.

Prummel, W. and W.A.B. van der Sanden. 1995. Runderhoorns uit de Drentse venen. Nieuwe Drentse Volksalmanak 112:84–131.

Pryor, F. 1991. Flag Fen – prehistoric fenland centre. Batsford, London.

Pryor, F. 1992. Current research at Flag Fen, Peterborough. Antiquity 66:439–457.

Rech, M. 1979. Studien zu Depotfunden der Trichterbecher- und Einzelgrabkultur des Nordens. Neumünster:Karl Wachholtz Verlag.

Roymans, N. 1996. The sword or the plough. Regional dynamics in the romanisation of Belgic Gaul and the Rhineland area. in: N. Roymans ed. From the sword to the plough. Amsterdam: Amsterdam University Press, 9–126.

Sanden, W.A.B. van der, 1990. ed. Mens en moeras; veenlijken in Nederland van de bronstijd tot en met de Romeinse tijd. Assen: Drents Museum,.

Sanden, W.A.B. van der. 1993–94. Early Iron Age ard shares from Drenthe, the Netherlands. Tools & Tillage 7:2–3:103–106, 118.

Sanden, W.A.B. van der. 1996. Through nature to eternity. The bog bodies of northwest Europe. Amsterdam: Batavian Lion International.

Sanden, W.A.B. van der. 1997a. Aardewerk uit natte context in Drenthe:het vroeg- en laat-neolithicum en de vroege bronstijd. Nieuwe Drentse Volksalmanak 114:127–141.

Sanden, W.A.B. van der. 1997b. Wagens, wielen en wieldelen uit de Drentse venen:de late ijzertijd en de Romeinse tijd. Nieuwe Drentse Volksalmanak 114:180–201.

Sanden, W.A.B. van der. 1998a. Veenvondsten in Drenthe (3): van maalstenen, wolkluwens, bronzen potten en veenlijken. Nieuwe Drentse Volksalmanak 115:103–106.

Sanden, W.A.B. van der. 1998b. Zware gaven – maalstenen uit natte context in Drenthe. Nieuwe Drentse Volksalmanak 115:107–130.

Sanden, W.A.B. van der. 1999. Wetland archaeology in the province of Drenthe, the Netherlands, in:B. Coles, J. Coles and M. Schou Jørgensen eds. Bog bodies, sacred sites and wetland archaeology. Exeter: University of Exeter, 217–225.

Sanden, W.A.B. van der and E. Taayke. 1995. Aardewerk uit natte context in Drenthe: 1100 v.Chr. tot 500 na Chr. Nieuwe Drentse Volksalmanak 112:149–186.

Schwabedissen, H. 1949. Die Bedeutung der Moorarchäologie für die Vorgeschichtsforschung. Offa 8:46–74.

Schwabedissen, H. 1951. Torfstiche mit Opfergefässen der Eisenzeit aus dem Rüder Moor, Kreis Schleswig. Offa 9:46–52.

Schwabedissen, H. 1965:Ein kaiserzeitlicher Moorfund von Büstorf, Kreis Eckernförde, in:Studien aus Alteuropa II. Cologne/Graz:Böhlau. 219–232.

Schwenzer, S. 1997. 'Wanderer kommst Du nach Spa..'. Der Opferplatz von Berlin-Spandau. Ein Heiligtum für Krieger, Händler und Reisende, in:A. Hänsel & B. Hänsel eds. Gaben an die Götter – Schätze der Bronzezeit Europas. Berlin: Museum für Vor- und Frühgeschichte und Seminar für Ur- und Frühgeschichte der Freien Universität. 61–66.

Tegenwoordige Staat van het Landschap Drenthe. Amsterdam/ Leiden/Dordrecht and Harlingen:J. de Groot, G. Warnars, S. en J. Luchtmans, A. en P. Blussé en V. van der Plaats, 1792.

Vebæk, C.L. 1945. Smederup – an Early Iron Age sacrificial bog in East Jutland. Acta Archaeologica 16:195–211.

Waals, J.D. van der. 1964. Neolithic disc wheels in the Netherlands. Groningen:diss. Universiteit van Groningen.

Waterbolk, H.Tj. 1950. Palynologisch onderzoek van de verster-king bij het Witteveen en de cultuursporen in het Bolleveen, beide bij Zeijen, gem. Vries. Nieuwe Drentse Volksalmanak 68:100–121.

Waterbolk, H.Tj. & W. van Zeist. 1961. A Bronze Age sanctuary in the raised bog at Bargeroosterveld (DR.). Helinium 1:5–19.

Zeist, W. van. 1958. De Valtherbrug. Nieuwe Drentse Volks-almanak 76:21–49.

13. North European Bronzes, Rock Art and Wetlands: Looking for Context and Relations. A Preliminary Study

John Coles

Archaeologists who study the later prehistory of north and west Europe generally concern themselves with rather standardised, traditional subjects such as settlements, cemeteries and metalworking. Less attention is paid to rock carvings other than by a few dedicated workers, and there is growing research on the landscapes of the period from 1800 to 500 B.C. – the classic Bronze Age; sometimes these subjects are treated as components in a world view, but more often they appear as isolated pieces of research. This paper is an attempt, and a very preliminary attempt at that, to address one of the problems in this abundance of evidence, that of context, and a particular context. My interests in most of this archaeology lie in three aspects – metalworking, rock carvings and wetlands.

Metalworking in the Bronze Age of the north of Europe is acknowledged by all prehistorians as a phenomenon *sensu stricto*. In northern Europe there were no native coppers or tin, or gold, that could be exploited; all of these metals, whether in the raw or as purified material, had to be imported into the north, to Denmark, southern Sweden and, in far less abundance, to more northern areas. Yet the quantity of bronze objects, and gold too, in southern Scandinavia is very considerable. And the quality of artifacts produced by local craftsmen or women is extraordinarily high. Imported bronzes do occur in the north, but the vast majority of objects were home-produced, that is, by native craftsmen and women. The mechanisms by which such metal was obtained from central Europe have for long taxed our explanatory faculties and will not be pursued here.

Among the many thousands of objects created in the northern Bronze Age are, of course, quantities of utilitarian tools and rather modest weapons for the chase – axes, spearheads, daggers, a range of industrial pieces including hammers and gouges, and a variety of ornaments for the

Figure 13.1. Sheet bronze shield from Nackhälle, Sweden. Diameter 623 × 595mm. National Museum, Stockholm.

human body. But there are also a considerable number of objects that demand attention, through size, technical achievement and seemingly – unusable character. They include beaten metal shields (Figure 13.1), elaborate battleaxes (Figure 13.2), metal helmets, embellished swords and a number of bronze figurines (see Broholm 1953). All of these have been shown or are otherwise considered to be non-functioning objects in terms of their physical ability to carry out their logical purpose – defending, hacking,

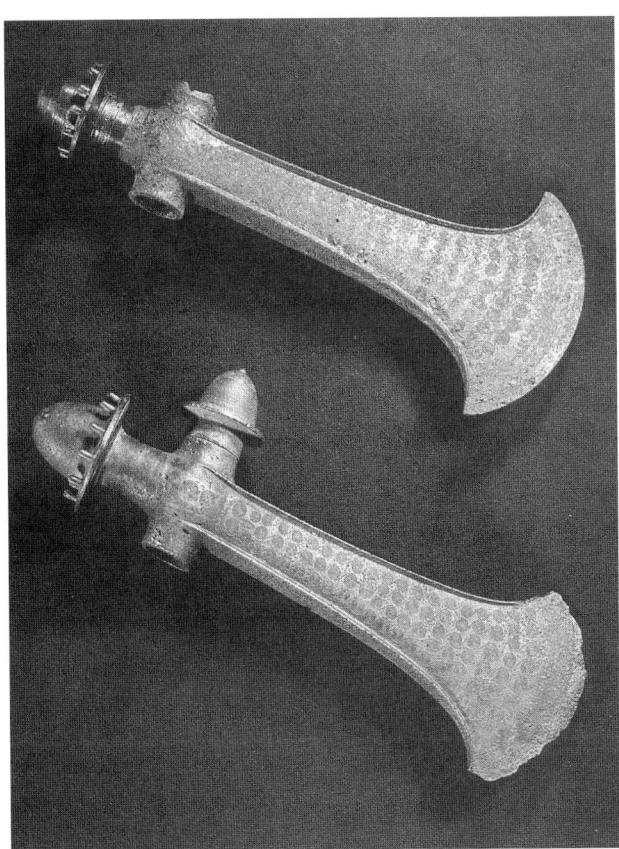

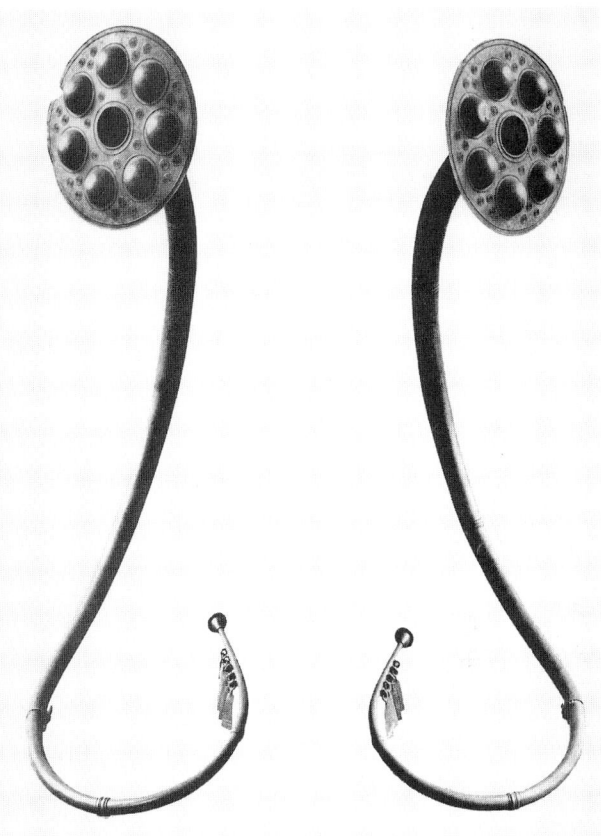

Figure 13.2. Bronze battleaxes from Egebak, Denmark. Lengths c. 400mm. (Jensen 1979).

Figure 13.3. Bronze lurs from Brudevaelte, Denmark. Diameter. Diameter of bell plate 275mm. (Broholm 1953).

attacking or whatever. A number of battleaxes, for example, are little more than a covering of metal over clay cores or are cumbersomely ill-balanced; sheet metal shields are so thin that they fall apart or bend double when hit by blade or axe (Coles 1962). Perhaps the finest example of craftsmanship in the European Bronze Age are the bronze lurs (Figure 13.3), long and evenly carved conical horns, cast in segments that were then cast together; these are noisemakers of course but are unwieldy, fragile and have limited repertoires of notes (Broholm *et al.* 1949).

The original purpose of these objects must surely be that of prestige and display, acknowledging power and wealth. They are not everyday objects either in appearance or in the places where they eventually came to rest. Barely any of these objects have been recovered from settlements, or from burials other than certain swords which by their condition show use; other objects may also exhibit traces of wear through handling but many of these pieces appear to be in as-new condition. The find-spots of most of these objects, shields, lurs, figurines and other prestige products are in wetlands of one sort or another, in peatbogs, in rivers

and streams, in ponds and small lakes; it is surely significant that they are not found on settlement sites, nor in burial grounds, nor in land that was dry – on valley slopes, hills – in the Bronze Age. There is a direct correlation between water and these value-laden objects.

Equally important must be the clear lack of association of these objects with other bronzes, or pottery, and in essence with anything at all. Most of these finds are singles, a single shield, or axe, or figure, although some are in pairs or, occasionally, multiples. Further, the condition of the objects is puzzling. Some appear to be new, barely used, and others seem to have been deliberately damaged – killed. Shields, for example, are very fragile and yet many are, or were, intact when deposited; a few others were probably damaged just before they were laid into the wet. Lurs are another example; some are more or less intact, while others have been rendered useless by breaking off their mouthpieces. These objects were valuable, all of them, valuable in metal, in technology and in appearance; yet they were deposited, discarded can hardly be correct, in wetlands and in waters, alone. We may presume that the act of deposition was public in some sort of ceremonial performance that brought together the community on an occasion, a commemoration or placation, but we can hardly know the circumstances; the context of deposition is unknown, as are their relationships with other objects.

There is another line of evidence that impinges on this difficulty. Much of northern Europe can be described as a land of rocks and water, and in the Bronze Age it was even more so; the land, formerly depressed by weight of ancient ice sheets, was rising but much of what we now see as fertile lowland was still beneath the Baltic waters. During the second and first millennium BC, land uplift exposed vast tracts and the dynamic landscape began its long evolution from sea-bound to marshland, to forests and grassland, and in many places on to the arable fields of today, much of the transformation natural but with human exploitation and manipulation at increasing intensities. The woodlands that colonised the newly-released lands were mostly cleared away during the later second and first millennia BC. The land of rocks, however, remained intact and the granites, gneiss, schist and quartzites provided extensive canvasses for Bronze Age artists.

Rock carvings of the Bronze Age were first identified over two centuries ago and discoveries continue to be made. Many thousands of sites are now known from southern Scandinavia, a vast majority located on or near their contemporary shoreline; debate continues about the precise relationships but it seems quite clear that water, or wetland, was one of the prerequisites in the selection of rocks for carving; orientation of the surface was also paramount, with carvings on rock sloping towards the west and north. As carvings were almost all designed to be viewed from downslope, the observer facing east, it would seem logical to suggest that the rising light was an important element in exposing and dramatising the carvings; in other words, an early morning event rather than that of a setting sun (see Coles 2000 for quantitative assessment).

From about 1800 BC, many rocks were carved with images of Bronze Age society, reflecting the symbols and values that different communities had come to deem important. Other than the ubiquitous cupmark, the dominant image carved in the rocks was the boat (Figure 13.4), elaborated in various ways but reflecting the overwhelming presence of water and the mechanism that societies had developed for communication, transport and colonisation (Coles 1993). Boats of the Bronze Age, insofar as we know them, were complex productions, valuable in wood and technology, but doubtless imbued with other significances that may have involved journeys into life and into death. Their elaboration on the rocks marked their importance to Bronze Age societies, and of course reflected their physical proximity to the life-blood of the north – the sea. Images of boats were often inscribed on small bronzes, such as razors

and knives (Kaul 1998), thus providing a generalised relationship between rocks and bronzework, but other carvings perpetuate images that more closely mirror the prestige metalwork of the north.

Here and there on the surfaces are carvings of just those objects that are found in isolation in the wetlands of the north. Humans holding, and apparently blowing, the lurs (Figure 13.5), humans armed with huge battle axes, humans defended by shields (Figure 13.6), and humans with other exaggerated objects such as spears and hammers occur on the rocks amidst other carvings (e.g. Coles 1994; Janson *et al.* 1989; Hagen 1990). More often, these artefacts are carved as single images, with no human form adjacent, but they almost always occur on surfaces where other symbols exist. So the wetland isolation of the real things is avoided here on the rocks – we are given a context.

The associations of carvings are extremely complex. A panel of images may consist of numbers of boats, many cupmarks, several axes and perhaps one or more humans. Another panel may have many discs, perhaps representing shields or in some cases quite detailed images of spoked wheels, another object found in wood and rarely in bronze in wetland isolations. More intriguing are images unmatched in reality, winged humans, armless humans, horsehead-decorated boat prows, gigantic battle axes dwarfing their human carriers (Figure 13.7), elkhead humans, and always the multiplicity of cupmarks.

The relationship between rock carvings and the actuality of bronze artefacts is a difficult process of disentanglement and correlation, and I will only discuss two examples where it seems that rock carvings provide some sort of context for wetland finds. The first concerns shields, the second is even more unusual.

From north and west Europe there are known perhaps 100 sheet bronze shields, fragile objects and richly decorated by ribs and bosses beaten into the thin sheet metal. Almost all the shields have been discovered in peat bogs or in other watery contexts, alone or in pairs or groups of shields (Coles 1962). Five or six, set upright in a circle, are reported from a Scottish site, and at Fröslunda in Sweden the quite astonishing total of 16 shields was recovered from a small peatbog in a field now arable but formerly a tiny inlet of a Bronze Age lake (Jankavs 1995); the shields had been laid in water, layer upon layer. A rock carving site just to the north-west of the findspot, and high over the former lake, consists of circular images, internally ringed, set in groups on the rock face (Svensson 1982) (Figure 13.8). To the east of the Fröslunda shield site, on a low rocky eminence that even now lies near the existing lake, is another rock carving with shield-like images (Jankavs 1999). It is tempting to envisage a direct association between these three sites, the reality of shields laid in water and their images perpetuated on the rocks - the one

Figure 13.4. Rock carvings at Backa, Bohuslän, Sweden. Painted. Scale one metre. (J. Coles photo 1991).

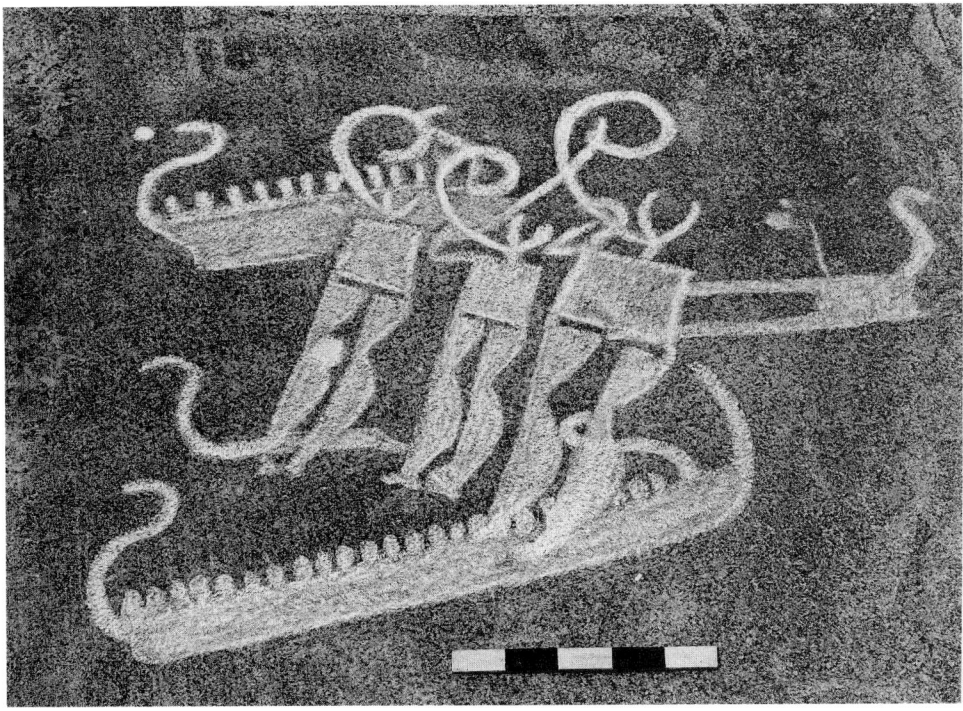

Figure 13.5. Rock carving at Kalleby, Bohuslän, Sweden. Chalked. Scale 25cm. (J. Coles photo 1976).

Figure 13.6. Rock carving at Hede, Bohuslän, Sweden. (Coles 1999).

lost to view forever (so it was thought), the others an everlasting reminder of the event which involved the deliberate loss and eventual destruction of objects that must have represented a huge wealth for the community who had commissioned the work or who had somehow obtained such a hoard of prestige objects. The wetland context is still to be worked out for this find, but the vital attraction of water, its particular depth and distance from the land, and indeed the position from which the shields were laid to rest, indicate the direct relationships that existed and the planning that went into the final operation.

Among the value-laden objects of the northern Bronze Age are several sets of bronze figurines, tiny in size but cleverly cast to reproduce particular shapes. Two sets of figures, found long ago in wetland context, contain horn-helmeted humans with battle axes (Figure 13.9), humans bent over backwards, doing a back flip (Figure 13.10), undulating snakes, and a horsehead. These two sets have often been reconstructed, in theoretical terms, as embellishing model wooden boats of which we have no evidence from the sites. They lack context by themselves yet rock carvings assign a certain relationship to them in Bronze Age symbolism. A number of rock carvings depict the horn-helmeted humans (Figure 13.5), snakes, horse-heads on boat prows (Figure 13.4), and the acrobats. Further, several carvings show elaborate boats with acrobats flipping backwards over them (Figure 13.11), in the way that humans are shown leaping over bulls in east Mediterranean Bronze Age paintings. Here on the rocks of the north we can see how the bronze figurines may have featured in ritualistic displays or performances, measuring and recording symbols that society demanded; under special circumstances the bronzes were consigned to the wetland but the carvings continued to perpetuate their former presence and their memory. What is exceptional here is the geographical separation of images from artefact, the bronzes mostly from wetlands in southernmost Scandinavia, their images carved on rocks some distance away to the north. Substitutes for the real thing, or commemoration in a convenient and accessible place?

Figure 13.7. Rock carving at Simris, Skåne, Sweden. Chalked. Height of male 450mm. (J. Coles photo 1980).

So far, we can show the connection in visual, and physical, terms of special bronzes, special rock carvings and the wetlands and waters that linked them in metaphysical terms. There are other clues that may help in our efforts to interpret these connections. The agencies of production, acquisition and presentation were of course the human communities of the Bronze Age. Within such societies were those whose prestige and control were maintained and enhanced by displays, whether of prowess or wealth, or indeed by the ability to organise presentations of one sort or another. Neither wetlands nor bronzes identify such people with any clarity, but rock carvings do provide some indication of individuals who were set apart, different, from the ordinary populace. Burials of the Bronze Age in the north are also clear indicators of differentiation within a society, although the provision of rich grave goods need not necessarily reflect an equivalence during life for particular individuals. The rock carvings provide quite explicit scenes, or associations at the very least, that offer insights into the patterns of behaviour.

Many carvings spread across the rocks show small unsexed humans, some armless and none with weapons or tools; many are in isolation, with no fellow passengers on the panels or carvings, while others are shown upon boats as crew or cargo (Figure 13.12). Sometimes set apart from these harmless and insignificant humans are larger figures, of males with erect penises, and armed with weapons of axe or spear (Figure 13.7). Some lack an obvious sexual identify and may wear horned helmets, body armour, and may carry and blow lurs (Figure 13.5). Some stand at stern or prow of boats, overseeing their cargoes of puny and

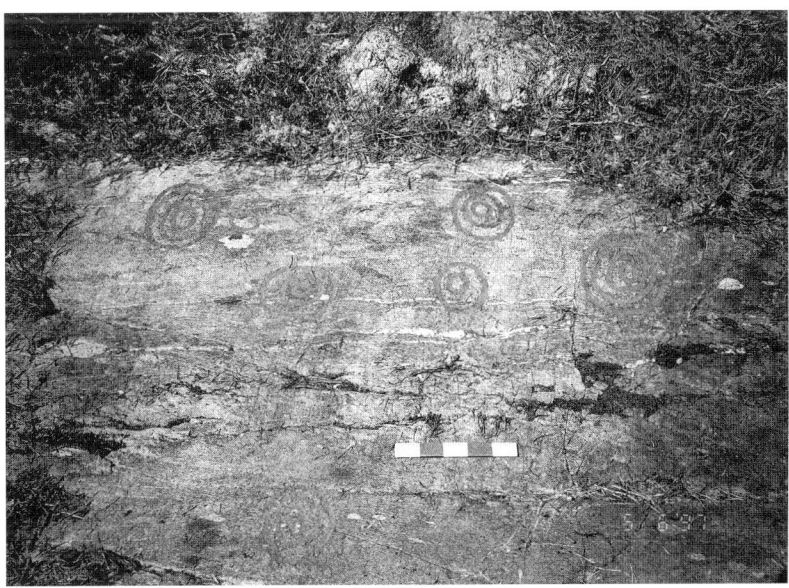

Figure 13.8. Rock carving at Knarrby, Dalsland, Sweden. Painted. Scale 25cm. (J. Coles photo 1997).

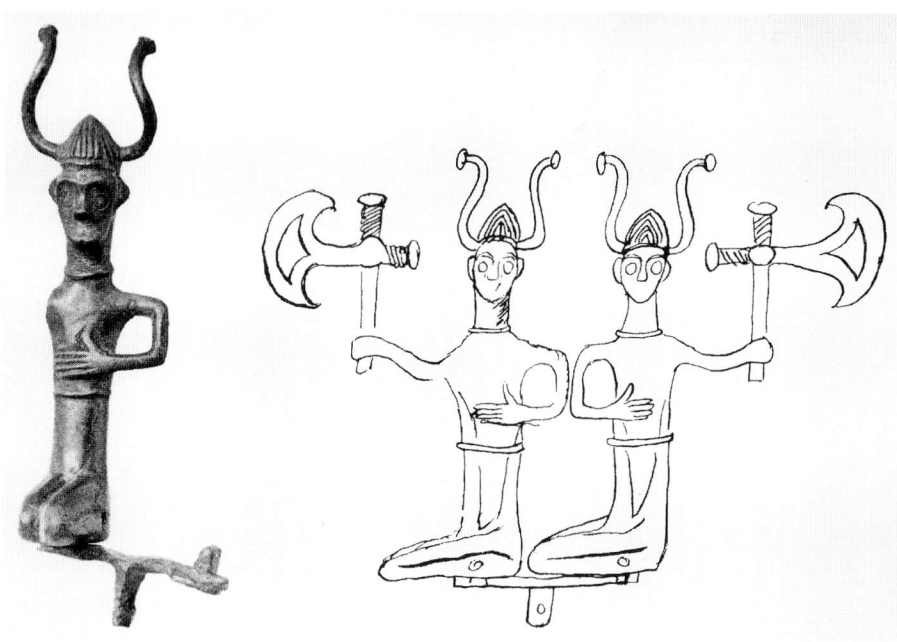

Figure 13.9. Bronze figures from Grevensvaenge, Denmark, mostly now lost. Top of helmet to knee 85mm. (Broholm 1953).

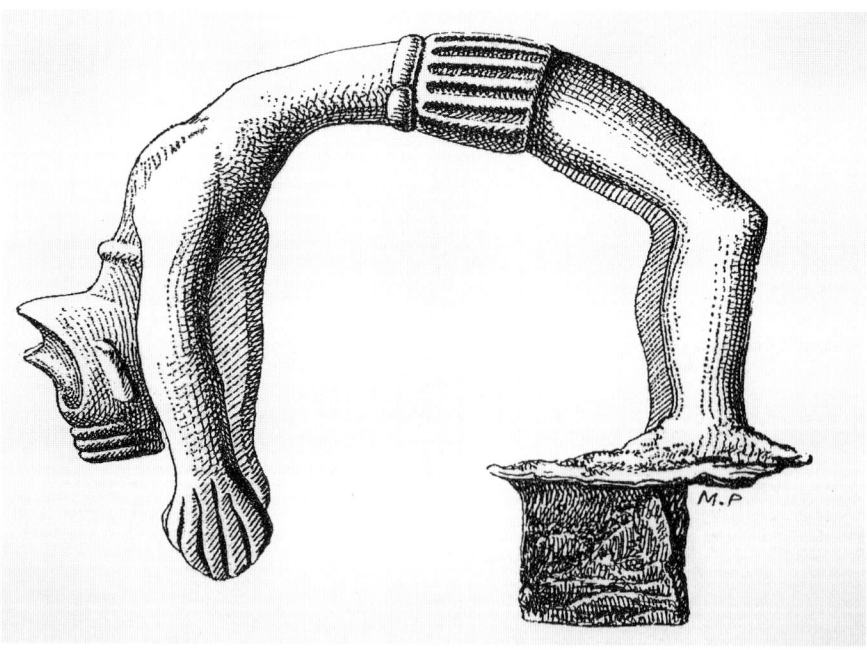

Figure 13.10. Bronze figure from Grevensvaenge, Denmark. Acrobat 85mm tall. (Broholm 1953).

Figure 13.11. Rock carving from Sottorp, Bohuslän, Sweden. Negative chalked. Scale 25cm. (J. Coles photo 1991).

Figure 13.12. Rock carving from Backa, Bohuslän, Sweden. Chalked. Scale 25cm. (J. Coles photo 1981).

Figure 13.13. Rock carving from Järrestad, Skåne, Sweden. Scale 25cm. (J. Coles photo 1995).

powerless humans lined along the gunwale (Figure 13.11). Some are carved on top of, overpowering, older carvings of boats or animals. There can be little doubt that these representations were meant to indicate power and prestige, an ability to control through representation the communities and their activities in the Bronze Age. If some of these human shapes were not of this world, their strength was enhanced by representation, and the position of the carvings near water and wetlands reflected the physical manifestation of prestige in the deposition of precious objects in the adjacent and distant wetlands, doubtless laid in obeisance to their powers for good and evil.

Conclusion

There are direct and indirect relations between north European Bronze Age metalwork, wetlands, rock carvings and the open waters of a dynamic landscape of land and sea. In some cases, rock carvings probably substituted for real metal in societies unable for whatever reason to acquire the precious copper, tin or gold, or to create those objects of the highest quality. In other cases, rock carvings reflected and perpetuated, in the visible world, the real metalwork made invisible by deposition in wetlands and water. In almost all cases, both metals and carvings were associated with water in some form or another. Unlike the single objects of metal deposited in bogs, marshes and pools, the rock carvings show them in association with other things – perhaps showing how they were worn, used, displayed and

even killed, and providing us with a glimpse of what performances or symbolic activities that may have taken place. In addition, the rock carvings are immobile, unlike metalwork, and we can thus be certain that these were the places, the platforms, for public or private performances, where assemblies were gathered, or where individual acts took place in propitiation, commemoration and aspiration. That many rock carvings were directly associated with water and wetlands must show that the dynamic, changing and yet predictable landscapes played a very significant part in social activities other than those of settlement, food acquisition, industry and burial. Many rock carvings consist of identifiable images that are miniature, small boats, small humans and other animals, but many sites contain full-size imprints of human feet (Figure 13.13); these might just represent the reality of human presence on the rocks, of those who came, stood and gazed downslope at the lower lands, the wetlands, and at the sea, the only real entry and exit to the Bronze Age world beyond, in more than earthly dimensions.

References Cited

Broholm, H.C. 1953. Danske Bronzealder. Yngre Bronzealder. Copenhagen: Gyldenalske Boghandel, 1953.

Broholm, H.C., W.P. Larsen, and G. Skjerne. 1949. The Lures of the Bronze Age. Copenhagen:Gyldenalske Boghandel, 1949.

Coles, J, 1962. "European Bronze Age Shields", Proceedings of the Prehistoric Society 28 (1962):156–190.

Coles, J. 1993. "Boats on the Rocks". In J. Coles, V. Fenwick and G. Hutchinson (ed) A Spirit of Enquiry (1993):23–31. Greenwich: National Maritime Museum.

Coles, J. 1994. Images of the Past. Uddevalla, Sweden: Bohusläns Museum, 1994

Coles, J. 1999. "Bronze Age shields in reality and reflection". In A. Gustafsson and H. Karlsson (ed) Glyfer och arkeologiska rum (1999):453–463. Göteborg, Sweden: Göteborg University.

Coles, J. 2000. Patterns in a rocky land:rock carvings in southwest Uppland, Sweden. Uppsala, Sweden. Uppsala University, 2000.

Hagen, A. 1990. Helleristningar i Noreg. Oslo: Det Norske Samlaget, 1990.

Jankavs, P. 1995. Långt borta och nära. Gudaoffer och vardagsting från bronsålderen. Skara, Sweden:Skaraborgs Länsmuseum, 1995.

Jankavs, P. 1999. "Rapsodi över hällbilder i Skaraborg". In A. Gustafsson and H. Karlsson (ed) Glyfer och archeologiska rum (1999):411–439. Göteborg, Sweden: Göteborg University.

Janson, S., E. Lundberg and U. Bertilsson. (ed) 1989. Hällristningar och hällmålningar i Sverige. Helsingborg: Forum, 1989.

Jensen, J. 1979. Guder og mennesker i bronzealderen. Viborg: Lademanns Danmarkshistorie, 1979.

Kaul, F. 1998. Ships on bronzes. Copenhagen: National Museum, 1998.

Svensson, K.P. 1982. Hällristningar i Älvsborgs Län. Älvsborg: Älvsborgs Länsmuseum, 1982.

14. South Scandinavian Wetland Sites and Finds from the Mesolithic and the Neolithic

Lars Larsson

Introduction

In prehistoric research on southern Scandinavia in general and the Stone Age in particular, wetland finds and sites play a very important role which manifests in different ways the general view of the past.

The oldest bog site, Mullerup on Zealand, was excavated in 1900 (Sarauw 1903) and since then several sites have been recognized and excavated (Larsson 1990). But there is a another aspect of bog finds. Scandinavian archaeologists have long interpreted collections of artifacts in wetlands as results of ritual acts. However, outside Scandinavia it seems to be more difficult to find acceptance of Stone Age depositions as votive depositions (Bradley 1990). A collection of artifacts is regarded as a cache for later use rather than the result of a ritual act.

That wetlands have been an important resource for information on prehistory is due to the large number of water basins from large lakes to small kettle holes formed during the retreat of the ice sheet, which in southern Scandinavia ended at about 11,000 cal. BP.

Most of the lakes were eutrophic with the conditions required for quick filling with organic material. Due to the geological structure of the bedrock and various moraine formations, the preservation of organic material in wetlands also varies. A small number of water basins were transformed into bogs already in Preboreal times, while a considerable number of lakes became bogs during Atlantic time. In the Subboreal, which equals the start of the Neolithic at 6000 cal. BP, many bogs dried out and were covered with forest. At the start of the Subatlantic period, which corresponds to the transition between the Bronze Age and the Iron Age at ca. 2600 cal. BP, there was increased precipitation, which meant that raised bogs began to form, in most cases on top of the former wetlands (von Post & Granlund 1926).

Sites and finds will be specifically exemplified here from Scania, the southernmost part of Sweden, but reference examples will be taken from Denmark, comprising the large peninsula of Jutland to the west and the islands in eastern Denmark, the largest ones being Funen and Zealand (Figure 14.1).

The reduction of wetlands

Dense population and early felling of forest caused shortages of fuel, and from the late 18th century peat cutting was introduced on a large scale. In order to gain more arable land many wetlands were drained, a process which started in the second half of the 19th century and is still going on. This process caused a radical change of the wetland landscape.

Thanks to cartographic sources that are extremely rich by international standards, we can gain a good idea of the extent of wetlands during this period preceding the large-scale drainage of the last two centuries. This picture does not agree entirely with conditions prevailing in the Stone Age, but the difference was probably not very great. If we compare the situation in the 18th century, when our map sources are particularly numerous and detailed, we see that very few wetlands have survived in today's landscape.

This is exemplified by a study of the run-off system of a small river in the western part of Scania. In the early 19th century wetland covered 29% of drainage system (Figure 14.2) (Wolf 1956). In the 1950s the wetland area was reduced to about 3% and today the figure is even lower. The wetlands that have survived are the lakes which would have caused more work in drainage than would have been economically acceptable. The information presented by Wolf was based upon research from the 1950s when few apprehended drainage as a problem but rather as an

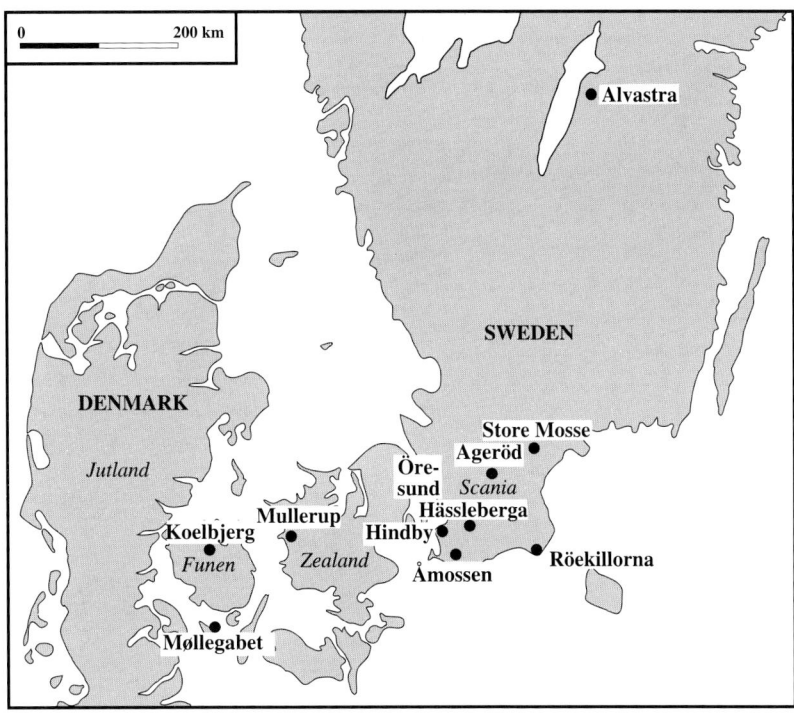

Figure 14.1. Southern Scandinavia with locations mentioned in the text.

important opportunity to enlarge and improve agriculture. It was not until recent years that the difficulties caused by a lowering of the groundwater level and an increasing outflow after melting snow and heavy precipitation have become evident.

Due to special environmental support, there are important attempts to restore wetlands today. In some cases this might save find-bearing layers of prehistoric age but in most cases such layers have already been more or less dried out and perishable artifacts severely damage or destroyed.

Depositions of artifacts in bogs were recognized during the 18th and 19th centuries, and some constitute the oldest finds in museum collections. During the late 19th and early 20th century a large number of artifacts were retrieved in southern Sweden, thanks to a campaign at the turn of the century, in which posters depicting the artifacts and describing their importance for science were distributed to peat-cutting farmers. The fact that pollen analysis was developed early in Sweden, with pollen dating of a significant number of bog finds as early as the 1930s, has stimulated interest in the cultural significance of wetlands. During the two world wars and shortly afterwards, peat cutting was intensive. In the late 1940s most of the cutting was done manually, which meant that artifacts and sites were easily

recognized. Today peat cutting is of minor extent and totally mechanized. Finds are rarely made except by archaeologists who are responsible for surveying certain areas where sites have been found earlier or where the environment indicates reasonable chances of finding prehistoric sites.

Finds in kettle holes

Settlements located in wetlands appear mainly during the Mesolithic (11,000–6000 cal. BP) and the early part of the Neolithic (6000–4000 cal. BP). However, recent finds show that remains of human activities in bogs can be traced back to the Late Paleolithic (13,500–11,500 cal. BP).

When a concentration of kettle holes at Hässleberga in south-western Scania was turned into crayfish ponds, bones and antlers were noticed in the unearthed material and were systematically collected (Larsson *et al.* 2000). Animals such as reindeer, wild horse, elk, and arctic fox were identified, and the radiometric datings show that the bones were accumulated in the small bogs during a long period from the late Allerød to early Preboreal (13,000–11,000 cal. BP). Cut marks and splitting of bones to extract marrow have been identified on bones from reindeer and horse. The finds at Hässleberga provide a very special type of winter kill

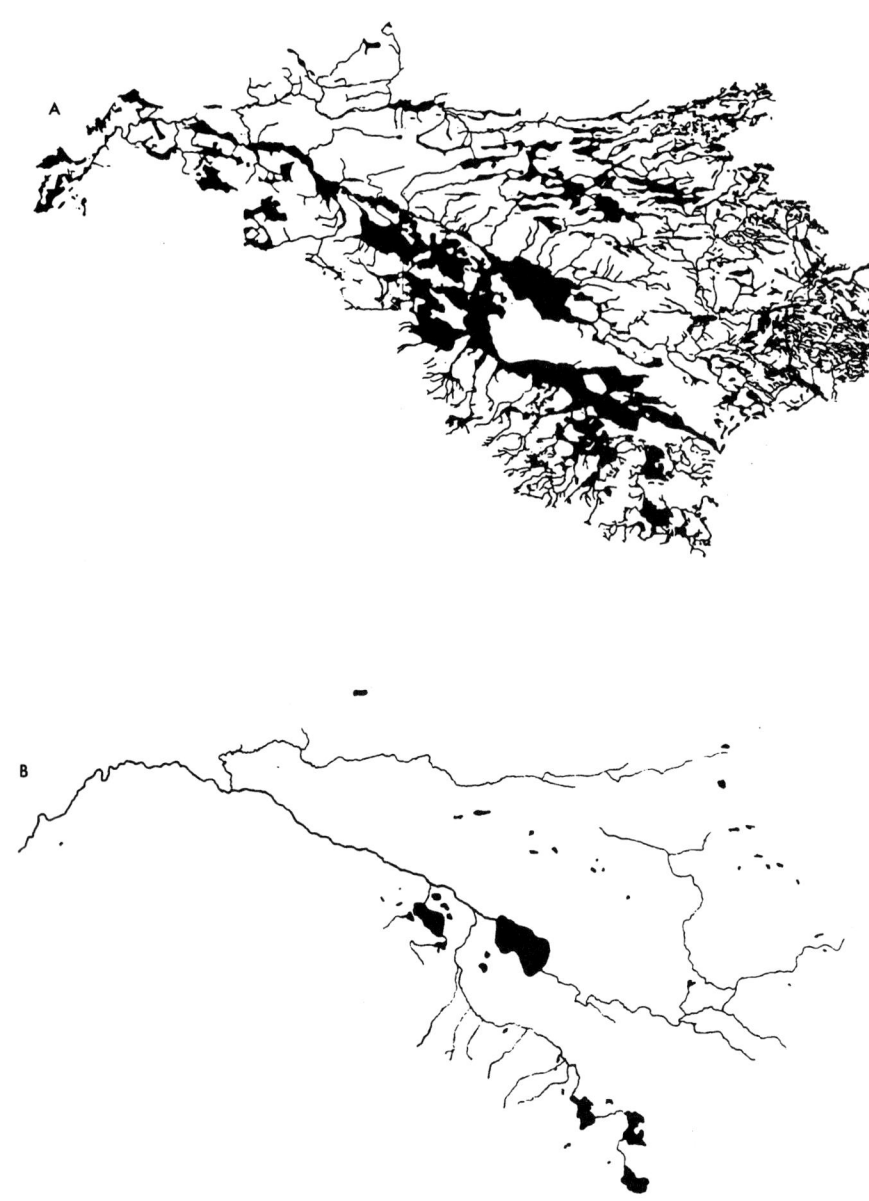

Figure 14.2. The run-off system of the small river Kävlingeån in western Scania, southernmost Sweden, with wetlands ca. 1815 (A) and 1953 (B).

sites on ice. When the ice melted in spring the remains sank down to the bottom of the kettle-holes, which were later transformed into small bogs. This kind of "settlement sites that sunk" might be much more common than these finds indicate, as antler and bones from reindeer causally identified during dredging and digging for clay have been registered in many kettle holes in southwestern Scania and Zealand (Degerbøl & Krog 1959; Liljegren 1975).

Bog sites

From the Preboreal and subsequently, bog sites appear which are particularly well known for the Mesolithic in southern Scandinavia – small camp sites including a hut structure, located on layers of gyttja and peat at the edge of a lake in the process of becoming overgrown. The area of settlement was often reinforced with small tree trunks, branches, and pieces of bark.

In some cases, constructional details such as small posts have been preserved, revealing that these were light structures intended for short-term use. A small number of examples from southernmost Sweden and their Danish counterparts demonstrate that the hut was rectangular or oval in shape (Larsson 1990). The size was between 3 × 3 m and 4 × 6 m. Roofs and walls were probably made of bark or reeds.

The reason why the bog sites have had such a dominant role in the study of Early Mesolithic settlement is that the good preservation environment has an unusually varied content, mostly of bone and antler but also of wooden artifacts.

The distribution of finds and the remains of the hearth provide a basis for a more detailed analysis of the function of the bog sites and the activities carried on there. The distribution of finds in relation to the postulated hearth has been interpreted as showing that the hut was divided into two different activity areas, perhaps a man's side and a woman's side (Grøn 1987; Blankholm 1991).

Bogs sites may be found in close connection. In some cases two or more huts may have been used contemporaneously, but in most cases a small number of people in some kind of family group used the same part of the shore for many years or even centuries. Most bog sites have been viewed as temporary camps of a seasonal character. This interpretation is supported by the numerous finds of shells in the refuse layer outside the hut floor, proving extensive collecting, mainly of hazel nuts but also of water chestnuts (Larsson 1983a).

From the Late Mesolithic few bog sites have been found, mostly because the previously attractive sites were now completely overgrown. The Early Mesolithic sites are clustered on the edge of the bog. The Late Mesolithic bog sites may be much more difficult to find as they are located in the deepest, most central section of the former large lake with the final stage of the filling in process. Some of these large infilled water basins were also transformed into raised bogs during a later part of prehistory, which led to thicker layers of peat. Commercial extraction of peat is thus in progress several years before the underlying sediments in the former lake are unearthed and sites found. In the bog Ageröds Mosse (Figure 14.1), central Scania, there is a site corresponding in scale to bog sites from the Early Mesolithic, with well-preserved material which also includes wooden artifacts such as bows, leisters, and fish traps (Figure 14.3–14.4). The site is dated to 7500 cal. BP.

During recent years several sites of about the same age have been excavated in an area but even further from the shore. On a island of organic litter about 700 meters from the firm land people camped for short periods with a duration of a few hours to some weeks. Based upon the distribution of flint, knapping spots as well as the de-limitation of huts have been identified. The finds also include a cash of more than hundred flint blades (Larsson 1999a).

As the lakes are filled they grow less attractive for settlement, and from the Early Neolithic there are just a few sites from the largest bogs in which areas with open water still exist (A. Fischer 1999).

The submerged Mesolithic landscape

Due to considerable transgressions during the Late Boreal and Atlantic periods, there is little evidence of coastal settlement during the Late Glacial and the early Post-Glacial period in southernmost Sweden. One has to consider that about one third of the land of southern Scandinavia during the Preboreal c. 10,000 BP was submerged because of the melting ice sheet and also as a consequence a rise of the land in northern Scandinavia during the Boreal. Especially in Late Boreal times, c. 9000 cal. BP, there was a global rise in sea level which brought the surface of the water from a position of more than 20 meters under the current level to only a few meters under this in just a few centuries (Christensen 1995).

To shed light on coastal settlement during this phase, attention may be directed to the strait Öresund, now the border between Denmark and Sweden. During the Early Mesolithic this sound included a land bridge as well as a deep and narrow bay of the North Sea with an archipelago (Larsson 1999b) (Figure 14.5). A rapid rise in the water level caused the bay to enlarge, at the same time as islands were submerged. The area contained several biotopes which made it attractive for settlement.

Due to the rises in sea level very little evidence remains of coastal settlement during the Early Post-Glacial in southern Scandinavia. It is only during a late part of the Boreal, about 8000 BP, that our knowledge of the southernmost part of Sweden improves. In order to obtain information on coastal settlement forms during the Early Mesolithic, marine archaeological investigations have been carried out on the Swedish side. The sites were concentrated on what is now a submarine furrow corresponding to the prehistoric course of a river (Figure 14.5). Along this river, both surveys and investigations have revealed a number of Late Mesolithic sites close to the present shoreline. During the investigations, at least four Early Mesolithic sites were recorded, the depths of which varied between 20 and 6 meters below surface level (Larsson 1983b; Larsson 1999b).

The best preserved site Pilhaken 4, which is partially layered in peat, is situated at a depth of 7 to 8 meters and has been dated to about 9000 BP (Figure 14.5). The muddy layer could be observed as a horizon in the steeply sloping submarine course of the river. The part nearest the course of the river channel is exposed to continuous erosion. With

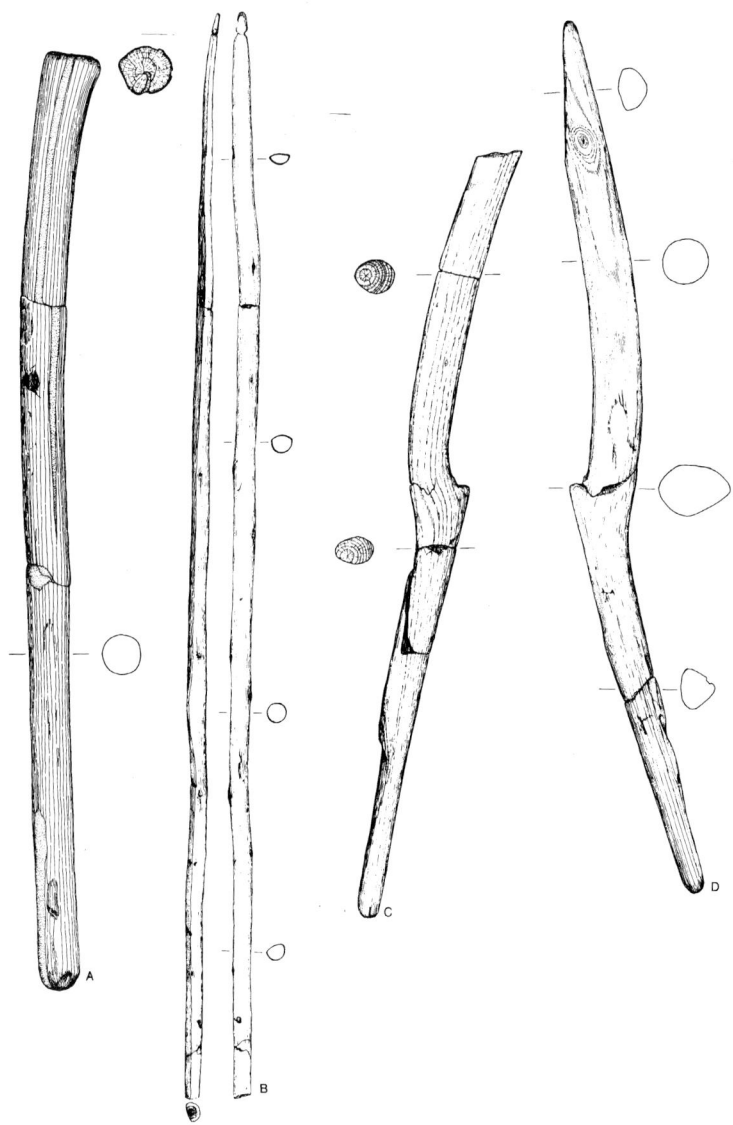

Figure 14.3. Wooden artefacts from the Mesolithic site Ageröd V, southernmost Sweden. A: axe shaft (c. 1:3), B: fragmentary bow (c. 1:5) and C–D: leister prongs (1:2).

large-capacity nozzles and the resulting back-suction, small trenches were dug. The stratigraphy consisted of alternate layers of mud and sand. The finds consist of flint artifacts from the late Maglemose Culture, and bones from roe deer, red deer, and aurochs were found (Larsson 1999b). The layers of peat were deposited in a comparatively well-protected basin, possibly in a part of the delta that may have included the former mouth of the river.

Surveys of the sea bed in the southern part of the sound,

in the vicinity of the location of the former land bridge between Denmark and Sweden, were carried out in 1992–94 in conjunction with prospecting for the bridge to be built over the Öresund. Flint artifacts were found in a significant number of test pits, pointing to the fact that the area had seen settlement (Larsson 1999b). None of these flints could be confirmed with certainty as *in situ* finds. Most also bore traces of rolling and salt water patination. Such traces were absent from a small number of flints,

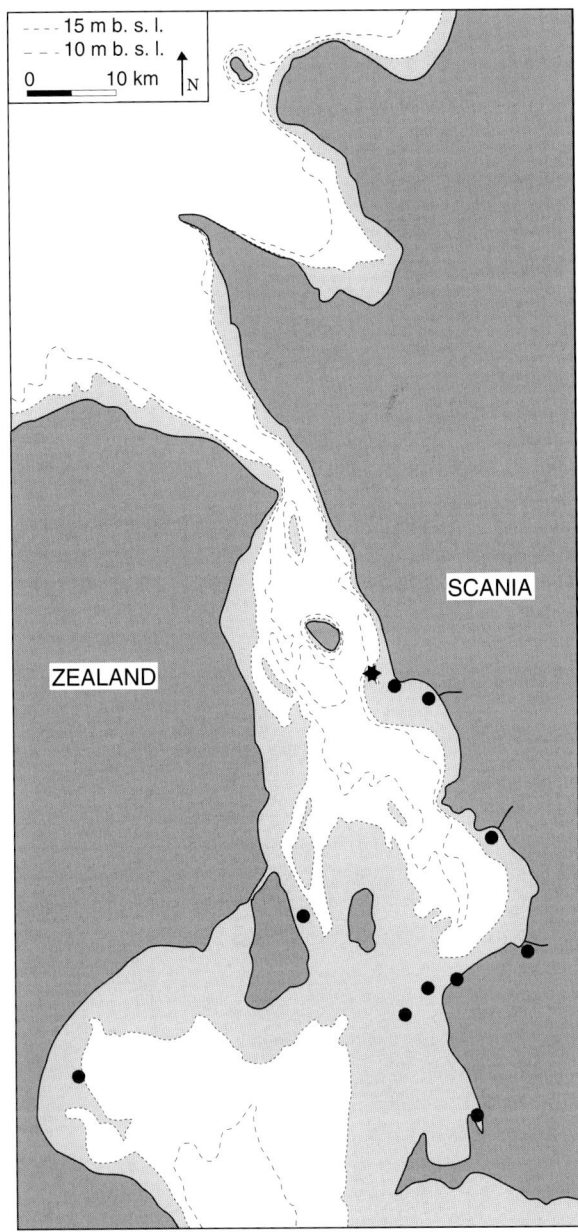

Figure 14.4. Submerged Mesolithic settlement sites (dots) in Öresund, the sound between present-day Denmark and Sweden. The sites are marked by a star.

however. These finds indicate that undisturbed find-bearing layers had been eroded quite recently, or that they may still exist in particularly well-protected areas. We were unable to identify any such find location, however, during the trial investigation.

Sites along a changing shoreline

The coastal settlement during the Boreal period was equally intensive as that which is well-proven for the hinterland. The late Mesolithic sites are located at river mouths and lagoons, indicating that the choice of settlement site during the Late Mesolithic was dictated by the same factors as those which governed Early Mesolithic man.

The great change in sea level during the Late Boreal and the Early Atlantic period must not be regarded as ecologically critical for the hunter-gatherer societies. New, abundant fishing environments were formed just as quickly as old ones disappeared. These changes were so drastic that their effects must have been clearly identifiable in the landscape during the lifetime of the inhabitants. The formation of new coastline environments attracted settlement. It is during the Late Mesolithic that large sites, often combined with cemeteries, are found in lagoons in southern Sweden and eastern Denmark (Larsson 1993). The sea level continue to rise for the rest of the Late Mesolithic, offering preservation in refuse layers just off the shore, later to form wetlands (Larsson 1988). At the same time the southern part of Denmark tilted, causing the coastal zone to submerge. During resent years surveys of this sea bottom landscape have resulted in the finding of several sites, some of them excavated as well (Andersen 1985). At the submerged site Møllegabet II on western Funen, the deceased was placed in the stern of a canoe which was wrapped in bark and placed below sea level close to the settlement (Grøn & Skaarup 1993).

Bogs and bog sacrifices in the Mesolithic

In connection with peat cutting, a large number of tools of bone and antler from the Mesolithic have been found (Figure 14.6). The reasons why they were found in bogs may vary. Some objects could derive from settlement sites which have not been noticed. Others may be traces of fishing. There are some examples in Denmark of a number of leister points being stuck into a former lake bed. Judging by the topographical situation, the place was a suitable fishing ground where the use of leisters was particularly effective. The fact that there are often finds of sticks of resin-rich fir with traces of burning at one end is evidence that leister fishing at night with the aid of torches was a frequent form of fishing (Larsson 1983a).

In other cases it is more difficult to find any practical explanation for the bog finds. One example is the finds of slotted bone points (Figure 14.6:4) in a small bog – Åmossen – in south-west Scania (Figure 14.1). A large number of slotted bone points have been found in a bog that is only about 200 meters in size, and about ten of these have been preserved, the majority intact. The bog is too small to have had the prerequisites for fishing. If the points had been used for fishing or hunting then they should have suffered extensive fragmentation, as we know from settle-

Figure 14.5. Deposition of microliths, some in fragmentary state, from the refuse layer of a bog site dated to c. 9000 cal BP at the bog Ageröds mosse, central Scania.

ment site finds with a large number of damaged points (Larsson 1978b; Larsson 1982). Yet this is not the case. It can hardly have happened that a batch of points fell in the water while being transported across it. The most likely explanation is that the bog was a sacrificial site where slotted bone points were deposited.

Dugout canoes and bows are types of artifact that occur surprisingly often at well-preserved find locations, whereas, for example, arrow shafts are extremely rare. In the Late Mesolithic world-view there may have been clear rules about how to handle tools that were no longer serviceable. We can take an example of this from archaeological investigations in connection with the building of a bridge from Funen to Zealand. Excavations at the bridge abutment uncovered several find spots located in a former bay of the sea (Pedersen *et al.* 1997; Christensen 1999). During a few centuries around 7000 cal. BP at least three dugout canoes, a fishing trap, an axe shaft, pointed sticks, and other objects were found, either whole or parts of them, beside a small but pronounced promontory in the fiord. Since there is no settlement site waste in the strict sense, the place is interpreted as a natural deposition for drifting wooden objects.

On the shore of a seemingly uninhabited island in the former lagoon, a further three Late Mesolithic dugouts were found. These had been retained on the shore by means of stakes driven into the ground. This find situation is interpreted as either a collection of special waste or what was left after a group of fishermen whose boat was driven by the wind and who were seized and killed by the people of the lagoon.

One cannot avoid wondering whether there may be more reasonable explanations than natural taphonomies or violent death. If the finds had belonged to the Neolithic it is fairly certain that the interpretation would have been in terms of some form of deliberate deposits – sacrificial sites! A considerable number of pots and other objects from this period were deposited in bays of the sea. There are several other Mesolithic finds from Southern Scandinavia which show great similarities to Neolithic votive acts (Larsson 1978a; Karsten 1994). In the refuse layer of a site dating to about 9000 cal. BP, at the edge of the bog Ageröds Mosse (Figure 14.1), a collection of more than thirty microliths, some in fragmentary state, were found (Larsson 1978:67–70) (Figure 14.7). By recording the position of each flint it was possible to study the position of the intact as well as the fragmentary pieces. The microliths had been stored in a receptacle and were broken before they were deposited. This find situation rules out the interpretation as a cache for future use.

Human skeletons of Mesolithic Age have also been found in bog. The oldest Swede is a young woman found in Store Mosse (Figure 14.1), a bog in southernmost Sweden (Nilsson *et al.* 1979). She has been dated to the Boreal period, ca. 9000 cal. BP. Judging by the find circumstances, the woman drowned. Likewise, the oldest Dane, a women,

Figure 14.6. The wooden platform from Alvastra, southern central Sweden, dated ca. 5.100 B.P.

was found in a bog at Koelbjerg on Funen (Figure 14.1). She has been dated to ca. 10,000 cal. BP (Bennike & Alexandersen 1997).

Neolithic wetland hoards

The role of the wetland as a place of contact with the spirit world was much more accentuated during the Neolithic.

The extensive drainage has had far-reaching conseq- uences for prehistoric remains in wetlands. In the 1980s

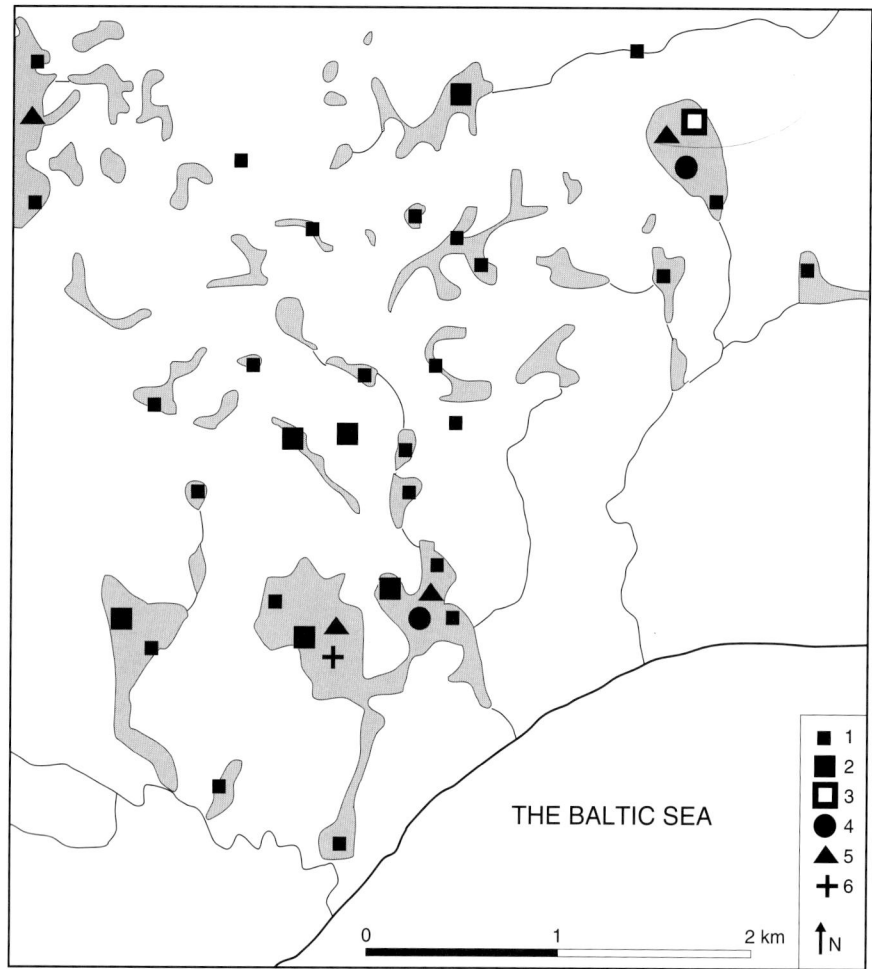

Figure 14.7. Votive depositions from the Neolithic within a small area of southernmost Scania, southern Sweden. Legend: 1: deposition of a single artifact, 2: deposition of at least two artifacts, 3: deposition of several artifacts, 4: deposition of artefacts made of antler or bone, 5: depositions from the Bronze Age and 6: depositions from the Iron Age. The wetlands are marked as hatched areas.

and 1990s the whole of Scania was the subject of a total survey. This comprised not just a survey of the landscape but also an attempt to document the thousands of farm collections that exist. It was found that almost all the Stone Age objects discovered in the last two decades were found in wetlands or have a patina showing that they were deposited in former wetlands. Earlier finds in farm collections were to large extent found on heights and lack a moss patina. This shows that the earlier found objects in farm collections are probably from graves while the later additions are from votive deposits in wetlands.

A study of the find circumstances of more than 600 Neolithic hoards in Denmark shows that 80% were found in wetlands (Nielsen 1985). The largest number of hoards

was registered in the museum collections during the late 19th and early 20th century as well as in the 1940s.

The Danish study is based on depositions including two or more flint or stone objects. But a large number of single finds have been found in wetlands, or the yellow to red patination strongly indicates that they have been preserved in a bog environment (Karsten 1994). In a study of the depositions in Scania 370 hoards have been identified, and the amount found in wetlands is similar to the figure for Denmark (Karsten 1994). However, in addition there are 928 single finds. Only a small number of these have not been found in wetlands.

Peat-cutting and drainage have uncovered a large number of objects, mainly of flint but also artifacts of bone,

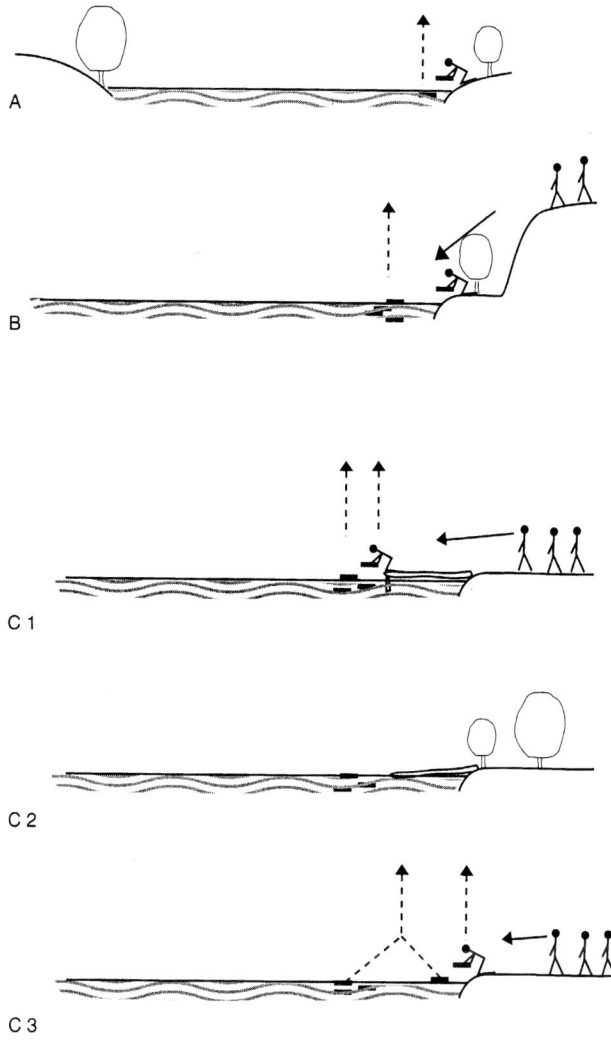

Figure 14.8. Votive deposition practices. A: deposition of a small number of finds as a private ritual by one person or a small number of persons, protected from other members of the society, B: votive depositions of public character by a person who was unknown to members of the society and C: votive depositions of public character by a person who was known to members of the society. At certain places votive depositions have been made over a long time, sometimes with intervals for centuries (C1–3).

antler, and amber. The most common category of find is flint axes, most commonly deposited in pairs. In most cases the deposition seems to be an isolated expression of a votive ritual, but in some instances a number of artifacts were deposited within a limited area during a time amounting to a hundred, in a few cases thousands of years. Entire tools are the most common finds but in some instances these finds might just represent the most obvious of a number of other depositions. An excavation of a small bog near Hindby

in southwestern Scania (Figure 14.8) presents this problem of representativeness (Svensson 1993). Remains of votive practice were found running through the Late Mesolithic, most of the Neolithic and into the Bronze Age (7000–3500 cal. BP). There are examples of axes deposited in pairs, but it is more common to find combinations of tools, sometimes broken up before deposition, as well as bones of animals or humans such as a deposition of a burnt fragment of an axe, two human bones and three pig eye-teeth.

The simplicity of the sacrifices makes it difficult for the layman to identify the objects as part of intentional votive depositions. It is probable that similar depositions in other bogs have not been observed apart from the easily recognized forms such as flint axes or daggers.

In the bog at Hindby a number of trunks were placed as a short trackway in order to facilitate access to the central part. Wooden constructions have been found in bogs in direct relation to deposition. In some cases, Quaternary geological investigations have shown that the bogs in which votive deposits were made were almost dried up when the sacrifices were made; they were only seasonally water-logged.

Among the bog sites of specific character is the structure in the Alvastra mire in south-central Sweden, dated to the middle of the Neolithic, about 5100 cal. BP (Figure 14.6) (Browall 1986). It consists of two square wooden platforms joined at an oblique angle. These two square parts are in turn divided into several rectangular "rooms", each with a hearth. Dendrochronological studies tell us that the structure was built on a single occasion, as the piles for the frame of the platform were all driven into the mire at the same time. Activities on the platform went on for a total of 42 years, but this included long periods when parts of the platform were unused. Large collections of burnt fruits and seeds have been found beside the hearths, such as crab apple, cereals, and hazelnuts. Human and animal bones also occur. The structure has primarily been viewed as an assembly place where each family had its own rectangular cell. Such a clear division may have been needed if two social groups, for example, two kin groups, used the same structure for activities of a sacral character.

Data from peat-cutting in Danish bogs indicate that wooden structures in conjunction with finds of a votive character were not as unusual as the finds from Alvastra might suggest (Rech 1979).

Large and small votive depositions

The frequency of Neolithic depositions is evident from a special study in the neighborhood of the author's residence on southern Scania. From my own field walks as well as from information gained concerning several farm collections, the find location of a considerable number of depositions is known. The distribution is shown on a map based upon information from the 1810s when drainage had not started (Figure 14.7). Find information about more than 40 flint artifacts, the majority consisting of axes, has been recorded; most of the flint artifacts have been found during soil preparation, a small number during digging for drainage (Figure 14.7). Judging by the yellow to red patination typical of the effect of the bog environment, an equal number of flint tools in the farm collections were found in wetlands, even though no information on the find circumstances is available. From the typology of the axes, it is clear that depositions were made during the entire Neolithic with a dominance during the late part of the Early Neolithic and the later part of the Middle Neolithic at about 5400 cal. BP and 4400 cal. BP respectively.

The larger wetlands also contain the largest number of tools, and most finds have been made close to the edge of the bog. In some cases finds have been made in small wetlands not recorded on the map from the 1810s. One of these bogs has a size of less than fifteen meters. On different occasion a flint axe, a fragmentary stone axe, a flint core with a shape resembling an axe, and a small polishing stone have been found. As the axes are of the same type, the artifacts might have been deposited simultaneously, and were later spread by soil preparation.

From a perspective comprising the entire Neolithic, almost every wetland was used for deposition.

Deposition of axes at settlement sites might sometimes appear in wetland conditions. In the excavation of a settlement from the Middle Neolithic (ca. 5000 cal. BP) at an ancient lagoon in southernmost Scania, two axes attracted special attention as the only intact examples. One was found in a former shoreline deposition covered by the occupation layer. By the shape it could be dated to the early part of the Neolithic at about 5900 cal BP. A few meters farther up the slope a fireplace was found below the occupation layer. According to a radiocarbon date it might be related to the axe, perhaps a fire lit in connection with deposition rituals (Larsson 1992). In the external part of the settlement an axe was found close to an unworked flint node. They were deposited just outside the shoreline which was marked by decaying stumps of bushes and small trees. Just these two objects were found in this part of the excavation trench.

These two axes represent two depositions with a time difference of more than a millennium. The older deposition belongs to a society with no site in the immediate neighborhood, while the younger one is a deposition contemporaneous with the settlement site. The latter is a kind of offering which is easy to miss if the excavation is just restricted to the occupation area.

The first intentional depositions of human beings in wetlands occur in the Early Neolithic (Stjernquist 1981;

Bennike 1999). In some cases the cause of death, such as strangulation with a rope or hard blows against the head, is visible. The way people were killed during the Early Neolithic does not differ from the practices in the Early Iron Age more than three thousand years later (C. Fischer 1999; Larsson 1998). It was not until a few years ago that votive deposition of humans from the Neolithic were recognized because of radiocarbon datings (Bennike & Ebbesen 1986).

In relation to Continental Europe, where depositions in rivers and lakes are well known, the number in southern Scandinavia is small. In connection with dredging work artifacts dated to the entire Neolithic – and the Bronze Age and Iron Age as well – were found at five different sites within a distance of four kilometers from the lower part of the river Segeå (Karsten 1994:141–144). However, this find category is probably much under-represented among the finds.

Deposition behavior

At least in certain respects the cosmology which is related to wetland offerings was active throughout most of the Neolithic and later periods as well. Some wetlands appear to have been surrounded by ideas of a sacral character throughout centuries, in some cases millennia. Such an example is Röekillorna (The Red Springs) in southeastern Scania (Stjernquist 1997). The name alludes to the high content of iron in the water, causing red sludge to be deposited on the bottom. In this well system depositions of artifacts were performed from an early part of the Neolithic until the Early Iron Age. The place was thus used for rituals for more than three thousand years.

In many cases the artifacts were deposited within a delimited area of the bog, even though individual finds may contain artifacts dating from a considerable time-span. Several of the wetland offerings may be seen as unfinished projects where the deposition period stretches over several generations. In this respect they may be given a similar interpretation to that suggested for Neolithic henges from southern England (Barrett 1994).

However, there seems to be a hiatus in Scania of almost a millennium when the wetland deposition sites were never or rarely used for depositions (Karsten 1994). In spite of a time interval there is a close spatial relationship between depositions dating from the Early Neolithic (c. 6000–5300 cal. BP) and depositions belonging to a late part of Middle Neolithic and a continuity into the Late Neolithic (4500–3800 cal. BP). This observation appears to be of major importance for understanding the world view of Stone Age societies and how it can be observed in material culture. Depositing artifacts within a delimited area means that knowledge of the ritual importance of the site survived for generations, including a period during which the site was not actively in use. Knowledge of the physical as well as the metaphysical components of the landscape was passed on during a long time without any visible manifestation in the material culture.

It looks as though there might be changes going on in the society when people had a need to establish links with much earlier societies. A detailed knowledge of the landscape included legends, tales, and stories which had been passed on from one generation to the next. During certain periods this relation to this wider environmental view is marked by acts of more or less ritual practice involving material culture. But for long intervals the knowledge was just passed on without any reaction by the society.

Taking the old offering sites into use once again during the later part of the Middle Neolithic might have been a way to reestablish contacts with earlier generations – a marker of a connection with societies of the past based on legends, as well as means of emphasizing a different value system from the society which had just been replaced.

The depositions in the landscape

Despite the special information obtained by the study of sites with accumulated deposition for a long time, one has to take into account that most of the wetland offerings are examples of single events when just a small number of artifacts were handed over to members of the metaphysical world. The size of the wetland used for votive depositions varies in size from large water basins to wetland of less than 20 meters. Even if exceptions do appear, the accumulated depositions over a long time seem to be made in larger wetlands and the single depositions in much smaller wetlands. As has been shown above, the number of votive depositions is considerable even in a local perspective. Depositions were made in practically every wetland. We are justified in saying that the landscape was sacralized.

Can we obtain any information about how the depositions were performed? In view of the fact that we are dealing with wetlands, it is conceivable that people stood on the bank and threw the objects out into the water. In those cases where a more detailed account is given about axes found close to each other, the finder has observed that they were carefully placed in a special arrangement in the wetland. Axes have been found close to, sometimes on top of each other, with the edges directed downwards or in a circle (Rech 1979:Abb. 2; Knutsson 1988:Fig. 40). This proves that artifacts were carefully placed in the water rather than being thrown out into the basin. As repeated depositions of artifacts took place within a limited area of the wetland which was shallow and in some cases seasonally dried up, there should have been a knowledge of the excellent status of tools in offerings which were several

generations old. The deposition or transformation of artifacts in water should involve a subtle quality of special value to the artifact. This must have been the primary intention behind wetland offerings.

The presence and absence of certain artifacts as well as the composition demonstrates that certain regulations existed as to what and how to deposit. The presence of axes is the most obvious. There is a variety in the frequency of deposition of different axe types within the Neolithic. Some axes are common in votive offerings while other types rarely appear in depositions despite a large number of axes recovered from locations other than wetland. During a late part of the Neolithic the number of axes decreases, replaced by flint daggers, which are also the most common grave goods.

Not only the selection of tools in a synchronic as well as a diachronic perspective can be distinguished but the chronological circumstances as well. During the Early Neolithic clay vessels are a very common deposition item on Zealand (Koch 1999). At the same time in Scania this form of deposition is very rare (Karsten 1994:61).

The reasons for votive deposition

Water is life-giving for all organisms. Wetland offerings might be related to an underground spirit connected with fertility, with the wetland being regarded as a point of bodily access to the hidden soul. Offering in wetlands might have been a regular practice of which the desired effects were long-lasting but not immediately noticeable. We can envision a habit whereby change is not visible unless the offerings cease.

Objects may have been deposited for several purposes. In general the sacrifices may be divided into three categories:votive gifts offered in the hope of a return gift; conciliatory offerings hoping for forgiveness; and communication offerings establishing community between the participants in the act and the envisaged recipients in a metaphysical world.

All categories of offering may have had a public character whereby it was important that the votive act could be observed by a considerable share of the members of the community (Figure 14.8:B–C). In some cases the physical conditions, with shores accessible to a large number of people, were particularly favorable for deposits of this type.

In other cases one almost has the impression that the deposition was a private ritual by one person or just a few people. Such depositions appear in a small bog or a part of a larger one, which in most cases was surrounded by bushes and trees. In such cases wetland offerings were closed acts in which a small number of people were involved.

It does not seem possible, however, to detect differences between large and small deposits as regards the composition of the objects. The reason for this may be that so few such votive sites, whether extensive or limited, have been excavated.

One has to keep in mind that deposition in wetland during the Neolithic is only one aspect of the relation between the societies and the metaphysical world. Depositions are found in dry ground, such as in connection with a large stone or a cliff. In these cases it is much more difficult to determine the cause of deposition. They may well be caches. In some cases deposition in wetland may have had the same purpose. As flint absorbs water and is easy to knap when water-soaked, the deposition of artifacts in water should result in better preservation for future use.

Depositions of votive character are also found in connection with causewayed enclosures such as on the bottom of the ditches and in pits within the enclosed area (Andersen 1997). Large depositions, mainly of pottery, appear close to the entrance of the megalithic tombs. These have been interpreted as offerings to the ancestors. In some cases the number of vessels deposited might have been as large as one thousand (Tilley 1996:295–257). Votive depositions are frequent in connection with settlements where axes are found, for example, in postholes for houses. Both small and large depositions of flint and stone artifacts damaged by fire are also documented (Larsson 2000).

References Cited

Andersen, N.H. 1997. The Sarup Enclosures. The Funnel Beaker Culture of the Sarup site including two causewayed camps compared to the contemporary settlements in the area and other European enclosures. Jutland Archaeological Society Publications XXXIII:1. Aarhus: Aarhus University Press.

Andersen, S.H. 1985. "Tybrind Vig. A Preliminary Report on a Submerged Ertebølle Settlement on the West Coast of Fyn." Journal of Danish Archaeology 4:52–69.

Barrett, J. 1994. Fragments from Antiquity. An Archaeology of Social Life in Britain, 2900–1200 BC. Oxford: Blackwell.

Bennike, P. 1999. The Early Neolithic Danish Bog finds: a strange group of people! Coles, B., J. Coles, and M. Schou Jørgensen, eds., Bog Bodies, Sacred Sites and Wetland Archaeology, 27–32. Warp Occasional Paper 12. Exeter: Department of Archaeology.

Bennike, P. and V. Alexandersen. 1997. "Danmarks urbe-folkning" (The aboriginal population of Denmark. National-museets Arbejdsmark:143–156.

Bennike, P. and K. Ebbesen. 1986. "The Bog Find from Sigersdal. Human sacrifice in the Early Neolithic." Journal of Danish Archaeology 5:85–115.

Blankholm, H.P. 1991. Aarhus: Aarhus University Press.

Bradley, R. 1990. The Passage of Arms. An archaeological analysis of prehistoric hoards and votive deposits. Cambridge: Cambrige University Press.

Browall, H. 1986. Alvastra pålbyggnad:social och ekonomisk bas. (The Alvastra Pile Dwelling:its Social and Economic

Basis). Theses and Papers in North-European Archaeology 8. Stockholm.

Christensen, C. 1995. The littorina transgressions in Denmark. A. Fischer, ed., Man and Sea in the Mesolithic. Coastal settlement above and below present sea level, 15–22. Oxbow Monograph 53. Oxford: Oxbow Books.

Christensen, C. 1999. Mesolithic boats from around the Great Belt, Denmark. Coles, B., J. Coles and M. Schou Jørgensen, eds., Bog Bodies, Sacred Sites and Wetland Archaeology, 47–50. Warp Occasional Paper 12. Exeter: Department of Archaeology.

Degerbøl, M. and H. Krog. 1959. The Reindeer (Rangifer tarandus L.) in Denmark. Zoological and Geological Investigations of the Discoveries in Danish pleistocene Deposits. Biologiske Skrifter. Det Kongelige Danske Videnskabernes Selskab 10. København: Munksgaard.

Fischer, A. 1999. Stone Age Åmose. Stored in museums and preserved in living bog. Coles, B., J. Coles, and M. Schou Jørgensen, eds., Bog Bodies, Sacred Sites and Wetland Archaeology, 85–92. Warp Occasional Paper 12. Exeter: Department of Archaeology.

Fischer, C. 1999. The Tollund Man and the Elling Woman and other bog bodies from Central Jutland. Coles, B., J. Coles, and M. Schou Jørgensen, eds., Bog Bodies, Sacred Sites and Wetland Archaeology, 93–97. Warp Occasional Paper 12. Exeter: Department of Archaeology.

Grøn, O. 1987. "Seasonal Variation in Maglemosian Group Size and Structure. A New Model." Current Anthropology 28, no. 3:303–327.

Grøn, O. and J. Skaarup. 1993. "Møllegabet – A Submerged Mesolithic Site and a "Boat Burial" from Ærø." Journal of Danish Archaeology 10:38–50.

Karsten, P. 1994. Att kasta yxan i sjön. En studie över rituell tradition och förändring utifrån skånska neolitiska offer-fynd.(To throw the axe in the lake. A Study of Ritual Tradition and Change from Scanian Neolithic Votive Offerings). Acta Archaeologica Lundensia 8:23. Stockholm: Almqvist & Wiksell International.

Knutsson, K. 1988. Making and using stone tools. The analysis of the lithic assemblages from Middle Neolithic sites with flint in Västerbotten, northern Sweden. Aun 11. Uppsala: Societas Archaeologica Upsaliensis.

Koch, E. 1999. Neolithic offerings from the wetlands of eastern Denmark. Coles, B., J. Coles, and M. Schou Jørgensen, eds., Bog Bodies, Sacred Sites and Wetland Archaeology, 125–132. Warp Occasional Paper 12. Exeter: Department of Archaeology.

Larsson, L. 1978. Ageröd I:B – Ageröd I:D. A Study of Early Atlantic Settlement in Scania. Acta Archaeologica Lundensia 4:12. Lund: Liber.

Larsson, L. 1977–1978. "Mesolithic Antler and Bone Artefacts from Central Scania." Papers of the Archaeological Institute University of Lund:28-67.

Larsson, L. 1978. "Mesolithic Antler and Bone Artefacts from Central Scania." Papers of the Archaeological Institute University of Lund 2:28–67.

Larsson, L. 1982. Segebro. En tidigatlantisk boplats vid Sege ås mynning.(Segebro. An early Atlantic site at the estuary of the river Sege å). Malmöfynd 4. Malmö: Malmö Museer.

Larsson, L. 1983a. Ageröd V. An Atlantic Bog Site in Central Scania. Acta Archaeologica Lundensia 8:12. Lund: Liber.

Larsson, L. 1983b. Mesolithic Settlement on the Sea Floor in the Strait of Öresund. Masters, P.M. and N.C. Flemming, eds, Quaternary Coastlines and Marine Archaeology. Towards the Prehistory of Land Bridges and Continental Shelves, 283–301. New York: Academic Press.

Larsson, L. 1988. The use of the landscape during the Mesolithic and Neolithic in southern Sweden. Archeology en Landschap. Bijdragen aan het gelijknamige symposium gehouden op 19 en 20 oktober 1987, ter gelegenheid van het afscheid van H. T. Waterbolk, 31–48. Groningen: Biologisch-Archaeologisch Instituut.

Larsson, L. 1990. "The Mesolithic of Southern Scandinavia." Journal of World Prehistory 4, No. 3:257–309.

Larsson, L. 1992. "Neolithic Settlement in the Skateholm Area, southern Scania." Papers of the Archaeological Institute University of Lund 9:5–43.

Larsson, L. 1993. The Skateholm Project:Late Mesolithic Coastal Settlement in Southern Sweden. Bogucki, P., ed., Case Studies in European Prehistory, 31–62. Ann Arbor: CRC.

Larsson, L. 1998. Prehistoric Wetland Sites in Sweden. Bernick, K., ed., Hidden Dimensions:The Cultural Significance of Wetland Archaeology, 64–82. Vancover: UBS Press.

Larsson, L. 1999a. Settlement and Palaeoecology in the Scandinavian Mesolithic. Coles, J., R. Bewley, and P. Mellars, eds., World Prehistory. Studies in Memory of Grahame Clark. Proceedings of the British Academy 99, pp. 87–106. London: Oxford University Press.

Larsson, L. 1999b. Submarine settlement remains on the bottom of the Öresund Strait, Southern Scandinavia. Thevénin, A., ed., L'Éurope des derniers chasseurs. Èpipaléolithique et Mésolithique. Peuplement et paléoenvironnement de l'Épipaléolithique et du Mésolithique, 327–334. Grenoble: CTHS.

Larsson, L. Axes and fire – contacts with the gods. Olausson, D. and H. Vandkilde, eds., Form – Function – Context. Material Studies in Scandinavian Archaeology. In print.

Larsson, L., R. Liljegren, O. Magnell, and J. Ekström. Archaeofaunistic Aspects of Bogfinds from Hässleberga, Southern Scania, Sweden. Bratlund, B., ed., Behaviour and Landscape Use in the Final Palaeolithic of the European Plain. Stockholm. In print.

Liljegren, R. 1975. Subfossila vertebratfynd från Skåne (Subfossile vertebrate finds from Scania). University of Lund. Department of Quarternary Geology. Report 8. Lund.

Müller-Wille, M. 1999. "Opferkulte der Germanen und Slawen" (Deposition rituals among Germanic and Slavonic tribes). Archäologie in Deutschland Sonderheft. Stuttgart: Theiss.

Nielsen, P.O. 1985. Neolithic Hoards from Denmark. Kristiansen, K., ed., Archaeological Formation Processes. The representativity of archaeological remains from Danish Prehistory, 102–109, København: Nationalmuseet.

Nilsson, T., T. Sjøvold, and S. Welinder. 1979. "The Mesolithic Skeleton from Store Mosse, Scania." *Acta Archaeologica* 49:220–238.

Pedersen, L., A. Fischer, and B. Aaby, eds. 1997. Storebælt i 10.000 år. Mennesket, havet og skoven. (The strait Storebælt during 10.000 years. Man, the sea and the forest). København: Nationalmuseet.

Rech, M. 1979. Studien zu Depotfunden der Trichterbecher- und Einzelgrabkultur des Nordens. (Studies of hoards from the Funnel Beaker culture and the Single Grave culture of Northern Europe). Offa-Bücher 39. Neumünster: Karl Wachholtz Verlag.

Sarauw, G.F.L. 1903. En Stenalders Boplads i Maglemose ved Mullerup, sammanholdt med beslægtede Fund. Aarbøger: 148–315.

Stjernquist, B. 1981. Näbbe mosse. A mysterious Stone Age lake. Florilegium Florinis Dedicatum, Striae 14:35–40.

Stjernquist, B. 1997. The Röekillorna Spring. Spring-cults in Scandinavian Prehistory. Regia Societatis Humaniorum Litterarum Lundensis LXXXII. Stockholm: Almqvist & Wiksell International.

Svensson, M. 1993. "Hindby offerkärr – en ovanlig och komplicerad fyndplats." (The bog for offering at Hindby – an unusual and complex site) Fynd no 1:5–11.

Tilley, C. 1996. An ethnography of the Neolithic. Early prehistoric societies in southern Scandinavia. Cambridge: Cambridge University Press.

von Post, L. and E. Granlund. 1926. Södra Sveriges torvtillgångar I. (The peat recourses of Southern Sweden). Sveriges Geologiska Undersökning Ser. C, no 2. Stockholm: Nordstedt.

Wolf, P. 1956. Utdikad civilisation (Drained civilization). Malmö: Gleerups.

15. Realizing the Archaeological Potential of the Scottish Peatlands: Recent Work in the Carse of Stirling, Scotland

Clare Ellis

Introduction

Over the past few years Historic Scotland has commissioned a series of research projects into the archaeology of AOC the Scottish peatlands (Ellis 1999a; 1999b; 1999c). The commissioning of these projects was part of a programme to develop a strategy for the comprehension, protection and management of Scotland's peatland resource in conjunction with other interested agencies (Hingley *et al.* 1999). However, the development of such a strategy is dependent upon a broad knowledge base of the potential of Scotland's peatland archaeology, which despite recent work (e.g. Clarke & Finlayson 1995; Clarke 1998; Ellis 1999a & b) lags behind that of neighbouring countries such as England and Ireland. The minimal quantity of cultural resource centred peatland research and our subsequent inadequate understanding and appreciation of this, is largely a consequence of lack of opportunity for invasive archaeological works. This is due to limited peatland industrialization (i.e. milling for fuel) and restricted wholescale destruction of areas of moss; although ironically many of Scotland's peatlands have been disturbed by widespread afforestation, the effects of which are potentially devastating upon buried organic archaeological remains (e.g. Ellis 1999b). This paper includes a brief discussion of the number and type of sites known from Scottish peatlands and a synthesis of the results of a wetland landscape research program that has advanced the understanding of the potential of the archaeological and palaeoenvironmental resource of Scotland's peatlands.

The database

The cultural resource of Scottish peatlands lies within over one million hectares of peat, comprising predominantly blanket bog, epitomized by the Flow Country of Caithness and Sutherland and more restricted lowland raised bogs, such as by Flanders Moss East in Stirlingshire (Hingley *et al.* 1999). In addition many of the soils in Scotland are classified as peaty and gleyed; these also hold the potential for the preservation of both archaeological remains and palaeoenvironmental data. Prior to the creation of a database of archaeological sites and artefacts found in all wetland types (Clarke 1997) the extent and nature of Scotland's peatlands archaeological resource was largely unappreciated. This wetland database was amended and utilised to quantify the current state of knowledge of archaeology within peatlands, to provide a greater understanding of specific wetland sites and to identify wetland sites worthy of further research (Ellis 1999a). The most promising wetland archaeological sites in Scotland were identified, their archaeological and palaeoenvironmental potential was assessed and each high potential site was categorized according to the perceived degree of risk to the archaeological remains (Ellis 1999a).

The database contained 6627 separate archaeological records and every record was classified according to its nature and type, with the aim of enabling a broad range of representative site types to be selected. The categories used were: landscape features, views, battle sites and oddities; wooden artifacts; other organic artifacts; organic structures; metal artifacts; stone artifacts; other inorganic artifacts; inorganic structures; organic and inorganic artifacts; organic and inorganic structures; and agricultural landscape remains. The use of these strict criteria to classify every entry in the database meant that certain very broad assumptions concerning the nature of the archaeology had to be made. Although the classification of some of the archaeological types may be contentious, the classification was employed in a systematic manner so that others could follow and reproduce the logic of the system. The soil type of each entry was also classified into a simplified series, again to enable a representative sample of soil types to be included in the sub-samples of high potential sites.

To facilitate an appreciation of the current state of knowledge and nature of wetland archaeology in Scotland every category of raw data was converted into a percentage of the total number of the records in the database. Some 70.86% of the archaeological records fell into the inorganic structure classification, whereas organic structures accounted for only 0.23% of the record. It is postulated that, despite the predisposition of the database to wetland zones, and therefore the presumption for the survival of organic matter, what was being observed was a significant bias to the survivability and recoverability of inorganic structures, as opposed to organic structures. This notion was enforced when the total percentage of inorganic artifacts (14.67%) was compared to that of organic artefacts (2.1%). It was considered that the mixed organic and inorganic category, be it artifactual or structural, held the highest archaeological potential. It was therefore encouraging that 7.95% of sites in the database fell into this category. Organic soils, predominantly peat, accounted for 53.69% of all the records, closely followed by peaty soils at 33.77%. Lochs accounted for approximately 5% of the records, with the majority of the sites described as crannogs. Beyond the broad relational patterns, details regarding absolute number of type-sites were particularly enlightening. There are only sixty recorded sites in Scotland located in peat from which exclusively organic artifacts have been recovered and eight sites from which solely organic structures have been recovered. However, there are significantly more mixed (comprising mixed organic and inorganic) artifacts (38) and structures (148), which bodes well for future Scottish wetland studies and research. However, an existing limitation to our further comprehension of the cultural resource of Scottish peatlands is the minimal description of each site entry and a lack of absolute dates for material recovered in the past; this issue is currently being addressed by the National Museum of Scotland dating program.

Initial assessment of the wetland database revealed that an inordinate proportion of the sites and artifacts on the database were derived from the valley of the Carse of Stirling. As part of Historic Scotland's peatland strategy, AOC Archaeology Group was commissioned to undertake an assessment of the archaeological and palaeoenvironmental potential of Blairdrummond Moss and Flanders Moss East. AOC Archaeology extended the boundaries of the study area to include all the raised mosses of the valley thus enabling a landscape approach to the assessment to be adopted. The project results demonstrated the potential of the cultural heritage resource of Scotland's peatlands and culminated in the excavation of a Neolithic wooden platform buried within peat deposits.

Case study: Carse of Stirling, a landscape approach

The initial project comprised a desk-based assessment of all known archaeological sites and find spots within the study area, a rapid walkover survey, geomorphological mapping and reconnaissance and reference coring of six of the raised mosses. Presented below is resume of the development of the landscape since the early Holocene, a review of known sites and artifacts and a summary of the rediscovery and excavation of Scotland's only known Neolithic wooden platform. Post excavation analysis is currently underway and it is anticipated that this will aid in the interpretation of the platform and augment an understanding of the local environment prior to and during the use of the structure.

The Carse of Stirling lies to the west of the city of Stirling within the Forth River Valley (Figure 15.1). The study area is bounded by the M9 on the east, the A84(T), the A873 and A81 to the north, the A81 to the west and the B835 and A811 to the south, covering approximately 117 km². The study area encompasses the Upper Forth River Valley (Vale of Menteith) and the extreme lower slopes of the Trossach Hills to the north, while the southern boundary is formed by the relic marine cliffs of the Touch and Gargunnock Hills.

Palaeoenvironment of the Upper Forth River Valley

The raised mosses and unconsolidated sedimentary deposits within the Carse of Stirling have received more attention than the archaeological remains contained within them (e.g. Durno 1956; Turner 1965). The onset of the Loch Lomond Stadial (11,000 years ago) saw the re-advance of ice as far as Arnprior depositing a stiff clayey (Mentieth) moraine that crosses the Forth valley, forming the topographic rise of Cardross and Parks of Garden. As post-glacial sea levels fell a series of buried raised beaches were formed (Laxton & Ross 1983). These post-glacial marine sediments form a plain some 4 to 5 kilometers wide and which slopes from about 14 m above OD in the west to 8 m OD in the east (Laxton & Ross 1983). Radiocarbon dates demonstrate that at about 8800 to 8500 BP a bed of peat (the sub-carse peat) started to form on the poorly drained surfaces of raised beaches (Sissons 1966; Ellis 1999b). During a post-glacial transgression much of the sub-carse peat was buried by marine carse clay, deposited between 8421 ± 157 and 5481 ± 130 BP (Francis *et al* 1970; Brooks 1971) (Figure 15.2). However, in some areas peat formation was not terminated, demonstrated by one of the deepest peat successions in Flanders Moss East which dates back to the early Holocene, 9405 ± 95 BP (AA-30340) (Ellis 1999b).

The original extent of peat cover in the Upper Forth Valley is difficult to estimate following the removal of vast tracts of peat in the eighteenth and nineteenth centuries, but

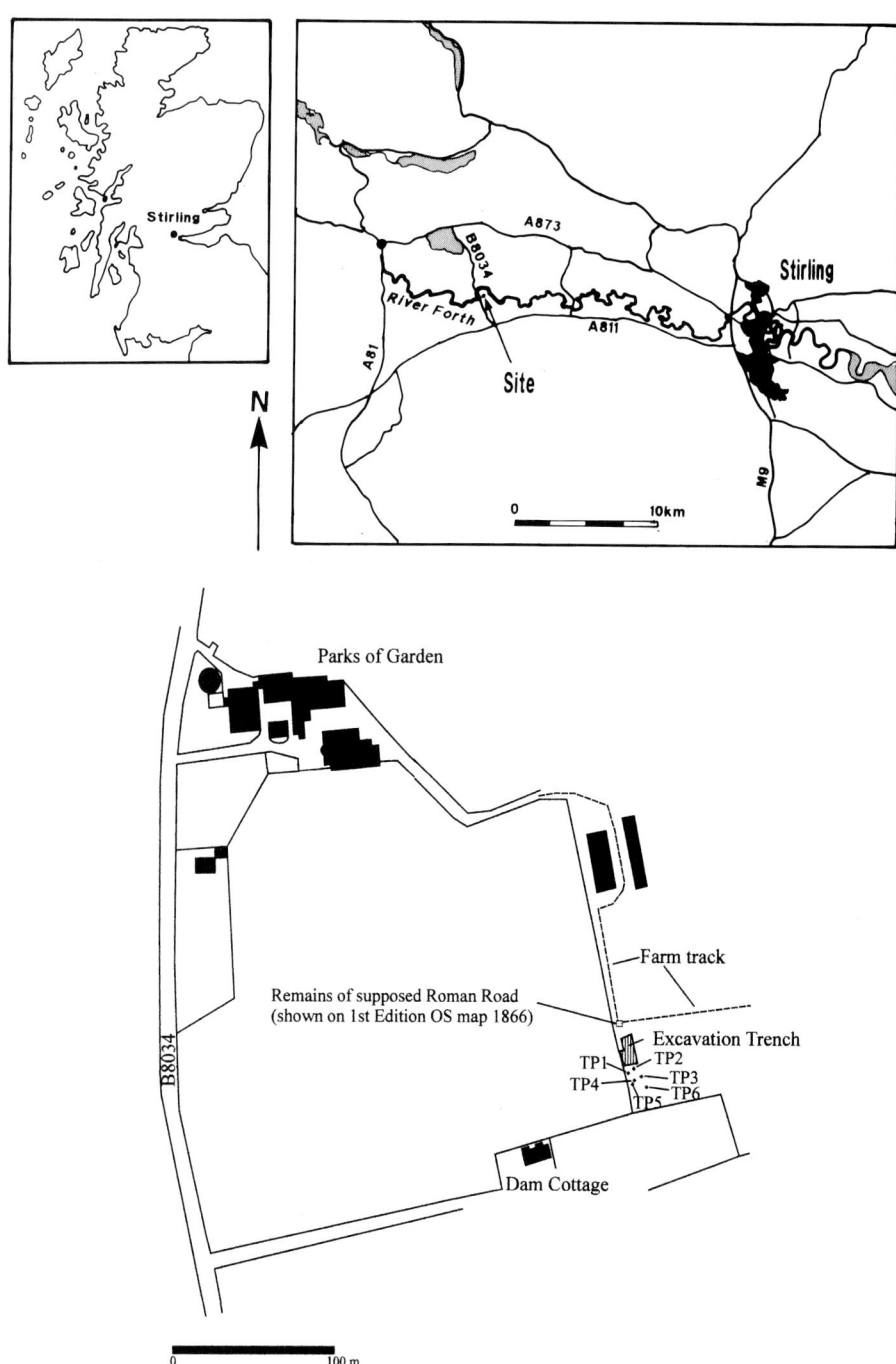

Figure 15.1. Location of excavation site.

it is reasonable to assume that Cadell's account of a 12 mile stretch of moss from 1 to 2 miles broad is fairly accurate (Cadell 1913). A basal date of 5840 ± 75 BP (AA-30333) has been obtained from the organic rich clay underlying Collymoon Moss (Figure 15.2) (Ellis 1999b). The date demonstrates that the moss started to form upon the drying

salt-marsh surface (the carse clay) following the final transgression of the Upper Forth Valley. The basal peat of Killorn Moss is dated to 4750 ± 70 BP (AA-30316), some 1000 years after that of Collymoon Moss (Ellis 1999b). The base of the peat from Ochertyre Moss, located at the eastern extreme of the study area, is dated 2550 ± 50 BP

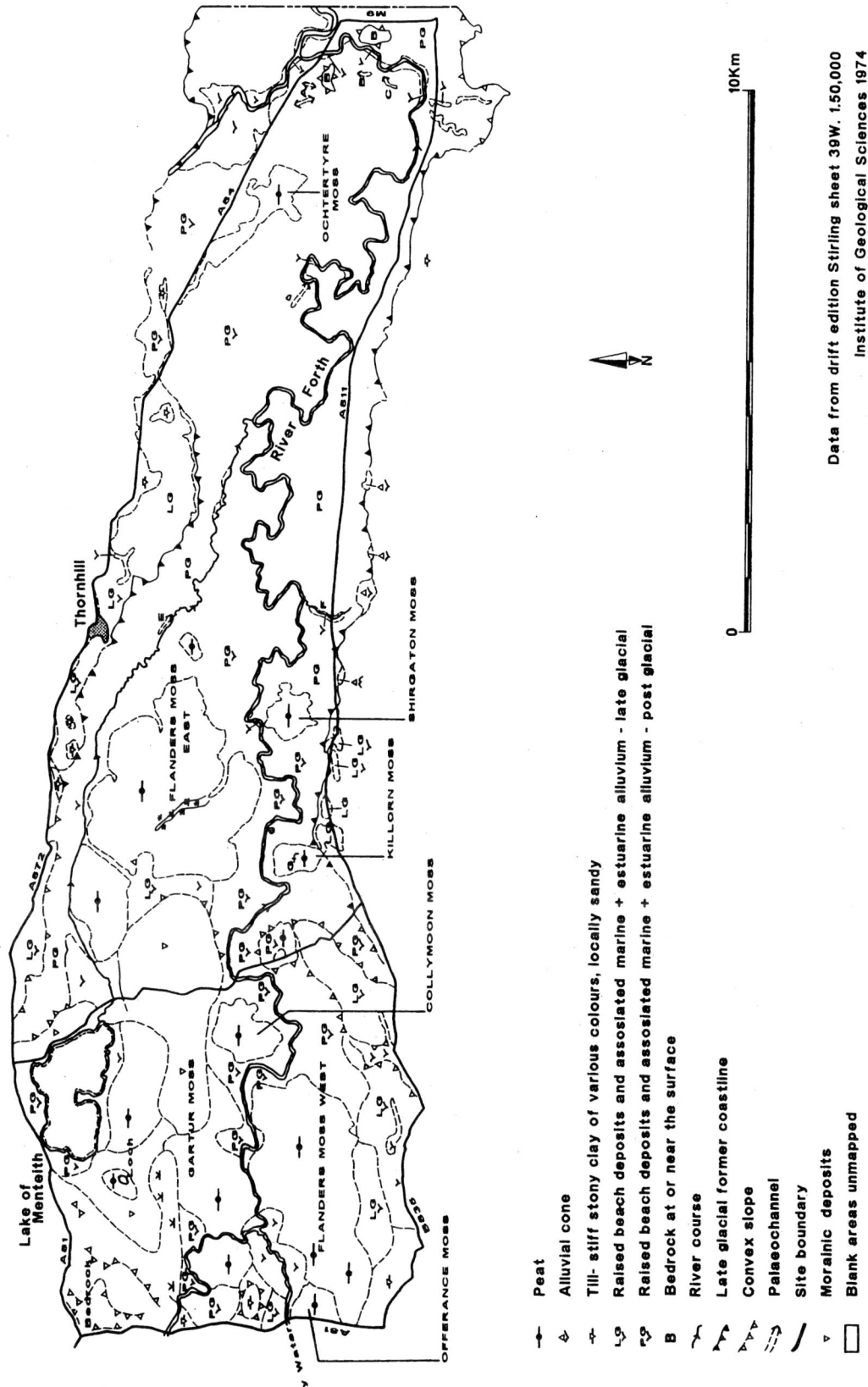

Figure 15.2. Geomorphology of the Carse of Stirling and location of raised bogs.

(AA-30320). These basal peat dates indicate that the retreat of the post-glacial sea was slow and geographically complex (Ellis 1999b). However, as a note of caution there may be a significant time-lag between the retreat of the sea and the development of peat due to the generally slow rate of paludification. It is very probable that between the mid 6th millennium BC to the 1st millennium BC estuarine mudflats acted as a natural barrier to north/south communication. During the last 2000 years north/south communication would have also been hampered by the lateral growth of the raised mosses of the Upper Forth River Valley (e.g. Cadell 1913).

The archaeological resource

The large quantity of archaeological material recovered from the Upper Forth Valley is largely due to extensive eighteenth and nineteenth century peat clearance, during which many organic and non-organic structures and artifacts were unearthed (Clarke 1997; Ellis 1999a & b). The Old Statistical Account gives a comprehensive description of the methodology followed in the removal of Blairdrummond Moss (OSA 1799). Prior to its removal Blairdrummond Moss covered a vast area of some 10,000 acres (OSA 1799), stretching from near the confluence of the River Forth and River Teith westward up the Forth River Valley for about 4 miles. Historical peat clearance also included the area of Poldar Moss located below Thornhill (Cadell 1913) and "improvements" along the northern margins of Flanders Moss East and around Flanders Hill (Johnson 1792).

As a consequence of this clearance a broad range of organic and non-organic artifacts, the fate of many unknown, have been recorded by various eighteen and nineteenth century antiquarian authors. Those from Flanders Moss East include a timber trackway uncovered near Pallabay Pow which may have been associated with some form of float or jetty and the remains of a log boat (Anderson 1967). Two swords were also found near Pallabay Pow reputedly lying in a cross position (Ordnance Survey 1862). Other finds from Flanders Moss East comprise a Romano-British glass bangle (Stevenson 1976) and a Late Bronze Age sword from Poldar Moss (Burgess & Colquhoun 1988).

The largest collection of finds results from the almost total clearance of Blairdrummond Moss. Probably the most important of these finds is a tripartite disc wheel dated to1255 to 815 BC (NMRS MS/735/1), which was reported to have been found 10 ft below the surface of the moss (Piggott 1959). This wheel was found in association with three other wooden wheels one of which was of ash; it may be supposed that these are the remains of a horse-drawn vehicle perhaps trapped in the peat as it made its way across

the moss. Access into and across Blairdrummond Moss appears to have been facilitated by at least one wooden road, the remains of which were uncovered on Blairdrummond Moss in about 1793 (RCHAMS 1979). It was described as being constructed from tree trunks lying full length on the surface of the carse clay with small pieces of timber crossing at right angles; the whole was covered with brushwood and measured some 4 ft in width (Tait 1794; RCHAMS 1979). A third wooden trackway was first recorded as "the remains of a supposed Roman Road" (Ordnance Survey 1862) located just a couple of kilometers to the west of Flanders Moss East. The road was described as "a causeway composed of the trunks of trees, with marks of bolts in the longitudinal sleeps" (Wilson 1878); it is at the approximate location of this site that the excavation reported below took place.

Further finds from Blairdrummond Moss include: a wooden mortar (Archaeology Society 1890); antler implements (e.g. Ordnance Survey 1862); arrow heads (Piggott 1959); several polished stone axes and maceheads (Wilson 1863); three flanged bronze axes, a socketed bronze axe, three socketed bronze spearheads, a bronze sword and bronze cauldrons (O'Connor & Cowie 1995); and a faience bead (Callander 1906). There are also two examples of Medieval homesteads located on the edge of the moss in the Upper Forth River Valley; one occurs on the western edge of Flanders Moss East and may have functioned as a hunting lodge.

The presence of prehistoric artefacts recovered from beneath and within the mosses, and the occurrence of sturdy wooden routeways within these demonstrates that the mudflats and later the moss was the focus of much human activity, but the absence, until recently, of any investigative excavation means that the nature of this activity is poorly understood.

Excavation of Parks of Garden Neolithic wooden platform

Parks of Garden, the site of this recent excavation is located on the east side of a low ridge stretching across the Upper Forth Valley (Figure 15.1). This ridge forms a line of high ground across the Forth River Valley from the village of Arnprior in the south, to the village of Menteith in the north and is the first high, dry crossing point across the valley to be reached travelling westwards from Stirling. It is clear from antiquarian accounts, geomorphological observations and the accounts of the current landowners that the peat represents the cultivated remnants of a once far more extensive moss, which would have stretched eastwards towards the River Forth. The remaining peat deposit is now a relatively thin, narrow wedge located between thick glacial moraine to the west and carse clay to the east. An archaeological evaluation carried out by AOC Archaeology in 1998 ascertained the presence and state of preservation of a wooden structure, described on the First Edition Ordnance Survey map (1862) as the remains of a Roman Road. The evaluation demonstrated that the structure was extremely unlikely to be a Roman road and unlikely to be a wooden track, but its function could not be ascertained. The structure lay within 0.40 to 0.60 m of the ground surface in actively oxygenating peat and the wooden members were soft and delicate. The poor state of preservation, coupled with the location of the structure within a narrow, relatively thin wedge of peat, made preservation *in situ* untenable and full excavation was recommended (Ellis 1999b).

The precise location of the site was ascertained through peat probing and non-invasive ground penetration radar (GPR) (Utisi *et al.* in prep), both methods enabling the detection (but not identification) of sub-surface anomalies within a peat profile. At Parks of Garden both methods produced evidence of a series of sub-surface anomalies. To determine the nature of some of these sub-surface anomalies a series of trial trenches were opened and archaeological remains were found in Trench A located at the break of slope of the glacial moraine.

The wooden structure was placed upon the damp *Polytrichum* peat surface (Context 137) (Figure 15.3 & Plate 15.1). A large *in situ* oak tree throw (Context 116) was utilized to form a stable northern edge and a smaller tree throw on the southern side (Context 170) was used for the same purpose (Figure 15.4). These two timbers created the frame of the trapezoidal substructure on which the main longitudinal timbers of the superstructure (e.g. Context 110, 113) were rafted (Figure 15.3). The tree throw substructure was strengthened by the central placement of a few near parallel roundwood timbers upon the surface of the moss (e.g. Context 163). Undressed roundwood logs and lengths of planking (e.g. Contexts 110, 111 and 113) were laid on top and perpendicular to the substructure (Figure 15.3). The strength and stability of the structure was ensured by the central placement of the majority of the timbers over the two parallel underlying substructure timbers. The northern end of these timbers lapped over the roots and moraine of the tree throw (Context 118 and 116) and were laid upon the silty peat (Context 158) overlying the moraine on the southern side of the structure. The central and eastern portions of the superstructure timbers were laid with their northern ends over the trunk of the tree throw (Context 116) and upon the *Polytrichum* moss surface. The central and southern portion of the platform eventually sunk into the peat on either side of the tree throw, primarily because many of the outermost timbers were short planking and consequently were not supported by the underlying substructure.

Where the wood of the superstructure touched the

substructure a thin brushwood layer (Context159) had been laid in places. It is postulated that the brushwood was used to protect the timbers and act as packing to prevent lateral movement of the superstructure. A thin layer of birch brushwood (Context 157) had also been placed under some of the planking. It is probable that this brushwood was deposited to create a level, firm and less damp ground surface on which to place the structure. No substantial buttressing posts, to prevent the lateral movement of the horizontal timber, were identified and only two possible *in situ* stakes were identified during excavation.

A series of split oak laths (Context 102) laid parallel to each other but perpendicular to the main timbers on the platform and appeared to have formed a "pathway" from the dry ground onto this platform (Figure 15.3). These laths could be traced as far back as the east facing section of the trench and were laid down the slope of the moraine and out onto the platform. The eastern end of these oak laths had been raised to create a level surface by the placement of coarse brushwood (Context 108) on top of the round wood timbers. On the southwestern side of the structure an area of partially burnt, parallel aligned timbers (Context 135) extended the southern boundary of the structure along the lower break of slope of the moraine. Evidently part of the structure, these timbers lay on a slightly different orientation (north-west south-east); their southern extent was not identified during the excavation.

The northeastern end of the structure was covered by an irregular layer of birch brushwood (Context 114) which merged into the twig rich, well humified peat of the surrounding moss (Context 115). It seems likely that this brushwood was deliberately placed over parts of the structure, although coverage was by no means uniform. The presence of a well-humified peat directly overlying the superstructure may represent its furnishing with moss, twigs and other organic matter to provide a soft and cushioned working or walking surface.

The long, relatively straight roundwood timber of the superstructure comprised both oak and alder (Crone 1999); in-field identification demonstrated that oak timber comprised the dominant species of the central and eastern portion of the superstructure. The wood species of the majority of the superstructure and substructure have yet to be identified. The *in situ* bog oaks display long straight boles suggesting that these were derived from a mature, dense oak forest (Crone pers comm). The source of wood (oak, alder and birch) used in its construction is very probably local (Hughes & Kenwood 1999). Two pieces of planking were identified during excavation (Plate 15.2); the western piece had been burnt on its underside prior to its incorporation into the superstructure. It is postulated that the cleft planks on the structure were the remnants of wood working activities elsewhere and these outer pieces

Figure 15.3. Plan of the superstructure of the wooden platform, Parks of Garden.

Plate 15.1. The superstructure of the wooden platform, Parks of Garden.

Figure 15.4. Plan of the substructure of the wooden platform, Parks of Garden.

Plate 15.2. Two re-used planks within the superstructure of the wooden platform, Parks of Garden.

of trunk had been stored and accidentally burnt prior to their inclusion within the structure. The presence of waste wood within the structure suggests anthropic activity and construction off site but presumably not at any great distance, perhaps in a clearing within the oak forest growing upon the drier ridge of glacial moraine.

The condition of the wood on the platform was poor. The roundwood and planking was extremely soft and fragile having lost much of the lignin, although the oak members were slightly better preserved (Crone 1999). The oak of the fallen trees was generally better preserved, especially those lower down in the peat profile (e.g. Context 116, 172). The highest oak trunk, Context 106 had been eroded by ploughing and weathering but exhibited a solid core. The presence of outlying oak trunks (Contexts 121, 106 and 136) probably served to protect the platform from destruction by ploughing and other agricultural activities (Figure 15.3 & 15.4).

Four separate wood elements of the structure have been previously dated (Ellis 1999b). The radiocarbon dates of these four elements are not statistically distinguishable and place the construction of the platform to the mid Neolithic, 4465 ± 40 BP (OxA-8124, the date derived from 5–10 year old birch brushwood). The radiocarbon date derived from the peat underlying the platform also falls within the quoted time-frame.

The function of the structure remains ambiguous, although it is hoped that post-excavation analyses may elucidate this problem. The 1st Edition Ordnance Survey map (1862) marks the "remains of supposed Roman road" in the general location of the site (Figure 15.1). The Rev. Wilson recorded a causeway composed of the trunks of trees, with marks of bolts in the longitudinal sleepers (1878), suggesting a corduroy road of the kind found elsewhere in Neolithic, Bronze Age and Iron Age contexts. Therefore one interpretation of the structure is as the remnants of a wooden trackway which would have led down from the drier high ground of the moraine onto the lower lying moss. However, the interpretation of the structure as a wooden trackway is problematic. The trapezoidal form of the structure and the lack of timber extending eastwards along the line of the tree throw (Context 116), northwards and southwards clearly demonstrates that the structure did not extend laterally in any one direction, as one would expect for a trackway. In general, excavated trackways tend to comprise either single or double planks with their long axes oriented to the direction of "traffic", such as the Sweet Track (Coles *et al.* 1984), or are constructed from shorter roundwood or planked timber laid transverse to the direction of traffic, a good example being the Iron Age Corlea Track in Ireland (Raftery 1990). In addition, no bolt marks were recorded in any of the timbers uncovered during the current excavation. It seems possible that what the Rev.

Wilson observed and the currently excavated structure are not one of the same.

It seems unlikely that the platform was used for domestic activities; there is no artifactual or ecofactual evidence to suggest such a use. The only foodstuffs recovered were a cache of hazelnuts and single hazelnuts scattered throughout the platform; as hazel wood has not been identified in this structure these must represent snack food brought to the site. Indeed, the complete absence of any artifactual material from the site militates against easy interpretation. The hunting platform at Star Carr (Mellars and Dark 1998) provides perhaps the closest analogy but there is no evidence at Parks of Garden for hunting activities.

The next pragmatic explanation is that the platform functioned as an assembly point for expeditions across the moss (possibly on foot or perhaps by dugout boats); a parallel would be the Neolithic Baker Platform (Coles *et al.* 1984). However, such an assembly point would have been better located on the higher and drier moraine immediately to the west of the wooden structure, or further north where the natural slope of the moraine is less marked. Understanding the nature of the moss surface at this particular location is vital to the interpretation of the function of the structure. The surrounding vegetation comprised woodland and *Polytrichum* moss with occasional pools of open water (Hughes & Kenward 1999) and scattered fallen oaks. Small craft may have been used to navigate between the pools and the fallen oaks could have been walked along. The presence of pools and trees may have outweighed the initial access problem down the steep slope of the moraine. Indeed it may be argued that this problem was overcome by the construction of the platform which could have functioned as a broad dry step.

The need to travel eastwards from the glacial moraine ridge also demands explanation. It is unlikely that peat covered the whole of the Upper Forth Valley during the period of platform use, but existed as discrete pockets (e.g. Flanders Moss East and Killorn Moss) within salt marsh (Ellis 1999b). Dating of the lower peat profile will aid in the unravelling of the rate and pattern of regression of the post-glacial sea. It is also more plausible that people would expend considerable energy building a wooden platform if access was to the retreating, resource-rich salt marsh and creeks of the Forth estuary, as opposed to entering a vast expanse of resource-poor raised moss. The platform would have been relatively flat at the time of its construction as its sloping nature is largely a function of post-depositional sink into the soft peat both during its use and post-abandonment. The east-west oak laths on the top of the superstructure may have acted as the main thoroughfare with room on both sides for the temporary storage of goods, etc.

The period of use of the platform has yet to be established. Continued peat and occasional tree growth continued after the abandonment of the platform. Initial macroplant and invertebrate analysis indicates that conditions became wetter with the progression of time (Hughes & Kenwood 1999) which may have forced the abandonment of the platform. The dense oak wood upon the edges and summit of the high ground to the immediate west of the platform appears to have continued to exist into this period, demonstrated by the presence of two oaks which fell off the moraine (Context 136 and 106). An up-rooted tree bowl (Context 133) overlaid the burnt timbers of the upper platform and it is likely that this oak was growing at the time of use of the platform and collapsed over its south-western edge during its use or shortly after abandonment.

Summary

The discovery of the excavated Neolithic platform serves to demonstrate the largely untapped archaeological potential of Scotland's peatlands. Prior to the excavation at Parks of Garden it was assumed that the wooden trackways of Antiquarian accounts that must be Bronze or Iron Age, primarily because of the preponderance of artifactual material of this period found in the Carse. The limitations of our current knowledge basis in Scotland and the enormity of the task before us in realizing the full potential of Scotland's buried cultural resource is all too apparent.

Further work is required to place the platform firmly within the socio-economic and ecological landscape. Preliminary palaeoenvironmental analyses of peat samples from the platform indicate that the immediate local environmental had not been radically altered by human activity. The moraine ridge between Arnprior and Menteith would appear to have been covered by a thick virgin oak wood in which opportunities for cultivation would have been limited. However, the ridge would have formed an important crossing point across the valley, facilitating trade and communication and would have also allowed relatively easy access to the retreating estuarine resources. It is postulated that the platform serviced hunting and gathering expeditions onto the mudflats with more permanent settlements located upon the higher ground of the Touch Hills and Trossachs. Indeed such wooden floats as that recorded in Flanders Moss East and the wooden track under Ochertyre Moss may have performed a similar function during the Neolithic period. Very few Neolithic dryland settlements on the edges of the Touch and Trossach Hills are known today, probably as a consequence of modern intensive farming practices and lack of field-based research.

The discovery and excavation of this Neolithic wooden platform within peat deposits is an exciting first for Scotland. The results of the survey and subsequent excavation demonstrate that the discovery and understanding of the function of such structures can be achieved only through a holistic landscape approach in which both wetland and dryland zones are incorporated and not studied in isolation.

Acknowledgments

The author is grateful to Mr Dalgleish (senior and junior) for their support, assistance and permission to intrude upon their land and to Murray Cook, Alan Duffy, Penny Johnson, John Gooder, Rob Engl, Nichola Radley, Martin Cook, John Bendicks and Anne Crone for their assistance in the field. Historic Scotland funded all the work described in this paper and contributed towards the costs of the author attending this WARP conference. The author is also grateful to Alan Braby for the production of Figure 15.2 and Sylvia Stevenson for the production of Figures 15.3 and 15.4.

References Cited

Archaeology Scotland. 1890. List of donations presented to the Society of Antiquities of Scotland. Archaeology Society. 5, 3 Appendix 11

Anderson, J. 1967. A history of Scottish forestry. 2v, Edinburgh. 1:67–70

Brooks, C.L. 1972. Pollen analysis and the main buried beach in the western part of the Forth valley. Transactions of the Institute of British Geographers. 55:161–170.

Burgess, C.B. and I. Colquhoun. 1988. "The Swords of Britain". Prahistoriche Bronzefunde, 4, 5, Muchen 41, 142.

Cadell, H.M.M. 1913. The Story of the Forth. Glasgow, James Maclehose & Sons.

Callender, J.G. 1906. Notices of (1) two stone cists from Oyne and Skene; (2) a late-celtic harness mounting of a bronze from Cusamond; (3) a stone mould for casting flat axes from Auchterles; and (4) two star-shaped beads from Aberdeenshire. Proceedings of the Society of Antiquaries Scotland. 40 (1905–6):37–38.

Clarke, C. and B. Finlayson. 1995. Scottish archaeological database for raised bogs. Edinburgh University Centre for Field Archaeology Report No 199, Edinburgh.

Clarke, C. 1997. Archaeological database for the Scottish wetlands. CSA Report No. 298. Unpublished report for Historic Scotland.

Clarke, C. 1998. Archaeological and palaeoenvironmental investigations at North Ballachulish Moss, highland, Scotland. (in prep) WARP Conference Proceedings.

Coles, J.M., B.J. Orme, and S.E. Rouillard. 1984. Somerset Levels Papers. No 10. Hertford, Stephen Austin & Sons Ltd.

Crone A. 1999. Report on the wood at Parks of Garden in C, Ellis 1999a. Archaeological assessment of the Scottish wetlands. Unpublished report for Historic Scotland.

Durno, S.E. 1956. "Pollen analysis of peat deposits in Scotland". Scottish Magazine:72 177–187.

Ellis, C. 1999a. Archaeological assessment of the Scottish wetlands. Unpublished report for Historic Scotland.

Ellis, C. 1999b. Wetland Archaeology. Carse of Stirling archaeological assessment. Unpublished report for Historic Scotland.

Ellis, C. 1999c. Parks of Garden. Data structure report. Excavation of a Neolithic wooden structure. Unpublished report for Historic Scotland.

Francis, E.H., I.H. Forsyth, W.A. Read, and M. Armstrong. 1970. The Geology of the Stirling District. Memoirs of the Geological Survey of Great Britain. Her Majesty's Stationery Office, Edinburgh.

Hingley, R., P. Ashmore, C. Clarke, and A. Sheridan. 1999. Peat, archaeology and palaeocology in Scotland. In B. Coles and M.S. Jorgansen (eds) Bog bodies, sacred sites and wetland archaeology: Proceedings of a conference held in Silkebourg, Denmark, September 1996. Prehistoric Society and Wetland Archaeology Research Project:105–114.

Huges, P. and H. Kenwood. 1999. Assessment of plant and invertebrate macrofossils from Neolithic deposits at Carse of Stirling, in Ellis, C. 1999a. Archaeological assessment of the Scottish wetlands. Unpublished report for Historic Scotland.

Johnson, T. 1792. Plan of Flanders Moss

Laxton, J.L. and D.L. Ross. 1983. The sand and gravel resources of the county of Stirling, Central Region. Mineral Assessment Report 131. Her Majesty's Stationery Office, Edinburgh.

Mellars, P. and P. Dark. 1998. Star Carr in Context: new archaeological and palaeoecological investigations at the Early Mesolithic site of Star Carr, North Yorkshire. McDonald Institute for Archaeological Research.

O'Conner, B. and T. Cowie. 1995. Middle Bronze Age dirks and rapiers from Scotland: some finds old and new. Proceedings of the Society of Antiquities of Scotland 125:349–358.

Ordnance Survey 1862. Perthshire sheet CXXX First Edition, 6

OSA. 1799. Appendix to the parish of Kincardine. In Sir John Sinclair (ed.) The statistical account of Scotland, drawn up from the communications of the ministers of the different parishes. 21, Edinburgh.

Piggott, S. 1959. A tripartite disc wheel from Blairdrummond, Perthshire. Proceedings of the Society of Antiquities of Scotland 23:120.

RCAHMS 1979. The Royal Commission on the Ancient and Historical Monuments of Scotland. The archaeological sites and monuments of Stirling District, Central Region. The archaeological sites and monuments of Scotland, series no. 7, Edinburgh.

Raftery, B. 1990. Trackways through Time: Archaeological Investigations on Irish Bog Roads, 1985–1989. Headline Publishing, Dublin.

Sissons, H.B. 1966. Relative sea-level changes between 10300 and 8300 B.P. in part of the Carse of Stirling. Transactions of the Institute of British Geographers 39:9–18.

Stevenson, R.B.K. 1976. Romano-British glass bangles. Glasgow Archaeology Journal 4:48.

Tait, C. 1794. An account of the peat mosses of Kincardine and Flanders in Perthshire. Transactions of the Royal Society of Edinburgh 3:266–278.

Turner, J. 1965. A contribution to the history of forest clearance. Proceedings of the Royal Society London, B161 (1965):343–354.

Utisi, E., C. Clarke, and C. Ellis, in prep. The use of ground penetration radar in Scotland. Antiquity.

Wilson, D. 1863. Prehistoric Annals of Scotland. 2v, London 1, 50.

Wilson, W. 1878. The History and Traditions of the Parish of Kippen. Unpublished lecture delivered at Kippen on the 12th of March 1878.

16. Medieval Children Coffins of Quimper (French Brittany)

Anne Dietrich and Véronique Gallien

In the XVIth century the courtyard of the Cathedral Saint-Corentin in Quimper (French Brittany) had been transformed into a place with a pillory and a Calvary (Map). Under this public place, a medieval cemetery was found. Its upper layers are dated to the XIIIth c. To this period belong the two graves of young children that have been exhumed. In both cases, a wooden coffin was laid inside a simple pit. This discovery was distinctive in that these coffins contain the wrapped bodies of two children, rather than the more commonly occurring skeleton.

History of the Children

Anthropological observations

Age, sex, stature, morphology, state of health and cause of death – for one of the two corpses – have been determined from our observations. The two skeletons are in a relatively good state of conservation and the complete bone system has been examined (Figure 16.1).

In both cases, the cerebrum has been conserved in the neurocranium in a very fragmentary state for one (S52) and complete for the second (S69) which was studied by a forensic pathologist.

Age estimation was carried out from the teeth eruption stages. The child S52 presents deciduous dentition in an embryonic state and age is evaluated between birth and 6 months, ± 2 months (Stages 1–2 on Schour & Massler tables, 1940). The child S69 presents developed first deciduous premolars, advanced mineralizing of dental root in the incisor tooth, canines and second premolars in growth, beginning of the mineralizing of the dental crown of the first molar. Age was estimated at between 18 months, ± 3 months, and 2 years, ± 6 months (Stages 5–6 on Schour & Massler tables, 1940).

The sex determination was tried through the morphoscopic method found by H. Schutkowski (1993). From investigations of the ilium, (great sciatic notch, arch criterion, iliac crest) and of the mandible (jaw, chin region, anterior dental arcade, gonion region) the sex recognition of these immature skeletons gave some results. This method was applicable only on the S69 child. Among these features, three were observed on the pelvis (angle of greater sciatic notch, left arch criterion and curvature of iliac crest) and one on the mandible (shape of gonion region). A predisposition to identify the child S69 as a girl was obtained.

The stature estimation has been done from the length of long bones according to G. Olivier's method (1963) and the first child, S52 was estimated to be between 50 and 57 cm of height, while the older was between 80 and 95 cm. Few other specific anatomic characters have been noted. The most unusual observation concerns an anomaly of deciduous teeth formation of the child S69. The first dentition of right maxilla possesses three incisor teeth and the two central incisor teeth of this maxilla have a special coalescence at the base of the crowns (in the collar region), whereas their crowns and roots are well separated. The dental arcades for the two teeth are completely formed up to the limit of fusion point between the two incisor teeth. The anomaly is unilateral.

The state of health of the infants at the time of their death can be pieced together from observations of stress indicators (dental enamel hypoplasia, Harris lines, [cribra orbitalia]) and traumatic injuries noted on the bones. The skeletons stress indicators were particularly well observed on the orbital roof of frontal bone (cribra orbitalia:porotic hyperostosis of orbital roof) and on long bones (Harris lines) as well as on dental enamel (hypoplasia). Their presence refers generally to deficiencies or growth disorders resulting from fevers, infectious diseases or

Map: Situation map of France and the city of Quimper.

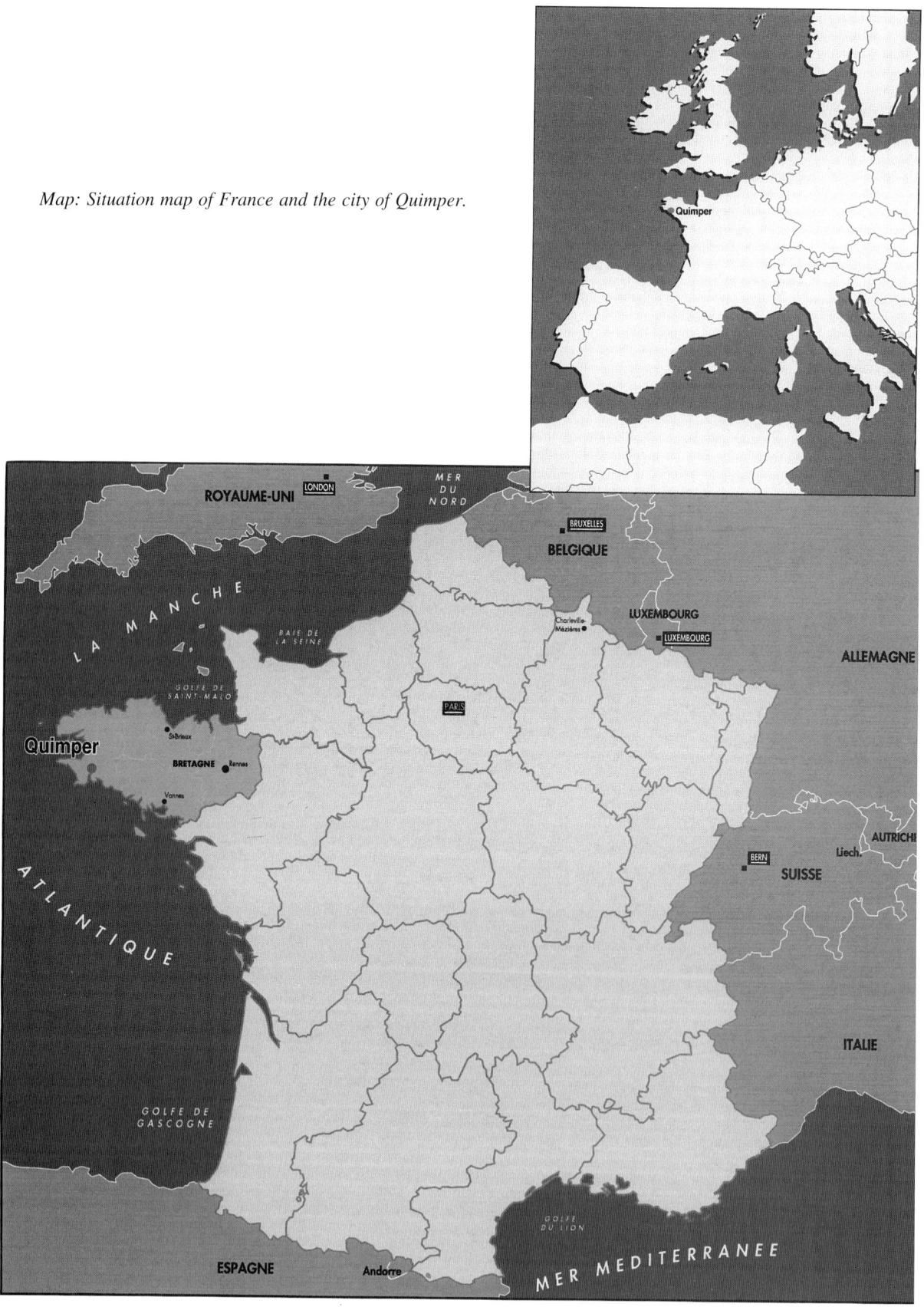

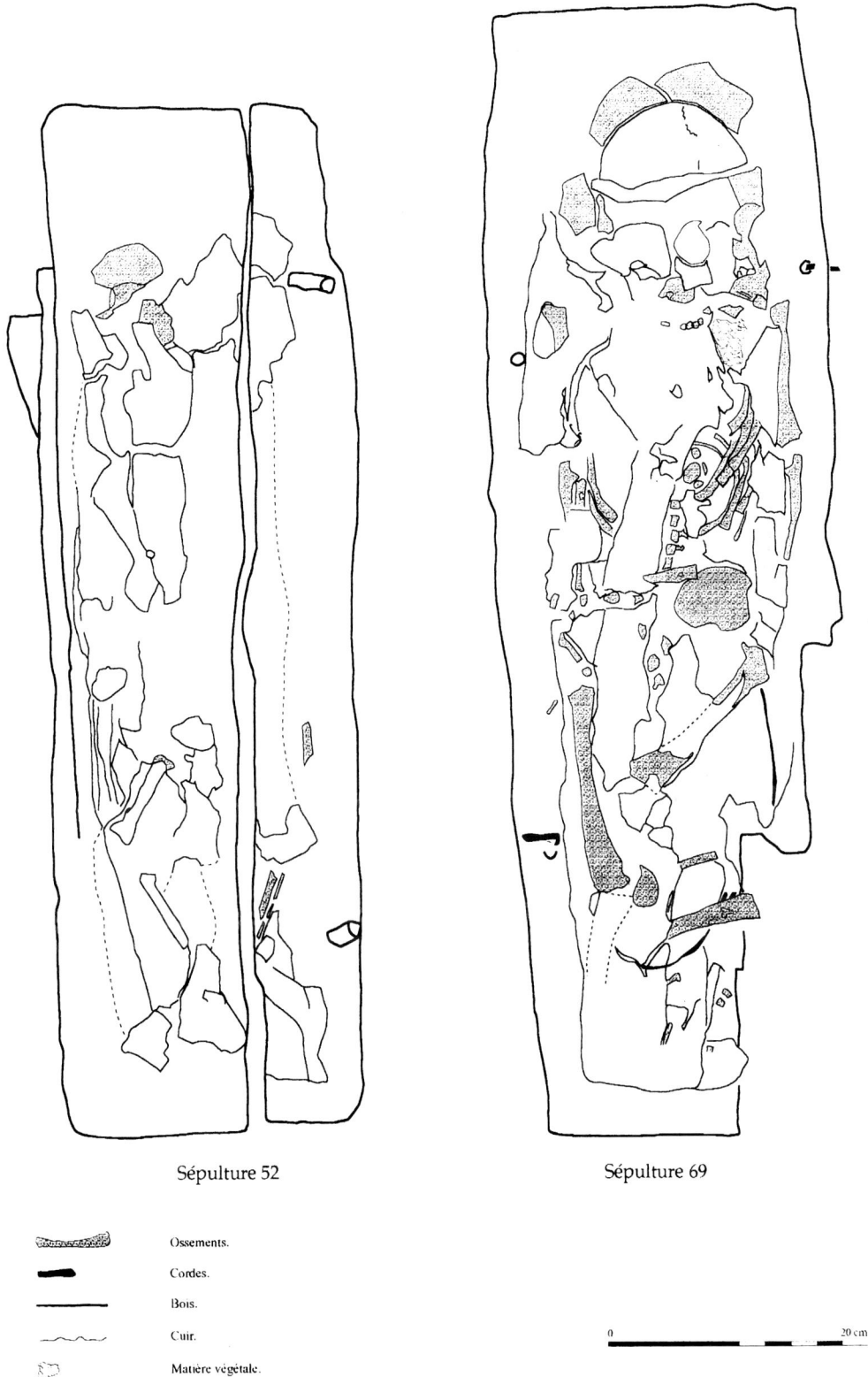

Sépulture 52   Sépulture 69

Ossements.

Cordes.

Bois.

Cuir.

Matière végétale.

0   20 cm

Figure 16.1. The skeleton of the two children.

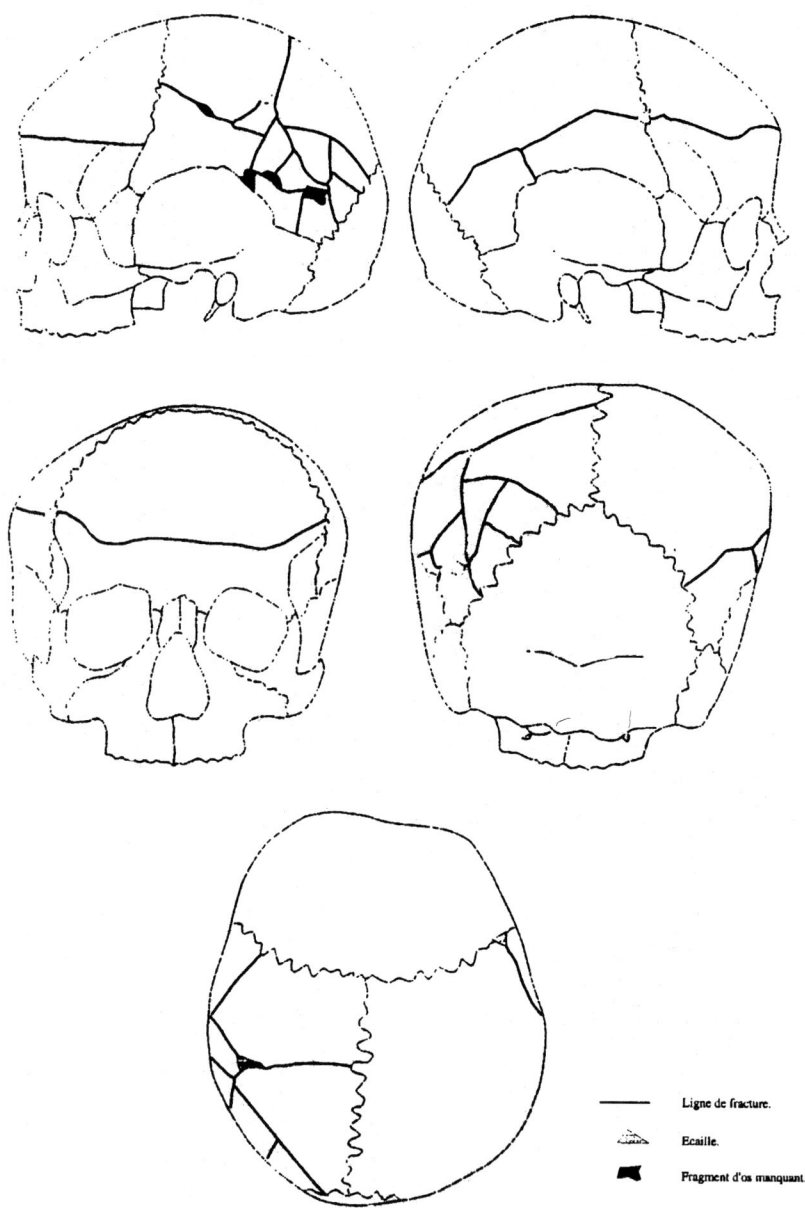

Figure 16.2. The cranium of S69.

nutritional deficiencies. The only disfunction indicator from the macroscopic and radiographic investigations, was the *cribra orbitalia* on S69, most often considered an iron deficiency anemia. This health problem probably also exists for the S52 child, but the mediocre state of bone surface conservation makes it difficult to confirm. The absence of dental enamel hypoplasia or Harris lines suppose that the two children benefited from satisfactory conditions of life during the short time of their existence.

The anthropological examination of the skeleton of child

S69 was completed by an attentive study of the cranial vault and cerebrum. Analysis of the cranium is related to the presence of a circular trauma with a comminutive zone on the left parietal bone. This lesion can be explained by a violent trauma, probably starting from an impact on the left parietal bone. Because no trace of fractured lip healing can be traced, we think the lesion occurred just before death (Figure 16.2).

Samples were taken from the edge of the brain, cerebrum and submitted to microscopic examination. They reveal the

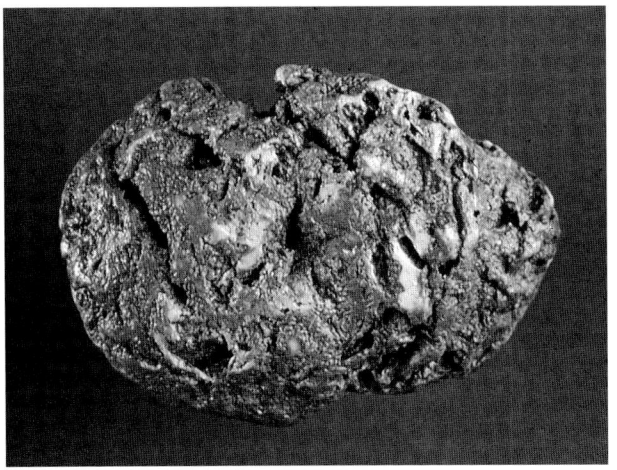

Figure 16.3. Photography of the brain of S69.

presence of abundant red blunt corpuscles zones and many internal brain hemorrhages on the periphery (space on all sides of flabby meninx). Finally, the concordant indications conduct us to diagnose a recent and diffuse cerebro-meningeal and leads us to specify with certitude the cause of the child's death being a traumatism (Figure 16.3).

Burial Procedure of the Children:

The aim of this section is to investigate the burial procedure in the XIIIth century. The funerals as religious ceremony and the procession are not evoked here. The coffins were prepared, the biers receive the corpses and the inhumation concludes the ritual ceremony.

The coffins

The smaller one is made of oak except for a side plank which is of beech. The second coffin is exclusively made of beech. Through the tree ring patterns we can assume that the wood was of forest origin instead of hedge exploitation. Dendrochronology (date obtained by V. Bernard, AFAN) dates the small coffin to 1286 AD. The sapwood was carefully eliminated and the only insect depredation concerns the bottom plank of the small coffin. The explanation is the thanatophagus living on the corps decomposition. Through the coleopteran family, the xylophagus are also thanatophagus insects.

The two coffins present the same shape on the 6 sides, all having a trapezoidal contour. The top is a narrow lath; the side slopes are made of single boards, as the bottom, the head and feet planks. The 6 month-old baby coffin measures 76 cm long, 25 cm wide and 13.5 high. The 18

Figure 16.4. Photography of the coffin S69 still closed.

month-old infant box has a volume of 91.5 cm × 25 cm × 16 cm (Figure 16.4).

The boards are quite heterogeneous, especially in thickness. The bigger coffin shows very thick (up to 3 cm) slopes. The conversion is always by radial splitting.

Observations on woodwork illustrate a rough craftsmanship while the fixing is rapidly done, appropriate to the short time of funeral rituals. The measurements seem approximately taken, but very well controlled and the grain follows the board axe instead of the board side. A marking line indicates the edge of the headboard of the bigger box. Each plank is unique and adapted to its place. The perforations seem to be very approximate and adjusted at the last moment. The tool-set is simply composed of a small axe, a drill and maybe a plane.

The fixing is quite different between the two coffins but both use peg-holes. The baby box is pegged with long and conical oak sticks. The side planks, head and feet, lie on grooves. While the four sides are set on the bottom, the top is adjusted and the whole box is assembled at once with pegs at each corner (Figure 16.5). The infant coffin has

QUIMPER
Place La'nnec

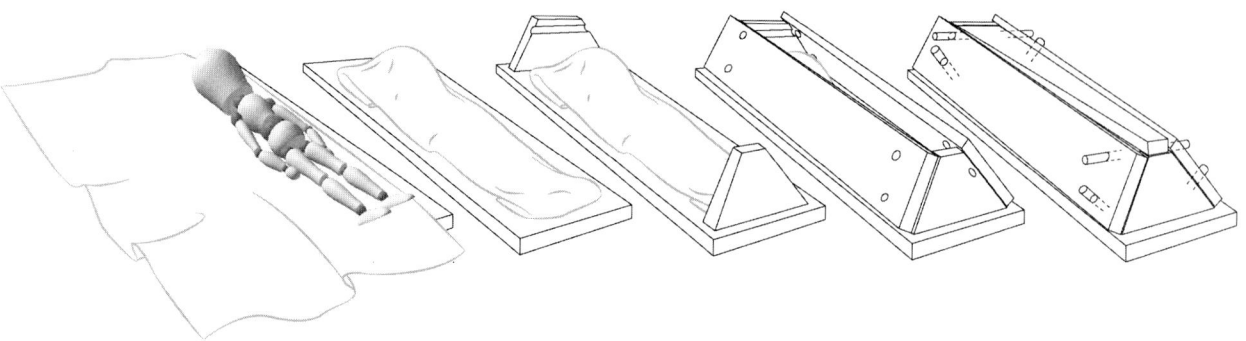

Figure 16.5. Reconstitution of the coffin S52 assemblage.

QUIMPER
Place La'nnec

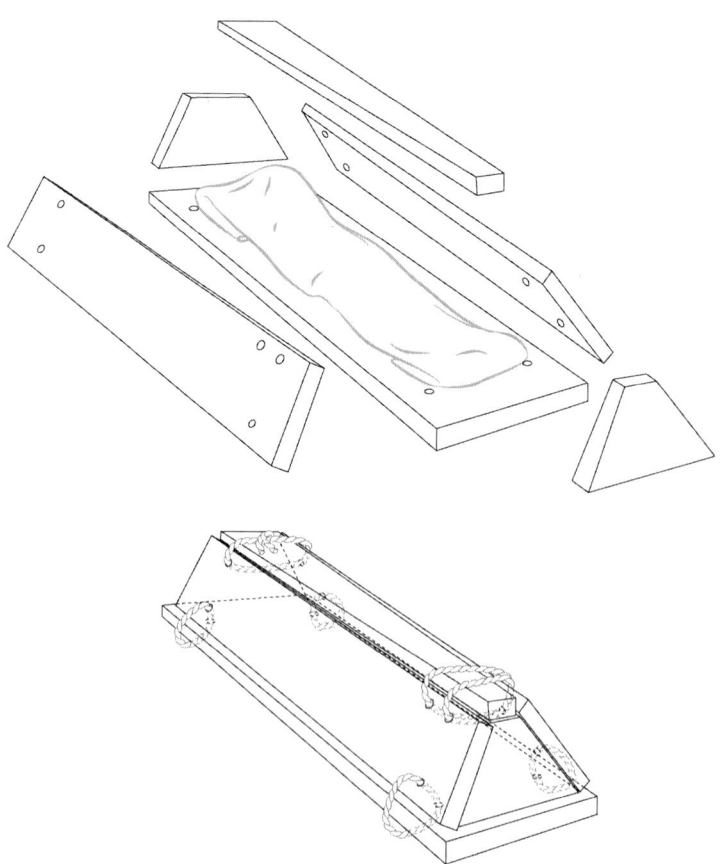

Figure 16.6. Reconstitution of the coffin S69 assemblage.

smaller holes, which allowed passage for cordage. Some of it remains and we were able to follow the fixing. The tread is made of willow "*Salix Sp.*", very roughly twisted in a Z position. Even if the various holes are not facing each other, the cordage is tied in the two box ends, at the head and at the feet. Only the topknots could be observed with simple knots (Figure 16.6).

In both cases the planks constitute a tension system strong enough to be transported and manipulated. The fabrication of these boxes contains an undeniable quality allowing their preservation but also implying some poor characteristics such as the absence of a clear conception, the irregular surface of the boards, the lack of nice joints, the probable speed of work. With such a configuration, we can hardly distinguish domestic work from a quick crafts-man production. It seems that the coffin maker knew how to work wood but ignored systematic steps of a good execution.

Other pegged wooden coffins were excavated in the same area, at Landevennec Abbey (Bardel, Perennec, to be published) for an earlier period, VIIIth–XIth century. They correspond to adult size. Outside of French Brittany, we know some examples for the XIth–XIIth century, like in the churchyard of Saint-Léger at Guebwiller (Brunel, 1988) or Saint-Mexme at Chinon (Husi *et al.* 1990), but also in Hull (Danielle, 1996) or York (Dawes & Magilton, 1980) England.

Installation of the Corpses

The bodies are laid down on the back. The arms are placed against the body according to the classical burial ritual of children in the Middle Ages. The skeleton of the younger individual (S52) was squeezed against a coffin wall; the body was probably knocked over during the descent into the grave.

Some vegetable twig balls support the neck of the child S69. These pellets consist of flattened fibres that seem to be grass, herbaceous or graminaceae. For the S52 body, only sparse fibres have been seen among shattered remains of ribs, vertebrae and mandible. The existence of a cephalic pillow is attested during the whole of the Middle Ages, as well as through archaeological records as by iconography and texts. As examples we can cite nettle, thyme and flowers found in some burials at Saint Victor of Marseilles or the walnut, mint and ambles under King Philippe the First's head at Saint Benoit sur Loir. In Fontevraud, as in our examples from Quimper, graminaceae have been collected (on child's feet). In the XVth century a fine iconography documentation is the "Triumph of death," painted by Bruegel the Elder (XVth century) with straw and rye pillows in tombs. The first role of the pillow is functional, to position the deceased's head facing the sky and heaven.

The odoriferous characteristics were very useful to cover the smell of putrefaction.

The bodies of both children were rolled in a leather shroud. The leather sheet or skin, probably rectangular, was laid down on the bottom plank. The body was installed and then the shroud was used to cover the dead body completely. The leather remains free on the right side and tucked under at the feet and around the head, without any pin, clasp or sewing. Although the use of animal skin is not totally unknown in medieval funeral practices, shrouds were thought to be in textile form such as linen, hemp, frieze or oilcloth.

Because of the excellent conservation of the organic materials, we can observe all the details of the attention given to the dead. We notice not only the preparations for the funerals (the making of the coffins, the head cushion, the cordage or the shroud) but we can also reconstruct the order and the way they were laid into the bier.

We assume the prefabrication of the various components as planks, pegs or ropes, shrouds. All of them are brought to the ritual place. The bottom plank of the coffin is put onto a table to ease the movements around the corpse. The leather shroud is put upon it. The dead body is positioned with its arms along the bust and the head maintained straight facing the sky with the herb cushion. The second half of the shroud recovered the body and is simply folded at the head and feet. Only then, the coffin is set up. The participants would start with the slopes, the head and feet boards in position, touching the corpse. The top closes the box in balance and rapidly, the pegs are forced in or the ties are knotted. The whole operation takes only minutes, with very simple gestures but the planks' position have to be strictly respected.

Now the coffin is ready to be transported by hand, maybe by only one person. Less is known of the burial in the cemetery. The pit is rough and contains no gifts such as a vase or charcoals to accompany the dead.

Conclusions

The first burial is of an infant. Its head has been carefully laid onto an herb cushion and the corpse is enclosed in a small trapezoidal coffin. The box is assembled with long and conical pegs. Oak and beech, cut on quarter are used. Dendrochronology dated the cutting of the tree to 1286 AD. The second tomb contains a small child whose age could be determined to 18 months through the examination of its bones and teeth, we assume the child could already walk. The head cushion and the leather envelope are identical to the previous one but with slightly bigger dimensions. The fixing of the beech planks are vegetal ties. The study of stress indicators shows that the two children benefited from satisfactory conditions of life. Although the skeleton of the second infant is still connected, its skull has been fractured by an impact starting at the left ear, crossing the front, reaching until the collapse of the opposite side. The observations of the cranial vault confirmed by the analysis of the brain remains have determined internal brain bleeding as the death cause (a cranial menigeal trauma, detectable by the neurocranion lesions and presence of accumulation of red blood corpuscles in the cerebrum).

In spite of the urgency of the rescue excavation work all the documentation and analyses were done with the co-operation of the archaeological team under the direction of Mr. J.P. Lebihan, xylologist A. Dietrich, anthropologists Mr. Y. Langlois and Mrs. V. Gallien and a forensic doctor Dc. F. Guillon. The extraordinary conservation of these organic remains presents us not only with a mortal accident, but gives us precise technical hints on the inhumation ritual before the general use of nailed coffins during the following century. The precise investigations of the coffins, the skeletons and the microscopic analysis of the brain shed new light on medieval funeral practices.

References Cited

Alexandre-Bidon D., and C. Treffort. 1993. A réveiller les morts, la mort au quotidien dans l'Occident medieval. Presses universitaires de Lyon, Lyon.

Alexandre-Bidon D. 1998. La mort au Moyen Age, XIIIème – XVIème siècles. Hachette, coll. La vie quotidienne, Paris.

Brunel P. 1988. En attendant le Jugement Dernier. Vivre au Moyen Age; 30 ans d'archéologie en Alsace, Strasbourg, 241–253.

Danielle Ch. 1996. Death and Burial in Medieval England. London, 126.

Dawes J., and J. Magilton. 1980. The Cemetery of Saint Helen-on-the-walls. The Archaeology of York the Medieval Cemetery no.12.

Dietrich A. 1998. Dégradations et effondrements de cercueils. In Dietrich A., Vertongen S. Rencontres autour du Cercueil. Bulletin GAAFIF 41–47.

Husi P., E. Lorans, and C. Theureau. 1990. Les pratiques funéraires à Sainte-Mexme de Chinon au Vème au XVIIIIèle siècle. Revue Archéologique du Centre, no. 29:131–168

Schour L., and M. Massler. 1918–1931. Studies in Tooth developement:the Growth Pattern of human Teeth. Journal American Dent. Ass. 27:1778–1793.

17. Neolithic watercraft:
Evidence from Northern Greek Wetlands

Christina Marangou

Introduction:geography and chronology

In a mountainous and partitioned country like Greece, mountains, steep slopes and water expanses have always, and particularly in prehistory, hindered inland communications between agricultural lands concentrated in restricted areas, isolated one from the other (Faugères 1989:88–89). Depressions and inner basins being still occupied by lakes and swamps during the Neolithic (Demoule *et al.* 1993), the importance of any means of transport, in particular across or along water, is obvious.

Two areas of Northern Greece, which have given evidence about Neolithic watercraft, are considered in this paper:Thessaly and Western Macedonia. Regarding the Neolithic phases of these regions, we follow here the subdivisions proposed by Demoule *et al.* (1993:366, fig. 2; *cf.* the slight differences of the chronological table in Andreou et al. 1996:538):The boundary between Early Neolithic (EN) and Middle Neolithic (MN) is situated around 5800BC calibrated (cal.), while the Late Neolithic (LN) starts about 5300BC cal., and the Final Neolithic (FN) begins about 4500BC cal. and ends around 3200BC cal.

Rivers, floodplains and lakes in Neolithic Thessaly

Surrounded by mountains attaining more than 2000 m, Thessaly, at the southern end of a system of mountains and valleys that stretches in a south-easterly direction from the Balkans into Central Greece (Demitrack 1994:37), contains the most extensive lowland basins of Greece (Johnson 1996:271) (Map 17.1). The Larissa (eastern) plain is about 1000km^2 and the Kardhitsa (western) basin about 2200km^2 (Johnson 1996:283). The chain of low hills separating the two basins do not hinder communication between them, neither did in prehistory (Gallis 1992:23).

In the Oligocene the Thessalian plain sheltered a shallow

sea and then a regional lake. The Larissa basin was formed during the Middle Pleiostocene. When the basin subsided, the ancient river Peneios eroded the Tempi gorges, retaining its outlet to the Aegean sea, while its southern drainage network turned to a large and shallow lake, Karla (ancient name:Viviis; in Greek:Βοιβηίς) (Demitrack 1994:37). A second lake, which disappeared several centuries ago, named Nessonis in ancient times, located to the north of Viviis/Karla, was not clearly separated from the latter, since there are no natural boundaries (Gallis 1992, 27).

The Peneios, flowing to the east across the two main basins, created wide Late Pleistocene floodplains, that were settled about 8,500 years ago by Neolithic farmers. According to recent research, in the Middle Holocene the rivers created an alluvium which covered the edges of EN and MN mounds, showing that the corresponding settlements were built on an active floodplain (Van Andel *et al.* 1995:131). Late in the MN the settlements acquired their present position 3–15 m above the river and the mounds rose above the plain (Van Andel *et al.* 1995:134). A later (MN to LN, 7000–6000 BP) alluviation may be related to human activity on the plain (Demitrack 1994:38). It appears in the lake Karla basin as well as on the floodplain and followed a period (between the EN and MN) in which many settlements on the marl slopes of the Middle Thessalian hills were abandoned in favour of plain sites (Demitrack 1994:38). In the LN new settlements were established in the alluvial plain, between earlier ones (Demoule *et al.* 1993).

Evidence from the site of Platia Magoula Zarkou indicated (Van Andel *et al.* 1995:140) that EN and MN farmers had settled in this active floodplain in order to take advantage of periodic flooding, for the cultivation of fresh, moist silt and spring-sown crops. Such early settlements must then have been occupied intermittently, perhaps even

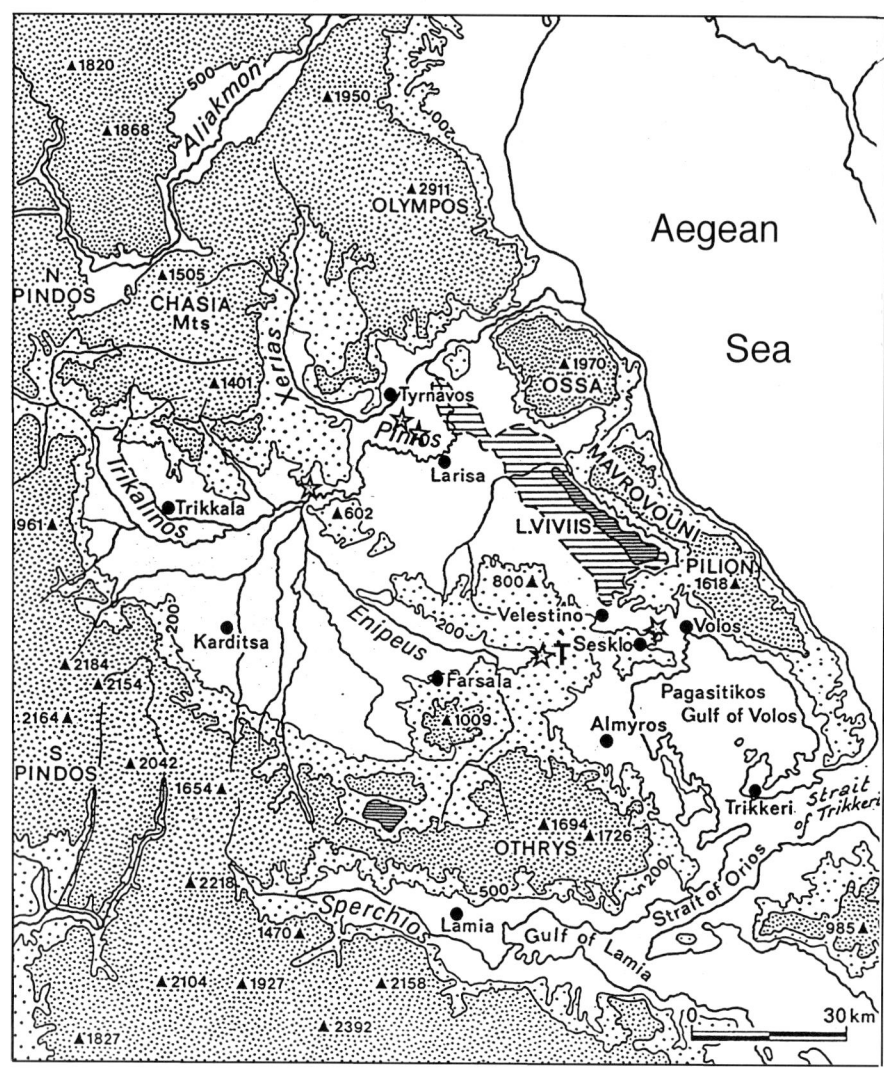

Map 17.1. Map of Thessaly in its Neolithic environment. Sites having provided watercraft models are indicated with an asterisk. T = Tsangli. Adapted from Marangou 1991, pl. VIa.

seasonally (Van Andel *et al.* 1995:141). This casts doubts on the relationship between the beginnings of farming and sedentism (Van Andel *et al.* 1995:141). According to Andreou *et al.* (1996:557), this model is based on the hypothesis of predictable regular flooding and presumes spring-sown crops, while the stratigraphic evidence from other Thessalian sites is not detailed enough to support the identification of periodic flooding. Demoule *et al.* (1993) argued that, in the EN and MN, uneven topographical distribution can be observed specially in regions drained by main rivers, but early sites were not restricted to the alluvial plain, but rest on different soil formations.

Up to recent years, the river and its tributaries flooded

every year and occupied most part of the Thessalian plain. A large part of the plain was covered temporarily or permanently by marshes created by the floods, and favouring mosquitos and malaria (Gerakis *et al.* 1996:206). The plain around the town of Larissa, in eastern Thessaly, and a larger area in the upper western plain have been much subject to floods (Jarman 1982:148). Moreover, in the Larissa plain the average annual precipitation is of 400–600mm, and a far greater rainfall in the surrounding mountains provides more water to the plain. The Kardhitsa basin receives an average annual rainfall of 600–1000mm (Johnson 1996:271).

Until recently, the lake Karla level and consequently its

limits depended on the flooding waters of Peneios, as well as on fluctuations in precipitation, since the plain slopes gently (Demitrack 1994:39). In the recent past the extent of the lake varied from 45,000 to 180,000 *stremmata* (one *stremma* = 1000m²; Gerakis *et al.* 1996:220). The location of the Neolithic lake-shore probably reflected variability of the lake level as well as in rainfall (Demitrack 1994:39). The prehistoric settlements identified till now probably followed the movement of the western lake-shore, while change of the lake level was not of much consequence for the eastern shore (Grundmann 1939, pl. 37; Gallis 1992, map 3). Besides shore settlements, some at least Neolithic sites were located on islands within the lake (such as Magoula Hadzimisiotiki:Grundmann 1939; cf. Gallis 1992:25).

Large draining schemes (1936–1969), isolated the lake from the flooding waters of the Peneios. The fauna disappeared or was reduced on the neighbouring mountains, and the chain along the air corridor of Eastern Greece for the migratory species was interrupted. Before the draining, the plain population near the lake Karla were agriculturers, and the mountainers fishermen. A very rich and varied bird population of 143 species, 430,000 water birds, were registered in 1954, and 600 tons of fish were captured per year. The negative effects from the floods did not disappear completely (Gerakis *et al.* 1996:220–226).

The ubiquity of water in this region in ancient times was also reflected in Greek mythology:it was a king of Thessaly, named Deukalion, who was believed to have survived after the flood, acting the part of the biblical Noah.

The lake and basin of Kastoria, Western Macedonia

In Western Macedonia (Map 17.2), the drainage basin of Kastoria (304km²) is situated in the valley of river Aliakmon (average altitude 200m), between two mountain chains, Pindos in the west and Verno in the east (Vafeiadis 1983:1). The river crosses towards the north-east a plateau of an average altitude of 600m, containing the lakes of Vegoritis, Petres and Heimaditis (Psychoyios 1992:23).

In such isolated basins of western Macedonia, at 500–600m of altitude, the forest may be thick. Early farmers would have met, in Epirus and Macedonia, a densely forested land of deciduous oaks, elms, ash, limes, hazel, and pines on the slopes (Demoule *et al.* 1993). Around 8000–4000 BP, in the area to the north-east of Pindos, flora included pine, beech and fir on the heights, associated to oak, birch, lime, juniper, pistachio, hazel and elm on the slopes, while typical aquatic plants thrived near the lakes (Faugères 1989:101, fig. 3, 103). Data about the Holocene climate, mostly from palynological cores from Northern Greece, suggest a Mediterranean climate, with winter rains and dry hot summers, but regional differences were important (Demoule *et al.* 1993).

Several prehistoric lake-sites or river-shore sites are attested in neighbouring areas, such as Maliq and Dunavec in the Korçë basin (Albania), and, among numerous settlements in the plains of Pelagonia (Hammond 1972:218, 223), Usta na Drim in Struga, by the lake of Ohrid (Former Yugoslavia). Ancient classical authors, e.g. Herodotus V.16, described lake settlements located in these regions, between the rivers Aliakmon and Strymon. The latter forms a natural frontier between Central and Eastern Macedonia.

The lake of Kastoria (30km², at about 620 m above sea level), located within the basin, was formed after the gradual subsidence of the basin in the middle Miocene. The lake surface decreased in the Pleiocene to its present form (Vafeiadis 1983:115–116). An open lake, its surplus waters directed to the river Aliakmon, it is surrounded by alluvial and karstic water bearing formations (Vafeiadis 1983:119). The climate is humid mesotherm, transitional between Mediterranean and continental, with a mean rainfall 671mm/annum (Vafeiadis 1983:116, 120).

The deepest point of the lake is 8.5m, at the northern extremity of the peninsula of Koritsa, where the town of Kastoria is built (Vafeiadis 1983:46). The lake level has lowered in the thirties and in the sixties, revealing large numbers of piles (Keramopoullos 1938; Moutsopoulos 1973–1974). It can till now be regulated artificially (0.4–0.6m; Vafeiadis 1983:46).

The prehistoric settlement of Dispilio, excavated since 1992 by a team under the direction of Professor George Hourmouziadis, University of Thessaloniki (Hourmouziadis 1996), is located 7 km south of Kastoria, by (and under) the lake. The topography of the lake shore on the south being almost horizontal, and because of the periodical fluctuations of the ground water, these areas were seasonally overflown with water (Vafeiadis 1983:46).

Communication across the lake of Kastoria was made preferably by boat till recently, before the construction of the modern road, in particular at Dispilio, as the shore was also easily accessible. The shortest crossing is 1400 m from Kastoria to the village of Mavrohori to the east (Tsolakis 1992, fig. 1).

Evidence about Neolithic watercraft in Northern Greece

In such environmental conditions as described above, both in Thessaly and Western Macedonia, some sort of watercraft would have been very useful for communication, fishing and waterfowl hunting, crossing, transport of heavy and bulky materials, animals and humans also during the Neolithic.

In Greece, the results of such an aquatic mobility,

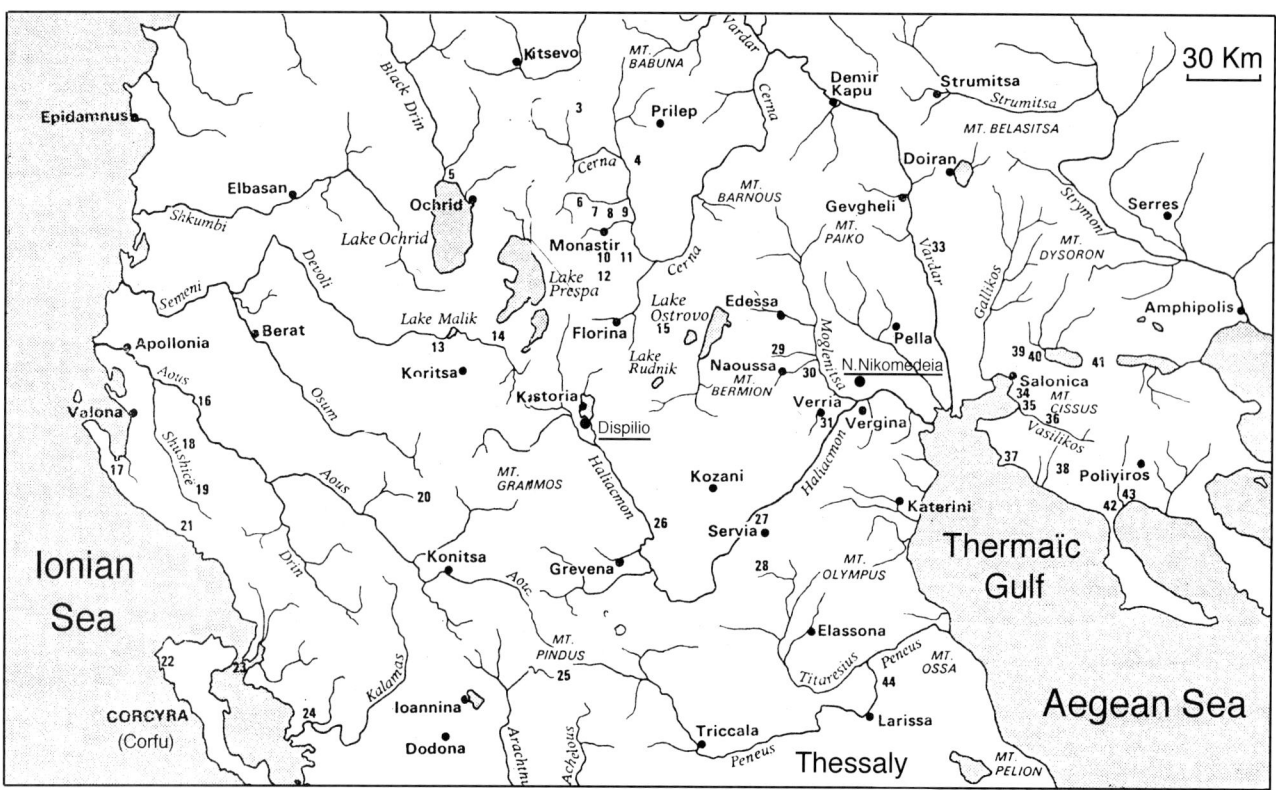

Map 17.2. Neolithic Western Macedonia and adjacent areas. Adapted from Hammond, 1972, 216–217, map 19.

including long distance transport of raw materials (obsidian) across the Aegean sea, and deep sea fishing, attested since the pre-Neolithic, are tangible and multiple (overseas raw materials and artefacts, fish bones, fishing gear). In the MN, besides lithics, stone and shell ornaments and stone vases made of material of foreign origins, with a low rate of circulation over long distances are involved (Demoule *et al.* 1993). In the LN, besides intensification of obsidian trade, among other manufactured products, *Spondylus Gaederopus* bracelets, marble figurines and grinding tools were transported overseas or from coastal sites to the hinterland (Demoule *et al.* 1993), probably along waterways. In the FN, utilitarian as well as prestige goods travelled over long distances, including from the islands to the mainland (Demoule *et al.* 1993).

Hints about the types of watercraft used come from two categories of evidence, which are boat-shaped:- Middle and Late Neolithic iconographic material, with all the expected difficulties of identification:it consists of clay models from various sites in Thessaly and from Dispilio; several others were found in Balkanic sites. As usually with models, we can not always be sure about the *scale* of reproduction. This means that sometimes the models might represent house equipment, such as wooden containers for bread preparation or troughs, for example:shape and raw

material are similar to those of dugouts (Marangou 1991:23–24). Also – boat gunwale outlines (Late Neolithic) are preserved in the soil at Dispilio.

Boat models as indicators of watercraft types and raw materials used

Logs

The dugout seems to be expected in inland waters in the Neolithic, depending on technological possibilities and availability of suitable tree trunks, all the more so, since preserved evidence from excavated examples from the Mesolithic onwards consist mostly of logboats (cf. lately Arnold 1995, 1996). The majority of recorded Neolithic boat models (Marangou 1991; 1996a; 1996b; 1998; Höck-mann 1996) also seem to represent dugouts.

Identified clay models date from the Middle, Late and Final Neolithic of Greece and corresponding periods of the Balkans, including the Karanovo, Gumelnitsa, Bakarno-Gummo and Vinca cultures, the Middle, Late and Final Neolithic of Thessaly (Marangou 1991, 1996a) and the Middle and Late Neolithic of Dispilio (Marangou 1996b, 1998).

Boat models, often fragmentary, occur in various sizes,

Figure 17.1. Dugout model from Dispilio, beginning of the Late Neolithic (length: 11.09 cm).

Figure 17.2. Traditional modern manoxylo, *moored at Dispilio (1993).*

their length varies from 6 to about 28 centimetres (mostly 10–25) (estimated original dimensions), and their breadth from 2.5 to about 14 cm, their height/depth from 2 to 7.2cm. As far as their often-fragmentary state permits to see, they may be subdivided in various categories, probably following a variety of types of the originals. They may be asymmetrical (example from Osikovo, Bulgarian Chalcolithic (FN); Frey 1991:197, fig. 2) or symmetrical, and have ellipsoid or approximately quadrangular, transversal sections. Their extremities are trapezoïdal or oval, occasionally bearing horizontal perforations, the sides more or less straight or tapering, and the length:breadth ratio (when the original dimensions are known or assumed) varies from 2 to 3.4 and even 5.8. They may be of the simplest type (e.g. example from Chalcolithic (FN) Drama in Bulgaria:Frey 1991:196, fig. 1.1) (*cf.* Figure 17.1). Another Bulgarian example (Telis-Redutite; Frey 1991, fig. 1.2), Chalcolithic (FN), shows a dugout with fitted transoms on both ends.

Technological possibility for woodworking in the Neolithic hardly needs to be stressed. Hundreds of suitable stone tools have been collected, among other places, from the Kastoria lake since several decades by local people and the excavations at Dispilio have yielded numerous tools (Hourmouziadis 1996:30, fig. 5b) and piles and post-holes of buildings (Hourmouziadis 1996:27, fig. 4; 34, fig. 7a; 35, fig. 7b). Several possible candidates for logs existed in Neolithic Dispilio, among others oak and pine, also used in the construction of houses. Oak was also used in other sites, such as Maliq (Albania) and Nea Nikomedeia (Western Macedonia), already in the Early Neolithic (Zohary *et al.* 1988:191). Modern flat, plank-boats of the area of Dispilio, called *manoxyla* or *monoxyla* (= logboats) (Figure 17.2), are made from pine, chestnut or elm (oral

communication (1993) by I. Kallinikos). Traditional boats of local types from other Greek lakes or lagoons are made from the trees available in the surroundings:oak, chestnut, pine, poplar or elm (Pantzopoulos 1989:42).

The exceedingly large breadth of some Dispilio models with a length:breadth ratio of 2:1, while other models from Thessaly have a ratio of up to 6, is particularly astonishing (see further). It has been suggested about balkanic examples, that two or three trunks or longitudinal elements had been joined together for the construction of large extended dugouts (Frey 1991:196, 197, fig. 2 and 199, fig. 4). Höckmann has proposed a similar interpretation of the Thessalian complex boat model from Tsangli (see further).

An exceptional find of the 1970s from Prodromos, situated at the edge of a large, marshy area periodically flooded, in the western Thessalian basin, with important changes of temperature, shows how far the EN technology could go. The wooden roof of a building (10 × 10 m), was made from branches and logs of trees, some of them, of a diameter of 30 cm, had been worked into planks or squared beams and were assembled by means of wooden pegs (Hourmouziadis 1971). Some evidence about possible use of planks is also available in the later phases of Dispilio. This means that advanced types of boats such as extended logboats with added elements are not excluded in the Neolithic. Such complex types are attested in Neolithic Denmark (Christensen 1990, fig. 11–12) and Italy (lake of Bracciano; *L'Archeologo subacqueo* I.3, September-December 1995:3).

Animal hide

Two models from Dispilio, dating from the end of the Middle Neolithic (Figures 17.3–17.4), do not look like usual dugouts, although they present a rather symmetric breadth shape and a more or less flat bottom. A flat bottom may as a matter of fact help for the stability of the model. Yet not only do they have exceedingly large breadth, with a length:breadth ratio of 2 (see above); they are also comparatively deep. Besides, they have one at least pointed end, slightly raised. This happens also on a third, Late Neolithic, smaller model. A Neolithic clay model from the Lake of Bracciano also presents pointed and slightly raised ends (illustrated in *L'archeologo Subacqueo* I, 3, 1995:3).

Since poplar is attested, it could have been used for expanded dugouts, which have pointed and raised extremities. Yet, the transversal hull section of the models is trapezoïdal, not elliptical, as would be the case for expanded dugouts (see for example Arnold 1995:150–151). Furthermore, if they were expanded, their sides should be low -the Dispilio models being, on the contrary, comparatively deep-unless, of course, strakes had been added to them.

Moreover, one Dispilio model bears an external lateral protrusion under the pointed extremity (Figure 17.3), although this could also be accidental, because of the rough modelling. Furthermore, a cavity can be seen in the interior of the same pointed end. On the second model from the same site, which also has a pointed, raised extremity, the other end shows in relief an external "rib" (Figure 17.4). On both models, the gunwale is irregular. It even consists of an applied, roughly modelled coil in one case, the sides being slightly curved inwards near the gunwale. The sides of the other model rise considerably amidships (Figure 17.4). Heights of stern and bow would be originally almost equal.

One might suggest, in these cases, a representation of the framework of a boat on which hide was extended, such as on a late 17th century A.D. drawing by Captain Phillip, of curraghs with a woven wickerwork and an external keel and stem element (McGrail 1998:fig. 10.2). Hornell (1938:35–7) criticised this drawing, since there was no corroborating evidence. Nevertheless, in the Middle Ages,

Figure 17.3. Boat model from Dispilio, end of the Middle Neolithic (maximum length: 20.2 cm).

Figure 17.4. Boat model from Dispilio, end of the Middle Neolithic (maximum length: ca 28 cm).

there are descriptions of curraghs with keels, and it is not excluded that before the nineteenth century British and Irish hide boats had an external keel like Phillip's boat or an internal one like the *umiak* (McGrail 1998:178, 182). In fact, the central longitudinal member of the *umiak*'s framework may be compared to a keel or a keelson, since it is inside the hide cover (McGrail 1998:182).

Skin boats had probably been used already in the Upper Paleolithic and Epipaleolithic (Ellmers 1986; Arnold 1996:36). Some Neolithic and Bronze Age rock carvings from Skandinavia may represent boats made from animal hide (Coles 1993; Greenhill 1995:93, fig. 83). An un-published possibly Neolithic burial discovered in the last century in northern Germany was made in a hide canoe, while an Early Bronze Age *coracle* used in a similar way is attested in Scotland (Höckmann 1996:41, and fig. 12.2). Besides, a piece of wood from a burial mound in Ireland of the 10th century AD has been interpreted as having belonged to the frame of a hide boat (Greenhill 1995:92; for a discussion about early use of hide craft see McGrail 1998:186).

Occasionally, some clay models have been interpreted as representing *curraghs*. They date from the Hungarian Early Neolithic (Höckmann 1996:37, fig. 9.2–4; 41, fig. 12), as well as from Bronze and Iron Age Ireland and Wales (Ellmers 1986:fig. on p. 31; McGrail 1998:186, 187, fig. 10.9). *Curraghs* and *coracles* (Hornell 1946:111–148; McGrail 1998:179, fig. 10.3, 180, fig. 10.4 and 10.5; Greenhill 1995, fig. 56, 80, 81), rounded and respectively long narrow boats, originally consisting of animal hide covering and a simple wooden and basket frame, have been used till recently in Wales or Ireland for example. Furthermore, an incised pattern on a Late Neolithic sherd from the Grabak cave (island of Hvar) in Dalmatia (Novak 1955:320, pl. 194) has been interpreted as a hide boat (Bonino 1983:66, fig.7B). A Late Neolithic incised ceramic bowl with an applied animal(?) head and a series of dots below the rim from Eastern Macedonia (Dikili Tash) could also represent a simple hide craft (Theocharis 1973, pl. 199). The preserved extremity of a boat (?) model fragment from Otzaki Magoula in Thessaly is modelled as an animal head with perforated eyes (Milojcic 1983, pl. 23, nr. 10).

Classical texts referring to the mythical beginnings of navigation often mention floats, rafts or boats of animal hide which would have been used for the first attempts to cross the sea in the north-east of the Aegean, after the flood. They are occasionally compared to hide floats used for crossing rivers, such as the Danube. The main source is Lycophronis Alexandra, 72–80:"...Atlas' daughter's diver son, who of old in a stitched vessel, like an Istrian [=Danubian] fishcreel with four legs, sheathed his body in a wineskin and, all alone [or:"with one oar"], swam like a petrel of Rhithymna, ... what time the plashing rain of Zeus laid waste with deluge all the earth" (translation:Lewis 1959:24).

Animal furs and hide were certainly used in the Neolithic. The floors of some LN Thessalian houses may have been covered with furs; bears are represented by feet bones only, which indicates the introduction of skins with distal extremities still attached (Demoule et al. 1993). In fact, in Thessaly, as well as in Macedonia, wild animals (deer, boar, auroch, fox, hare, beaver, birds and fishes) have been found in LN sites in very small proportions. They must have been exploited mostly for their furs, feathers or antlers rather, than as food supplements (Demoule *et al.* 1993). Hide would also be available from domesticated animals from the beginning of the Neolithic, since sheep, goats, pigs, cattle and dogs are present, sheep being predominant

in number of remains, and *Bos* in terms of meat yields in most sites (Demoule *et al.* 1993). At Dispilio, not only real bones of large animals, but also, among a number of animal figurines, a large horned animal head (preserved length ca. 20cm) was found. It has been argued that sites located by lake Karla would have the possibility to keep bovines for tillage, a great advantage counterbalancing the drawback of a flooded area (Halstead 1977). In the recent past, herding was also practised near the lake of Kastoria, while bears are scarcely perceived on the surrounding mountains.

Among other sites, Prodromos (Hourmouziadis 1971) gave large quantities of bones of domesticated and wild animals, as well as bone tools specialised for the work of soft materials, such as hide, besides stone tools for woodworking, but few tools related to food processing.

Willow and Pistacia could provide branches suitable for a frame, possibly covered by hide.

Reeds and basket

The same branches could be used as a frame for a wickerwork/basket boat, covered with waterproof material. Numerous imprints of basket under the base of vases during all phases of the Neolithic (Treuil 1989:151) show the use of basketry, in that case for supporting unbaked vases during pottery making.

A model from Eridu (5200 BC), constructed from clay coils, with inward curving of the sides towards the gunwales, probably represented a reed bundle boat, coated with bitumen. Protuberances on the outer surface may represent "throughbeams" (Qualls 1985). Later iconographic Mesopotamian evidence shows crescent-shaped craft, either reed boats or wooden boats imitating the reed craft form (Vosmer 1996:225). Representations of *papyrus* boats from Predynastic Egypt (mostly 4th millennium BC), used for fluvial navigation, engraved on rocks or painted on pottery (Basch 1987:33–34, fig. 65–66; 49, fig. 76; 50, fig. 78), are particularly abundant.

Besides, ethnography and experimental archaeology give evidence for the possible use of reeds for boat construction in the Aegean Sea since the Mesolithic (9000BP), when obsidian was brought from the island of Melos to Franchthi in the Peloponnese (Southern Greece). The Hellenic Institute for the Preservation of Nautical Tradition crossed the sea from Central Greece to this island in 1988 (Tzalas 1995), in a flat, paddled craft, with inward-curving pointed ends. This consisted of a wooden (cypress) frame, on which bundles of *Scirpus lacustris L. ssp. Lacustris* were fastened with lashings of vegetal fibre rope (Tzalas 1995:445–446, 453–454, 456, note 8). The craft imitated a traditional raft (McGrail 1998:164) or boat (*papyrella*) used till recently in the shallows off the island of Corfu (north-western Greece).

Wicker-work or reed bundle boats, covered with waterproof material may have been used in the sea around the middle of the 3rd millenium BC, in the Arabian gulf and western Indian Ocean (Ra's Al Jinns, Oman). Some pieces of bitumen with impressions of bundled reeds lashed together, basket weave, and wooden planks lashed, stitched or sewn together, bear remains of barnacles on the surfaces opposite the impressions. Therefore, it was concluded that these surfaces were immersed in seawater and consequently that the bitumen had coated the hulls of vessels (Vosmer 1996). As a matter of fact, reeds were the main material worked with flint tools at Mesolithic Franchthi (Jacobsen 1999). Reed structures are attested at Dispilio, among others, possible standing fishing weirs, and convenient raw material should be available by the lake in the Neolithic.

Combinations of various materials for boat construction in prehistory have also been assumed:it has been suggested that an Early Dynastic boat model from Egypt represented a papyrus boat covered with animal hide (Nibbi 1993).

Bark

The intriguing ribs, the pointed raised ends, and indeed the raised sides amidships of some models, may also recall bark canoes (Adney et al. 1964, for example, figs. 106, 115, 118; Marshall 1985:9; McGrail 1998:89, fig. 7.1). Bark boat originals have exceptionally been proposed for some Hungarian Late Neolithic examples (Höckmann 1996:37 and fig. 9.1).

Bark boats are attested mostly at certain latitudes, in North America, Siberia, Scandinavia, Chile and Australia (McGrail 1998:88, table 7.1). The most suitable tree species for bark in North America (35° – 65° N) is birch, but second-rate trees may also be used, such as elm and chestnut (Adney *et al.* 1964:14–15; McGrail 1998:89–90, table 7.2). Use of bark in the area of Dispilio is not as impossible as that, since, according to palynological evidence, birch did exist around 4000–6500 BP, in the north-east of the Pindos mountains, at an altitude of 800–1100 m (Faugères 1989:101, fig. 3, 103), which is, 150 to 450 m higher than the Kastoria lake. Elm and chestnut are indeed used for modern plank boats of Dispilio, and willow or hazel could do for the frame (cf. Adney *et al.* 1964:17; McGrail 1998:90, table 7.4).

Thus evidence shows the possibility of a parallel existence and use of different boat types made from various raw materials, even on the same site. This co-existence of varied types has also been suggested concerning rock-carved boats from Bronze Age Scandinavia (Coles 1993).

One could argue that some constructional traits of boat building originated in other building traditions, as it happened with carry-overs from skin boat to plank boat traditions (Crumlin-Pedersen 1972) and that some at least

examples simply copied features characteristic of one material, transposing them in a different material. All this remains of course hypothetical, as long as preserved finds of hide/reed/bark – real boats cannot be added to the numerous dugouts.

Boat gunwale outlines; from models to real boats

Precisely, about these "real" boats:If the boat models are only attested for the moment from the older phases of Dispilio (B2–B3/C1), when the presence of water seems to be more palpable, the boat outlines date from the latest phase (A0).

In the latest Neolithic of Dispilio, in the upper strata (where wood is not preserved), three outlines (*cf.* Crumlin-Pedersen 1991) of boat gunwales have appeared till now, as a colour variance in the soil, in two contiguous trenches. The outlines are 3.3, 3.0 and 3.5 (preserved) meters long and 0.80, 0.73 and 1.40 m wide (maximum breadth).

The first outline (A) (Hourmouziadis 1996, fig. 12; Marangou 1999, fig. 6), probably of a dugout (Figure 17.5) is asymmetrical, with a pointed and a larger, more or less trapezoïdal end. It has not yet been excavated. Traditional wooden *manoxyla* (see above) of similar dimensions and asymmetrical shape are still used in this particular area of the lake for fishing (Figures 17.2, 17.6).

The second outline (B) (Marangou 1996b) has the same south-east – north-west orientation and approximately the same size, but it has two more or less pointed ends. The outline, about 4 centimetres wide, seemed to "move" towards the west during its excavation up to a depth of some centimetres, when it disappeared. This could suggest that it consisted only of a wooden frame, and, consequently, that it was a hide-boat gunwale frame, but could also be due to the bad preservation of the bottom of a dugout, reversed or not.

Both outlines may be related to a number of postholes, with several alternative interpretations (Marangou 1996b). One large (30×22 cm) posthole at the northern end of outline B could have been a mooring post; today manoxyla are fastened with a chain to wooden posts (Figure 17.2).

Besides, pairs of small postholes are repeated with a constant span (60–70 cm) between them on the external limit of the eastern side of both outlines. These postholes could be supports for a fishing trap or signs over water level of the location of sunken boats (Marangou 1996b).

The boat outlines would confirm the aquatic environment of Neolithic Dispilio. They are situated to the east of a "channel"-like linear feature, which surrounds the excavated part of the settlement on the east and the north. They might be in a fishing location (*cf.* Andersen 1985:55), or they might have sunk or been abandoned in the periphery of the settlement; changes of water level and periodic floods

Figure 17.5. Boat gunwale outline (A) in Dispilio, end of the Late Neolithic. Photograph kindly provided by Professor G. Hourmouziadis.

could be involved. However, the excavation has not been extended to the east of the outlines, and the outlines' connection with the "channel" is not yet clear. Moreover, the houses of this phase consist of land structures of clay bricks and floors and the precise location of the lake then is still unknown.

A fishing and water-fowling function of the boats seems probable, as is their use for the transport of reeds and tree trunks for building. Fish vertebrae and fishing gear (Hourmouziadis 1996:44, fig. 13), even a fish-shaped stone pendant have been discovered at Dispilio. Among other bird bones, possible swan or eagle bones were occasionally used for the construction of bone flutes (Hourmouziadis 1996:fig. 17; Malea *et al.* 1997).

The third outline (C) appeared two strata below B, in a stratum following a stratum with possible standing fishing weirs. It has a pointed end, but the other end disappears in the non-excavated margin of the trench. The port and starboard sides were not completely joined at the northern pointed end (bow?), which therefore was probably not completely closed. The gunwale was preserved to a maximum depth of 4 cm. Besides, there is a transversal thinner (2 cm wide) linear element near the preserved end. If not the transom of a dugout, this could well be the transverse member – a thwart? – of a hide boat with its gunwale frame preserved. Moreover, the ratio length:breadth of this outline -2.5 – implies a craft of quite different proportions than the first two (length:breadth about 4.1), and consequently of different type, unless this is the result of its circumstances of preservation.

Contextual information

Not only real watercraft, but also models occur in groups. Most Dispilio models come from an area of two or three neighbouring trenches, dating from the end of the MN and the beginning of the LN. The settlement in this period was probably located in a marshy environment, the post-houses had floors consisting of layers of clay, wooden beams and stone structures.

A number of boat models and oval ("boat-shaped") vases were found in a restricted sector, in 2 or 3 successive strata. Some of them were situated in relationship to a possible storage or cooking space, as well as several vases of different types (jars, "fruitstands", saucers and bowls), and miniature vases and figurines. Burnt and disintegrated wood remains from piles, structures and floors, a lot of animal- and some fish-bones, as well as a number of postholes, possibly from light structures, complete the picture (Marangou 1996b).

In such a context, the role of boat models, as that of figurines, could have been protective or magic, for instance. This does not necessarily exclude a practical function of some examples (such as lamps, "incense/odoriferous substances -burners" or simply containers), while their sizes vary considerably. No analyses have been made yet, so the last point cannot be resolved for the moment.

Discovery of boat models in a Neolithic domestic context is not unique. An exceptional find from a Late Neolithic house in Battonya (Hungary) consisted of seven possible boat models (Höckmann 1996:38). A model from Chalcolithic Cascioarele (Romania) was also found in domestic context, as well as several anthropomorphic and zoomorphic figurines and miniature vases, and various tools (Stefan 1925:142–143, 164). Another example comes from Middle Neolithic Selevac in Serbia, at some hours from the rivers of Morava or Danube, where fish consumed on the site must have been caught. A concentration of pottery, tools, figurines and miniature objects, including a boat model with perforated ends, was found in association with

Figure 17.6. Traditional fishing manoxyla *moored among the reeds near Dispilio (1993).*

the floor of a post-hole house, related to an oven (Tringham *et al.* 1990:336, pl. 10.5, no (02–1178); 373). Again, it is not excluded that the boat model had a (parallel?) practical function, for example, as it is suggested in the publication, that of a loom piece or weight (see comments in Marangou 1996a). Similar ideas have been discussed concerning a possible functionality of boat models from early 5th millennium Eridu (Strasser, 1996; J. Bourriau and J. Oates, 1997).

It is possible that a comparable to Selevac situation, a scene of several figurines and a boat(?) with pointed upturned ends, upper surface decorated with transversal parallel incisions and quadrangular section was represented in miniature, in an early Late Neolithic model of the interior of a house. This was found under the floor, near the hearth of a real Thessalian house, in Platia Magoula Zarkou, located at a distance of 800 m from the Peneios now. Earlier Neolithic layers were contemporary to flooding, although it is not possible to estimate its frequency. It has been argued that Neolithic occupation in Thessaly was intermittent (see above) and took place only outside the flood season, while during the flood dry sites were occupied (Van Andel *et al.* 1995:138). In the 19th and 20th centuries, the flood season lasted from December to May. After the Peneios flood in spring 1982, a large area near Platia Magoula Zarkou turned into a shallow lake (Van Andel *et al.* 1995:138).

Boat models, as well as figurines and other models (Marangou 1998), often come from house floors and garbage:one at least of the Dispilio fragmentary examples comes from a probable garbage space and models from

Otzaki and Vucedol were found in pits (Marangou 1991:29). Boat models would thus be connected to limited in time everyday activities. They would be related to few households, since they occur rarely, while occasional concentrations in certain domestic sectors are difficult to explain. This might suggest specific economic or social activities of some social groups, possibly restricted in specific areas of a settlement. Of course, evidence is not sufficient yet and such hypotheses need to be corroborated by further finds and study.

From the inland waters to the sea

For the moment, evidence about Neolithic watercraft only comes from inland sites, in spite of the intensive navigation in the Aegean Sea. However, relatively few Neolithic coastal sites have been excavated till now, several of which did not give any other clay models either. Besides, as this material has been identified relatively recently, unrecognised specimens may still lay in Museum deposits together with other miscellaneous clay fragments.

Even in this inland corpus of material, some of the original boats could have been seaworthy, at least more than basic dugouts. A unique model from Tsangli in southeastern Thessaly (Giannopoulos 1910:63, fig. 3; Wace-Thompson 1912:124, fig. 74 c; Marangou 1991, pls. IV, VIIb–IX; 1996a, fig. 2–3) (Figure 17.7), dating possibly from the MN, shows such an advanced type of hydrodynamic shape. The identification of the original is however not easy:it could be an extended dugout, but its

Figure 17.7. Middle Neolithic boat model from Tsangli (length: 10.2 cm).

length:breadth ratio is only 1.5; it is too large. This could be explained by an assemblage of longitudinal elements of several stems (Höckmann 1996, fig. 4.2) (cf. above). If it were a hide boat (of *curragh* or *umiak* type?), then its "keel" would represent an element of a hide boat frame. Nevertheless, the morphology of the interior rather recalls hollowed wood.

In fact, the model is divided transversally in two compartments by what seems to be a bulkhead left in the solid, or a fitted transom. Basic log-boats may be partitioned in order to separate functional spaces on board, to secure the load, fish or fishing gear, or transported animals, or to provide seats. The inner division and large breadth of the Tsangli model may suggest that the boat was well adapted as a cargo.

Another advanced type of craft probably constitute some clay models from Final Neolithic sites, in a region of lakes and rivers around Pelagonia (Former Yugoslavia), Romania and Albania (Marangou 1991), near the Greek border (Map 17.2). Communication among sites located in the Pelagonian closed valley is easy (Simoska *et al.* 1975; *cf.* Hammond 1972:41–46), easier towards the south than towards the north, and close cultural contacts existed among several of these sites (group Bakarno Gummo-Suplevec-Grnobuki). A number of these models at Maliq, Vucedol and Bitola have been identified as representing double dugouts, much more stable in open water and more roomy for transport. Some of them have projections on one or both ends, which are occasionally perforated. Paired logs, among other places, were attested in the beginning of the 20th century in Albania for the crossing of rivers and transport of animals (Marangou 1991, pls. VIb, VIIa). Some evidence about prehistoric paired logs is available from Italy. In the 16th century A.D. monks of the Athos monas-

teries used them indeed in the Aegean sea by the Chalkidiki peninsula (Central Macedonia) for fishing (Marangou 1991:27).

As about the complex Tsangli model, Tsangli, a particularly large Neolithic settlement (the mound measures about 200×210 m:Decourt 1990:51, note 18), is located in the centre of a valley which runs along a small tributary of the Enipeus river (Map 17.1, T). Half an hour to the south, on the hills, there are the vestiges of the Greek city of Eretria (Wace *et al.* 1912:86). A group of seven Neolithic sites located closely to each other have been identified, at low altitude, between the valley and the slopes (Tsangli is at 204masl; Decourt 1990:55). They are situated near a much-frequented passageway, which leads across the low mountains from the Aegean (the Volos or Pagasean gulf), to the interior plain (Decourt 1990:50–51, 127–128). In fact, the low hills to the south and the north do not hinder relations with the Almyros plain and the Peneios basin (Wace *et al.* 1912:241–242). This was also a later communication axis from the coast to the Western Thessalian plain (Decourt 1990:67). This is one of the alternative passages of the Roman army under Flamininus in 197BC:he would have crossed through the hills, near the historical town of Eretria, close to Tsangli, coming from the east towards the Enipeus valley (Decourt 1990:109 and fig. 127). Eretria was located on the summit of a trapezoidal hill close to the ridge, which permits to descend precisely to the plain of Almyros (Decourt 1990:103).

As a mater of fact, the mountains separating the large inner Thessalian basins from the coast let only the small plains of Volos and Almyros in direct contact to the sea networks (Map 17.1). These smaller eastern coastal plains have climatic conditions resembling south-eastern Greece (Johnson 1996:271). Not only Tsangli was not really isolated from the seacoast, but it even occupied a privileged position between the sea and the hinterland communication axes.

Besides, sea was closer to Neolithic sites, which are now at some kilometers from the bay of Volos. The important in the Late and Final Neolithic site of Dimini lies on the western edge of the coastal plain of Volos, at 3km from the present coast. The site has been reconstructed recently as a coastal settlement (Zangger 1991). A major episode of alluviation in the 4th millennium BC has formed the plain, pushing the coast away from Neolithic Dimini (Andreou *et al.* 1996:543). The site seems to have been occupied as long as the coast was close, meaning that the location and may be the function of the site were closely related to the environmental setting (Zangger 1991). In the Bronze age, when the sea moved away, another site was selected (Petromagoula) on the coast, at a distance of a few kilometres (Zangger 1991).

A similar situation occurred in the eastern part of

Western Macedonia, in the Yannitsa plain and marshes and the Thermaic Gulf (Map 17.2). The riverine alluvium in the deltas of the important rivers Aliakmon, Axios and Gallikos had not yet moved the coast away, to its present position. The site of Nea Nikomedeia, inhabited since the EN, was located near the shore of the Thermaic Gulf around 7500 BP. From then on, competing lake/lagoon/marsh and marine influences succeeded one another (Faugères 1989:106–107; Psychoyios 1992:23, 25–27).

Towards the end of the Middle Neolithic, the sea level rose in various areas of the Aegean Sea, while in plains humidity increased and some parts of them were submerged by the sea. The sea level continued to rise in the first part of the Late Neolithic (Psychoyios 1992:27). The climate became the warmest and most humid after the Glacial period about 5900BC and remained like that till about 5600BC. The climatic optimum coincided with the first part of the Late Neolithic (around 5000BC), when the temperature was about 2 degrees Celsius warmer than today and rainfall increased (Psychoyios 1992:28).

The significance of the omnipresence of water in the regions considered and the relative proximity of the Aegean coast shed new light on the multiple clues about overseas relationships of the Thessalian and Macedonian hinterlands in the Neolithic period.

Incidentally, the naval expedition of the Argonauts for the Golden Fleece started, according to the myth, from the Volos bay. However, the precise location of the Late Bronze Age site, the Homeric Iolkos, from where Jason departed, can not be identified with certainty (Andreou *et al.* 1996:550–551 and note 91).

Conclusions:variety and multifunctionality of Neolithic watercraft?

A means of aquatic mobility and acquisition of alternative food resources in an insecure Neolithic environment is an important instrument for survival by the water. This significance can also be assumed from the fact that watercraft was even represented in clay.

At this stage of investigation, in the regions considered, a relatively limited number of material has been identified, mostly dated from the last part of the Greek Middle Neolithic and the Late Neolithic. Lack of Early Neolithic and early Middle Neolithic boat representations, if not fortuitous, might also be a repercussion of contemporaneous customs. Their occurrence later could be attributed to a pronounced interest in aquatic mobility in the end of the Middle Neolithic and in the early Late Neolithic. One could correlate this interest to changes in the aquatic environment, such as increase of humidity, floods of rivers, extension of lakes, or submersion of coastal plains by marine transgression, as well as to increased involvement in nautical

matters. The latter trend could have been influenced by evolution in networks of communication and exchange across or along water, in the frame of more diversified economic and broader social interests. Improvement of technical capabilities and therefore of watercraft performances facilitating access to new territories could have a further bearing on this trend, particularly in areas offering more challenges, for example in sites located within view of the sea.

In spite of the fact that navigation is more intensified in the Final Neolithic, scarcity or absence of models of boats from Northern Greek sites, contrasting to the situation in other regions (e.g. Chalcolithic Bulgaria, Albania and Pelagonia), could again be due to excavation or identification hazards. The case of Dispilio according to present evidence is striking:real boat outlines are found precisely in the later strata, while boat models come from the older phases. Nevertheless, this absence or scarcity coincides with a shift in three-dimensional art concerning choice of represented subjects, morphology and probably function towards the end of the Neolithic period in Thessaly; among other mutations, not only models of boats, but also models of other structures are now extremely rare.

A variety of watercraft types, some of which at least were not simple dugouts, are assumed to have been used in the Neolithic, all the more so, if we bear in mind the probability of non-preserved evidence and of non-identified specimens. Different types of the original watercraft had been represented and may co-exist even in our restricted corpus of models, as they do among the limited number of preserved gunwale outlines.

The local environmental conditions must have been consequential for the choice not only of raw material, but also of boat types. Navigation in large lakes and rivers as well in the sea waters would necessitate compliance to more demanding requirements than fishing and water fowl hunting in marshes, or crossing shallow and short water expanses. In addition, different boat types had presumably been used for distinct functions in the same environment. Therefore, several alternative solutions for selecting raw material and shapes for the construction of Neolithic watercraft are conceivable. The choice may have depended primarily on local, environmental and topographic conditions, but also on economic requirements, specialised functions, social needs, and even symbolic concerns.

Acknowledgements
The author is grateful to Professor Barbara Purdy for supporting participation to the WARP 1999 in Florida. Special thanks are due to Professor George Hourmouziadis for entrusting me with the study and publication of a part of the material of his excavations at Dispilio, as well as for helpful discussions. Ioannis Kallinikos, last constructor of

the traditional boats of the Kastoria lake, *karavia* and *manoxyla*, provided invaluable information. Discussions with Maria Mangafa[+] and Maria Dinou, who supplied information on the botanical and anthracological material, as well as with other colleagues from the Dispilio excavations proved very useful. Many thanks are due to Anne Dietrich for fruitful discussions on wood and bark and to Lucien Basch for helpful comments about primitive boats. Finally, warmest thanks are addressed to Ole Crumlin-Pedersen for precious advice and encouragement during a stay at the Centre for Maritime Archaeology in Roskilde in April 1997.

References Cited

Adney, E.T., and H. Chapelle. 1964. The bark canoes and skin boats of North America. Washington, D.C.: Smithsonian Institution, Museum of History and Technology.

Andersen, S. 1985. "Tybrind Vig. A preliminary report on a submerged Ertebølle settlement on the West coast of Fyn". Journal of Danish Archaeology 4:52–69.

Andreou, St., M. Fotiadis, K. Kotsakis. 1996. "Review of Aegean Prehistory V: The Neolithic and Bronze Age of Northern Greece". American Journal of Archaeology 100:537–97.

Arnold, B. 1995. Pirogues monoxyles d'Europe centrale. Construction, typologie, évolution (Dugouts of Central Europe. Construction, typology, evolution), vols. 1–2, Archéologie neuchâteloise 20–21. Neuchâtel: Musée cantonal d'archéologie.

Basch, L. 1987. Le Musée imaginaire de la marine antique (The imaginary museum of ancient ships). Athens: Institut Hellénique pour la préservation de la tradition nautique.

Bonino, M. 1983. "Le imbarcazioni monossili in Italia" (The dugouts in Italy). Bolletino del Museo civico di Padova LXXII:51–77.

Bourriau, J. and J. Oates. 1997. "Spinning or sailing?:the boat models from Eridu". Antiquity 71:719–721.

Christensen, Ch. 1990. "Stone Age Dug-out Boats in Denmark: Occurrence, Age, Form and Reconstruction". Experimentation and Reconstruction in Environmental Archaeology, D.E. Robinson (ed.), Oxbow Books Monograph 5. Oxford: Oxbow Books: 119–141.

Coles, J. 1993. "Boats on the Rocks". A spirit of Enquiry. Essays for Ted Wright, WARP Occasional Paper 7, J. Coles, V. Fenwick and G. Hutchinson (eds). Exeter: WARP, Nautical Archaeology Society and National Maritime Museum: 23–31.

Crumlin-Pedersen, Ole. 1972. "Skin or wood? A study of the Origin of the Scandinavian Plank-boat". Ships and Shipyards, sailors and fishermen, Introduction to maritime ethnology. Ol. Hasslöf, H. Henningsen, A.E. Christensen jr. (eds), Copenhagen University Press: 208–234.

Crumlin-Pedersen, Ole. 1991. "Bådgrave og gravbåade på Slusegård" (Boat-graves and grave-boats at Slusegård). S. Andersen, B. Lind, Ole Crumlin-Pedersen, Gravformer og gravskikke Bådgravene, Slusegårdgravpladsen III. Forfatterne: 93–263.

Decourt, J.C. 1990. La vallée de l'Enipeus en Thessalie. Etudes de topographie et de Géographie antique (The Enipeus valley in Thessaly. Studies of ancient topography and geography). Bulletin de Correspondance Hellénique, supplément XXI. Paris:Ecole Française d'Athènes.

Demitrack, A. 1994. "A dated stratigraphy for the Late Quaternary in Eastern Thessaly and what it implies about landscape changes". La Thessalie. Quinze années de recherches archéologiques, 1975–1990. Bilan et perspectives. Actes du colloque international Lyon, 17–22 avril 1990, volume A. Athens: 37–40.

Demoule, J.-P., and C. Perlès. 1993. "The Greek Neolithic: a new review". Journal of World Prehistory 7:355–16.

Ellmers, D. 1986. "Fellboote, Einbäume, Schiffe. 11000 Jahre Schiffbau" (Hide-boats, dugouts, ships. 11000 years of ship construction). Schiffsarchäologie in Deutschland 2:28–37.

Faugères, L. 1989. "Introduction. A. Le cadre géographique" (Introduction. A. the geographical framework). R. Treuil, P. Darcque, J.-Cl. Poursat, and G. Touchais, Les civilisations égéennes du Néolithique et de l'Âge du Bronze. Paris: Presses Universitaires de France: 81–109.

Frey, O.-H. 1991. "Varna –ein Umschlagplatz für den Seehandel in der Kupferzeit?" (Varna – a transhipment centre for the maritime trade in the Chalcolithic?). J. Lichardus ed., Die Kupferzeit als historische epoche, Symposium Saarbrücken und Otzenhausen 1988, Teil I, Saarbrücker Beiträge zur Altertumskunde, Band 55. Bonn: Rudolf Habelt: 195–201.

Gallis, K. 1992. Atlas of the prehistoric settlements of the eastern Thessalian Plain. Larissa.

Gerakis, P.A., and E.Th. Koutrakis. 1996. Greek wetlands. Goulandri Museum of Natural History, Greek Centre of biotopes-wetlands. Athens: Commercial Bank of Greece.

Giannopoulos, N. 1910. "Prähistorische Funde aus Thessalien" (Prehistoric finds from Thessaly). Athenische Mitteilungen 35:61–64.

Greenhill, B., with J. Morrison. 1995. The archaeology of boats and ships. An introduction. London: Conway Maritime Press.

Grundmann, K. 1937–1939. "Magula Hadzimissiotiki. Eine steinzeitliche Siedlung im Karla-See" (Magula Hadzimissiotiki. A Stone Age settlement in the Karla lake). Athenische Mitteilungen 62–62:56–69, plates 31–37.

Halstead, Paul. 1977. "Prehistoric Thessaly:the submergence of civilisation". Mycenaean Geography. British Association for Mycenaean studies: 23–29.

Hammond, N. 1972. A History of Macedonia, I:Historical geography and prehistory. Oxford: Clarendon Press.

Höckmann, O. 1996. "Schiffahrt in der Steinzeit" (Navigation in the Stone Age). Omaggio a Dinu Adamasteanu, a cura di Marius Porumb. Clusium (Cluj-Napoca):25–60.

Hornell, J. 1946. Water transport. Origins and early evolution. Cambridge University Press.

Hornell, J. 1938. "Curraghs of Ireland". Mariner's Mirror 24:5–39, 148–175.

Hourmouziadis, G. 1971. Two new Early Neolithic settlements in Western Thessaly. Archaiologika Analekta ex Athinon (4):164–175.

Hourmouziadis, G. 1996. Dispilio (Kastoria) the prehistoric lakeside settlement) (bilingual edition). Thessaloniki: Codex. 1996.

Jacobsen, T.W. 1999. "Maritime mobility in the Prehistoric Aegean". Tropis V. Proceedings of the Vth international Symposium on Ship Construction in Antiquity (Nauplia, 26–29 August 1993), H.Tzalas (ed.). Athens:Hellenic Institute for the Preservation of Nautical Tradition.

Jarman, M.R., G.N. Bailey and H.N. Jarman. 1982. Early European agriculture. Its foundations and development, Papers in Economic Prehistory, volume 3, in honour of Eric Higgs. Cambridge: Cambridge University Press.

Johnson, M. 1996. "Water, animals and agricultural technology: a study of settlement patterns and economic change in Neolithic Southern Greece". Oxford Journal of Archaeology 15.3:267–295.

Keramopoullos, A. 1938. Researches in Western Macedonia in Praktika tis Archaiologikis Etaireias: 58–61. L'archeologo Subacqueo I, 3 (1995):3.

Lewis, Naphtali. 1959. Samothrace:the ancient literary sources. Samothrace. Excavations conducted by the Institute of Fine Arts of New York University, Karl Lehmann (ed.), vol. 1. London: Routledge and Kegan Paul.

McGrail, S. 1998. Ancient boats in North-West Europe. The archaeology of water transport to AD 1500. London and New York: Longman.

Malea, E., P. Papageorgiou, G. Hourmouziadis, G. Panagiaris. 1997. Bone made aulos from the prehistoric lake side settlement of Dispilio, Kastoria:Technology of construction and arising conservation problems. Ancient Greek Technology, Proceedings of the 1st International Conference. Thessaloniki: 525–530.

Marangou, Chr. 1991. "Maquettes d'enbarcations:les débuts" (Watercraft models:the beginnings). Aegaeum 7, Thalassa. L'Egée préhistorique et la mer. Actes de la 3e Rencontre Internationale de l'Université de Liège, Station de recherches sous-marines et océanographiques, Calvi (Corse), 23–25 April 1990, Robert Laffineur ed. Liège: Université de Liège: 21–42, pls. II–IX.

Marangou, Chr. 1996. "From Middle Neolithic to Early Bronze Age: Consideration of early boat models". In Tropis IV, Proceedings of the 4th International Symposium on Ship Construction in Antiquity (Athens, 29 August –1r September 1991), H. Tzalas (ed.). Athens: Hellenic Institute for the Preservation of Nautical Tradition: 277–293.

Marangou, Chr. "More evidence about neolithic inland craft (Dispilio, Kastoria)". Tropis VI, Proceedings of the VIth international Symposium on Ship Construction in Antiquity (Lamia, 28–30 August 1996), H. Tzalas (ed.) Athens: Hellenic Institute for the Preservation of Nautical Tradition (in press).

Marangou Chr. "Three-dimensional clay representations from Dispilio, Lake of Kastoria, Northern Greece". Wetland Archaeology Research Project Conference in Ireland, Dublin, 26–29 August 1998 (in press).

Marangou, Chr. 1999. "Evidence about a Neolithic dugout (Dispilio, Kastoria) (preliminary report)". Tropis V, Proceedings of the Vth international Symposium on Ship Construction in Antiquity (Nauplia, 26–29 August 1993), H. Tzalas ed.. Athens: Hellenic Institute for the Preservation of Nautical Tradition: 275–282.

Marshall, I.C.L. 1985. Beothuk Bark Canoes:An Analysis and

Comparative Study, Canadian Ethnology Service paper no. 102, A Diamond Jenness Memorial volume, National Museum of Man Mercury series. National Museums of Canada.

Milojcic, Vl. 1983. Die Deutschen Ausgrabungen auf der Otzaki-Magula in Thessalien III. Das späte Neolithikum und das Chalcolithikum. Stratigraphie und Bauten. (the German excavations on the Otzaki Magoula in Thessaly, III. The Late Neolithic and the Chalcolithic. Stratigraphy and architecture) Bonn: BAM.

Moutsopoulos, N. 1973–1974. Kastoria. History-Monuments-Ethnography. Since its creation to the 10th c. a.C. Prehistoric, Historical and Early Christian periods. Epistimoniki Epetiris Panepistimiou Thessalonikis 6 (1973–1974):261–432; English summary: 433–474.

Nibbi, Al. 1993. "An Early Dynastic hide-covered model papyrus boat". Revue d'Egyptologie (44):81–101, plates 4–7.

Novak, G. 1955. Prehistorijski Hvar (Prehistoric Hvar). Zagreb.

Pantzopoulos, Y. 1989. The boats of the Greek lakes and lagoons (in Greek, with an English summary), in Archaiologia (Athens) 32:41–45.

Psychoyios-Smith, O. 1992. Space and natural environment of Macedonia). I. Aslanis The Prehistory of Macedonia. I. The Neolithic period. Athens: Kardamitsas: 21–35.

Qualls, C. 1996. "Early Mesopotamian boats and the Gulf", unpublished paper presented to the Bahrain Historical and Archaeological Society, 1985, quoted by Vosmer: 225.

Simoska, D., and V. Sanev. 1975. "The neolithic settlement Veluska Tumba at Bitola. A report of the protecting excavations in 1971–1972". Macedoniae Acta Archaeologica 1 abstract: 85–88.

Stefan, G. 1925. "Les fouilles de Cascioarele" (The excavations at Cascioarele). Dacia 2:138–197.

Strasser, T. 1996. "The boat models from Eridu:sailing or spinning during the "Ubaid period?". Antiquity 70:920–925.

Theocharis, D. 1973. Neolithic Greece. Athens: National Bank of Greece.

Tringham, R., and D. Krstic. 1990. Selevac. A neolithic village in Yougoslavia. Monumenta Archaeologica 15. Los Angeles, California: University of California, Institute of Archaeology.

Tsolakis, P. 1992. The boats of Kastoria, their history and construction (in Greek, with an English summary). Thessaloniki: University Studio Press.

Tzalas, Harry 1995. "On the obsidian trail. With a papyrus craft in the Cyclades". Tropis III, Proceedings of the 3rd International Symposium on Ship Construction in Antiquity, Athens 1989, (ed.). H. Tzalas. Athens: Hellenic institute for the Preservation of Nautical Tradition: 441–469.

Vafeiadis, P. 1983. Hydrogeological study of the Kastoria basin, with three additional maps, doctoral thesis, submitted to the geological section of the School of Physics and Mathematics of the Aristotle University of Thessaloniki. Thessaloniki.

Van Andel, Tjeerd H., K. Gallis and G. Toufexis. 1995. "Early Neolithic farming in a Thessalian river landscape, Greece". Lewin, J., M.G. Macklin, and J.C. Woodward, (eds). Mediterranean quaternary river environments. Rotterdam/ rookfield: A.A. Balkema: 131–143.

Vosmer, T. 1996. "Watercraft and navigation in the Indian Ocean:an evolutionary perspective". The Prehistory of Asia and Oceania XIII International Congress of Prehistoric and Protohistoric Sciences, Colloquium XXXII, Trade as a subsistence Strategy. Post Pleistocene Adaptations in Arabia and Early Maritime Trade in the Indian Ocean, Forlì, Italia, September 1996, vol. 16. Forlì: ABACO: 223–242.

Wace, A.J.B., and M.S. Thompson. 1912. Prehistoric Thessaly. Cambridge University Press.

Zangger, Eberhard 1991. "Prehistoric coastal environments in Greece: The Vanished Landscapes of Dimini Bay and Lake Lerna". Journal of Field Archaeology 18, 1:1–15.

Zohary, D., and M. Hopf. 1988. Domestication of plants in the Old World. The origin and spread of cultivated plants in West Asia, Europe and the Nile Valley. Oxford: Oxford University Press.

18. Reconstruction of the Woody Vegetation at the Acheulian Site of Gesher Benot Ya'aqov, Dead Sea Rift, Israel

Ella Werker and Naama Goren-Inbar

Introduction

Over the past ten years intensive research has been carried out at the site of Gesher Benot Ya'aqov. The many and varied aspects of this research contribute much to the reconstruction of paleoenvironments and hominid behavior in an Acheulian site located in the Dead Sea Rift System, along what has been termed the "Levantine Corridor" (Thomas, 1985; Tchernov, 1988, 1992) This corridor is considered a possible route through which biotic exchange took place and hominids dispersed from one paleogeographical domain to others. The "Out of Africa" model, which views the phenomenon of *Homo erectus* as spreading from the African continent to Eurasia, considers this corridor to be one of the main potential dispersal routes.

It is indeed in the Levantine sector of the Dead Sea Rift that the most ancient non-African archaeological evidence exists. At Gesher Benot Ya'aqov, this material is assigned to the Acheulian Industrial Complex and is bedded in a Pleistocene formation, consisting of a sequence of archaeological occurrences located on a lake shore and its vicinity. The Gesher Benot Ya'aqov sites are dated to 0.78 mya (Verosub *et al.*, 1998; Goren-Inbar *et al.*, in press) and the exposures comprise a sequence of 34 m thick (Feibel *et al.*, 1998) in which at least seven archaeological horizons are present (Goren-Inbar, 1998). Each of these sites yielded thousands of stone artifacts *in situ* within a thoroughly investigated stratigraphical sequence.

The meagre hominid osteological evidence from Ubeidiya is unsatisfactory in that it does not enable definite species identification (Tobias, 1966). At Gesher Benot Ya'aqov the hominid remains were found out of context and the elements do not permit a definite conclusion regarding the bones' antiquity (Tchernov and Geraads, 1983).

Excavations at the site of Ubeidiya have contributed an extraordinary wealth of data on Lower Pleistocene fauna.

The extremely diversified and abundant material (Tchernov, 1986) includes hundreds of species originating in different paleogeographical regions, as well as many endemic ones. Excavations at Gesher Benot Ya'aqov also yielded paleontological data, much of it still undergoing study (Goren-Inbar *et al.*, 1994; Goren-Inbar *et al.*, in press). Though the large mammals present at each site differ and are assigned to different biostratigraphical units, the existence of an extensive biomass in the Dead Sea Rift as documented at both sites necessitates a thorough paleoenvironmental assessment.

The vegetation of the Hula region during Lower and Middle Pleistocene times had to be plentiful enough to sustain large herbivorous animals. The present ecological conditions could not have provided the quantities of food needed for these ancient animals. The change is in great part due to the intervention of modern man. However, changes in climate and ecological conditions may also have been responsible for differences in vegetation, which may have been both qualitative and quantitative. Plant species that are extinct today might have grown there in Pleistocene times and new wild species may have been introduced since then into the region. Direct examination of plant remains provides at least partial answers to these questions.

The Pleistocene vegetation consisted of both woody and herbaceous plants. In general, however, only plant parts which possess thick endurable cell walls can, under the right conditions, be preserved for long periods of time. Such plant parts include pollen grains, seeds and fruits, phytoliths and wood. The representation of plant types, as well as the distance of their site of growth from the excavation, differ between the groups. Pollen grains, seeds and fruits may represent both woody and herbaceous plants. Pollen grains are usually well preserved due to their very hard durable coat principally composed of a highly resistant component, sporopollenin. However, they can be carried over great distances due to their microscopic size. Seeds and fruits may be either hard-coated, with lignin or other materials impregnating the cell walls or soft coated, large or small, and therefore may be represented and dispersed to various degrees. Phytoliths are crystal bodies found in various tissues of certain plant taxa, the formation of which is sometimes affected by environmental conditions and other factors (Warnock, 1998 and literature cited therein). Phytoliths may also constitute single organisms with silicified cell walls. Wood generally has relatively thick, lignified cell walls, though the thickness of the walls and the degree of lignification vary between species. Wood, however, represents only woody plants. Small shrubs and herbaceous plants are generally not represented. Trunks and branches are usually found *in situ* if not carried by water or transported by man. Each of the above groups, therefore, can provide a certain type of information. Clearly, any attempt to reconstruct the paleoenvironment and behavior of the Acheulian hominids in a specific region should be based on the integration of data from both micro and macrobotanical assemblages.

Interpretations of the behavior of early hominids (Acheulian) are based largely on present-day African environments and to a much smaller extent on fossil data retrieved by excavations. These models are specific to the African continent. However, the "Out of Africa" models which are concerned with the dispersal of hominid groups require data beyond that continent. The Gesher Benot Ya'aqov plant remains which are discussed below, contribute much to our understanding and to the attempt to reconstruct the specific paleoenvironmental conditions within the Dead Sea Rift ecological zone.

A site like Gesher Benot Ya'aqov can indeed furnish detailed paleoenvironmental data, as substantial parts of the geological formation in which the site is bedded are waterlogged and furnish ideal conditions for plant preservation. Both macro and micro plant assemblages exist in the sedimentary sequence exposed at the site and are undergoing study (Melamed, 1997). Since the microbotanical analysis is still in its preliminary stages, only some preliminary results of the macro remains will be presented here; the authors are well aware of the partial nature of the data.

The Gesher Benot Ya'aqov Site
Geography

The Gesher Benot Ya'aqov site is located in the northern sector of the Dead Sea Rift, in the northern Jordan Valley. In this geographical region, the Hula Valley (Figure 18.1) is bordered by the rift escarpments in the east (the Golan

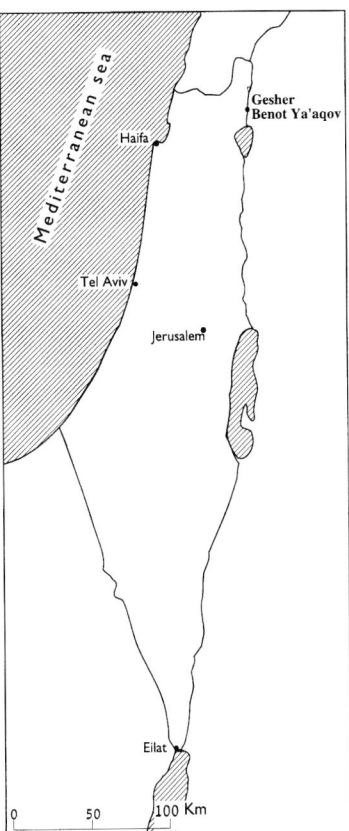

Figure 18.1. Location map of the site (Hula Valley).

Heights, up to 800–1000 m above msl) and west (the Naftali Mountains) and two landform barriers, the Hermon range (up to 2800 m above msl) in the north and by the elevated basalt block of Korazim (up to 300 m above msl) in the south.

The Hula Valley is a flat, elongated and narrow structure (25 km long and 6–8 km wide). Its southern sector was occupied, until it was drained in the 1950s, by the shallow Hula Lake. The central part of the valley is covered by peat/lignite deposits deriving from marshes, while the northern part, slightly more elevated, is covered by alluvial soils and crossed by several streams that supply water to the whole Jordan Valley (Belitzky, 1987; Horowitz, 1979; Karmon, 1953)

Climate

The climate of the Hula Valley is warm Mediterranean due to its relative low elevation (*ca.* +70 m above msl). The maximum summer temperature is 40˚C, while during winter the temperature drops to below 0˚C. The mean annual temperature is 21˚C (Gat and Paster, 1974) and the mean annual precipitation 600–800 mm (Gat and Paster, 1975).

History of Research

Early archaeological surveys and excavations were carried out at the Gesher Benot Ya'aqov site between 1935 and 1968 by Garrod (Goren-Inbar and Belitzky, 1989 and references therein) Stekelis (1960) and Gilead (1968; 1970). These excavations encountered numerous difficulties, as well as intensive destruction of the northern parts of the site (Goren-Inbar and Belitzky, 1989 and references therein). Geological observations assigned the site's deposits to the Benot Ya'akov Formation (Horowitz, 1979 and references therein). The site's extent was estimated to be some 1.5 km along the River Jordan and its water-saturated deposits were considered to result from the proximity of the river. The influence of a shallow Pleistocene lake, the Paleo-Hula Lake, which preceded the River Jordan in this geographical area was not mentioned. Paleontological and lithic assemblages were described (Goren-Inbar and Belitzky, 1989 and references therein), but no mention was made of preservation of plant material in these Pleistocene deposits.

The Renewed Excavations

Geology

Lacustrine sediments and Lower Pleistocene basalt flows along the north-south-trending structural line known as the Jordan Lineament (Belitzky, 1987) have been affected by tectonic activity. Folding and faulting of these units have resulted in the formation of a narrow embayment (the Benot Ya'aqov Embayment) in the south-eastern part of the Hula Basin. The limnic-fluvial sediments of the Benot Ya'akov Formation were deposited in the embayment as well as in the Hula Basin *senso stricto*. Additional tectonic activity, including left-lateral movement on the north-north-east-trending structural line, has produced folding and faulting of these deposits. Continued subsidence of the Hula Basin has left the Benot Ya'aqov Embayment as the only location where the Benot Ya'akov Formation is exposed (Belitzky, 1987; Goren-Inbar and Belitzky, 1989; Goren-Inbar et al., 1992a).

The Benot Ya'akov Formation deposits are remnants of a freshwater lake system which existed in the Hula Valley and are now assigned to the Lower/Middle Pleistocene (Feibel *et al.*, 1998; Verosub *et al.*, 1998; Goren-Inbar *et al.*, in press). The deposition in the area currently under study took place on the southernmost part of the paleo-Hula Lake, in an embayment of the Hula Basin (Goren-Inbar *et al.*, 1992a,b; Feibel *et al.*, 1998).

The discovery of previously unknown exposures of the Benot Ya'akov Formation (Goren-Inbar and Belitzky, 1989; Goren-Inbar *et al.*, 1992a) initiated a multi-disciplinary research project which is aimed at the excavation, analysis and publication of all the Pleistocene evidence encountered and collected during seven field seasons (1989–1991; 1995–1997) (Belitzky *et al.*, 1991; Feibel *et al.*, 1998; Goren-Inbar *et al.*, 1991; Goren-Inbar *et al.*, 1992a; b; 1994; Goren-Inbar and Saragusti, 1996; Goren-Inbar, *et al.*, in press). During the course of these excavations (Figure 18.2) a wealth of plant material was recovered, comprising wood, bark, seeds, fruits and pollen grains.

The present study is concerned only with the macro-botanical remains. It is focused on preliminary data resulting from the analysis of wood and bark remains retrieved during the current multidisciplinary research of the Gesher Benot Ya'aqov site.

The Wood Assemblage of Gesher Benot Ya'aqov

Location and Stratigraphic Assignment of the Gesher Benot Ya'aqov Wood

Pieces of wood were found in the following contexts: 1) Eroding from sediments (Benot Ya'akov Formation) underlying the River Jordan. 2) In the Holocene floodplain deposits of the River Jordan. 3) In sediments which were mechanically quarried out of the geological trenches dug into the Benot Ya'aqov Formation. 4) In archaeological horizons, excavated from *in situ* contexts.

None of the wood segments originating from the Jordan flood plain deposits were kept as they might be mixed with Holocene material. Wood pieces originating from the mechanical excavation (JCB) of six geological trenches in the study area (Figure 18.3) were anatomically analysed, although it was frequently impossible to assign them to a specific layer among those exposed in the trench. *In situ* material recovered from the archaeological excavations and the walls of the geological trenches was mapped into the detailed spatial and stratigraphic framework of the site.

Excavation, Registration and Conservation of Wood

A grid, aligned with the Israel Grid, was applied to the study area to serve as a reference system for the spatial locations of finds originating in the excavation. A grid of one-meter squares was suspended over the area intended for excavation. The archaeological horizons, tilted by tectonic activities, were exposed along the strike and dip of the layer. Once the artifact-bearing horizons were laterally exposed, the finds, including the wood segments, were mapped and photographed, and their exact position recorded, before removal. The standard unit of excavation was the projection of a one-meter square on a tilted surface.

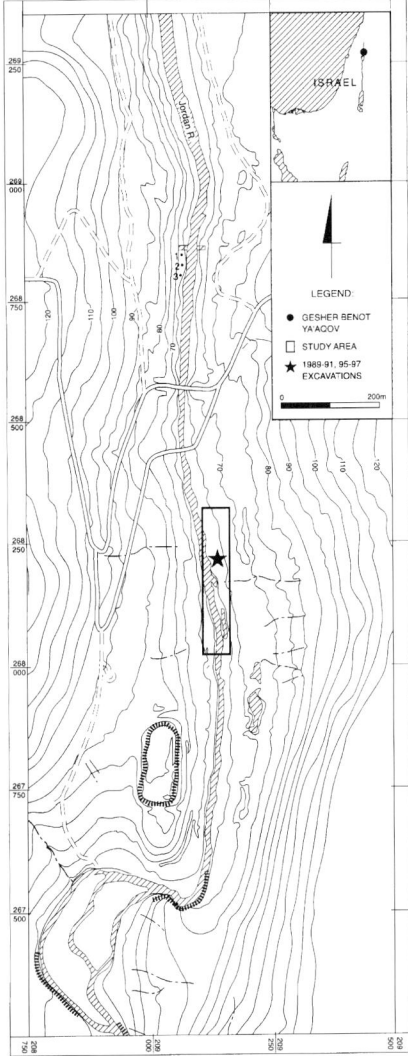

Figure 18.2. Location map of past excavations and present study area.

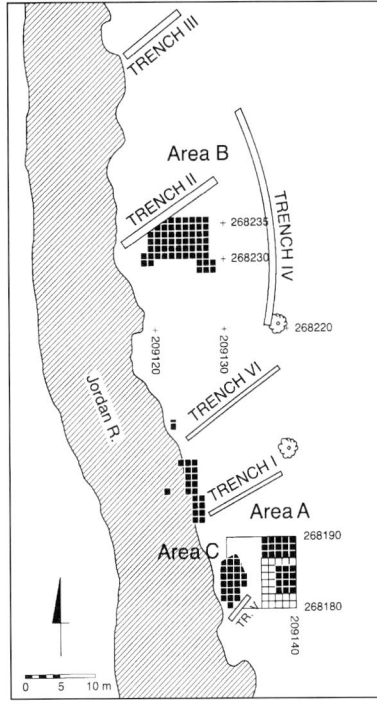

Figure 18.3. Detailed location map of recent excavations.

Each unit was further subdivided into four quadrants or subsquares. Excavation was carried out in each sub-square to an average depth of 5 cm. Coordinates were taken for all items larger than 2 cm, and each was also drawn on the maps.

Utilizing elevation controls (benchmarks) established along with the site grid, elevations above msl were determined for the various surfaces and for the excavated items, using a combination of surveying instruments and line levels. At the beginning of each phase of excavation, the exact elevation of the sub-square to be excavated was registered, with readings taken on two opposite points of the tectonically inclined area, one at the highest point (on the northeast) and the other at the lowest (on the southwest).

This procedure was carried out in order to record the tilt of the layers.

Pieces of wood were excavated with varying tools and methods. In coarse-grained sediments, wood was exposed with icepicks and dental tools and with the constant application of brushes. In finer-grained sediments, water was applied as an aid to excavation, though frequently dental tools were also used. It should be noted that the Gesher Benot Ya'aqov sediments are characterized by the presence of molluscs, including layers which consist of coquina and do not permit a single excavation method using very fine tools. The presence of hard materials within the soft deposits dictated the application of alternative strategies corresponding with the sediment type.

Of the organic material, only fragments of wood longer than 2 cm were documented (drafted, coordinated and sometimes photographed). The other organic items were usually invisible during excavation. On exposure of an archaeological surface, the finds, especially the basalt artifacts and the pieces of wood, were watered constantly to prevent deformation due to shrinkage of the woods, exfoliation and destruction of some of the basalt artifacts. On exposure, the wood samples were placed in plastic bags filled with water in order to prevent desiccation. Descriptive data on each wooden piece (location, size, layer, *etc.*) were written on Mylar paper with Mylar pencil and placed in the water together with the wood. The same descriptive data were also catalogued. Wood cleaning took place in the laboratory, as well as insertion of special tags carrying the serial catalogue number for each piece or group of pieces. The longest wood pieces had to be treated differently in the field, since their transportation in water proved to be destructive, causing many fractures and splitting of the brittle ancient pieces. The packing procedure included placing each piece (after moisturizing) inside a long plastic sleeve together with a plastic irrigation pipe tilted upwards and the serial catalogue number. The top of the sleeve and the pipe were closed together with masking tape and placed inside an elongated cardboard box. Styrofoam was then sprayed on one face of the sleeve, and after drying the package was turned over and the same procedure carried out on the other side. A catalogue number was inserted into the foam. After drying the package was ready for transportation and needed watering (through the pipe) only once every few days.

Wood pieces immersed in water and kept in plastic bags and those cocooned in Styrofoam were left untouched until anatomically sampled and ready for the beginning of the long conservation treatment.

Preparation of Wood for Anatomical Examination

Wood segments originating from the excavations and the trenches were examined anatomically. The wet segments were sectioned by hand with a razor blade, in cross, longitudinal radial and longitudinal tangential directions, and the thin sections were examined under the light microscope. Whenever needed the segments were cleared with a solution of 4% NaOH. The anatomical features of the wood segments were compared with slides of present-day wood and with wood catalogues (Fahn *et al.*, 1986; Schweingruber, 1990).

Results

The wood sample originating in the Gesher Benot Ya'aqov excavations consists of all the segments recovered during the seven field seasons (1989–1997). Table 18.1 presents the inventory of wood segments and the number of identified and unidentified segments. The total number of segments counted during sampling procedure for wood anatomy analysis was sometimes higher than that encountered during the process of excavation and initial cataloguing. This difference results from fragmentation caused by fissures and breaks, both visible and invisible, during exposure. The fissured/broken segments had been "glued" together, forming a single item, by fine clays and silts which disintegrated into many fragments when immersed in water for storage and conservation purposes. Thus, after anatom

Table 18.1. Number of wood segments excavated and those botanically analysed.

Type of segment	N
Segments recovered at GBY excavations	1568
Segments after conjoining	1403
Segments botanically examined	916
Segments botanically identified	638
Segments botanically identified as bark	97
Segments botanically unidentified	181

ical sectioning, many small pieces that could not be analysed were discarded.

A substantial number of wood fragments were conjoined, either in the preliminary registration phase or during that of anatomical sampling. Wood segments were conjoined from different pieces, resulting in a different sum of segments from that initially recorded in the field.

The following data, analyses, results and discussions are limited only to those segments which underwent anatomical analysis.

A total of 916 specimens were anatomically examined. Of these, 181 wood segments (excluding bark) could not be identified for the following reasons (Table 18.1):

a. Desiccation of the waterlogged wood at some period before sectioning, after excavation due to unfavorable preservation conditions or at some time in the past before excavation. The extent to which the pieces of wood were shrunk and distorted varied in accordance with their anatomical features. Wood with thin cell walls, for example, is more liable to shrinkage, while the direction of the cellulose microfibrils in different cells and cell wall layers may be responsible for twisting.

b. The extent of wood preservation also varied between species independently of the effect of desiccation. This could have resulted from the wood's response to other external conditions like pressure or biological agents, which depends on the wood structure and chemical composition of the walls and lumen of cells.

c. In some of the specimens roots of contemporary plants have penetrated into the waterlogged wood, partially obliterating its structure. In all the above cases gross structure or more delicate details like fine spiral thickenings, the type of pits or the presence of crystals may be hard to distinguish. Identification therefore may become impossible or uncertain.

d. Some of the specimens were young branches with very few growth rings. The wood structure in the first growth rings may be different from that of more mature wood, and therefore uncharacteristic.

e. In ring-porous wood, narrow growth rings consisting only of large vessels of early wood are often produced in years of drought. If the section includes only such rings, the wood may appear diffuse-porous. Identification may therefore be erroneous unless a wider growth ring with narrow vessels can also be distinguished in the section. Wood identification according to its anatomy and frequencies of the identified taxa are presented in Table 18.2.

A total of 27 genera of trees and shrubs and a few woody climbers were identified. They comprised the bank vegetation and dryland vegetation in the Hula Plain and adjacent regions in the Pleistocene. The bank vegetation includes *Fraxinus syriaca*, *Salix* spp., *Populus euphratica*, *Ficus carica*, *Hedera helix*, *Myrtus communis*, *Nerium oleander*, *Ulmus* and *Vitis*. The remaining genera belong to the dryland vegetation. Almost all the identified genera are also found in the region today, though some of them in very different proportions. Thirteen segments of one type of wood could not be identified, either by comparison with available wood samples and consulting wood atlases or by the GUESS computerized database program (Wheeler *et al.*, 1986). This appears to be an extinct species, at least in the region, not found in Europe, the Middle East and North Africa.

Discussion

The vegetation of the Hula Plain has undergone drastic changes through the intervention of man and does not represent the primary vegetation that prevailed during the Pleistocene. The greater part of the area today consists of cultivated and disturbed areas, and the vegetation accordingly consists mainly of segetal and ruderal plants, between which remnants of the primary vegetation are dispersed.

Among the wetland vegetation only scattered trees of *Fraxinus* and *Ulmus* are found today, *Fraxinus* in the Upper Galilee, Hula Plain, Beit Shean Valley and Golan, and *Ulmus* in the Lower Galilee. Their relatively high percentage in the excavation, especially that of *Fraxinus*, further confirms the prevailing assumption that in the past these two species constituted the primary park forest on wetland in the Hula Valley (Zohary, 1959). Zohary suggested that these species penetrated from the north in one of the more humid periods of the Quaternary.

Among the dryland vegetation of the Hula Valley and on the flanks of the Golan and Gilead, on the basalt of Korazim, and reaching up to the Dan Valley there are variously dispersed remnants of *Quercus ithaburensis* and *Pistacia atlantica* (Weisel *et al.*, 1978). Zohary (1959) suggested that these two species constituted a park forest as the primary climax vegetation in these regions.

Lower Palaeolithic paleobotanical data originating from Levantine sites are scarce, sporadic, accidental and mainly comprise pollen and fruits [Ubeidiya, Umm Qatafa, Tabun

Table 18.2. Distribution of identified wood and unidentified bark.

Taxonomic Identification	Common name	N	%
Amygdalus (korschinskii?)	Almond	1	0.15
Cedrus	Cedar	2	0.29
Cerasus	Cherry	1	0.15
Cerasus?		1	0.15
Crataegus	Hawthorn	4	0.59
Ficus carica	Fig	2	0.29
Ficus carica?		1	0.15
Fraxinus syriaca	Ash	245	36.08
Fraxinus?		28	4.12
Hedera	Ivy	1	0.15
Jasminum	Jasmine	1	0.15
Jasminum?		1	0.15
Juniperus	Juniper	3	0.44
Lonicera	Honeysuckle	5	0.75
Lonicera?		2	0.29
Lycium	Box-thorn	1	0.15
Lycium?		3	0.44
Myrtus	Myrtle	7	1.03
Myrtus?		2	0.29
Nerium?	Oleander	1	0.15
Olea europaea	Olive	45	6.63
Olea?		15	2.21
Periploca?	Silk-vine	1	0.15
Pistacia	Pistachio	4	0.59
Pistacia?		3	0.44
Pistacia atlantica		25	3.68
Pistacia (atlantica?)		2	0.29
Pistacia atlantica/vera		8	1.18
Pistacia (palaestina?)		1	0.15
Pistacia vera		3	0.44
Pistacia (vera?)		2	0.29
Populus	Poplar	4	0.59
Populus?		1	0.15
Pyrus	Pear	6	0.88
Pyrus?		2	0.29
Quercus calliprinos	Oak	27	3.98
Quercus (calliprinos?)		3	0.44
Quercus ithaburensis		26	3.83
Quercus (ithaburensis?)		2	0.29
Quercus ithaburensis/ calliprinos		11	1.62
Rhus pentaphylla/ tripartita	Sumac	2	0.29
Rhus pentaphylla/tripartita?		1	0.15
Rhus?		1	0.15
Rosaceae, Prunoideae	Rose family	1	0.15
Salicaceae?	Willow family	2	0.29
Salix	Willow	26	3.83
Salix?		8	1.18
Ulmus	Elm	14	2.06
Vitis	Vine	2	0.29
Ziziphus/Paliurus	Jujube/Christ-thorn	2	0.29
Unknown tree		13	1.91
Bark	bark	97	14.28
Bark?		7	1.03
Total		679	99.98

(Jelinek *et al.*, 1973; Horowitz, 1979; Neuville, 1951)].

The paleobotanical wood assemblage of Gesher Benot Ya'aqov is unique and hence sheds light on aspects not previously treated. It is the first record of unfossilized wood material that is assigned to the Lower/Middle Pleistocene times, in addition to yielding other paleobotanical data unmatched at any other site in Israel and adjacent countries. The existing data are not comparable at present to the record of any other archaeological locality in terms of paleo-climate, paleomagnetic chronology and depositional sequence.The data originating in the site are consequently isolated and discontinuous, and additional data are needed before an attempt at wider correlations and conclusions can be carried out. Nevertheless, the unique data from Gesher Benot Ya'aqov present an opportunity to gain a better understanding of the woody vegetation at the site and its vicinity and the type of environment in which the Acheulian hominids existed.

This paper attempts, for the first time, to present a paleoenvironmental reconstruction of the Lower/Middle Pleistocene vegetation in the Levant. The foundations of this reconstruction lie in the results of a study of wood anatomy as revealed by the Gesher Benot Ya'aqov plant assemblage. The 27 genera of trees, shrubs and woody climbers are the primary data set for a basic understanding of the paleobotany of the Dead Sea Rift zone and its adjacent areas. Evidently, these results are partial and should be supported by studies derived from complimentary fields of paleobotany such as those of fruits and seeds, pollen and phytoliths. In order to gain a better understanding of the paleoenvironment, these botanical aspects should be integrated with additional data derived from other disciplines such as paleontology, geology and prehistoric archaeology. The partial and preliminary results discussed above, however, contribute much to our understanding of the type of vegetation and its associations during the Acheulian occupation of the Israeli northern Rift Valley. We consider this data set to be both informative and thought-provoking. The continued presence of most of the assemblage genera since Lower/Middle Pleistocene times until the present reflects stability of the woody *Flora palaestina* and its resistance for many thousands of years. Moreover, it indicates that this assemblage was not affected by the continued presence, during hundreds of thousands of years, of the hunter-gatherer groups that populated these parts of the eastern Mediterranean region.

Future studies of the Gesher Benot Ya'aqov site should be concerned with the implications of the wood, shrub and climber plant data and should attempt to gain a better understanding of the meaning of this continuity and almost lack of change (except for the extinction of one genus) from the perspective of the hominids' behavior .

Acknowledgments

The field work of this study was supported by the LSB Leakey Foundation and the National Geographic Society. Laboratory analysis and conservation were supported by the Irene Levi-Sala Care Archaeological Foundation, the Israel Science Foundation and the Hebrew University of Jerusalem. We thank the GBY staff and participants for their dedication and are especially grateful to O. Cohen, I. Saragusti, H. Taub, O. Marder, M. Wiseman, and P. Enamorado-Rivero. Figure 1 was drawn by G. Hivroni and Figure 2 by C. Douzil.

We thank Prof. B. Purdy for making the conference possible and inviting us to participate.

References Cited

Belitzky, S. 1987. Tectonics of the Korazim Saddle. M. Sci. Thesis, Hebrew University, Jerusalem.

Belitzky, S., N. Goren-Inbar, and E. Werker. 1991. A Middle Pleistocene wooden plank with man-made polish. Journal of Human Evolution 20:349–53.

Fahn, A., E. Werker, and P. Bass. 1986. Wood Anatomy and Identification of Trees and Shrubs from Israel and Adjacent Regions. Jerusalem: The Israel Academy of Sciences and Humanities.

Feibel, C.S., *et al.* 1998. Gesher Benot Ya'aqov, Israel: new evidence for its stratigraphic and sedimentologic context. Journal of Human Evolution 34.3:A7.

Gat, Z., and Z. Paster. 1974. Agroclimate of the Golan Heights: The rain. Bet Dagan: Israel Meteorologiscal Service.

Gat, Z., and Z. Paster. 1975. Agroclimate of the Golan Heights: Temperatures and Relative Humidity. Bet Dagan: Israel Meteorological Service.

Geraads, D., and E. Tchernov. 1983. Femurs Humains du Pleistocene Moyen de Gesher Benot Ya'aqov (Israel). L'Anthropologie 87.1:138–141

Gilead, D. 1968. Gesher Benot Ya'aqov Hadashot Archeologiot 27:34–35.

Gilead, D. 1970. Handaxe industries in Israel and the Near East. World Archaeology 2:1–11.

Goren-Inbar, N. 1998. Gesher Benot Ya'aqov: the Acheulian Cultural Sequence. Journal of Human Evolution 34.3:A8.

Goren-Inbar, N., and S. Belitzky. 1989. Structural position of the Pleistocene Gesher Benot Ya'aqov site in the Dead Sea Rift Zone. Quaternary Research 31:371–376.

Goren-Inbar, N., and I. Saragusti. 1996. An Acheulian biface assemblage from the site of Gesher Benot Ya'aqov, Israel: indications of African affinities. Journal of Field Archaeology 23:15–30.

Goren-Inbar, N., Z. Lewy, and M.E. Kislev. 1991. The taphonomy of a Jurassic bead-like fossil from an Acheulian occupation at Gesher Benot Ya'aqov. RAR 8.2:133–136.

Goren-Inbar, N., *et al.* 1992a. New discoveries at the Middle Pleistocene Gesher Benot Ya'aqov Acheulian site. Quaternary Research 38:117–128.

Goren-Inbar, N., *et al.* 1992b. Gesher Benot Ya'aqov – the "bar": an Acheulian assemblage. Geoarchaeology 7.1:27–40.

Goren-Inbar, N., *et al.* 1994. A butchered elephant skull and associated artifacts from the Acheulian site of Gesher Benot Ya'aqov, Israel. Paléorient 20.1:99–112.

Goren-Inbar, N., *et al.* Technological Innovation and Biotic Exchange in an Acheulian Site on the Out-of-Africa Corridor. (in press).

Horowitz, A. 1979. The Quaternayt of Israel. New York: Academic Press.

Jelinek, A., *et al.* 1973. New excavations at the Tabun Cave, Mt. Carmel, Israel, 1967–1972; a preliminary report. Paléorient 1.2:151–183.

Karmon, Y. 1953. The settlement of the Northern Huleh Valley since 1838. Israel Exploration Journal 3:4–25.

Melamed, Y. 1997. Reconstruction of the landscape and the vegetarian diet at Gesher Benot Ya'aqov archaeological site in the Lower Paleolithic period. M.Sc. Thesis Bar-Ilan.

Neuville, R. 1951. Le Paleolithique et le Mesolithique du desert du Judee.

Schweingruber, H.F. 1990. Anatomy of European Woods. Stuttgart: Haupt.

Stekelis, M. 1960. The Paleolithic deposits of Jisr Banat Yaqub. Bulletin of the Research Council of Israel G9:61–87.

Tchernov, E. 1986. Les Mammiferes du Pleistocene Inferieur de la Vallee du Jordain a Oubeidiyeh. Memoires et Travaux du Centre de Recherche Francais de Jerusalem. Paris: Association Paleorient.

Tchernov, E. 1988. The paleobiogeographical history of the southern Levant. The Zoogeography of Israel. Eds. Y. Yom-Tov and E. Tchernov. The Hague: Dr. W. Junk Publishers. 159–250.

Tchernov, E. 1992. Biochronology, paleoecology, and dispersal events of hominids in the southern Levant. The Evolution and Dispersal of Modern Humans in Asia. Eds. T. Akazawa, K. Aoki and T. Kimura. Tokyo: Hokusen-sha: 149–188.

Thomas, H. 1985. The early and middle Miocene land connection of the Afro-Arabian plate and Asia: A major event in hominoid dispersal? Ancestors: The Hard Evidence. Ed. E. Delson. New York: A.R. Liss: 42–50.

Tobias, P. 1966. A member of the genus homo from Ubeidiya. Jerusalem: Israel Academy of Sciences and Humanities.

Verosub, K.L., *et al.* 1998. Location of the Matuyama/Brunhes boundary in the Gesher Benot Ya'aqov archaeological site. Journal of Human Evolution 34.3:A22.

Warnock, P. 1998. From plant domestication to phytoliths interpretation. The history of paleoethnobotany in the Near East. Near Eastern Archaeology 61:238–252.

Weisel, Y., G. Pollak, and Y. Cohen. 1978. The Ecology of Vegetation of Israel. Tel-Aviv: University of Tel-Aviv.

Wheeler, E.A., *et al.* 1986. Computer-Aided Wood Identification. North Carolina Agricultural Research Service, North Carolina State University.

Zohary. 1959. Geobotany. Merhavia: Sifriat Hapoalim (in Hebrew).

19. Ancient Wooden Objects and Structures in Oxbow Peat Bogs of the European Northeast (Russia)

Grigory M. Burov

Introduction

Wetland sites have inventories of organic materials which are incomparable with those of sites where often only ceramics and stone are present. In some cases one wet site can give more information than hundreds of typical terrestrial sites concerning the life of prehistoric peoples.

The present paper is devoted to specific wetland sites situated in the Northern Dvina basin: Vis I and Vis II, Yavronga I and Marmugino. There are no other known wet sites there. To the east of the Ural mountains, in the Shigirsky and Gorbunovsky peat bogs, wooden artifacts of the Neolithic and Eneolithic Age were discovered in the early 20th century.

Neolithic peat bog settlements have been excavated in the Lake Onego region (pile dwellings on the Modlona River) (Bryusov 1951), in the Pskov district (pile dwellings Usvyaty IV and Naumovskoe) (Miklyaev 1969), in Latvia (Sarnate, Abora I) (Vankina 1970) and Lithuania (Sventoji A, B, 2B, 3B, 9, 23, Zemaitiske I and II (Rimantiene 1979). Mesolithic artifacts made of organic materials are known from Nizhnee Vertetye (Foss 1941) and Nizhnee Veretye I (Oshibkina 1989) in the Lake Onego region. Wooden structures and objects have been found in excavations of Medieval towns of the 8th to 15th centuries at Staraya Ladoga, Novgorod (Kolchin 1968), Beloozero (Golubeva 1973), and David-Gorod (Earwood 1993). Badly preserved wooden artifacts have also been discovered in barrows of the East European steppe belt.

Materials of the oxbow peat bogs have been introduced into archaeology as a result of the author's investigations of 1960–1968 (Komi branch of the USSR Academy of Sciences). Yavronga I, Vis I and Vis II can be considered as both terrestrial and as wet sites. People lived on a sandy dry hillock but artifacts penetrated into adjacent oxbow deposits. Only Marmugino (oxbow peat bog) is an exception as a place of fish weirs outside of dwellings.

The materials from Vis II, Yavronga I and Marmugino are practically unknown in publications of Western Europe and North America, while those of Vis I have appeared in many books published throughout the world. In this paper, the principal data about organic finds at sites of the European Northeast are summarized.

Settlements with Oxbow Peat Bogs

A Method for Prospecting of Ancient Wooden Objects and Structures

Organic artifacts were found by using a specific strategy to locate ancient wooden objects and structures in peat bogs formed by infilled oxbow lakes. Wooden objects illustrated in publications throughout the world were found mainly in settlements which are typical wet sites, often peat bogs. Such settlements seldom leave surface traces and are usually discovered by industrial workers. Prospecting for wooden artifacts can yield good results at settlements on dry sandy hillocks and terraces. Such sites may be discovered easily. They often include oxbow lakes which have become or are in the process of becoming peat bogs. Ancient people preferred to settle along rivers. Thus, when abandoned meanders ("cut offs") gradually filled in and were converted into bogs, organic objects inevitably found their way into these back swamps where they remained intact owing to the preservative properties of water, sapropel and peat.

The locations of oxbow peat bogs are not difficult to find in most cases due to the fact that they occur in river floodplains. Outcrops of oxbow deposits rising at least partially above the water level can often be observed in river cliffs. In the absence of outcrops, prospecting along small and medium-sized rivers with oxbow meanders is simpler than prospecting on large rivers with oxbow lakes of a different genetic type called "channals". A concave edge of a terrace points to the presence of an oxbow peat bog. Such peat bogs can be delimited by digging or augering along the concave edge.

When wooden objects find their way to the bottom of an oxbow, they are preserved in the sapropel and peat. Some objects not carried away by the current may sink before a meander is cut off. Parts of fish weirs, of course, remain there because they have been driven into the bottom.

There will be no cultural layer in the oxbow deposits, however, if people settled on the terrace after the peat bog was covered by mineral overburden, or if a settlement had ceased to exist before the river migrated into its location. Success in finding sites is more likely if a settlement which existed intermittently over a long period is explored. But the duration of occupation is of great importance also; people who have left only a few flints or sherds of pottery would hardly have left enough wooden objects to be found in the course of exploratory excavations.

From a practical point of view, oxbow peat bogs containing an abundance of wooden objects situated near lake outflows are the most suitable for prospecting because there is very little floodplain alluvium and no overburden to hamper access to the ancient deposits.

General Characteristics of the Oxbow Peat Bogs with Cultural Remains

An ideal place for locating prehistoric wooden artifacts is the area of Lake Sindor (Vychegda basin, Knyazhpogost region). Two bow-shaped peat bogs relating to the rivers Vis and Simva were discovered in 1960 and 1962 (Figure 19.1). Vis is an outlet of Lake Sindor. The oxbow peat bog at Vis I was discovered on the first day of prospecting excavations. We were going to stop work at this place after three days yielded no materials that were unquestionably accepted as artifacts. Then four nice objects came to light almost simultaneously.

An abundance of wooden artifacts have been found in Vis I and Vis II peat bogs. The bogs are located at the concave edges of the higher sandy ground (hillocks) on which two multi-period settlements existed for a long time. The peat bogs are 16–20 m wide; the length of the former is 150 m, and that of the latter is 250 m; their maximum depths are 1.6 m and 3.5 m respectively. During excavations of 575 m² at the peat bog of Vis I from 1960 to 1967, a stone industry, 168 organic objects (wood, bark, and sedge), and 39 pieces of worked wood were found. These finds date back to the period 8300–7000 BP. Five uncalibrated radiocarbon dates for the bog were obtained: 8080±90 (LE–776), 7820±80 (RUL–616), 7150±60 (LE–684), 7090±80 (LE–685), and 7090±70 (LE–713) BP. When the first wooden objects were recovered from the peat bog, the existence of a Mesolithic complex at this site was not suspected. That is why the radiocarbon dates surprised us greatly. Later, we obtained additional proof for the great age of the bog.

A number of fish weirs were discovered in the peat bog of Vis II together with a large amount of Vanvizdino culture pottery (AD 400–500), birch bark right-angled sheets, wooden and bark objects as well as divers bone artifacts and glass beads (Burov 1996b). This peat bog differs from that of Vis I. Vis I contains artifacts both in sapropel and peat buried under clayey deposits (1.1–1.3 m thick) associated with a transgression of Lake Sindor (Figure 19.1). Vis II has cultural remains only in sapropel – mainly pieces of fish weairs above the different objects. It is possible that modern artifacts might still be incorporated into peat of Vis II but the sandy hillock of Vis II has not been inhabited since the 6th century AD.

In 1966 an infilled oxbow lake in the village of Marmugino (Veliky-Ustyug region, Vologda district) on the Yug River revealed Neolithic fish weirs. However, in 1966–1967 we failed to dig close to the concave edges of the sandy terraces where the Eneolithic settlements of Marmugino and Kiroksa (Kholmogory region, Arkhangelsk district) are situated (Burov 1988: Fig. 1; 1990b: Fig. 1). In both cases peat bogs were discovered but they are of quite different times and therefore without artifacts.

In 1968 along the bow-shaped edge of the dry sandy hillock which belongs to the site of Yavronga I (Pinega basin, Karpogory region, Arkhangelsk district), an oxbow peat bog containing sticks with worked ends (apparently from fish weirs similar to those at Vis II) were found along with a fragment of a simple bow dated by C14 to the Iron Age. A 26 m² area was excavated (Burov 1974:55–58). Besides remains of fish weirs, there were practically no wooden objects recovered because ancient people lived on the sandy hillock only during the Stone Age.

Mesolithic-Neolithic wooden artifacts were found in oxbow peat bogs as well as in other areas of Europe: Plekhanov Bor (Burov 1988:157) and Zamostye II, at Podzorovo and Sventoji 9, at Pobiel 10 in Poland (Bagniewski 1990:160–166) and Noyen in France (Mordant & Mordant 1992). So these bogs deserved to be found and studied.

Vis I

Wood, bark, and sedge artifacts of Vis I include objects of different kinds.

Bows, Arrows, Spears and Throwing Clubs

Hunting equipment predominates in the wooden inventories of Vis I and Vis II. In a series of 30 bows from Vis I (Burov 1981), only one example is made of wood of a deciduous (birch or other leaf-bearing) species. It has a narrowing and

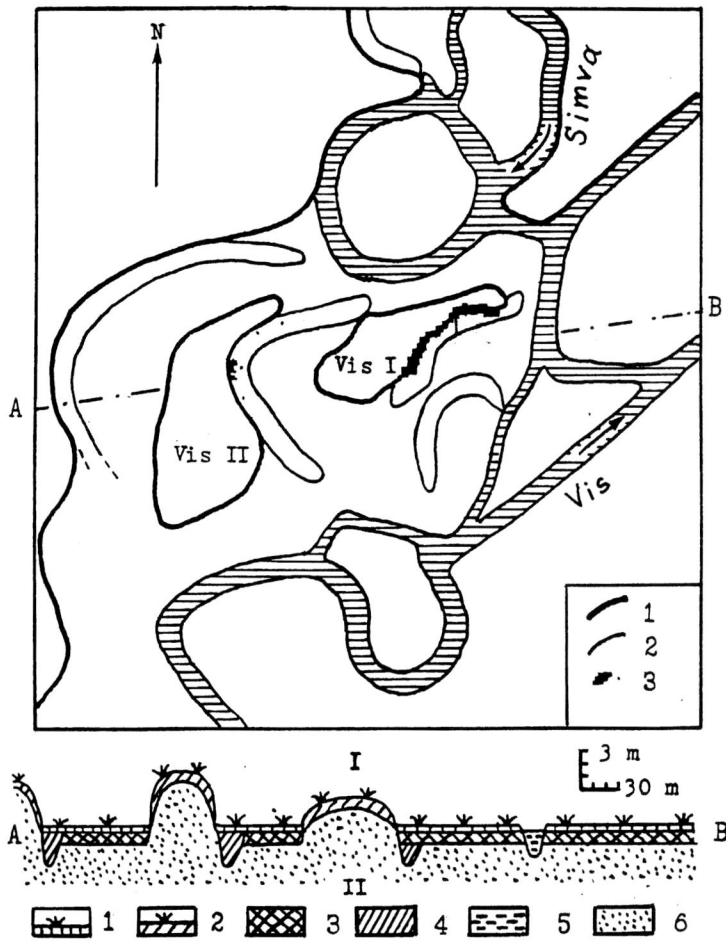

Figure 19.1. Sites of Vis I and II. Vis I plan: (1) Outline of sandy hillock, (2) outline of oxbow peat bog, (3) trench and digging. Vis II profile: (1) peat, (2) cultural layer, (3) clayey lake deposits, (4) oxbow peat bog, (5) water, (6) sand.

thickening in the middle part (Figure 19.2:1) and seems to be similar to the Mesolithic so-called flat bows from Holmegaard IV in Denmark. The rest are made of softwood and are asymmetrical, massive and usually have an opening on the upper end for fastening and regulating a string. They join the group of long bows and form the Vis type in manual and path variants. The manual ones have a carved projection on the upper end and as a rule one or two side cuts on the middle portion for passing arrows (Figure 19.2:2,3). These bows, 134–156 cm long in the straightened position, have no archaeological parallels but are similar to materials of the 17th century AD from Brazil (Meyer hnd). This fact speaks of the technical expediency of such a type. The bows of the second variant are made from young trees without branches and are ca 350 cm long with an opening, notch (cut around the circumference) or head on one end (Figure 19.2:4). It may be supposed that they were placed on animal trails as cross bows. Big, strong bows were used

in this way in Siberia (Popov 1937:165). Our experiments with a specially made copy of a gigantic bow from Vis I testify that such bows were suitiable for shooting heavy arrows for a distance of tens of meters.

Arrows sometimes have a conical wooden head. There are also in the collection an arrow of one piece of wood and a similar dart (the head is spindle-shaped) as well as plenty of shafts made of laths and beams and having a diameter between 0.7–1.5 more often than between 1.5–3.1 cm. Some shafts have a sharpened or split end for a point or with double grooves for flint inserts (Burov 1993a).

The spear is made completely of wood with the back part thinner than the fore part. The L-shaped and straight type throwing clubs are, according to ethnographic data, weapons used to hunt waterfowl (Franz 1928). There is also in the inventory a short dart or throwing club which had a skew attached flint or bone head.

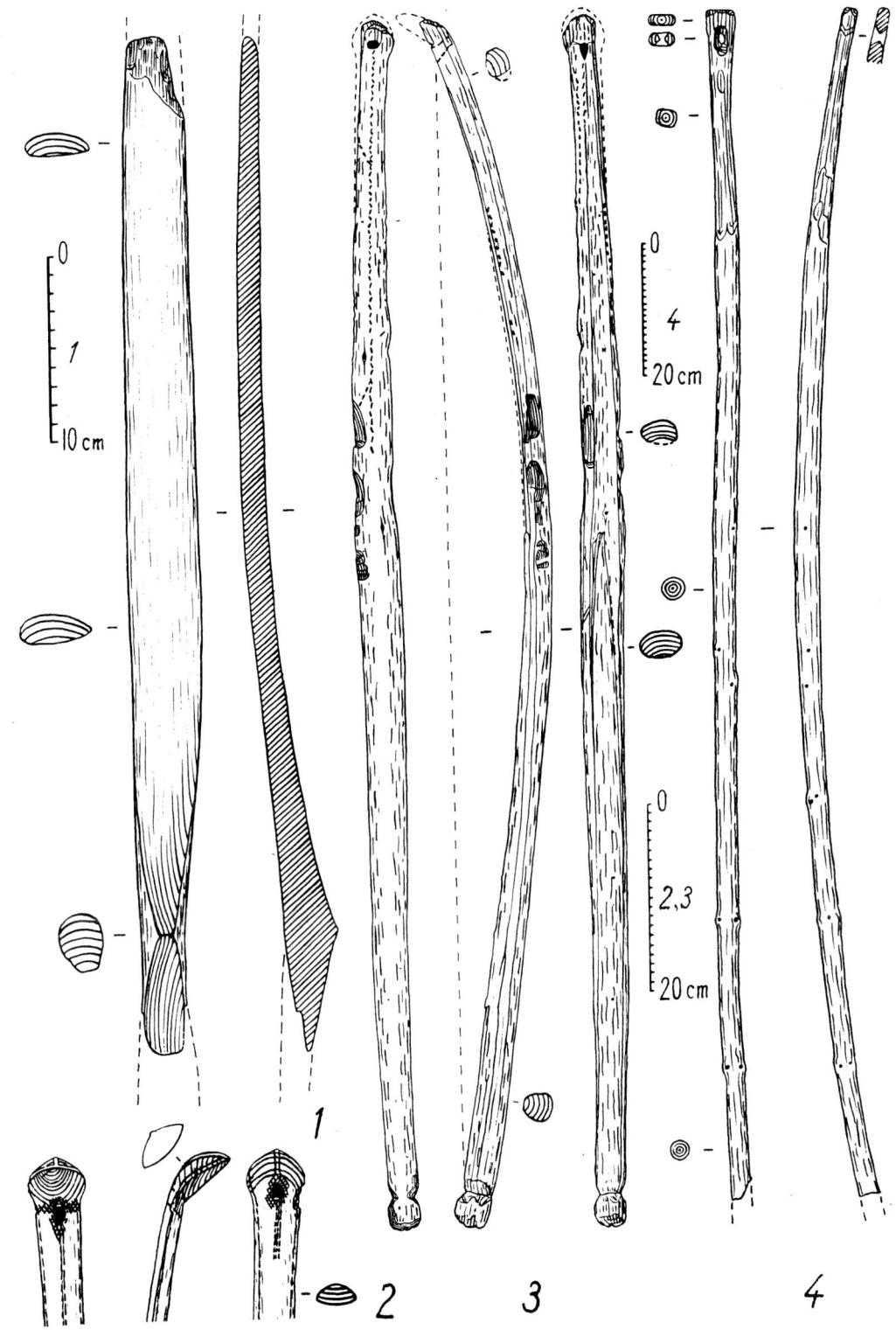

Figure 19.2. Vis I. Bows of the Holmegaard (1) and Vis (2–4) types (nos. 164, 97, 106, 26). The concentric circles show disposition of tree rings but not their quantity. The inventory number points to the stratigraphic position of the object (Burov 1990a:Abb.3).

Fishing Gear

Equipment needed for fishing consists of grass netting, a fishing net float of pine bark, a wooden disc for a fish driver, a fragment of a fish trap (?), a series of hoops (Burov 1992), and three pieces of specific leisters (prongs). The netting was made of a sedge fiber cord (Figure 19.3:6, 6a). The distances between the knots are 4.5–5.5 cm because the net was intended to catch pike. The only similar net is from the Karelian Istmus at the site of Antrea-Korpilahti, where remains of a complete fishnet, 27 m long and 3 m wide, were recovered. Judging from the finds at Antrea-Korpilahti, the float has a hole on one end and is very oblong (Figure 19.3:5). Fish were driven into the net with the help of a splashing stick in the form of a disc with an opening in the center for the handle. Such discs have been found at sites of many periods from the Mesolithic to Medieval times (Figure 19.8:9).

An object (Figure 19.3:7) made of thin (0.1–0.2 cm) pine laths 0.5–0.8 cm wide was tied by plant cross strips (plaits) (Burov 1998:62) can, of course, be a mat, but the small size of the laths and short distances between separate cross strips (1.0–1.5 cm) allow us to assume that this object was a fish trap.

In the collection there are fragments of eight hoops with long slanting cuts at the ends; the cuts have been tied together by two (Figure 19.4:1). Half of them were perforated. Their only analog comes from the Mesolithic site of Nizhnee Veretye (Burov 1992: Fig. 2:1). These objects were probably intended for making landing nets or fish traps of netting, and the holes served for fastening the net with a cord inside the hoop in order to prevent rubbing the cord against the edges of an ice hole or a fish surround. Such a practice is known by the Finns on the river Kymi (Sirelius 1906:161, Fig. 229).

The leisters consisted of three pieces of wood. Their S-shaped side prongs, with a barb and, as a rule, flat surface of the upper portion from within (Figure 19.3:1), have analogies at Mesolithic-Neolithic sites of Denmark (Andersen 1955:57–59, Fig. 19; Fig. 26), northern Germany, Sweden, Latvia (Vankina 1970: Plate II:3–5), and Lithuania (Rimantiene 1979: Fig. 18:3–5). A portion of such a leister of three wooden sticks lashed together by cord has been found at the underwater site of Skjoldmaes, Denmark (Skaarup 1995: Fig. 3). Judging from similar Eskimo fish spears (Greenland) Brinkhuizen 1983: Fig. 3:B), the middle prong in the shape of a point (attached to the lower, usually wide, portion of the shaft) served to run through a fish while the side prongs were intended for keeping it in fixed position with the help of their barbs. But such leisters were fit for fishing only on river sections with silt or on soft sea floors. Originally described side prongs were considered by scholars as a kind of thorwing club.

Means of Transportation

A paddle (Figure 19.3:4) was made of softwood (Burov 1996a). The blade has parallel sides, distinct shoulders in the place where the blade passes into the handle, and a pointed end. The object could be used not only for paddling but also for pushing off when moving upstream in shoals. The paddle is analogous to that from Nizhnee Veretye but synchronous paddles of Western Europe are of quite another type; they have an oval blade. A canoe of the 8–7th millennium BP was found in France (Mordant & Mordant 1992) and leads one to believe that the paddles were used in boats.

At the site of Vis I, fragments of runners from approximately 19 sledges belonging to two types were found (Burov 1999). The runners of the first type were trough-shaped with a slight flange along the left or right side and two pairs of horizontal or skew holes at the head and at some distance from the rear end. The type was named Heinola because such a Mesolithic runner came to light previously near Heinola in southern Finland. It seems the sledge consisted of two runners tied together with straps in such a way that the flanges were situated on the outside of each (Figure 19.5:1).

The second (Vis) type includes flat, asymmetrical artifacts with a flange at one of the sides and several pairs of vertical holes for planks connecting two runners forming the sledge The toboggan of the Canadian Indians presents a parallel example (Lips 1955) (Figure 19.5:2,3). The sledge of the Vis type represents one of the highest technological achievements of the Mesolithic.

The skis also are divided into two types, Vis and Veretye, based on the fore parts which were found (Burov 1989b:393–397). The objects of the first type, made from hardwood, appear to have a gradually tapering projection (Figure 19.4:2) and there are flanges along both edges on the upper surface. The absence of holes in the fore part, symmetry of the artifacts, and the flat lower surface allow them to be interpreted as skis rather than sledge runners. These objects have no analogies and are the oldest skis known from archaeological records.

The fore part of a Veretye type ski, made from softwood, is not sharpened and has a concavity occupied by a carved projection which is situated mainly on the lower surface. The front portion has a pair of holes which probably served to attach a belt similar to a horse's bridle and used when going downhill to keep the skies on the feet. An Arabian traveler of the 12th century, Abu-Hamid, wrote about such a belt on skis he had seen in the area of the northern Urals. The ski of the Veretye type has no flange along the sides but was provided with a wave ridge along the middle of the upper surface.

The carved projection, bevelled backwards, was in the

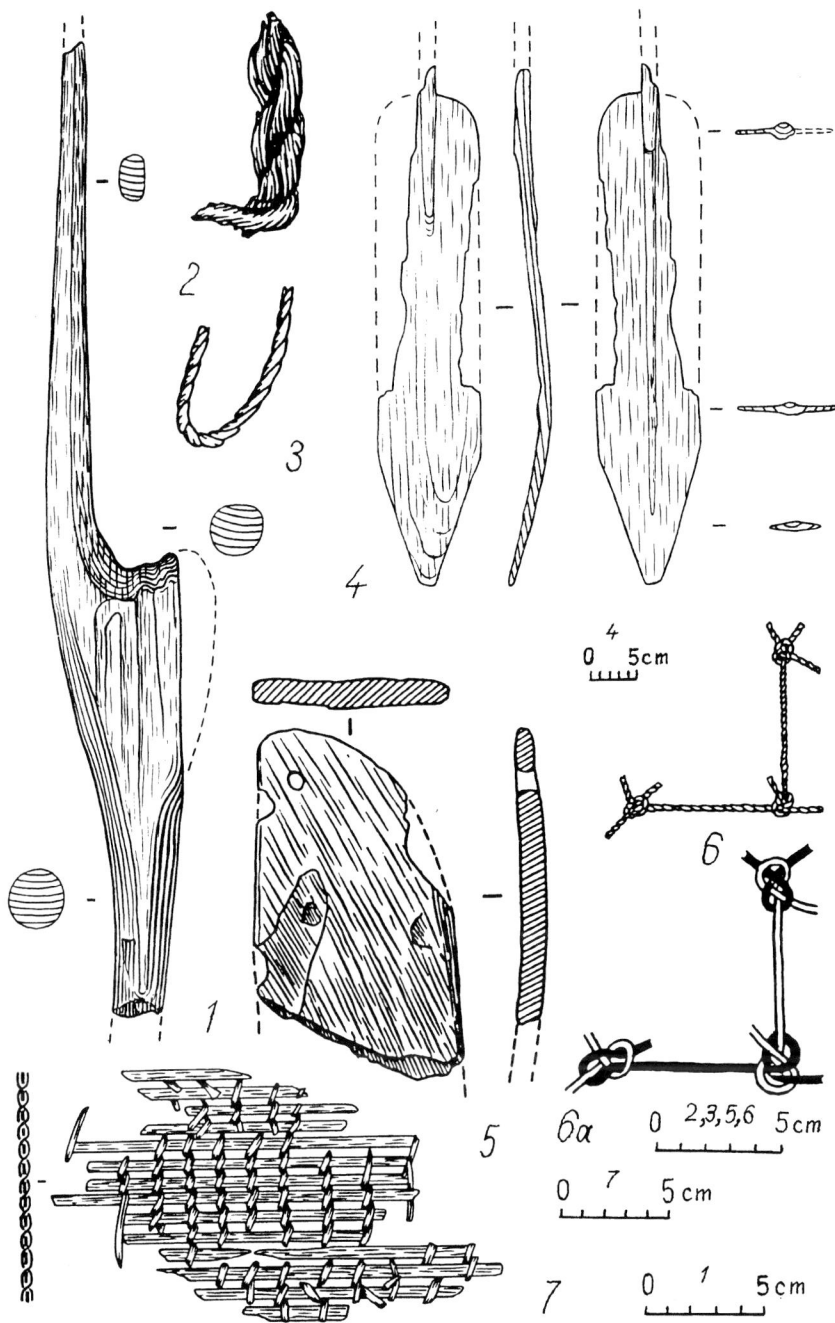

Figure 19.3. Vis I. Wooden (1,4), grass (2,3,6,6a), pine bark (5) and composite (7) object of wooden laths and plant strips. (1) throwing club (No. 41); (2,3) cords (No. 69, 118); (4) paddle (No. 137); (5) float (No. 111); (6,6a) netting (No. 118); (7) fishtrap/mat (No. 34).

form of a female elk's head. Being lower than the gliding surface of the ski, the projection would have prevented the reverse movement of the ski over packed snow. According to ethnological data, Silberian peoples covered the lower ski surface with fur for the same purpose (Antropova 1953).

One runner of Vis type also has a projection like that (Figure 19.5:2). It probably served as a "parking brake" to avert movement of the sledge on a slope. The described ski has an indisputable analog at Nizhnee Veretye.

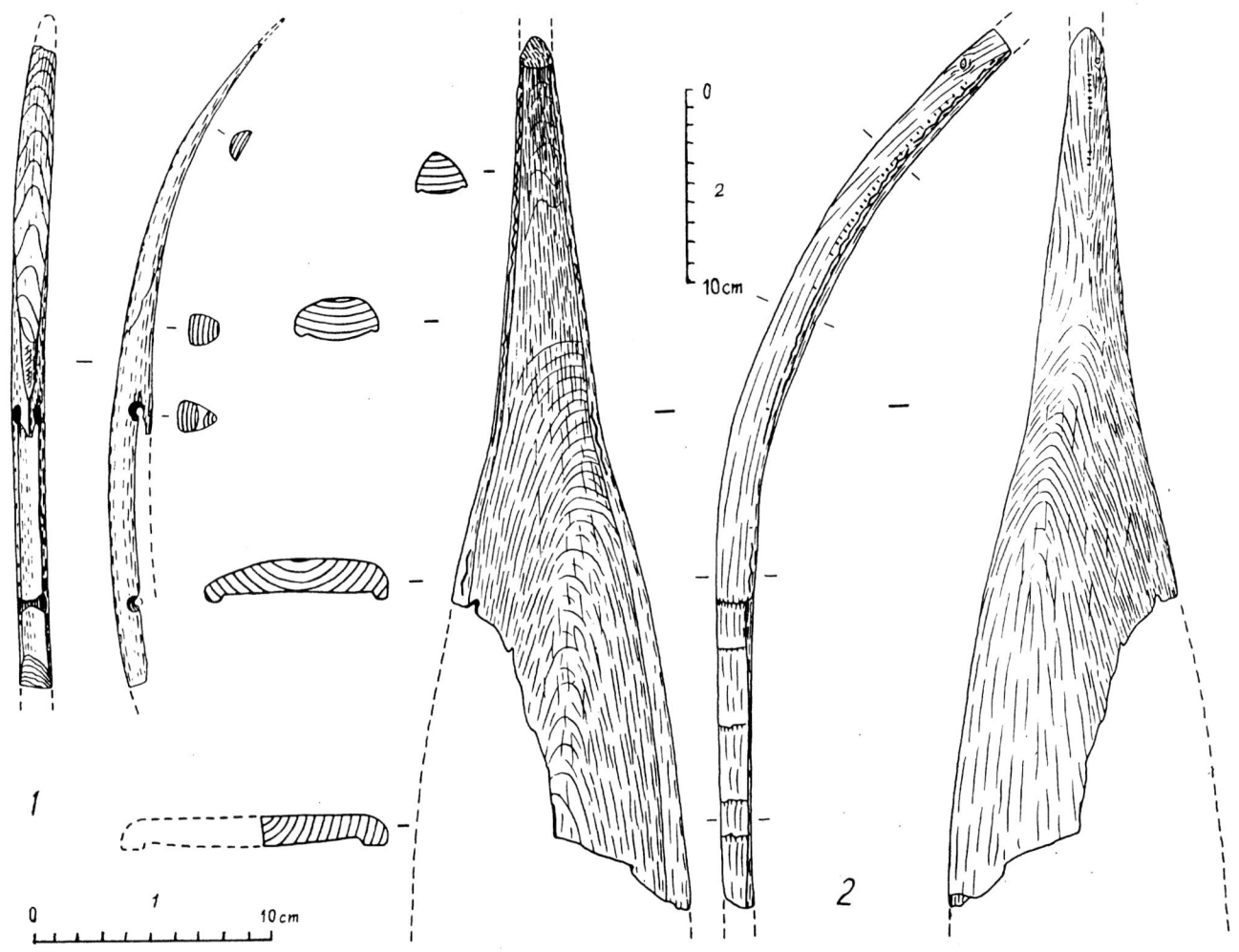

Figure 19.4. Vis I. Wooden objects. (1) hoop (No. 99); (2) ski of the Vis type (No. 168).

Processing Organic Materials

As far as possible, the natural (but barkless) surface of a trunk was used when making wooden objects while its squared side was smoothed by an indented tool which left parallel striations on runners, skis, and paddles (Burov 1998:53–54), (Figure 19.3:4). An experiment pointed out that this instrument was an arched scraper with a wooden shaft and inserted flint microliths which looked like small teeth when they became worn. The inhabitants of Vis I were familiar with techniques of log splitting – even into thin laths (0.1–0.2 cm) – and perforating. The holes are usually biconical, having been bored from both sides of the objects and range in diameter from 0.1–3.5 cm.

Bow-shaped artifacts, called "small bows" in the collection, were not intended for hunting. They have a more or less curved profile with almond-shaped terminals on both ends (Burov 1989b:397–400). The objects widen and thicken gradually toward the middle. They can be divided into two groups according to their sizes of about 55 cm and 26–27 cm long. The differences between the small bows and hunting bows are in the symmetry of the former, their shortened proportions, small size, and the absence of a device for passing of arrows. Thus, the conclusion has to be drawn that the tools were used for boring and making fire (a rotating pivot having been fitted in the loop of a string). Fungi of the Polyporaceae family (tinder fungus) found in the Vis I peat bog recall similar finds at the Mesolithic site of Hohen Viecheln in Eastern Germany; this form of tinder fungus was used to keep fire alive.

Mesolithic populations appear to have preferred white spruce to pine in the manufacture of fishing hoops, hunting and small bows. White spruce is almost as durable as pine and more supple than pine (Burov 1998:54–58). At the same time, pine is better for splitting and was mostly used

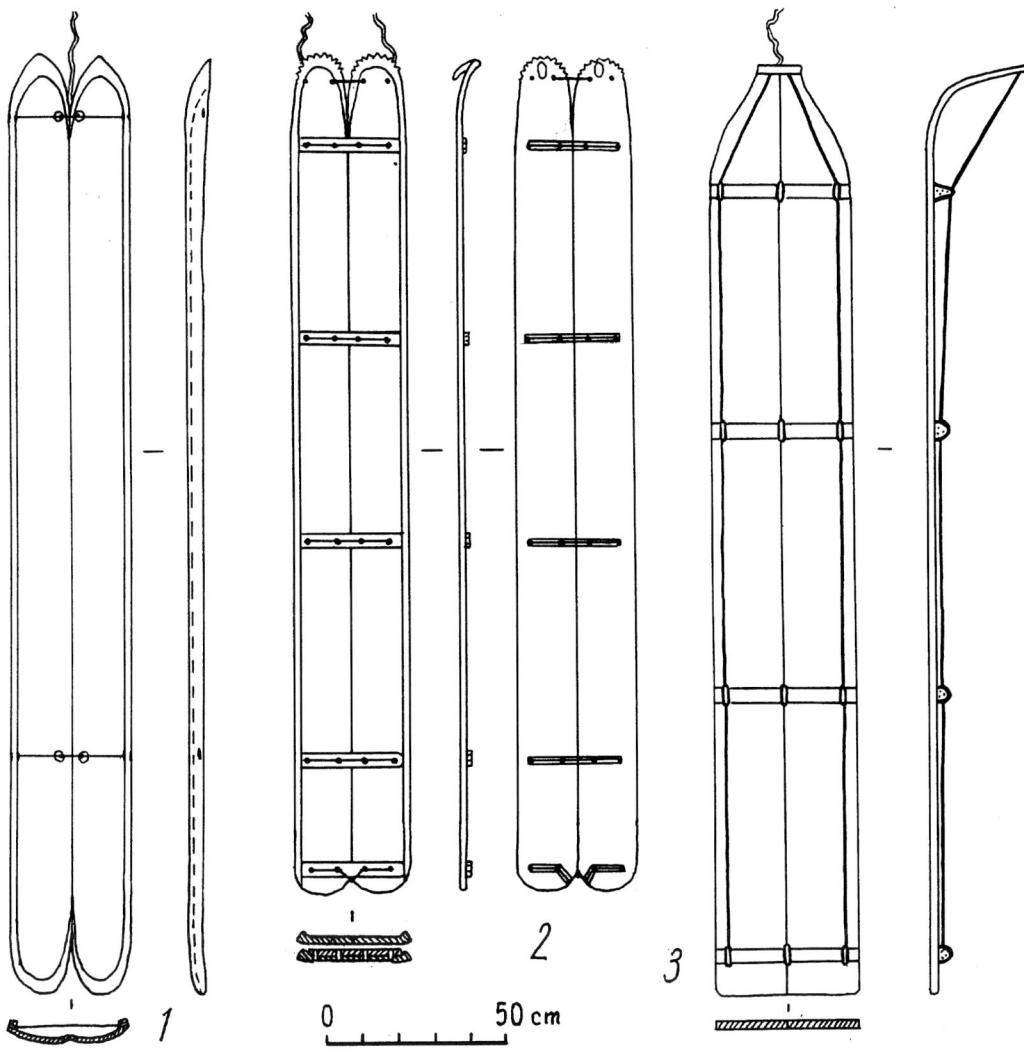

Figure 19.5. Vis I. Reconstructions of sledges of Heinola (1) and Vis (2) types; (3) Canadian toboggan, after J.E. Lips.

in the Neolithic and until the 1st millennium AD when massive curved objects went out of common usage (Burov 1993b).

A container from Vis I is almost square in planview and is made of a rectangular sheet of birch bark (Burov 1998: 58-62). The container was fastened with thin wooden pegs. It is the most ancient birch bark vessel known by the author but the existence of such artifacts in the Mesolithic was suggested by E. Vogt. Containers which were made before the invention of pottery could have been intended for both storage and cooking. Ethnological information about birch bark coopers testify to this conclusion. Boxes made from single cross-shaped sheets of birch bark suitable for carrying stone tools but unfit to hold liquids existed in the Mesolithic of northeastern Europe at Nizhnee Veretye I (Oshibkina

1989: Fig. 2:7). It represents a period earlier than Vis I and, at that time bark containers intended for liquid probably had not been invented yet.

The net at Vis I was made of a two-strand Carex cord 0.15–0.20 cm thick, but cords 0.4 cm and about 1.0 cm in diameter were found also (Figure 19.3:2,3). When making the net, the cord was knotted with a sheet-bend knot which is very strong under tension and continues to be used even today in the manufacture of synthetic nets.

Art

On many of the wooden artifacts, there are sculptured and engraved decorations (Burov 1989a). The elk sculpture on the ski from Vis I is a masterpiece of Mesolithic art and has many parallels in prehisoric art on antler, bone, stone, and metal. Among these, is the pommel of a bronze dagger decorated with a female elk's head image from the Seyma burial ground in the Volga-Oka region (4th millennium BP). This dagger provides a clue to the significance of the wavy ridge decorating the upper surface of the ski. On the hilt of the Seyma site dagger, a snake is stretching to the head of an elk. This and other objects allow us to regard the wavy ridge as a relief image of a reptile. The worship of both elk and snakes was widespread in the Mesolithic and continued throughout the Neolithic, Bronze Age and even into Medieval time. In some cases the female elk and snake were involved in a dual cult. A number of facts testify that the snake was connected with the cult of fruitfulness and served as a bearer of the male principle (Burov 1997:59). Fourteen objects (hunting and small bows, and the scraper) had almond-shaped terminals (Figure 19.2:2,3) that varied in cross section. They had some practical use such as fixing the string of the bows, preventing bows with holes from breaking, and making it easier to hold small bows and scrapers which were two handled. But we may speak also of their aesthetic and magic meaning. Apparently, such decorations are stylized carvings of the heads of snakes. Engraved ornamentation has been found on tools and transportation artifacts (Figure 19.2:2,3; Figure 19.4:2). It consists of zigzags, straight lines, skew nets or crosses, and rows of V-shaped signs or notches. The latter motif was seen on edges only. Such a phenomen was not recorded in the Neolithic and Bronze Age when usually only ceramics, and art and cult objects were decorated.

Periodizaton

The rich wooden inventory of Vis I gives us a rare opportunity to detect three periods in the history of Mesolithic populations at this site on the basis of organic artifacts (Burov 1990a). The objects were found in a peat bog at varying depths (from 1.1–2.9 m below the modern surface) as well as in the upper stratum of sand lying below the peat deposits. Although the thickness of the peat bog varies in different places, we have evidence to believe that the accumulation of the deposits took place almost simultaneously. For example, two parts of a manual bow, later glued together, were recovered at a distance of ca 50 m from each other. They were discovered in places of varying thickness of the peat bog but at the same relative depth.

The earliest period (Figure 19.2:1; Figure 19.3:1,3,6,7; Figure 19.4) is the time when ca 40% of the organic artifacts were made of hardwood (Figure 19.2:1; Figure 19.4:2). Later it was used very seldom. The second period is represented by an abundance of bows and hoops with openings (Figure 19.2:2–4). while in the third (most recent) period, we recorded the widespread use of wooden artifacts

with surfaces smoothed by an arched scraper (Figure 19.3:4,5). In the first period (earliest), there was probably no intentional selection of wood species. The transition to the predominant use of softwood may be explained by the fact that the people began to choose wood types consciously. Confiers have very straight boles; their wood has a high elastity, and is also soft and light (Perelygin 1963: 95, 117–144, 252).

Typical of the first period (earliest) were the skies of the Vis type, runners of the Heinola type not worked by an arched scraper, some bows (manual of Holmegaard type, of Vis type without a hole or with a ledge on the middle part, and path bows with a head), and the straight and L-shaped clubs. Bows provided with openings for regulating string, hoops for fastening netting, and the spears were typical of the second Mesolithic period of Vis I. In the middle of this period, runners (of Heinola and Vis types), skis (of Veretye type) and paddles appear with surfaces worked by an arched scraper. Such predominate in the third (most recent) period. This periodization permits us to date Nizhnee-Veretye (by comparison) as a site from the second period. The western analogies of Vis I Mesolithic artifacts testify to Baltic area connections of Vis I culture.

Vis II

Nearly the same inventory of artifacts exists at this site as at Vis I (Burov 1996b).

Weapons

The bows are made from softwood and are either simple or composite (Burov 1983). The simple bow handles are narrow but thick as a rule; the nock pegs for attachment of the string are chiefly without a head. These flatbows have elliptical, rectangular, or segmental cross sections (Figure 19.6:1). They are only 52–55 cm long and some of them were children's toys. The single bows were identified tentatively, and a few of these may belong to other categories (Figure 19.8:8).

There are 42 convex flat rods from the composite bows which are about 100 cm long and 1.0–4.0 cm wide. The rods often have small grooves and cuts on the flat side (Figure 19.6:4,5) because they were glued together in pairs. The objects belong to types with the ends cut off (Figure 19.6:9) and with the nocks having no peg heads (Figure 19.6:5). The composite bows of the first type probably had nock parts of bone. Bows of this type are like those from Medieval Novgorod (Medvedev 1966: Table 1). The origin of the composite bow in the European northeast seems to be connected with an influence from areas to the southeast of Vis II. The earliest evidences of its existence were found in Mesopotamia, Siberia, and Japan (Rausing 1967:148).

The wooden arrows – for hunting squirrels and forest

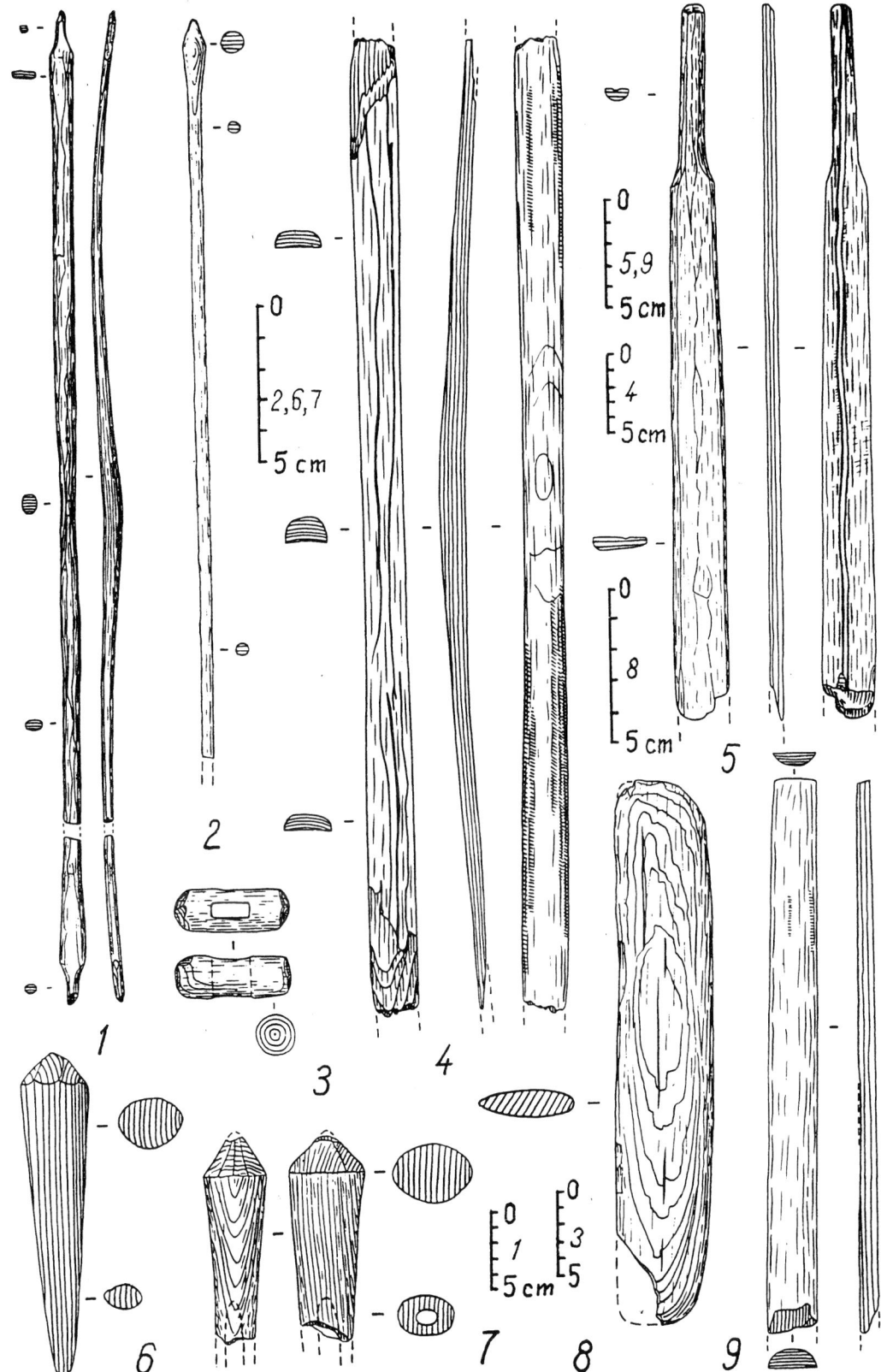

Figure 19.6. Vis II. Wooden artifacts: (1) simple bow; (2) arrow; (3) cross piece; (4,5,9) parts of composite bows; (6,7) arrowheads; (8) scraper.

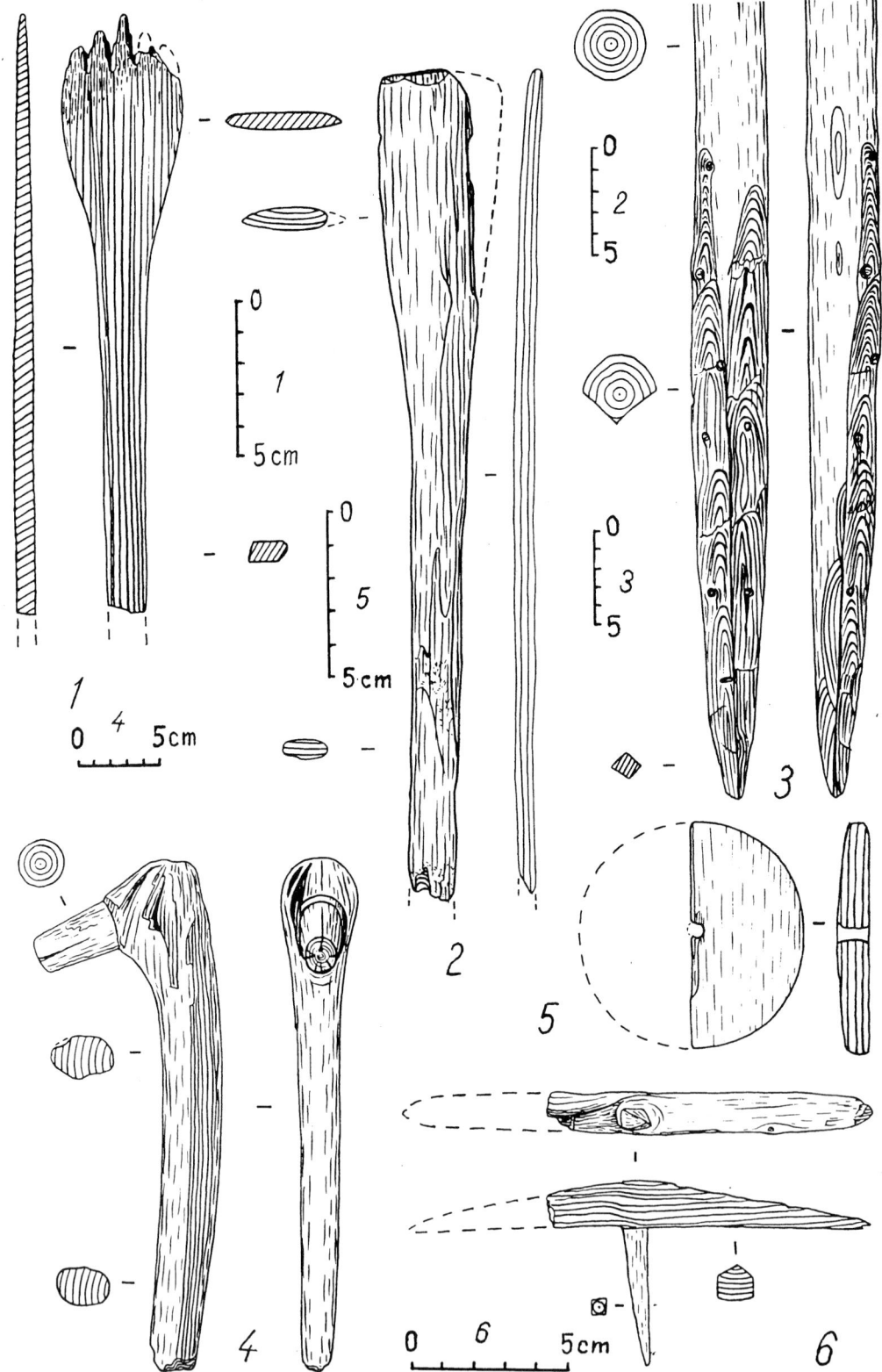

Figure 19.7. Vis II. Wooden objects: (1) comb; (2) spatula; (3) stake; (4) axe handle; (5) whorl; (6) pin.

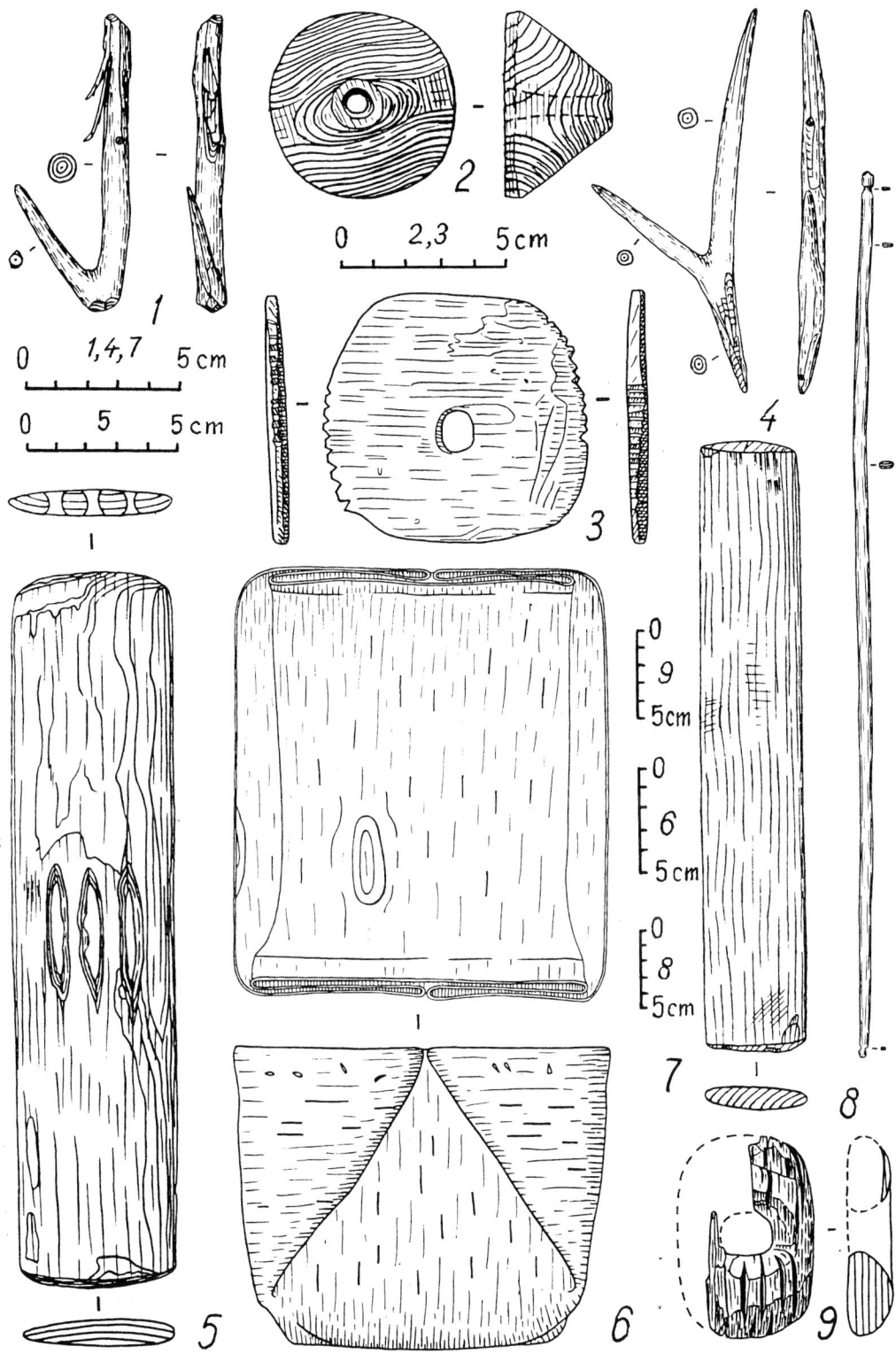

Figure 19.8. Vis II. Wooden objects: (1,4) fish hooks; (2) whorl; (3) tool to decorate ceramics; (5) plank for softening straps; (6) birch bark container; (7) measuring stick of net maker; (8) toy bow or piece of loom; (9) fish driver.

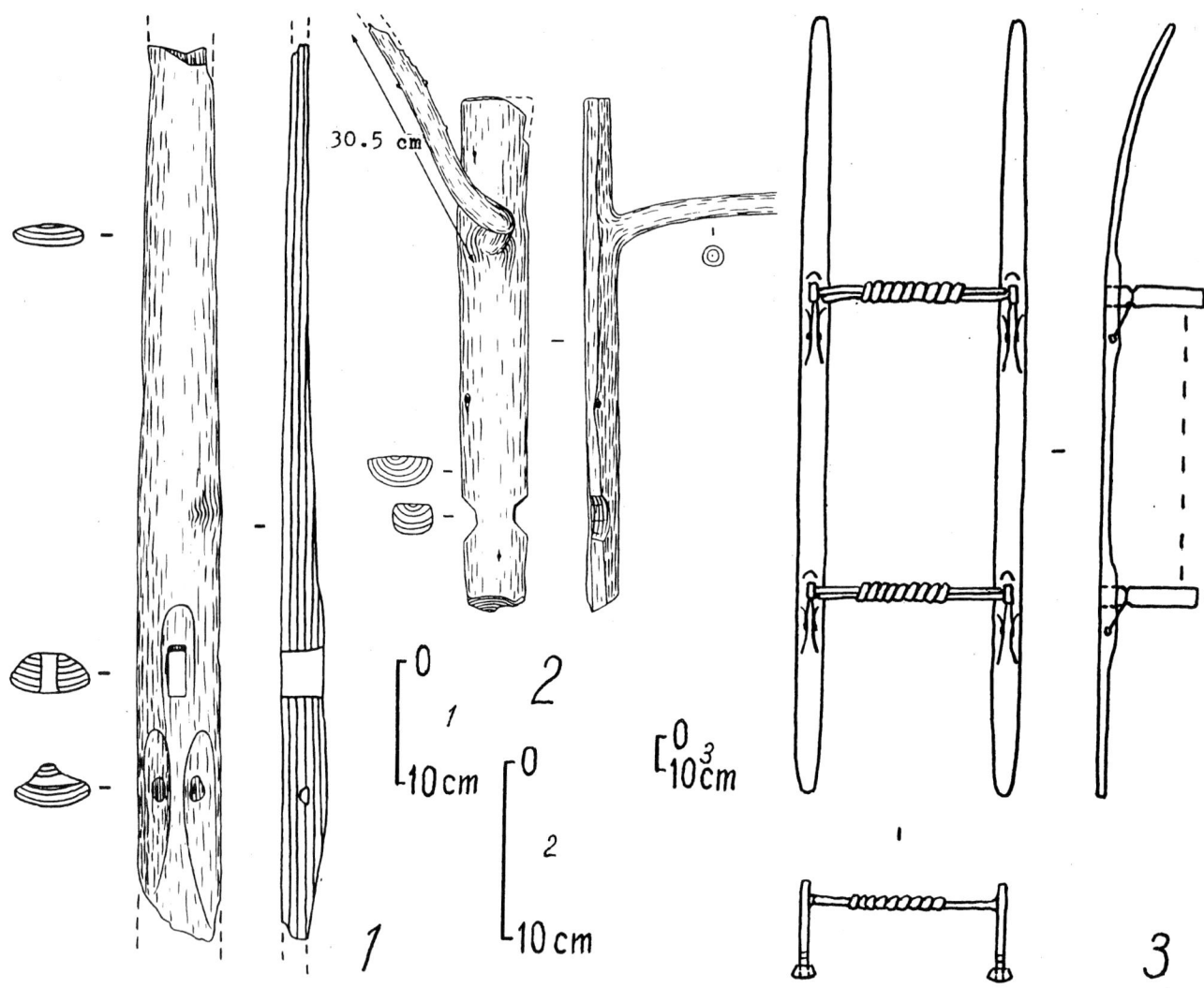

Figure 19.9. Vis II. Sledge with poppets of Novgorod type: (1) runner; (2) poppet; (3) reconstruction.

birds – have generally a biconical head (Figure 19.6:2) and mainly were made of separate pieces. The arrowhead was tanged (Figure 19.6:6) or had a hollow for the attachment of the shaft (Figure 19.6:7). In the inventory, there is also a composite weapon which looks like a wooden dagger with slots to receive three inserts held in place by a clayey paste. The inserts were not preserved and were probably made of iron (Burov 1996b: Fig. 2:4).

Fishing and Digging Tools

The settlement of Vis II subsisted to a considerable degree by fishing and, therefore, existed for a long time. In the sapropel below the peat, at depths of 0.7–3.5 m, were discovered numerous fish weirs made of stakes and young conifers driven vertically with the thick end at the bottom.

A portion of one side of the stake is pointed (Figure 19.7:3). The weirs resemble similar constructions of the Komi and Urgrians of the river Ob (Sirelius 1906). The fishing equipment includes a rectangular measuring stick used for net making (Figure 19.8:7); loops of twigs, probably from nets; elliptical fish driver plates 10.9–14.3 cm long with an opening in the center for the handle (Figure 19.8:9); angling hooks made of thin switches with a twig and usually having three pointed ends (Figure 19.6:4) but sometimes only one barb (Figure 19.8:1). All of these artifacts are matched in medieval Novgorod (Kolchin 1968) and ethnological materials (Brinkhuizen 1986: Fig. 8:9). The people also hunted and gathered, had cattle breeding and, from the presence of wooden mattocks, it appears that agriculture was probably of some significance.

Paddles and Sledges

The paddles were pointed and had leaf-shaped blades (Burov 1984:162–165). The top of the shaft had, apparently, a bulge or a cylindrical cross piece held in place by a rectangular mortise (Figure 19.6:3). The sledge had two runners with mortises in which poppets supporting a platform were fitted (Burov 1995). There were two models at Vis II. The first one (of hardwood) had poppets with long horizontal branches (Figure 19.9). They were tied together in pairs and formed a support for the remaining of the platform. Such poppets are known also in medieval Novgorod (Kolchin 1968:51–55). The second model (of softwood) (Figure 19.10) is similar to sledges of Mangazeya (a transpolar Sibertian town of the 16–17th century AD). Every poppet had a mortise for fixing a transversal beam (with end tenons) supporting the platform (Belov *et al.* 1981:70–73).

Woodworking Tools

Instruments to process wood were iron axes and adzes, the wooden handles of which were preserved. The axes had a tube for the conical projection of the handle; the adzes were fitted with a slot. Axe handles (Figure 19.7:4) are longer (29.5–50.5 cm) than the adze handles (22.2–23.0). The axes are analogous to the finds at Hallein, Austria (Vouga 1923: Pl. XLII), La Tene, Switzerland, and Nydam, Denmark. The adze handles look similar to those from medieval Russian towns (Staraya Ladoga, Novgorod). A beetle (a piece of a log with a handle) was apparently used for splitting wood. Small bows, probably used for drilling, had a hole and one or two pairs of shoulders under it on one end for regulating a looped string.

The master woodworkers of the Vanvizdino culture used some Mesolithic handicraft achievements. In particular, hardwood (Figure 19.6:6; Figure 19.8:9; Figure 19.9:1,2) was seldom used. But the Mesolithic population did not appear to have made wooden objects by using mortise and tenon joints, sometimes with pegs, or by fire tempering. Such methods were recorded for the time of Vis II; for example, burnt wood (Figure 19.6:8; Figure 19.7:1) and holes for pegs (Figure 19.10:1,2) are known. Also, people of the Vanvizdino cultures glued objects together, a technique not known in the Mesolithic. In addition, the wooden inventory of Vis II is more varied than that of Vis I and consists mainly of smaller objects (Burov 1993b).

Other Handicraft Tools and Domestic Objects

Skin working was done with flint scrapers and D-shaped wooden tools (Figure 19.6:8). T-shaped pins (Figure 19.7:6) seem to have been produced for fastening skins on the ground before they were worked. A plank with three parallel slits (Figure 19.8:5) was used to soften straps, like those of the Khants of Western Siberia (Manninen 1932:Abb. 229). Sinews were worked by combs with short teeth and long shafts (Figue 19.7:1). Conical and disc-shaped spindle whorls occur (Figure 19.7:5; Figure 19.8:2). Some small spatulae with remains of clay on their surface (Figure 19.7:2) seem to be intended for pottery materials. A tool in the form of a small wheel with teeth may have been intended to decorate ceramics (Figure 19.8:3). For the same purpose, a two-strand grass cord of Carex and Calamagrostis was preserved within a lump of clay.

Roofs of dwellings were probably covered with birch bark sheets while windows might have blinds of perforated laths like medieval Novgorod (Kolchin 1968:85). Birch bark containers, similar to those of Vis I, were fastened with plant strips (Figure 19.8:6). A small wooden ball might be a child's toy, and a sculpture of a woman (6.3 cm long) with a pointed hat may be a cult object.

Marmugino Fish Weirs Site

Fishing Structures of Marmugino

In the village of Marmugino, three Neolithic fish weirs of different periods were discovered in oxbow sediments of the River Yug ca 3.5 m from the surface of its floodplain. The rest of the structures lay at depths of 1.25–1.35, 0.80–0.95, and 0.55–0.65 m below the horizontal surface of the buried oxbow peat bog. According to C14 dating, they dated from 4700±60 BP (LE–711, weir No 1) to 4510±50 (LE–703, No 2) as well as to intermediate times (No 3). The weirs consist of screens made of pine wood laths 0.5 to 2.5 cm wide and driven into the river bed (oxbow bottom) to a depth of 0.3–0.4 m (Figure 19.11). The screens from weir N° 1 were of laths 2.2 m long, while those of weir No 2 were made of laths having a length of only 1 m, because the oxbow was infilled by sediments and became shallow. The screens were discovered mainly in a horizontal position. One of them was rolled up (Figure 19.11:1:A). In one case, the weir split in different directions (Figure 19.11:1:B); another one fell in one piece (Figure 19.11:2:C). Stakes also were found.

According to ethnological data, such screens were intended to build fish surrounds consisting of weirs (wings) and a collection chamber that looked like a kidney in planview. Sometimes, however, the screens were used to isolate fish in an oxbow lake connected with a river during spring flood.

It was necessary to excavate the Marmugino site in September and October even when it was snowing. In the spring, water and ice destroyed a wide belt along the cliff of the floodplain. Therefore, the structures would not have been preserved for even one year. We do not know how

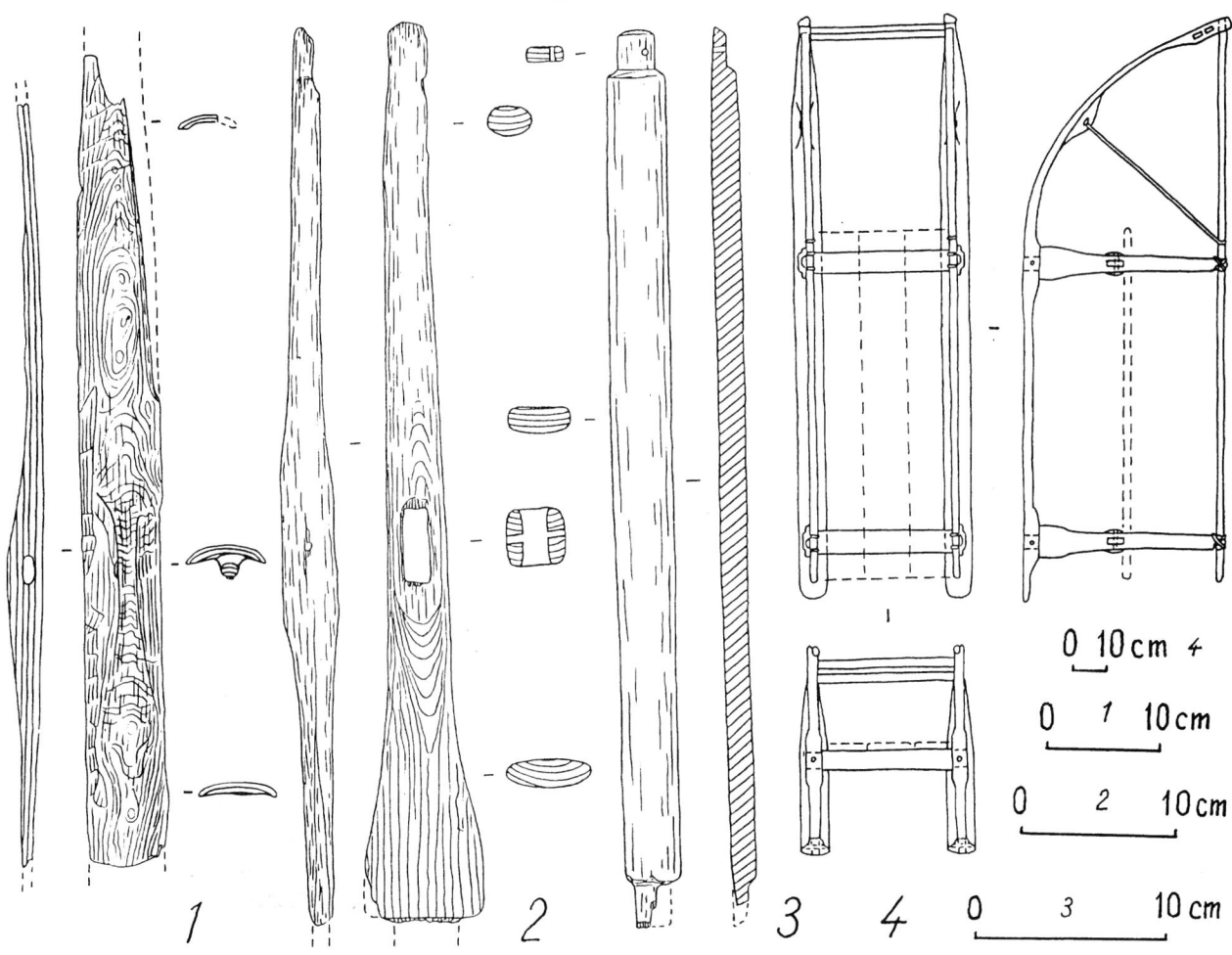

Figure 19.10. Vis II. Sledge with poppets of Mangazeya type: (1) runner; (2) poppet; (3) fore cross plank; (4) reconstruction.

large the cut off portions of the Marmugino weirs were but it was ascertained that the oxbow lake where they were built had been ca 80 m wide. The unexcavated parts of weirs which occurred in the floodplain deposits might not be more than 10–15 m in length because investigations were conducted near the bank of the ancient lake. The structures stretched at acute angles to the bank and this has analogies to ethnological materials (Brinkhuizen 1983:Fig. 16:C). According to them, a screen could not be more than 3–5 m long. Thus, the recovered portion of weir No 1 included remains of two to three screens while that of weir No 2 consisted of one or two such objects. The distance between weirs No 1 and 2 in the cliff was 14 m.

East European Prehistoric Structures of Lath

Fish Traps or Screens

The Marmugino fish weirs have parallels (Figure 19.12) at Mesolithic-Neolithic/Bronze Age sites of the Voronezh basin (Podzorovo) (Levenok 1969), Upper Volga area (Zamostye II, Ivanovskoe III and VII, Berendeevo IIa, Sakhtysh I) (Kraynov 1991; Lozovsky 1999), Eastern Baltic area (Sventoji 2B and 1A, Zemaitiske I in Lithuania, Zvidze, Sarnate and Abora I in Latvia, Kuuselankangas and, apparently, Nivala and Tervo in Finland) (Rimantiene 1979; Rimantene 1983; Vankina 1970; Loze 1986; Valonen (1952); Nunez 1995) and in Sweden at Skedemosse on Oland Island (Hagberg 1967). At the latter site, dated by C14 to the third quarter of the 4th millennium BP, there was a complete surround among the three structures made of laths. A part of such a structure without wings, dating to the Neo-Eneolithic, was investigated in 1878 by I.S.

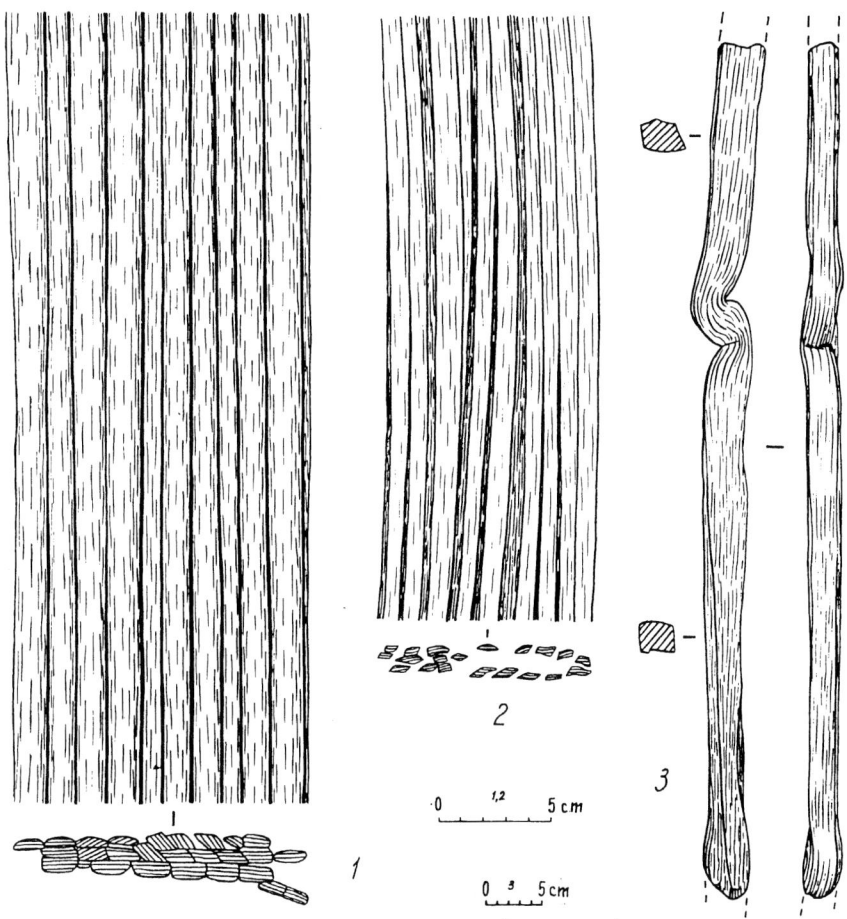

Figure 19.11. Marmugino. Plans of fishing structures I (No. 1) and II (Nos. 2 and 3): (1) outline of excavation area; (2) cliff; (3) laths; (A) rolled up screen (rouleau); (B) line of weir; (C) the same of weir No. 2.

Polyakov at Plekhanov-Bor on the Oka (Burov 1988:157). Thus, inhabitants of northeastern Europe from the 8th to the 4th millennium BP frequently used fishing structures constructed of screens made of laths.

However, in the opinion of archaeologists who excavated several sites mentioned above, the excavated structures are "fish traps having a concial form." But there is no proof for such a supposition. Of course, some "traps" (Sarnate, Abora I, Sventoji 2B, Sakhtysh I and others) were found close to prehistoric dwellings but they are screens which, according to ethnological data, were kept rolled up in rouleaus. No frameworks were excavated. The laths that were used are 1–2 cm wide and ca 0.5 cm thick while ethnological fish traps consist of flexible laths ca 0.4 × 0.2 cm because the trap has a cylinder-conical form and must be lightweight. That is why there are intervals of 4 cm and more between laths. Finally, at sites where cross strips were preserved, the distances between the cross strips are too large (Sarnate and Abora I) to consider these objects as fish traps. A series of Stone Age traps of this type that were found in Denmark, the Netherlands, and other countries, were manufactured of thin twigs. From the finds in eastern Europe, only the object of thin laths recovered at Vis I can be interpreted as a fragment of a fish trap.

The fishweirs dated to 8000–4500 BP were known long ago in Denmark. They consist of wicker screens ca 5 m long in the shape of a wattle fence (Pedersen 1999). On territory of the former USSR, Stone Age fish weirs, in places where they were used, were first discovered by V.P. Levenok in 1965 and by the present author who devoted a special publication to ancient fishing structures (Burov 1988). Another publication about this topic is a paper by Lozovsky (1999) who has not used my paper. He has made some mistakes in facts and interpretation but gave a description of an interesting Mesolithic fishing structure at Zamostye II. It and some structures discovered earlier at sites of the Upper Volga area suggest that weir fishing already occurred in eastern Europe in the Mesolithic.

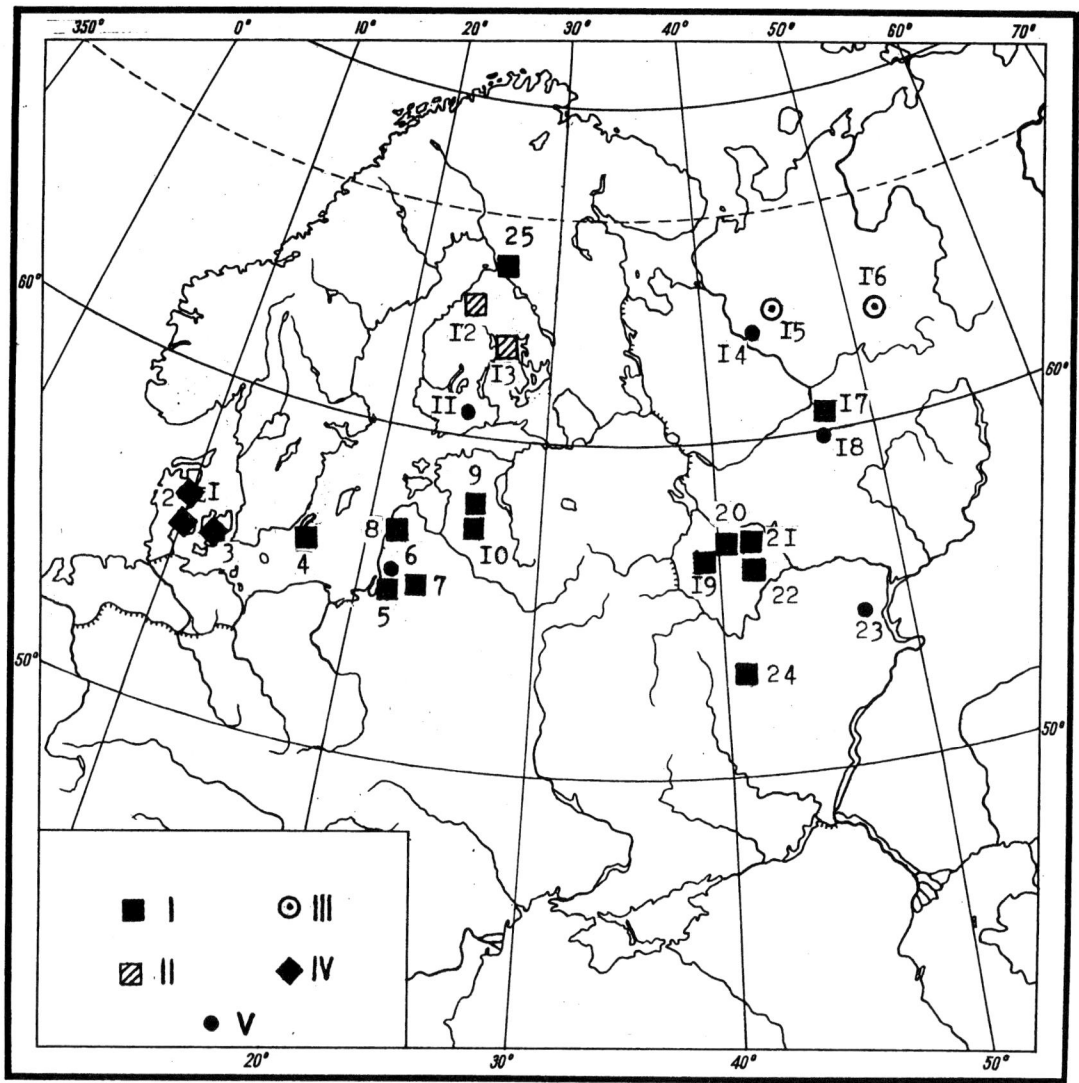

Figure 19.12. I: structures of lath of the 8–4th millennia BP including screens found in dwellings; II: those of laths not dated; III: weirs of stake and poles (Iron Age); IV: structures of wattles. (1–3) Oleslyst and other sites in Denmartk; (4) Skedemosse; (5) Sventojo ia and 2b; (6) Sventoji 9(?); (7) Zemaitiske I; (8) Sarnate; (9) Abora; (10) Zvidze; (11) Kyrkslätt; (12) Nivala-Sarjankylä; (3) Tervo-hiideman; (14) Pingisha; (15) Yavronga I; (16) Vis II; (17,18) Marmugino; (19) Zamostye II; (20) Sakhtysh I; (21) Ivanovskoe III and VII, Berendeevo Iia; (22) plekhanov-Bor; (23) Lugovoe; (24) Podzorovo; (25) Kuuselankangas.

V. Lozovsky (1999:142) groundlessly doubts the interpretation of finds at Podzorovo and Sakhtysh I as fishing structures. But screens of long laths which lie one parallel to another are obviously artificial. There is no way that a complete piece of pine roundwood can split "under the pressure of sediments ... into a number of thin splinters." In lake and bog sediments, pinewood preserves excellently. The fact that, for the most part, the lath construction has no "binding or other links between the splinters" is caused by unfavorable conditions for their survival. The cross strips, which must be strong, probably were usually made of skins.

But in Latvia, bast, and in Finland, birch bark, cross strips were used. Such a peculiarity speaks of eventual chronological similarity between the three Finnish sites (Figure 19.12), one of which (Kuuselankangas) has a radiocarbon date of 4770±130 BP, very close to those of Marmugino and Podzorovo.

Built by People or by Beavers?

V. Lozovsky has also touched upon the activity of beavers. He did not know that their dams had been already eliminated from the list of known ancient fishing structures of eastern Europe (Burov 1988:Fig. 10). A construction in the district of Kyrkslatt (southern Finland) defined by G. Topelius as a weir is, in fact, a beaver's product (Rimantene 1983:75). A comparison of wood from oxbow peat bogs investigated in 1966–1969 (1) near the Eneolithic site (settlement) of Marmugino, (2) on the river Pingisha, and (3) near the village of Lugovoe (Burov 1972:34–41) (Figure 19.12) with pieces of wood gnawed by modern beavers indicated their identify. An object excavated by R. Rimantene (1983) as a "fishing structure" of Sventoji 9 looks very similar to that at Lugovoe. Unfortunately, there are no descriptions of woodworking techniques and no corresponding illustrations.

The presence of some prehistoric artifacts does not prove that the structure was also made by the people. At Plekhanov Bor, a room of laths was discovered together with beaver dams. At the site of Zvidze, four "fish traps" were excavated which could be considered as screens forming fish weirs. As to the neighboring "fish weirs" of "stakes" and "poles," it is impossible to recognize its artificial origin because of the lack of information in Loze's publication. The same must be said of Zamostye II. The beaver dams include rows of thin log parts 2 m long or less, put almost verticaly (Panov 1990:69). Consequently, not only are there no indisputable fish traps of the Stone Age but also no Mesolithic-Neolithic fish weirs made of stakes have yet been found in northeastern Europe.

B. Coles (1998) wrote that "people would have taken advantage of the beavers' product." In this connection it is possible that people built their weirs and surrounds using beavers' dams. It is very interesting to note that a board with a sharp burned end was found among pieces of a beaver's dam at Lugovoe during our excavations in 1972. This board was probably used at night as a torch while spearing fish that were hiding among the branches of the dam.

Conclusion

Since organic artifacts are extremely rare in sites dating back to the Mesolithic anywhere in the world and, since Vis I has the largest collection of wooden artifacts from this time period, it is of great scientific interest. With regard to Vis II, this site – including plenty of organic objects and fish weirs – has no parallels in eastern Europe where no synchronic wet sites have been found. The collection from Vis II adds considerably to our knowledge about material culture of the early centuries AD in the European northeast. The significance of this site is well demonstrated by the fact that only 100 m² were dug but 228 unique and unusual objects of botanical materials were found. It is desirable to excavate more than 2000 m². As to the Marmugino struc-

tures, they belong to the most expressive patterns of the Neolithic fish weirs in eastern Europe.

References Cited

Andersen, S.H. 1995 Coastal adaptation and marine exploitation in Late Mesolithic Denmark – with special emphasis on the Limfjord Region. In: Fischer A. (ed.), Man and Sea in the Mesolithic. Oxford: Oxbow Books: 41–66.

Bagniewski, Z. 1990. Obozowisko mezolityczne z doliny Baryczy. Warsaw – Wroclaw: Panstwowe wydawnictwo naukowe.

Belov, M.I. and others. 1981. Mangazeya: material'naya kul'tura russkikh polyarnykh morekhodov i zemleprokhodtsev XVI–XVII vv. Moscow: Nauka.

Brinkhuizen, D.C. 1983. Some notes on recent and prehistoric fishing gear from northwestern Europe. Palaeohistoria 25:7–53.

Bryusov, A.Ya. 1951. Svaynoe poselenie na r. Modlone i drugie stoyanki v Charozerskom rayone Vologodskoy oblasti. Materialy i issle dovaniya po arkheologii SSSR 20:7

Burov, G.M. 1972. Arkheologicheskie pamyatniki Verkhney Sviyagi. Ul'yanovsk: Irivolzhskoe knizhnoe izdatel'stvo.

Burov, G.M. 1974. Arkheologicheskie kul'tury Severa evropeyskoy chasti SSSR (Severodvinskiy krey). Ul'yanovski: Pedagogicheskiy institut.

Burov, G.M. 1981. Der Bogen bei den mesolithischen Stämmen Nordosteuropas. In: Gramsch, B. (ed.) Das Mesolithikum in Europa. Veröffentlichungen des Museums für Ur-und Frühgeschichte Potsdam 14–15. Berlin: Deutscher Verlag der Wissenschaften: 373–388.

Burov, G.M. 1983. Luki i derevyannye strely V–VI vv. n.e. s poseleniya Vis II v Privychegod'e. Kratkie soobshcheniya Institute arkheologii 175:55–62.

Burov, G.M. 1984. K izucheniyu rybolovstva na Evropeyskom Severo-Vostoke v seredine I tysyacheletiya n.e. Voprosy istorii Evropeyskogo Severa: 147–165.

Burov, G.M. 1988. Zapornyy lov ryby v epokhu neolita v Vostochnoy Evrope. Sovetskaya arkheologiya 3:145–160.

Burov, G.M. 1989a. Mesolithic art from the European north east (USSR). Mesolithic Miscellany 10(1):27–30.

Burov, G.M. 1989b. Some Mesolithic wooden artifacts from the site of Vis I in the European north east of U.S.S.R. In: Bonsall, C. (ed.). The Mesolithic in Europe. Edinburgh: John Donald. 391–401.

Burov, G.M. 1990a. Die Holzgeräte des Siedlungsplatzes Vis I als Grundlage für die Periodisierung des Mesolithikums im Norden des Europäischen Teils der UdSSR. In: Vermeersch, P.M. and P. Van Peer (eds.). Contributions to the Mesolithic in Europe. Leuven: Univ. Press. 335–344.

Burov, G.M. 1990b. Poselenie Kiroksa i marmuginskiy tip eneoliticheskikh poseleniy v Severodvinskom krae. In: Nagovitsyn, L.A. (ed.) EneolithLesnogo Urala i Povolzh'ya. Izhevsk: Udmurt institute: 105–119.

Burov, G.M. 1992. Nesolithic fishing in the European northeast (Russia). Mesolithic Miscellany 13(2):1–9.

Burov, G.M. 1993a. Derevyannye orudiya okhoty u mezoliticheskikh plemen Kraynego Severo-Vostoka Europy. Rossiyskaya arkheologiya 3:149–164.

Burov, G.M. 1993b. Obrabotka dereva i beresty u plemen vanvizdinskoy kul'tury. Voprosy istorii Evropeyskogo Severa: 129–144.

Burov, G.M. 1995. Sani rannego srednevekov'ya s posseleniya Vis II v basseyne Vychegdy. Rossiyskaya arkheologiya 3:184–192.

Burov, G.M. 1996a. On Mesolithic means of water transportation in northeastern Europe. Mesolithic Miscellany 17(1):5–15.

Burov, G.M. 1996b. Wooden objects and constructions of the fifth century AD at the site of Vis II, north east European Russia. News WARP 20:27–31.

Burov, G.M. 1997. Sistema znakov na keramike srubnoy istoriko-kul'turnoy obshchnosti: opyt interpretatsii. Drevnosti Stepnogo Prichernomor'ya i Kryma 6:48–63.

Burov, G.M. 1998. The use of vegetable materials in the Mesolithic of Northeast Europe. In: Zvelebil, M. and others (Eds.) Harvesting the Sea, Farming the Forest. Sheffield: Academic Press: 53–63.

Burov, G.M. 1999. Die mesolithischen Schlittenformen in Nordosteuropa. In: Cziesla, E. and others (eds.) Den Bogen spannen. (Festschrift für B. Gramsch). Weissbach: Beier & Beran: 117–136.

Coles, B. 1998. Beaver works. NewsWARP 23:40–41.

Earwood, C. 1993. The Medieval town of David-gorod, Belarus. Antiquity 67:534–547.

Foss, M.E. 1941. Kostyanye I derevyannye izdeliya stoyanki Veretye. Materialy i issledovaniya po arkheologii SSSR 2:212–235.

Franz, L. 1928. Alteuropäischen Wurfhölzer. In: Festschrift. P.W. Schmidt. Wien: 800–808.

Golubeva, L.A. 1973. Ves' i slavyane na Belom ozere X–XIII vv. Moscow: Nauka.

Hagerg, U.E. 1967. The archaeology of Skedemosse. Vol. I. Stockholm: Almqvist & Wiksell.

Kolchin, B.A. 1968. Novgorodskie drevnosti: derevyannye izdeliya. Moscow: Nauka.

Kraynov, D.A. 1991. Rybolovstvo u neoliticheskikh plemen Verkhnego Povolzh'ya. In: Gurina, N.N. (ed.) Rybolovstvo i morskoy promysel v epokhu mezolita – rannego metalla v lesnoy i lesostepnoy zone Vostochnoy Evropy. Leningrad: Nauka: 129–152.

Levenok, V.P. 1969. Novye raskopki stoyanki Podzorovo. Kratkie soobshcheniya Instituta arkheologii 117:84–90.

Lips, J.E. 1955. Vom Ursprung der Dinge. Leipzig: Brockhaus.

Loze, I.A. 1986. Rybolovnyy zakol epokhi neolita na poselenii Zvidze. Kratkie soobshcheniya Instituta arkheologii 185:78–82.

Lozovsky, V. 1999. Archaeological and ethnographic data for fishing structures from northeastern Europe to Siberia and the evidence from Zamostje 2, Russia. In: Coles, B. and others (eds.). Bog bodies, sacred sites and wetland archaeology. Exeter: Short Run Press: 139–145.

Manninen, I. 1932. Die finnisch-ugrischen Völker. Leipzig: Harrasowitz.

Medvedev, A.F. 1966. Ruchnoe metatel'noe oruzhie (luk i strely, samostrel) VII–XIV vv. Moscow: Nauka.

Meyer, H.n.d. Bogen und Pfeil in Zentral-Brasilien. Leipzig:w.n.

Miklyaev, A.M. 1969. Pamyatniki usvyatskogo mikrorayona, Pskovskaya oblast' – Arkheologicheskiy sbornik (Leningrad) II:18–40.

Mordant, D. and Mordant, M. 1992. Noyen-sur-Sein: A Mesolithic waterside settlement. In: Coles B. (ed.) The wetland revolution in prehistory. Exeter: Short Run Press: 55–64.

Nunez, M. 1995. Recent wetland finds from Yli-Li, northern Finland. NewsWARP 18:29–31.

Oshibkina, S.V. 1989. The material culture of the Veretye-type sites in the region to the east of Lake Onega. In: Bonsall, C. (ed.) The Mesolithic in Europe. Edinburgh: John Donald: 402–413.

Panov, G.M. 1990. Bobry. Kiev: Urozhay.

Pedersen, L. 1999. Fishing structures in wetlands. In: Coles, B. and others (eds.). Bog bodies, sacred sites and wetland archaeology. Exeter: Short Run Press: 185–190.

Popov, A.A. 1937. Okhota i rybolovstvo u dolgan. In: Meshchaninov, I.I. (ed.). Pamyati B.G. Bogoraza. Moscow-Lenigrad: Izdatel'stvo AN SSSR: 147–206.

Rausing, G. 1967. The bow: some notes on its origin and development. Lund: Gleerup.

Rimantene, R.K. 1983. Rybolovnoe sooruzhenie na beregu Baltiyskogo morya (Shvyantoyi 9). In: Krizhevskaya, L.Ya. (ed.) Izyskaniya po mezolitu i neolitu SSSR. Leningrad: Nauka: 73–78.

Rimantiene, R. 1979. Sventoji. Narvos kulturos gyvenvietes. Vilnius: Mokslas.

Sirelius, U.T. 1906. Über die Sperrfischerei bei den finnisch-ugrischen Völkern. Helsingfors: Scoiété finno-ougrienne.

Skaarup, J. 1995. Hunting the hunters and fishers of the Mesolithic: twenty years of research on the sea floor south of Funen, Denmark. In: Fischer, A. (ed.) Man and sea in the Mesolithic. Oxford. Oxbow Books: 397–401.

Valonen, N. 1952. Geflechte und andere Arbeiten aus Birkenrindenstreifen. Vammala: Suomen Muinaismuistoyhdistys.

Vankina, L.V. 1970. Torfyanikovaya stoyanka Sarnate. Riga: Zinatne.

Vouga, P. 1923. La Tène. Leipzig: Hiersemann.

20. Perishable Technologies and Invisible People: Nets, Baskets and "Venus" wear ca. 26,000 B.P.

O. Soffer, J. M. Adovasio and D. C. Hyland

Introduction

The hunter-gatherers who occupied the Old and New Worlds between ca. 50,000 and 10,000 years ago left a voluminous material record that is extraordinarily rich in technological diversity. Although these technologies have informed our reconstructions of Paleolithic and Paleoindian lifeways, these insights were gained almost exclusively from the study of durable materials – stone, ivory, antler, and bone.

This has occurred in spite of ample ethnographic evidence that perishable organic technologies form the bulk of hunter-gatherer material culture even in arctic and sub-arctic environments (e.g., Damas 1984; Helm 1981). Over 30 years ago, Clark (1968) hypothesized that what was preserved in the Old World archaeological record constituted only about 15% of what was actually used. New World archaeologists working with materials recovered from environmental contexts with optimal preservation not only confirmed Clark's estimate, but documented an even more pessimistic reality. Taylor (1966:73), for example, noted that in most dry caves, fiber artifacts outnumber those made of stone by a factor of 20. Croes (1997:536) reported that wet sites often yield inventories where >95% of prehistoric material culture are made of wood and fiber. Collins (1937) confirmed these findings on sites in the Alaskan permafrost.

However, even a cursory examination of these ratios among published Late Paleolithic and Paleoindian assemblages (especially from open sites), indicates nearly 100% of material remains are produced in durable media. This inversion clearly results from a number of factors, among which preservation bias and inadequate recovery techniques figure most prominently. The devastating effect of post-depositional factors that erase any evidence of perishable technologies was recently and amply demonstrated in excavations of Paleolithic wet sites such as Ohalo II in

Israel (Nadel *et al.* 1994) and Paleoindian sites like Monte Verde in Chile (Dillehey 1997). The inadequacy of recovery techniques used in excavating most Paleolithic or Paleoindian sites – most notably the very infrequent use of flotation techniques as well as simple inattention to the presence of charred organics – and the resulting non-recovery of organic perishable remains were again dramatically demonstrated by the recovery of charred plant remains at Dolní Vestonice II (Mason *et al.*1994) and at El Jyjo (Freeman *et al.* 1988).

As many scholars have observed, the wealth and diversity of perishable implements, which likely existed in Upper Paleolithic-Paleoindian times, as well as the past failure to recover these items strongly biases the understanding of these economies and technologies. Additionally, this discrepancy conceals the inventories made and used by the majority of late Pleistocene people – namely women, children, and older individuals (Adovasio 1999, 2000; Adovasio and Hyland 2000; Conkey 1991; Kehoe 1990, 1991; Owen 1996, 2000; Soffer *et al.* 2000a, 2000b, 2000c). This situation obtains because technologies used by females and, by extension, children and older individuals, are usually far more perishable than those used by males – an observation amply confirmed by cross-cultural ethnographic data on the division of labor by sex and the concomitant implements associated with the different tasks (e.g., Mason 1910; Murdock 1937; Murdock and Provost 1973; Watanabe 1968).

This contribution begins to redress these past omissions by focusing on the widespread evidence for perishable organic technologies extant in Europe between at least 30,000 B.P. and the close of the Pleistocene, ca. 10,000 years ago. Specifically, by using data derived from textile impressions, from the recognition of bone, ivory, and antler tools used to weave, construct baskets and nets, as well as

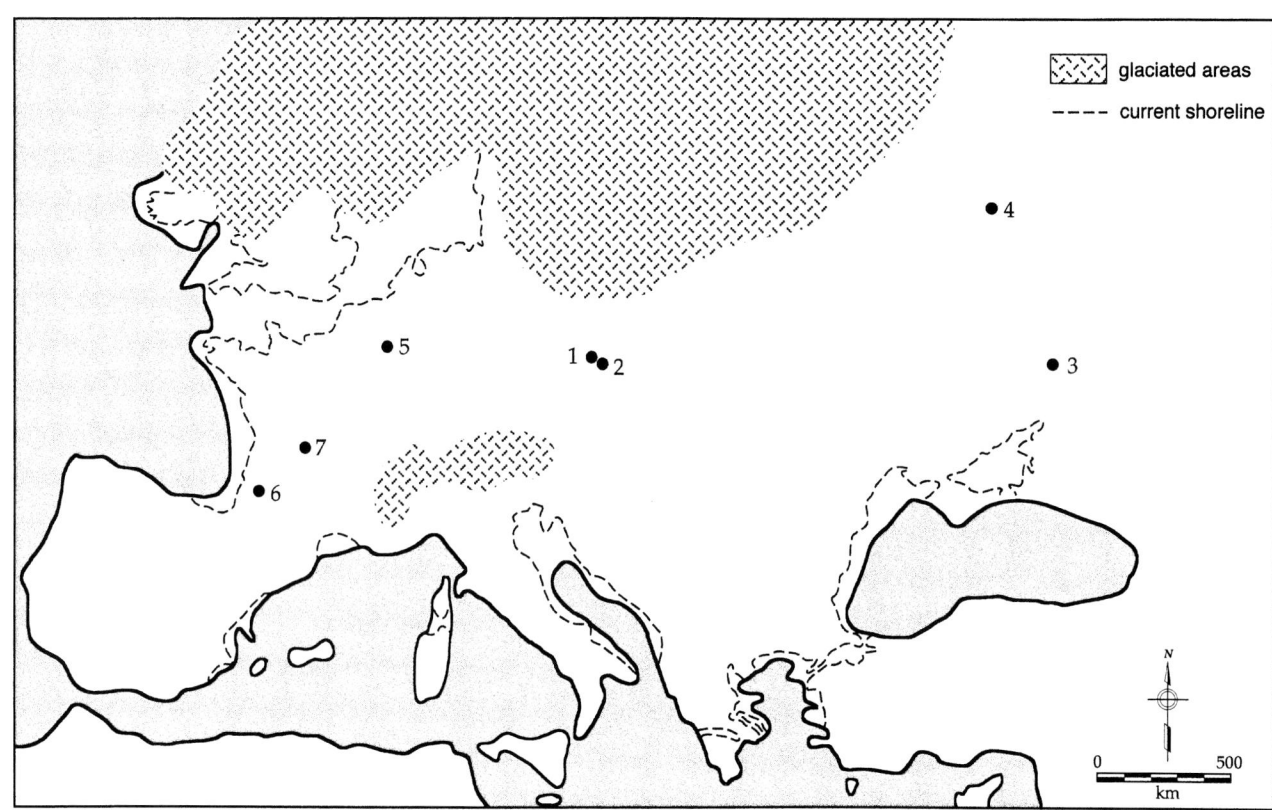

Figure 20.1. Upper Paleolithic European sites that have produced textile impressions: (1, 2) Dolní Vestonice I and II, Pavlov I; (3) Kostenkî I; (4) Zaraisk; (5) Gönnersford; (6).

iconographic evidence from the "Venus" figurines, the authors address the social significance of basketry, textiles, cordage, and cordage byproducts such as nets.

The Impressions

Data from Moravia

The recently completed collaborative research on textile impressions found on small-sized fragments of fired clay from the sites of Dolní Vestonice I and II and Pavlov I in Moravia, Czech Republic, documents the existence of diverse and sophisticated textile and cordage technologies in Europe by ca. 28,000 B.P. (Adovasio *et al.* 1996, 1998, 1999; 2000; Soffer *et al.* 1998) (Figure 20.1). These technologies include the production of cordage and nets, the making of baskets, and the twining and loom weaving of cloth. During the analysis, 79 impressed fragments were identified and examined from a Gravettian ceramics inventory numbering >10,000 pieces. This ceramic assemblage was recovered from the above mentioned sites and is assigned to the Pavlov culture – a local variant of the Gravettian technocomplex (Svoboda *et al.* 1994, 1996).

The presence of these impressions in Upper Paleolithic Moravia raises the question of whether weaving and basket making were unique to the Pavlov culture, an entity replete with other technological innovations including ground stone technology and ceramics production (Adovasio *et al.* 1999; 2000; Soffer 2000), or whether, more importantly, these perishable technologies were present elsewhere in Eurasia. Ongoing research by the authors in collaboration with Russian, German, and French colleagues is intimating that these industries are, indeed, far more widely distributed in the Old World.

Impressions from Russia

Recent fieldwork with materials excavated at a number of Russian sites is producing new evidence that weaving, net making, and basketry were practiced far outside of Moravia. For example, excavations at the well-known site of Kostenkí I-2, a Gravettian-age site dating to ~21,000 B.P., yielded a number of clay fragments, one of which is impressed with cordage (Praslov 1991; Soffer *et al.* 2000c). Similarly, ongoing excavations at the Gravettian-age site of Zaraisk produced impressions of knotted netting in fired clay from a layer dating to ca. 19,000 B.P. (Soffer *et al.* 2000c; Trusov 1998).

Evidence from Germany

The preliminary examination of bone and antler inventories recovered from Upper Paleolithic sites in Germany identified at least one item, a worked basal fragment of antler from Gönnersdorf, that bears impressions of cordage on its flattened side (Soffer *et al.* 2000c). The site, a well-known multi-seasonal residential locale dating to the Magdalenian (ca. 15,000 B.P.), yielded numerous bone needles as well as other implements that are probably associated with textile production (Bosinski 1995; Tinnes 1994). While it may seem unlikely that antler or bone can, indeed, be thus impressed, this specimen is not unique. Sakharov (1952) and Bader (1952) reported netting impressions on a poorly dated human cranium recovered from a stratigraphic exposure along the Skhodnin River in Russia. Based on the stratigraphy, the cranium was assigned to the late Pleistocene. Sinel'nikov (1952) conducted a number of actualistic studies to account for this phenomenon and concluded that humic acids could indeed etch the design on bone. This impression is also currently under study.

The French Data

The production of cordage in Upper Paleolithic France was first documented by Abbé Glory (1959), who discovered a fragment of rope in one of the adjacent galleries at Lascaux. Unfortunately, the rope did not survive, but recent examination by the authors of both its positive and negative casts (curated at the IPH in Paris) confirms the use of rope at least 15,000 years ago, albeit one with a six-ply construction rather than the previously reported three-ply formula (Leoroi-Gourhan and Allan 1979; Soffer *et al.* 2000c). Not an isolated phenomenon, there are corroborative examples of organic perishable technology known from late Pleistocene France. Cheynier (1967) published, albeit in an anecdotal fashion, a textile impression from the Solutrean level at Badegoule in France (dating between ca. 21,000–18,000 B.P.). Additionally, recent examination of Cheynier's collections revealed actual burned fragments of textile fabric adhering to pieces of flint (Soffer *et al.* 2000c).

To summarize, impressions of cordage, netting, textiles, and of baskets are apparently ubiquitous across Upper Paleolithic Europe. Furthermore, ongoing research on impressions of ceramic fragments from the first true pottery, dating to ca. 13,500 B.P., recovered from the Russian Far East, is producing further evidence for the wide distribution of perishable technologies across the furthest reaches of Eurasia (Derevianko 1997; Hyland 2000; Zhushchik-

hovskaya 1996). Clearly, specimens analyzed to date contain impressions of cordage as well as twined basketry or textiles. The absence of perishable technology from synthetic publications about Upper Paleolithic lifeways and text books on prehistory reflects a failure to recognize these items rather than evidence of their absence from the late Pleistocene record.

The Diversity of the Impressed Perishables

Technological Diversity

The Moravian Upper Paleolithic perishable inventory, augmented by new findings from Russia, Germany, and France, reveals that an extraordinarily wide range of perishable items was produced in late Pleistocene Europe by a broad array of addititve methods (Adovasio *et al.* 1999, 2000; Soffer *et al.* 1998). As noted in Table 20.1, the inventory includes single-ply, multiple-ply, and braided cordage, knotted netting, plaited wicker-style basketry, as well as a wide variety of non-heddle, loom-woven textiles, including simple and diagonal twined pieces, plain woven, andtwilled objects. A number of these pieces exhibit intentional structural decoration (Soffer *et al.* 1998), while others evidence conjoining of two pieces of fabric by a whipping stitch to produce a seam (Adovasio *et al.* 1999, 2000). This observation denotes sewing – a production technique confirmed by needles and the use of tailored clothing as is discussed further below.

All of these impressions, which range in age from ca. 28,000 B.P. to ca.13,000 B.P., represent well-made items. The typological heterogeneity coupled with the general regularity and narrow gauge of the warp and weft elements used in all of the textile/basketry types identified to date, suggest a high level of standardization and antecedent development, both for these specimens and the fiber industry at large. As noted elsewhere, younger Mesolithic/Archaic and Neolithic/Formative perishable assemblages usually exhibit a far more restricted array of types with a clear preference for certain warp and weft manipulations as well as preferred initial spin and, especially, final twist directions of both cordage and yarns (Adovasio *et al.* 1999, 2000). Possibly, this later homogeneity reflects a stabilized technology practiced over millennia while the observed Upper Paleolithic heterogeneity may reflect developed but relatively new technologies. The observed Upper Paleolithic heterogeneity may also indicate idiosyncratic production at the level of the household as well as the functional nature of the Pavlovian sites in Moravia (Adovasio *et al.* 1999, 2000). Specifically, it is suggested that since the Moravian sites appear to have been aggregation loci where a number of independent social units gathered seasonally, weaving and basket making at these sites should evidence

Table 20.1. Recovered Perishable Fiber Technology from Upper Paleolithic Europe, by Class.

Type	Pavlov I	Dolní Vestonice I	Dolní Vestonice II	Gönners-dorf	Kostenkí I	Zaraisk
			Textiles: Twining			
Open Simple, Z-Twist Weft	2	–	–	–	–	–
Close Simple, Z-Twist Weft	2	1	–	–	–	–
Open Simple, S-Twist Weft	3	3	–	–	–	–
Close Diagonal, Z-Twist Weft	1	–	–	–	–	–
Open Diagonal, Z-Twist Weft	4	–	–	–	–	–
Close Diagonal, S-Twist Weft	2	–	–	–	–	–
Open Diagonal, S-Twist Weft	3	1	–	–	–	–
Close Simple, S-Twist Weft	–	1	–	–	–	–
Close Simple, Z and S-Twist Wefts	–	1	–	–	–	–
Open and Close Simple, Z and S-Twist Wefts	–	1	–	–	–	–
Close Simple, Unknown-Twist Weft	2	–	–	–	–	–
Open Unknown, Z-Twist Weft	2	–	–	–	–	–
Unknown Simple, S-Twist Weft	1	–	–	–	–	–
Unknown	7	1	1	–	–	–
			Textiles: Plain Weave			
1/1 Balance Plain Weave	–	5	–	–	–	–
			Basketry: Plaiting			
2\2 Twill	–	4	–	–	–	–
Unknown	–	6	–	–	–	–
			Cordage			
Single, One-Ply, Z Spun	1	–	–	–	–	–
Multiple, Two-Ply, S Spun, Z Twist	2	1	–	–	–	–
Multiple, Two-Ply, S Spun (?), Z Twist	–	1	–	–	–	–
Multiple, Two-Ply, Z Spun, S Twist	4	–	–	–	–	–
Multiple, Two-Ply, Z Spun (?), S Twist	–	3	–	–	–	–
Compound, Two-Ply, Z Spun, S Twist	1	–	–	–	–	–
Braided, Three-Strand	1	1	–	–	–	–
Z Twist	3	–	–	–	1	–
S Twist	8	6	–	1	–	–
Unknown	–	1	–	–	1	–
			Miscellaneous: Knotted Netting			
Weaver's Knotted	4	–	–	–	–	1

greater technological variability than at locales where multiple social units co-resided together on a more permanent basis.

The assemblages also contain several impressions of knotted cordage (Adovasio *et al.* 1999). Depending on its precise configuration, ethnographically and archaeologically known knotted cordage often represent fragments of netting. Although several varieties of knotted netting exist, only one type of knotted netting is thus far inferred for the specimens from the studied sites – a weaver's knot or one of its variants such as a fishnet knot.

Potential Raw Materials

The textiles, basketry, and cordage specimens from Upper Paleolithic sites in Europe were clearly made of plant rather than animal fiber, though at present an explicit identification of species is impossible. Pollen analyses from the Moravian sites indicate a predominately open landscape with bast-bearing and other plants available at the sites (Adovasio *et al.* 1999, 2000). A number of these taxa are suitable for construction material. Mason and her colleagues (Mason *et al.* 1994) suggest that the fibrous bark of both alder (*Alnus* sp.) and yew (*Taxus* sp.) were locally available and that the herbaceous flora may have included milkweed (*Asclepias* sp.) and nettle (*Urtica* sp.), all of which have well-documented ethnographic and prehistoric uses as perishable production media (Barber 1991). Additionally, nettle has a long history of use as a weaving fiber in Europe (Barber 1991; Hald 1941).

Perishable Inventories

Since the studied impressions are highly fragmentary, quite small, and little more than miniature "negatives" of the original products, it is impossible to specify with confidence the original form or size of any of the items. It is certain, however, that both woven textiles and plaited baskets are represented. As argued elsewhere (Adovasio *et al.* 1996, 1998, 1999, 2000; Soffer *et al.* 1998), it is highly likely that the plaited items represent baskets or mats. Similarly, the relatively wide range of textile gauges and weaves suggest mats, perhaps wall hangings, blankets, and bags, as well as a wide array of apparel including such items as shawls, shirts, skirts, and sashes. The extremely narrow gauge of some of these (e.g., specimen DVI-#1 strongly argues for fine, woven clothing. Additionally, the presence of sequentially spaced knots on some of the impressions suggests the production of netting, while the identification of whipping stitch seams indicate the textile sewing of more complex structures such as clothing and bags (Adovasio *et al.* 1999, 2000; Soffer *et al.* 1998).

Tools to Produce Nets, Make Baskets, and Weave Cloth

The extensive evidence for weaving, as well as net and basket making in the Upper Paleolithic, necessarily raises the question of whether specialized tools were utilized by the makers. To answer this, the senior author initially examined ethnographic collections from hunter-gatherer cultures curated at the Illinois State Museum, the Smithsonian Institution, and the American Museum of Natural History with the purpose of identifying the diagnostic morphology of such tools as well as of delimiting unique wear patterns which result from such use. Armed with this information, a preliminary survey of European Upper Paleolithic bone, ivory, and antler implements was conducted to ascertain if specific weaving, netting, or basket making tools could be definitively identified. Although this research is necessarily at a preliminary stage, these initial forays into the pertinent literature have yielded the following observations.

First, it should be noted that Upper Paleolithic textiles *are* coeval with the implements associated with sewing, weaving, as well as with net making. As noted elsewhere, eyed needles make their first appearance during the Gravettian period *sensu lato*, and are reported in all parts of Europe, from France (de Baune 1993) to Sungir' in Russia (Bader 1998). Although some of these, such as the large needle from Predmostí in the Czech Republic (Klíma 1990:Figure 28; Valoch 1982:Figure 4), may have been used in net making, its subsequent and much smaller ivory equivalents, ubiquitously found in later Upper Paleolithic sites, attest to extensive sewing and possibly to embroidery – a hypothesis initially suggested by Cheynier (1967). While these needles are traditionally associated with the sewing of leather and hide garments to make tailored clothing (Bader 1998; Sterdeur-Yedid 1979), the authors

suggest that the size of many of the needles is far too small for this and, more likely, reflects working with woven textiles and/or accessory stitching or embroidering rather than conjoining animal hides.

Similarly, many Upper Paleolithic inventories contain implements, previously identified as hunting weaponry or decorative "art" objects, which may have been associated with textile production (e.g., the bone "spear head" from Predmostí [Klíma 1990:Delte3; Valoch 1982:Figure 1] or its analogue from Abri Blanchard [White and Breitborde 1992:Figure 7] which may be a net spacer; the "figurative fork" from Predmostí which suggests its use as a netting spacer [Figure 20.2]). Additionally, it is hypothesized that the sitting anthropomorphs made of mammoth phalanges from Predmostí (Klíma 1990:Figures 23–24) and their equivalents from Avdeevo in Russia (Gvozdover 1995: Figure 154) possibly served as loom weights. Moreover, the enigmatic "rondelles," including the cut-out ivory circular objects from Sungir' (Bader 1978:Figure 114), likely were spindle whorls, as were the perforated mammoth bone disks from Mezhirich (Figure 20.3). Similarly, the

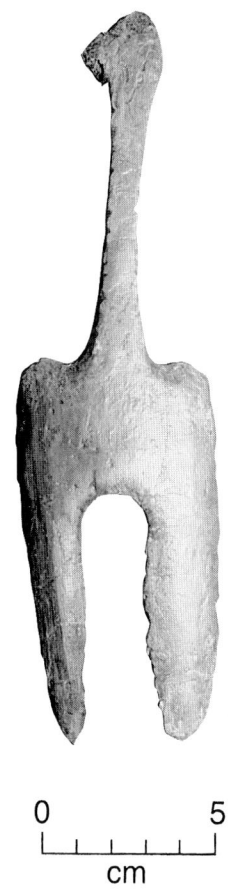

0   5
cm

Figure 20.2. Zoomorphic fork-like piece from Predmostí likely used in weaving and looping (photo O. Soffer).

Figure 20.3. Circular disk made of mammoth bone from the Upper Paleolithic site of Mezhirich, likely used as a spindle whorl (photo V. Suntsov).

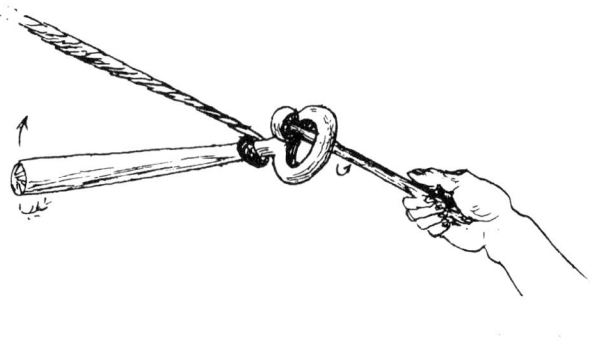

Figure 20.4. Perforated implement used in Portugal for twisting of cordage (after Lacorre 1960:Figure19).

engraved foot-shaped "pendant" made of fossil ivory and found at Kniegrotte, Germany, conceivably was used as a fiber comb (Feustel 1974:Figure XXVII).

Finally, at least two scholars, Heite (1998) and Lacorre (1960) using ethnographic analogies, suggested that the bone, ivory, or antler "shaft straighteners" (known in Europe as *batons de commandement*) were probably associated with organic perishable technologies. Specifically, Heite (1998) demonstrated that similarly shaped objects, termed "madmen," are presently used in Iceland for spinning yarn, while Lacorre (1960:Figure 19) pointed out that similarly shaped instruments were used historically in Portugal to produce cordage (Figure 20.4).

Preliminary survey of osseous implements in European Upper Paleolithic collections from sites from the Atlantic

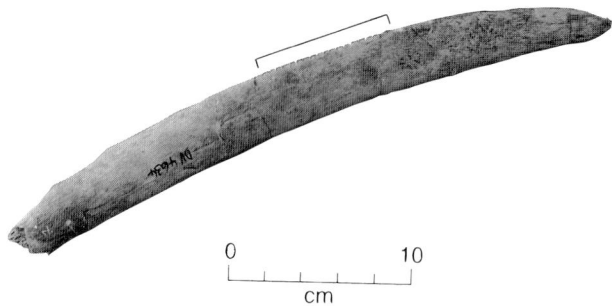

Figure 20.5. Fragment of a batten made in ivory from the Upper Paleolithic site of Dolní Vestonice I (photo O. Soffer).

to the Urals identified a number of net spacers or gauges, weaving sticks, and battens such as the batten from Dolní Vestonice (Figure 20.5). These materials are ubiquitous in the inventories and their existence corroborates the mounting evidence that extensive weaving, net making, and basket making, indeed, occurred during the Upper Paleolithic.

In summary, it is clear that not only are impressions of textiles and baskets made of organic materials present in the European Upper Paleolithic record, but so also are the actual tools used to produce them.

Upper Paleolithic Iconography:The "Dressed Venus"

The third source of information concerning Upper Paleolithic perishable technologies derives from iconography – specifically from the items of clothing depicted on the "Venus" figurines. As argued elsewhere, prior research on these items of late Pleistocene material culture paid extraordinary and exclusive attention to the sexual characteristics of the images (Soffer *et al.* 2000a, 2000b, 2000c) often overlooking entirely the clothing depicted on these females. A detailed examination of these representations, which number over 200 specimens just for the Gravettian period (ca. 30,000–20,000 B.P.), reveals variability in the clothing depicted, and the detailing indicates that this clothing was made of plant rather than animal fibers (Soffer *et al.* 2000a, 2000b, 2000c).

The Dressed "Venus"

A recent study of "Venus Wear" demonstrates that a significant number of these figurines are depicted dressed in headwear, belts, bandeaux, and bracelets (Soffer *et al.* 2000a, 2000b, 2000c). These include the well known "Venus" of Willendorf (ca. 25,000 B.P.) shown wearing a radially stitched hat (Figure 20.6) as well as analogous

0 5

cm

Figure 20.6. Side view of the Venus of Willendorf with diagnostic coiled headwear (photo S. Holland).

pieces from Kostenkí and Avdeevo (ca. 21,000 B.P.), some of which show radial centers while others depict plaited start basket hats. All of these specimens exhibit such exquisite detailing of the production sequence used to make the headwear, that one can identify the number of rows as well as splices and other modifications employed in shaping the item. Although more abstract, examples from Western Europe, such as the "Venus" of Brassempuoy, the "Venus with the Gridlike Head" from Laussel, as well as the "Negroid head" from Grimaldi (also dating to the Gravettian period), display netted headwear in the form of snoods (Soffer *et al.* 2000a, 2000b, 2000c).

One form of the depicted bodywear is the bandeau, with or without straps, worn above the breasts. The most realistically detailed of these is on a marl figurine recovered from Kostenkí I–2 (ca. 21,000 B.P.). Not only is the weft selvage depicted but also the conjoining of the straps to the bandeau, presumably by sewing (Figure 20.7).

The third item of clothing found on the "Venuses" are woven belts (Soffer *et al.* 2000a, 2000b, 2000c). As discussed in detail elsewhere, these belts in Eastern Europe

are depicted worn on the waist, while in Central and Western Europe, they ride low on the hip. At times these hip belts are highly abstracted, as in the case of the "Venus" of Dolní Vestonice (Marshack 1991:Plate 4 a-b), while in other cases, such as on the large Kostenkí I-2 body fragment made of marl (Soffer *et al.* 2000b:Figure 9), they are quite detailed. Based on the the sole example of the "Venus" of Lespugue (Soffer *et al.* 2000b:Figure 10), the belts, in Western Europe, were conjoined to string skirts as demonstrated in exquisite detailing on the back of the figurine.

When Upper Paleolithic dressed female images are depicted, equal attention is paid to the detailing of their clothing as is to the delineation of their primary and secondary sexual characteristics (Soffer *et al.* 2000a, 2000b, 2000c). This directly implies that weaving and basket making were socially important skills whose highly perishable intricacies were made permanent in ivory and stone. It is also noted that the headwear of the clad "Venuses" is accented while their faces, if outlined at all, are not. This strongly suggests that appropriate headwear was socially and ideologically more important than individual identity.

Notably, it is only the female images that are displayed dressed. The depictions of Upper Paleolithic males, as well as of unsexed anthropomorphs, are not. This patterning suggests that weaving and basket making were apparently associated with women and, predicated on the limited occurrence of clad "Venuses," these perishable technologies were likely associated only with specific categories of women.

Perishable Technologies and Upper Paleolithic Daily Life

The discovery that diverse and sophisticated organic perishable technologies were widespread across Europe by at least 25,000 years ago offers a number of important revisions to the current reconstructions of Paleolithic lifeways. These, specifically, include subsistence practices, social organization, and the gendering of labor.

Hunting in the Upper Paleolithic

According to the received wisdom, daily life in the late Pleistocene was replete with brave male hunters stalking, intercepting, and successfully dispatching an impressive range of such large sized animals as mammoths and mastodons. This received wisdom, however, ignores the reality that the largest number of faunal remains at many European Upper Paleolithic residential sites are comprised of small-sized taxa such as hares and marmots (Adovasio *et al.* 1999; Soffer 1984; Soffer *et al.* 1998) as well as the limitations of hunting weaponry before some 20,000 B.P. (Soffer, 2000).

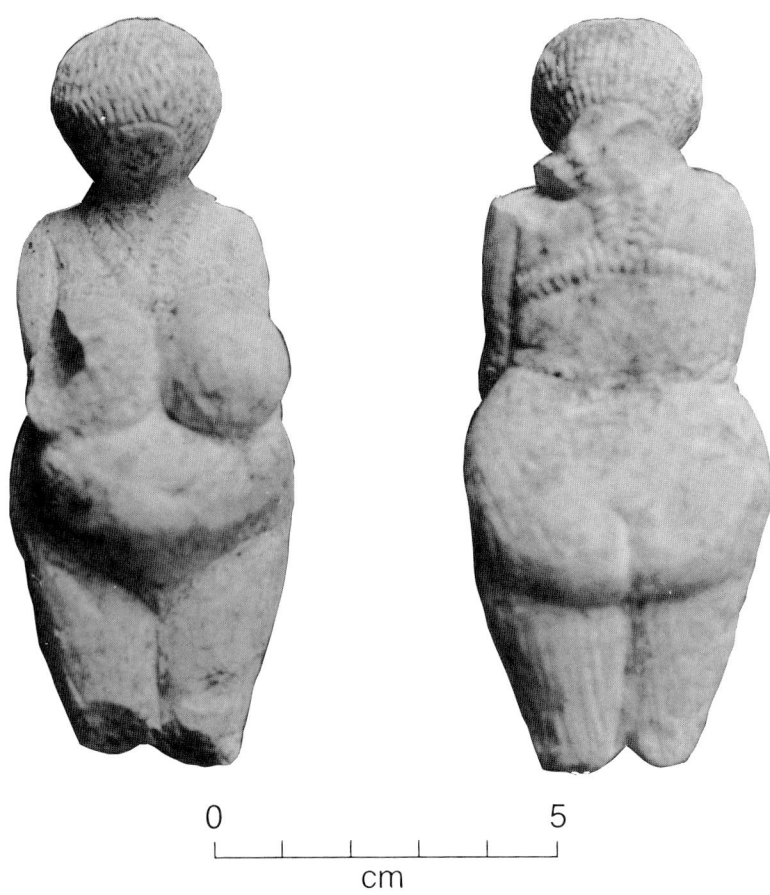

Figure 20.7. "Venus" figurine made of marl from Kostenkí 1–2.

For example, the abundance of microliths and geometric microliths in the Moravian Gravettian inventories herald the presence of complex multi-component weaponry – most likely lethal barbed throwing and thrusting spears and lances. These short-range weapons limited prey choice to large sized animals and hunting methods to ambush. On the basis of cross-cultural ethnographic data, Churchill (1993) established that long distance hunting weaponry (i.e., spear throwers, bows and arrows) considerably increases the effective range of hunting implements as well as allows the effective hunting of medium and small sized animals. Additionally, he demonstrated that the absence of long-distance hunting weaponry favors co-operative hunting (Churchill 1993). The Gravettian-age European record contains no spear throwers, and while some scholars have argued that some of the microliths may have served as arrows – and, thereby, suggesting that bow and arrow hunting came into practice some 26,000 years ago – evidence for this is equivocal and in need of empirical demonstration. This fact, combined with the abundant and

varied faunal remains at the Moravian Upper Paleolithic sites, has led a number of Czech scholars to argue for collaborative hunting (e.g., Klíma 1964; Musil 1994, 1997; Svoboda *et al.* 1996).

The identification of net making in Upper Paleolithic sites in Central and Eastern Europe suggests that nets were likely used for hunting (Adovasio *et al.* 1999, 2000; Soffer *et al.* 1998). To date, the mesh size of the identified nets is quite small, which would have made them suitable for the hunting of only small sized prey. This observation concurs with the high number of fur-bearers like hares and foxes recovered from the Moravian sites (Musil 1994, 1997). It also agrees with cross-cultural ethnographic data which cite the use of nets to capture small terrestrial fauna, including fur-bearers, throughout the world (Andrews and Adovasio 1996; Satterthwait 1986, 1987; Steward 1938; Roscoe 1990, 1993). While the ethnographic literature further documents the widespread use of nets for fowling and fishing, the lack of fish remains and the paucity of avian elements at the Moravian sites (Musil 1994, 1997) and at other sites in Central and Eastern Europe suggests that fish and birds were not an important component of the Upper Paleolithic diet. This observation corroborates Oswalt's (1976) hypothesis that net hunting of terrestrial game likely preceded fishing and fowling with nets.

The fine gauge of the netting identified to date strongly suggests that, perhaps, the larger gauge fraction of this perishable industry has not been preserved or recovered. Ethnographic data show that nets, in addition to capturing small-sized mammals in the 3–20 kg (9.1 lb) weight range, were successfully used to capture a wide range of larger-sized herbivores from kangaroos and horses in post-contact times in Australia (Satterthwait 1986, 1987) to antelope, deer, and mountain sheep in North America (Andrews and Adovasio 1996; Frison *et al.* 1986). This, in turn, suggests that, if practiced, net hunting possibly was used to procure larger-sized taxa in Upper Paleolithic Europe as well.

The possibility of net hunting carries important implications. Cross-cultural research indicates that net hunting is a communal effort. Due to the relative lack of expertise necessary for success as well as the minimal danger involved in such a non-confrontational harvesting technique, net hunting requires few skilled hunters. Consequently, it can and does utilize the labor of the entire co-residential social unit (Anell 1969; Frison 1987; Satterthwait 1986, 1987; Steward 1938; Wilkie and Curran 1991). It is, therefore, the one hunting method strongly associated with the labor of women, children, and the elderly (Murdock 1937; Murdock and Provost 1973). In contrast to other confrontational methods of prey capture, net hunting does not require excessive physical strength nor the elaboration of stone tool kill weaponry. Immobilized animals can be dispatched in relative safety and security at close range by a wide variety of methods such as stabbing and clubbing. Furthermore, it is likely that net hunting was more frequent in the past than documented in the ethnographic record, where it is rare but widespread (Manhire *et al.* 1985; Satterthwait 1986, 1987). Summarizing, nets are a form of tended facilities, which, as Oswalt (1976) denoted, substitute human labor through collaborative action for technological elaboration.

It is important to emphasize that net-hunting is also associated with large harvests in short periods of time and, thus, with the production of a surplus (Satterthwait 1986, 1987). Although such surplus in some ethnographic cases is associated with participation in a market economy (e.g., the Ituri forest [Wilkie and Curran 1991]), in other cases, such as in Aboriginal Australia (Satterthwait 1986, 1987) or in New Guinea (Roscoe 1990, 1993), it is associated with large gatherings, feasting, and ceremonialism.

In brief, the study of seemingly insignificant weavers knots, usually invisible or unnoticed in the archaeological record, permit us to consider the daily conduct of the, heretofore, unseen majority of Upper Paleolithic people and to examine the subsistence pursuits of not only prime-aged men but those of women, children, and older individuals.

Women at Work

The identification of weaving and basket making across Europe, combined with the iconographic association of these technologies with women, permit us to escape such two-dimensional androcentric tropes as "Ice Age Hunters" and to address issues of gender, agency, labor, and the value placed on that labor.

The authors recognize that it is very difficult to unambiguously associate any technology with specific social groups in archaeology – be it men, women, or children. However, as discussed in detail elsewhere, there are cogent and compelling reasons for associating weaving and basket making with Upper Paleolithic European women (Soffer *et al.* 2000a, 2000b, 2000c). These arguments are summarized here by noting that this association is justified not only by the iconographic data discussed above but also by analogy to division of labor documented in simpler societies in the recent past.

The cross-cultural ethnographic record firmly associates women with plant harvesting and processing (Kehoe 1990, 1991; Murdock 1937; Murdock and Provost 1973; Owen 1996, 2000; Watson and Kennedy 1991). Additionally, it associates women with transforming plant matter into more complex perishable structures such as baskets, mats, and cloth. As numerous scholars assert (Barber 1990, 1994; Hald 1980; King 1991; Schneider and Weiner 1989), it is the women who usually are the weavers and basket makers in all simpler societies where textiles and basketry are made for household and communal needs. It is acknowledged that ethnographic accounts indicate that occasional perishable organic items, such as woven footwear, cordage, rope, and certain type of nets, are also made by males. These exceptions, however, are just that and not the norm—an observation especially valid for textiles and textile production where male involvement is usually minimal except in some advanced tribal and chiefdom level contexts.

In the case of Upper Paleolithic Europe, iconographic depictions of woven and fabric-based clothing clearly associate these products with women. Furthermore, the finely detailed "Venus-wear" indicates that the figurine creators were either extensively guided by the knowledge of expert weavers and basket makers or that the weavers and the basket makers were the actual artists (Soffer *et al.* 2000a, 2000b, 2000c).

Perishables as Precious Products

The mere fact that the garments worn by Upper Paleolithic "Venuses" are so accurately depicted is clearly significant. Perishable weaving and plaiting skills were apparently important enough in late Pleistocene European cultures to be immortalized in such durable media as stone, ivory, and bone (Soffer *et al.* 2000a, 2000b, 2000c). Furthermore, such iconographic transformation suggests that the women who wove the textiles and made the baskets likely held positions of marked status in their societies.

Further insights into Upper Paleolithic ideologies are gained by considering the fineness of the produced textiles. As discussed in this chapter and elsewhere, the extremely narrow gauge of many of the textiles imprinted on the ceramic fragments from Dolní Vestonice and Pavlov I reflects a highly refined textile industry (Adovasio *et al.* 1997, 1998, 1999, 2000; Soffer *et al.* 1998). Indeed, the finest of these, such as the plain weave from Dolní Vestonice I, is comparable to cloth produced not only in the Neolithic but also in the later Bronze and Iron Ages (see Delte). Fine weaving, as Barber (1990, 1994) and Brumfiel (1996) demonstrated, is a very labor intensive activity and generally associated, both prehistorically and historically, with intensification of female labor. The existence of such fine weaving during the Upper Paleolithic, thereby, suggests a similar pattern. This observation concurs with the marshaling of the labor of women, children, and the elderly as postulated in the previous discussion of communal net hunting.

While the value placed on women's hunting skills during the Paleolithic may never be known with certainty, iconographic evidence permits raising this question in terms of women's weaving and basketry. In his discussion of prehistoric valuables, Clark (1986) identified ivory and shell objects as likely "symbols of excellence" during the Upper Paleolithic. Although in doing so he restricted himself to a consideration of items made in durable media, Clark noted that in simpler societies such valuables are often made of highly perishable materials. This observation, which is amply confirmed by hunter-gatherer ethnographies, raises a strong possibility that Upper Paleolithic textiles and basketry functioned in an equal capacity and served as part of prestige economies (Soffer *et al.* 2000a, 2000b, 2000c) as they clearly did in much later state level societies (Barber 1994). While, in all likelihood, it will remain impossible to unequivocally identify the specific agents who made the durable valuables, iconography conveys that women definitely used Upper Paleolithic textiles and basketry. Parsimony as well as ethnology argue that women were the producers of these perishable valuables and, by extension, possibly the durable icons.

Conclusions

In summary, this contribution suggests that the following observations may be offered about late Pleistocene lifeways:

1. A variety of textile, basketry, and netting items were produced in Upper Paleolithic Europe by at least 27,000 B.P., constituting the earliest evidence for these perishable technologies in the world (Adovasio *et al.* 1998, 1999; Soffer *et al.* 1998). This, in turn, illustrates that fiber artifact production is millennia older than previously envisioned and implies that these technologies were part and parcel of the technological repertoire of the first Americans.

2. This Paleolithic perishable inventory undoubtedly included items intended for both household and hunting needs. Specifically, the widespread evidence for nets across Europe suggests that communal net hunting was practiced, which likely yielded food surpluses needed at such aggregation sites as Dolní Vestonice I and Pavlov I.

3. Upper Paleolithic communal net hunting conceivably involved the participation of all members of the coresidential units and its advent likely signals increased demands on the labor of women, children, and older individuals.

4. The ubiquitous Upper Paleolithic clad "Venus" figurines associate the wearing of woven and twined clothing with a particular category of women. This clothing, which likely depicts ritual wear, includes twined wear as well as non-heddle, loom woven plain weaves (Soffer *et al.* 2000a, 2000b, 2000c).

5. Upper Paleolithic iconography as well as ethnographic and ethnohistorical analogs associate the production of textiles and baskets with women. This permits not only a discussion of women's work in the remote past but also reveals one of the principles used to gender Upper Paleolithic people.

6. The association of textiles and basketry products with a limited number of the "Venuses" (Soffer *et al.* 2000a, 2000b, 2000c), suggests that a variety of roles were available to Upper Paleolithic women in addition to purely gender-based ones. Some of these roles, such as those of weavers and basket makers, were clearly associated with positions of status.

7. The extraordinary fineness of some of the weaves and the iconographic attention devoted to this technology suggest that these organic perishables were "symbols of excellence," serving as important signifiers in prestige economies and in status demarcation.

8. Finally, and perhaps most importantly, it is these organic perishables, and not the inorganic durables, that allow a social perspective on Pleistocene lifeways

and reveal the previously unseen decision making process and the deployment of labor. Ideally, both social theory as well as evolutionary ecology enable us to perceive decision making at the individual level.

While the authors freely admit that the Pleistocene record is likely too coarse for consistent interpretation, the studies summarized above demonstrate that it is not too coarse to reflect the behavior of different constituencies which comprised past hunter-gatherer groups – e.g. the women, the children, the young and the old. It is, thus, quite tragic that not only is there a general failure to develop stronger methodologies for the recovery of organic perishables from all archaeological contexts but their likely existence in our reconstructions of the past is, for the most part, ignored. Clearly, prehistoric life was facilitated by far more than stone – would that we remember this!

Acknowledgments
We gratefully acknowledge the assistance of many individuals with the research as well as with the production of this chapter. Jeff Illingworth conducted the preliminary identifications of the new impressions from Russia and Germany as well as assisted in the production of this paper. The artwork was produced by Steve Holland. Discussions with numerous colleagues on both sides of the Atlantic, including Francoise Audouze, Elizabeth Barber, M.D. Gvozdover, Joyce Marcus, Alexander Marshack, Linda Owen, Mary Ann Owoc, Jiri Svoboda, and Randall White, greatly contributed to the formulation and refinement of the ideas presented here. This contribution was edited by J. E. Thomas and produced by D.R. Pedler.

Olga Soffer wishes to acknowledge the financial support of her research on Upper Paleolithic adaptations in Central and Eastern Europe by the International Research and Exchange Board with funds provided by the U.S. State Department (Title VIII program), the U.S. National Academy of Sciences, the National Endowment for the Humanities, the Wenner-Gren Foundation for Anthropological Research, the Woodrow Wilson Institute, the American Museum of Natural History, and the Research Board of the University of Illinois. The views and statements expressed above and any errors or biases found within this document are solely the responsibility of the authors and in no way should be attributed to any of the above-mentioned funding agencies or organizations.

References Cited
Adovasio, J.M. 1999. Perishable Artifacts, Paleoindians, and Dying Paradigms. Paper presented at the Clovis and Beyond – Peopling of the Americas Conference, Santa Fe, New Mexico. October 28–31.
Adovasio, J.M., O. Soffer, D.C. Hyland, B. Klíma and J. Svoboda, 1998. Perishable Technologies and the Genesis of the Eastern Gravettian. Anthropologie XXXVI/1–2:43–68.
Adovasio, J.M., O. Soffer, D.C. Hyland, B. Klíma, and J. Svoboda. 1999. Textil, 'Kosakarstvi a Site v Mladém Paleolitu Moravy. Archeologické rozhledy LI:58–94..
Adovasio, J.M., O. Soffer, D.C. Hyland, J.S. Illingworth, B. Klíma, J. Svoboda. Perishable Industries from Dolni Vestonice I: New Insights into the Nature and Origin of the Gravettian. In Proceedings of the Masaryk Conference, edited by J. Malina and D. Sosna, Masaryk University, Czech Republic. (in press)
Adovasio, J.M., O. Soffer and B. Klima. 1996. Upper Paleolithic fibre technology: Interlaced woven finds from Pavlov I, Czech Republic, c. 26,000 years ago. Antiquity 74:526–534.
Andrews, R.L. and J.M. Adovasio. 1996. The Origins of Fiber Perishables Production East of the Rockies. In: J.B. Peterson, ed., A Most Indispensable Art: Native Fiber Industries from Eastern North America, The University of Tennessee Press, Knoxville, pp. 30–49.
Anell, B. 1969. Running down and Driving of Game in North America. Studia Ethnographica Upsaliensis XXX.
Bader, O.N. (ed). 1998. Pozdnepaleoliticheskoe Poselenie Sungir'. (The Late Paleolithic Site of Sungir'). Moscow: Nauchnyj Mir.
Bader, O.N. 1952. O drevnosti Skhodninskoj chereprnoi kryshki i o kharaktere ee naruzhnoi poverkhnosti. (On the antiquity of the Skhodnin calvarium and on the nature of its exterior surface.) Uchenye Zapiski MGU. Vyp. 158:155–157.
Bader, O.N. 1978. Sungir'. Verhnepaleoliticheskaia stoianka. (Sungir'. An Upper Paleolithic site). Nauka, Moscow.
Barber, E.J.W. 1991. Prehistoric Textiles. Princeton University Press, Princeton, N.J.
Barber, E.J.W. 1994. Women's Work: The First 20,000 Years. New York: W.W. Norton and Co.
Bosinski, G. 1995. Gönnersdorf'. In Quaternary Field Trips in Central Europe, edited by W. Schirmer, vol 2, pp. 906–910. Munchen: Verlag Dr. Fredrich Pfeil.
Brumfiel, E.M. 1996. Figurines and the Aztec State:Testing the Effectiveness of Ideological Domination. In Gender and Archaeology. ed. R.P. Wright. pp. 143–166. Philadelphia: University of Pennsylvania Press.
Cheynier, A. 1967. Comment Vivait l'Homme des Cavernes. (How Cavemen Lived). Paris. Robert Arnoux.
Churchill, S. 1993. Weapon Technology, Prey Size, Selection, and Hunting Methods in Modern Hunter-Gatherers: Implications for Hunting in the Paleolithic and Mesolithic. In G.L. Peterkin, H.M. Bricker, and P. Mellars, (eds). Hunting and Animal Exploitation in the Later Paleolithic and Mesolithic of Eurasia. Archaeological papers of the American Anthropological Association no. 4:11–24.
Clark, D.L. 1968. Analytical Archaeology, Methuen, London.
Collins, H.B. Jr. 1937. Archaeology of St. Lawrence Island, Alaska. Smithsonian Miscellaneous Collections vol. 96. Smithsonian Institution, Washington, D.C.
Conkey, M.W. 1991. Contexts of Action, Contexts for Power: Material Culture and Geder in the Magdalenian. In J.M. Gero and M.W. Conkey, (eds). Engendering Archaeology. Blackwell, Oxford: pp. 57–92.
Croes, D.R. 1997. The North-Central cultural dichotomy on the

Northwest Coast of North America: Its evolution as suggested by wet-site basketry and wooden fish-hooks. Antiquity 71:594–615.

Damas, D. (ed). 1984. Arctic. Handbook of North America Indians vol. 5. Smithsonian Institution, Washington, D.C.

de Baune, S.A. 1998. Les hommes au temps de Lascaux. (Man during the times of Lascaux). Paris: Hachette.

Derevianko, A.P. (ed). 1998. University of Illinois Press, Champaign, Illinois.

Dillehay, T.D., (ed). 1997. The Archaeological Context and Interpretation, Monte Verde: A Late Pleistocene Settlement in Chile, vol. 2. Smithsonian Institution Press, Washington, D.C..

Feustel, R. 1974. Die Kniegrotte. (The Kniegrotte). Weimar: Hermann Bohlaus Nachfolger.

Freeman, L.G., J. Gozáles Echegaray, R.G. Klein, and W.T. Crowe. 1998. Dimensions of Research at El Juyo:An Earlier Magdaleian Site in Cantabria Spain. In H.L. Dibble and A. Montet-White, (eds). Upper Pleistocene Prehistory of Western Eurasia. The University Museum, University of Pennsylvania, pp. 3–40.

Frison, G. 1987. Prehistoric, Plain-Mountain Large-Mammal Communal Hunting Strategies. In M.H. Netecki and D.V. Netecki, (eds). The Evolution of Human Hunting. Plenum Publishing Corp., New York, pp. 177–224.

Frison, G.C. 1998. R.L. Andrews, J.M. Adovasio, R.C. Carlisle, and E.A. Edgar. Late Paleoindian Animal Trapping Net from Northern Wyoming. American Antiquity 51:352–363.

Glory, Abbé. 1959. Débris de corde paleólithique à la Grotte de Lascaux. (Remains of a Paleolithic cord from the Cave of Lascaux). Mémoires de la Société Préhistorique Française 5:135–169.

Gvodover, M.D. 1995. Art of the Mammoth Hunters. The finds from Avdeevo. Oxbow Monograph 49. Oxford.

Hald, M. 1942. The Nettle as a Culture Plant. Folk-Liv 6:28–49.

Heite, L. 1998. Spear Straightener or Spinning Tool? Mammoth Trumpet 13, no. 3, pp. 18–19.

Helm, J. (ed). 1981. Subarctic. Handbook of North America Indians. vol. 6. Smithsonian Institution, Washington, D.C.

Kehoe, A.B. and S.M. Nelson. 1990. Points and Lines. In Powers of Observation:Alternative Views in Archaeology. Archaeological Papers of the American Anthropological Association no. 2:23–37.

Kehoe, A.B. 1991. The Weaver's Wrath. In:The Archaeology of Gender. Proceedings of the Twenty-Second Annual Conference of the Archaeological Association of the University of Calgary. Eds. D. Walde and N.D. Willow. pp. 430–435. The University of Calgary, Archaeological Association.

King, M.E. 1991. The Perishable Preserved:Ancient Textiles from the Old World with Comparisons from the New World. The Review of Archaeology 13:1, pp. 2–11.

Klíma, B. 1963. Dolní Vestonice. Nakladatelství Ceskoslovenské Akademie Ved. Praha.

Klíma, B. 1990. Lovci mamutu z Predmostí. (The mammoth hunters from Predmosti). Academia, Praha.

Lacorre, F. 1960. La Gravette. Impremerie Barneoud S. A., Laval.

Leroi-Gourhan, A. and J. Allain. 1979. Lascaux Inconnu. (The Unknown Lascaux). Paris:CNRS.

Marshack, A. 1991. The Female Image: A Time-factored Symbol. A Study in Style and Aspect of Image Use in the Upper Palaeolithic. Proceedings of the Prehistoric Society 57, part 1:17–31.

Mason, O.T. 1910. Woman's Share in Primitive Culture. D. Appleton & Co. New York.

Mason, S.L.R., J.G. Hather and G.C. Hillman. 1994. Preliminary Investigation of the Plant Macro-remains from Dolní Vestonice II, and Its Implications for the Role of Plant Foods in Palaeolithic and Mesolithic Europe. Antiquity 68:48–57.

Manhire, T., J. Parkington, and R. Yates. 1985. Nets and fully recurved bows: rock paintings and hunting methods in the Western Cape, South Africa. World Archaeology 17:161–174.

Murdock, G.P. 1937. Comparative Data on the Division of Labor by Sex. Social Forces 15:551–553.

Murdock, G.P. and C. Provost. 1973. Factors in the Division of Labor by Sex:A Cross-Cultural Analysis. Ethnology 12:203–226.

Musil, R. 1994. The fauna. In: J. Svoboda, (ed). Pavlov I. Excavations 1952–1953. ERAUL 66:181–210.

Musil, R. 1997. Hunting Game Analysis. In J. Svodoba, (ed). Pavlov I –Northwest. The Dolní Vestonice Studies. Vol.4. Institute of Archaeology, Academy of Sciences of the Czech Republic, Brno: pp. 443–468.

Nadel, D., A. Danin, E. Werker, T. Schick, M.E. Kislev and K. Stewart. 1994. 19,000–Year-old Twisted Fibers from Ohalo II. Current Anthropology 35:451–457.

Oswalt, W.H. 1976. An Anthropological Analysis of Food Getting Technology. John Wiley & Sons, New York.

Owen, Linda R. 1996. Der Gebrauch von Pflanzen in Jungpaläolithikum Mitteleuropas. (Plant products in the Upper Paleolithic of Central Europe). Ethnographisch-Archäollogische Zeitschrift 37:119–146.

Owen, Linda R. 1999. Questioning Stereo-Typical Notions of Prehistoric Tool Functions. Ethno-analogy, Experimentation and Functional Analysis. In Ethno-Analogy and the Reconstruction of Prehistoric Artefact use and Production. Edited by L.R. Owen and M. Porr. (in press).

Praslov, N.D. 1991. O keramike epokhi paleolita. (About Paleolithic ceramics). In V.M. Masson, (ed). Drevnie Kul'tury i Arkheologicheskie Izyskania. Materialy k plenumu IIMK 26–28 noiabrya 1991 g. IIMK, Sankt Peterburg: pp. 47–50.

Roscoe, P.B. 1990. The Bow and Spreadnet:Ecological Origins of Hunting Technology. American Anthropologist 92:691–701.

Roscoe, P.B. 1993. The Net and the Bow in the Ituri. American Anthropologist 95:153–155.

Sakharov, V.V. 1952. Geologicheskie uslovia zalegania Skhodnenskoj chereprnoi kryshki. (The geological context of the Skhodnin calvarium). Uchenye Zapiski MGU. Vyp. 158:151–154.

Satterthwait, L. 1986. Aboriginal Australian Net Hunting. Mankind 16:31–48.

Satterthwait, L. 1987. Socioeconomic implications of Australian Aboriginal net hunting. Man (N.S.) 22:613–636.

Schneider, J., and A.B. Weiner. 1989. Introduction. In A.B. Weiner and J. Schneider, (eds). Cloth and Human Experience. Smithsonian Institution Press, Washington, pp. 1–15.

Sinel'nikov, N.A. 1952. Ob obrazovanii tkanevidnogo rel'efa na Skhodnenskom fragmente cherepa. (About the formation of textile impression on the Skhodnin cranial fragment). Uchenye Zapiski MGU. Vyp. 158:175–179.

Soffer, O. 1984. The Upper Paleolithic of the Central Russian Plain. Orlando: Academic Press.

Soffer, O. 2000. Gravettian Technologies in Social Contexts. In Hunters of the Golden Age, edited by W. Roebroeks, M. Mussi, and J. Svoboda. Leiden:The University of Leiden Press, (in press).

Soffer, O., J.M. Adovasio, D.C. Hyland, B. Klima, and J. Svoboda. 1998. Textiles and Basketry in the Paleolithic – What then is the Neolithic? In: Historical-Cultural Contacts Between Aborigines of the Pacific Coast of Northwestern America and North Eastern Asia. (ed). A.R. Artem'ev, Institute of History, Archaeology and Ethnography of the Peoples of the Far East, Russian Academy of Sciences Far Eastern Branch.Vladivostok. pp. 311–317.

Soffer, O., J.M. Adovasio, and D.C. Hyland. 2000a. The Well-Dressed Venus:Women's Wear ca. 27,000 BP. Archaeology, Ethnology, and Anthropology of Eurasia. 1:37–47.

Soffer, O., J.M. Adovasio, and D.C. Hyland. 2000b. The "Venus" Figurines:Textiles, Basketry, Gender and Status in the Upper Paleolithic. Current Anthropology 41:511–537.

Soffer, O., J.M. Adovasio, J.S. Illingsworth, Kh. A. Amirkhanov, M. Street, and N.D. Praslov. 2000c. Paleolithic Perishables Made Permanent. Antiquity (in press).

Sterdeur-Yedid, D. l979. Les Aiguilles á Chas au Paleolithique. (Paleolithic Eyed Needles). XII Supplement á Gallia Pre-histoire. Paris: CNRS.

Steward, J. 1938. Basin-Plateau Aboriginal Sociopolitical Groups. Bureau of American Ethnology Bulletin 120.

Svoboda, J., V. Lozek, and E. Vlcek. 1996. Hunters between East and West. Plenum Publishing Co. New York.

Taylor, W.W. 1966. Archaic Cultures adjacent to the Northeast Fronteers of Mesoamerica. In: G.E. Ekholm and G.R. Willey, (eds). Handbook of Middle American Indians. vol. 4. University of Texas Press, Austin, Texas: 59–94.

Tinnes, J 1994. Die Geweih-, Elfenbein-und Knochenartefakte der Magdalénienfundplätze Gönnersdorf und Andernach. (The Antler, Ivory, and Bone artifacts from the Magdalenian sites of Gönnersdorf and Andernach). Inaugural-Dissertation zur Erlangung des Doktorgrades der Mathematisch-Natural-wissenschaftlichen Fakultät der Universität zu Köln, Germany.

Trusov, A.V. 1998. Kremnevyj kompleks Zaraijskoj paleo-liticheskoj stoyanki (The lithic complex from the Zaraisk Upper Paleolithic site). In:Vostochnyj Gravett, (ed). Kh. A. Amirkhanov. Moscow:Nauchnyj Mir. pp. 279–298.

Valoch, K. 1982. Die Beingeräte Von Predmistí in Mähren (Tschechoslowakei). (The ivory inventory from Predmistí in Moravia (Czechoslovakia). Anthropologie XXI:57–69.

Watson, P.J., and M.C. Kennedy. 1991. The Development of Horticulture in the Eastern Woodlands of North America: Women's Role. In: Engendering Archaeology. Edited by J.M. Gero and M.W. Conkey, pp. 255–275. Oxford: Blackwell.

Watanabe, H. 1968. Subsistence and Ecology of Northern Food Gatherers with Special Reference to the Ainu. In: R.B. Lee and I. DeVore, (eds). Man the Hunter. Aldine de Gruyter, New York:pp. 69–77.

White, R. and L.R. Breitborde. 1992. French Paleolithic Collections in the Logan Museum of Anthropology. Logan Museum Bulletin (new series), vol. I, no.2.

Wilkie, D.C. and B. Curran. 1991. Why Do Mbuti Hunters Use Nets? Ungulate Hunting Effiiency of Archers and Net-Hunters in the Ituri Rain Forest. American Anthropologist 93:680–689.

Zhushchikhovskaya, I.S. 1996. Current Data on the Early Pottery of Primorie Region, Russian Far East. Journal of Korean Historical Society. 21:301–312.

21. The Conservation of Wetland Archaeological Sites in New Zealand/Aotearoa

Dilys A. Johns

Introduction

Human occupation by the indigenous people of New Zealand, the Maori, began between 800 and 1000 years ago.

The organic remains that have been deposited in wet sites since human arrival provide a particularly valuable component of the archaeological record, as pottery and metals were completely absent from pre-European (i.e. pre-eighteenth century) Maori culture. The lack of metals meant that the tools of the pre-European Maori were manufactured from stone, shell, obsidian, fiber and wood. Wood played a central role in everyday Maori life, and this is reflected in the skill demonstrated in the manufacture, of not only domestic items, houses and canoes, but also in the remarkable carved artworks.

Archaeologists are able to study this rich aspect of Maori material culture largely though chance finds or, more rarely, through the excavation of wet sites.

Wetland archaeological sites are found throughout the country. They occur in a variety of places, including swamps, lakes, streams, river margins, and the ocean bed. Wetland legislation in New Zealand is covered by nine Acts of Parliament (Scott 1966). Nearly all of these Acts are primarily concerned with the ecology of wetlands, rather than with cultural aspects.

Despite the regulatory role of Regional and District Councils, the Historic Places Trust, and the Department of Conservation, no specialized protection for wet sites exists, even though it is well-known that the composition of these fragile systems requires that they are kept environmentally stable. Protection often occurs after development rather than as the result of any systematic programs (Allen 1998).

Maori involvement and ethics

Maori regard the landscape as embodying their ancestors, and this extends to carved artifacts. While the Maori conservation ethic often requires places and artifacts to remain undisturbed *in situ*, even if subject to natural decay, Maori are concerned about destruction caused by human activities such as drainage.

Maori, *tangata whenua* (literally "people of the land"), have the responsibility to protect and manage the physical and spiritual resources of their land.

Over the last two decades, the archaeological community has become increasingly aware of the need to be more active in establishing links with Maori.

It is now generally accepted that *iwi* must be involved in the interpretation, custody, protection, and display of their sites and the artefacts associated with them, and that this can be achieved only with a strong and effective partnership in place.

There are long-term benefits for all parties when relationships are strengthened between archaeologists, conservation and *iwi*.

Differing uses of wetlands and site conservation strategies.

Maori life was often centered around water, and to the Maori, wetlands were sources of food, defensive possibilities and transport. They knew the preservative qualities of swamps, and many artifacts were deliberately buried in them, either for preservation, or for concealment during times of strife or warfare. The use of swampy land for safekeeping of artifacts included the hiding of prized carvings in swamps when a village was being hastily left for long or short periods. The Te Atiawa *pare* (Figure 21.2) is a good

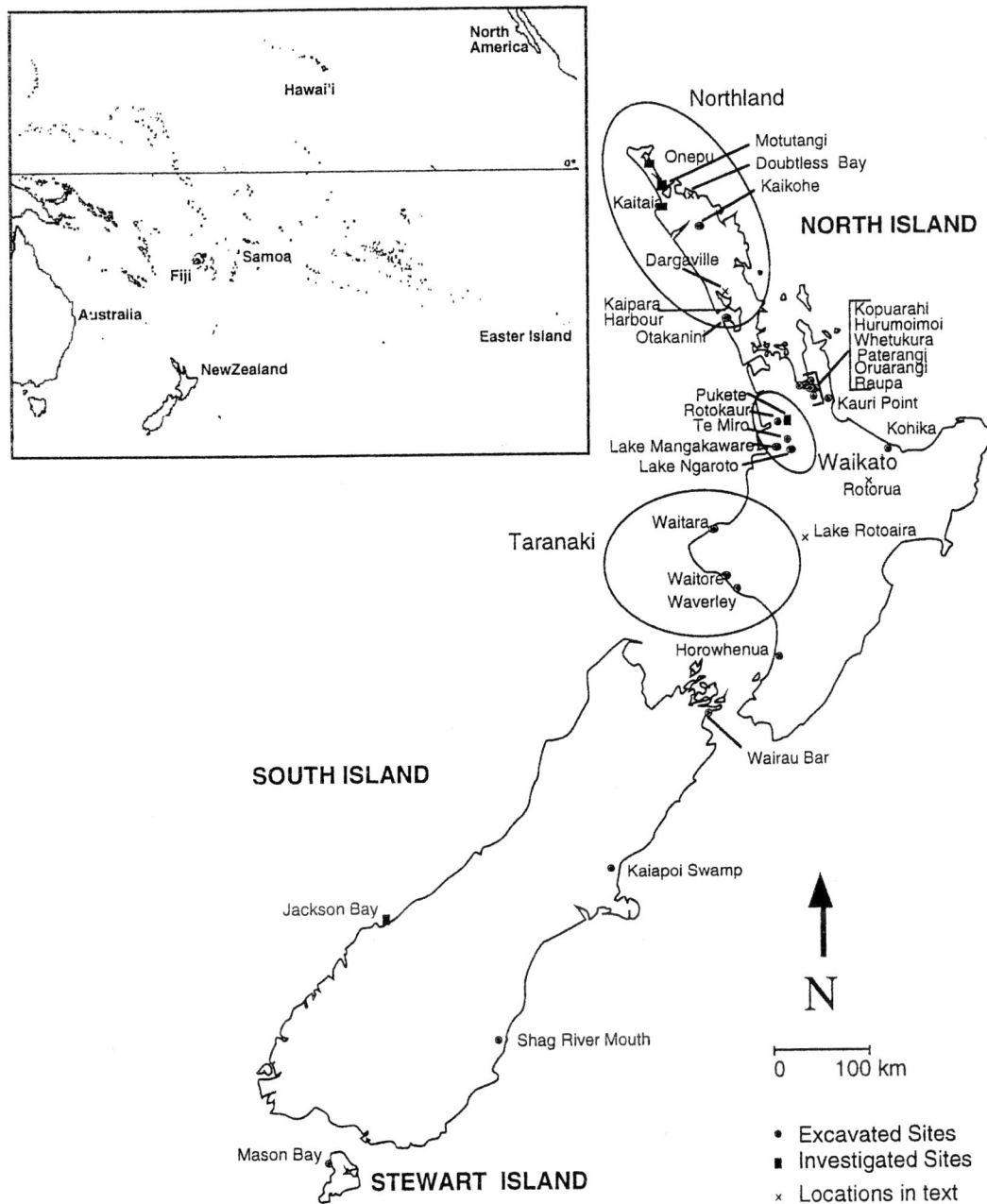

Figure 21.1. Locations mentioned in text and wetland archaeological sites investigated in New Zealand.

illustration of a find from this type of activity (Houston 1954).

For a large part of this century, the New Zealand Government subsidised the draining of lowland, peaty, wet areas to provide farmland.

Draining of swamp land on farms has revealed water-logged archaeological deposits. Sometimes this has been accompanied by the raising of funds, either by a local museum or by a central Government body, to excavate and document the site. Kohika, in the Bay of Plenty (Irwin 1975), and Te Miro, in the Waikato (Edson 1979), are examples of sites whose excavations were initiated by the exposure of artifacts during drainage projects.

However, and more importantly, the draining of swamps has had an extremely detrimental effect on waterlogged deposits, ensuring that both site strata and artifacts are

Figure 21.2. Te Atiawa pare (door lintel) or paepae (doorsill of a raised food storehouse) found in the swamp near Waitara. Taranaki Museum.

irreversibly damaged. Many sites in the Taranaki, Waikato and Northland regions (Figure 21.1) have been destroyed as a result of swamp draining and farming activities.

In New Zealand, unlike the situation in England, where a national program of wetland survey and evaluation has been implemented (Corfield 1998), basic knowledge of our wetland resource is still incomplete.

There is an urgent need for thorough surveys of the wetland sites in Taranaki, Waikato and Northland, areas, which were known to have been heavily occupied in pre-European times. Two wetland projects which attempt to redress some of the gaps in this knowledge have been started recently, and they are discussed below under current projects.

Different wetland sites in New Zealand

In 1989, Barr noted that New Zealand's wetsites could be divided into five main categories:

- historic sites
- ditch features
- drowned dryland sites
- caches
- island *pa*/wetland habitation sites

Historic wetland sites were often associated with land use by European settlers such as flax milling around swampy margins, and gum digging near Dargaville (Figure 21.1).

The most frequently-discussed examples of ditch features, which were probably used for gardening, occur in the far north around Motutangi, Onepu and Kaitaia. The famous Kaitaia carving (Figure 21.3) was found near ditch features in this area during a drainage project. Ditch features have also been recorded near Wairau Bar.

Examples of drowned dryland sites can be found in the central North Island near Rotorua, notably at Ohinemutu and Lake Okataina. Both these sites were inhabited when dry, and subsequently became waterlogged as a result of

the levels of the lakes on which they were located rising through volcanic or seismic events.

This paper concentrates on cache sites, which are often also chance finds, and island *pa*/wetland habitation sites, as these site types have produced most of the artifacts which have been conserved at the University of Auckland.

The highest concentration of cache sites found recently has been in the Taranaki region, where there are extensive wetlands, and land is intensively farmed

Island *pa* are artificial island sites, which were nearly always fortified and built mainly in the swamps of the North Island towards the end of the pre-contact period. They were concentrated mainly in the Waikato area (Bellwood 1978, Cassels 1972, Shawcross 1968), although other significant swamp pa which have been excavated are in the Bay of Plenty (Irwin 1975) and on the Hauraki plains (Furey 1996) (Map 21.1).

Settlements were also built on the margins of streams and lakes. At least three of the sites (recorded but not yet excavated) on Lake Ngaroto are examples of this type.

Past wetland archaeological investigations

Wet archaeological sites and their significance were recorded in the early part of this century. As early as the 1920s a surveyor drew careful maps of pre-European drains near Kaitaia (Wilson 1921).

One of the reasons that wet sites were generally un-excavated, even though they were known to be a rich resource for understanding pre-contact life in New Zealand, was the difficulty and expense of excavation and subsequent conservation. This was exacerbated by the lack of know-ledge and facilities for the conservation of any recovered artefacts.

The earliest wetsite excavations were those at Kauri Point (Shawcross 1963), Lake Ngaroto (an undefended wet habitation site) (Shawcross 1968), and Lake Mangakaware (Bellwood 1971 and 1978). Lake Mangakaware comprised a fortified village built on a small promontory on the shores of the lake.

Later excavations were at Waitore, in southern Taranaki, which had been a fifteenth-century fishing camp in the swamp of a coastal sand dune (Cassels 1979), and Kohika, a swamp *pa*, located within the once-vast Rangitaiki swamp (Irwin 1975, Lawlor 1979).

It was the Kohika excavations which demonstrated the need for competent conservation, to deal with the wealth of material being unearthed.

Current projects

Recently, Dr Harry Allen and I, together with several colleagues from the University of Auckland and elsewhere,

Figure 21.3. The Kaitaia carving found during draining in swampland near Kaitaia in Northland. Te Rarawa. Auckland Museum.

have been successful in securing funds from the Royal Society of New Zealand to conduct a three-year research project on the wetlands of Taranaki. This project is being conducted with the full co-operation of the *tangata whenua,* and we have developed an on-going association with Ngati Mutunga tribe as partners.

Taranaki, on the west coast of the central North Island, was heavily populated during New Zealand's prehistory, and a wide range of archaeological sites is represented, including many types of wetsites. Many have been drained for farming and industry during this century.

For this project a multi-disciplinary team has been assembled which consists of archaeologists, an ethnographer, a dating specialist, a conservator, a palynologist, and a Maori researcher.

The project proposes integrating Taranaki's wetland and adjacent dry-land sites using archaeological conservation and paleoenvironmental studies, and incorporating this data with a knowledge of waterlogged artefact find spots and Maori traditions concerning the swamps, within a landscape archaeology approach. Work will begin in 2000, with collection of Maori traditional, historic and local knowledge of wetlands, their use and cultural values, followed by field studies including surveying, coring and test excavations.

I have been involved in developing a further project during 1999 with contract archaeologists Warren Gumbley and Kevin Jones. It is being funded by the New Zealand Department of Conservation, Science and Research Unit.

This applied research project aims to establish guidelines for the development of conservation strategies, monitoring and ultimately conserving wetland archaeological sites in New Zealand.

It will result in recommendations for the management, protection and maintenance of three important previously-investigated wetland sites: Lake Mangakaware, Lake Ngaroto, and Kauri Point (Bellwood1978, Shawcross 1968, and Shawcross 1963).These sites, although located in the countryside, have been subject in the past to the requirements of the forestry industry and are currently influenced by the varied activities of farming

The May 1999 issue of the WARP newsletter mentions this project.

Artifact assemblages

Pre-European Maori knew the properties of different native wood species, and they chose particular types of wood to manufacture the different items they required, including tools, carvings, canoes, houses, and gardening and domestic implements. Wallace (1983) carried out a survey of artifacts held in five New Zealand museums, and identified the species of wood of which they were made. What he found was that generally, although there were exceptions, kanuka *(Leptospermum ericoides)* was favored for digging sticks, kauri *(Agathis australis)* branch heartwood for fernroot beaters, matai *Podocarpus spicatus)* for bowls, rimu *(Dacrydium cupressinum)* for combs, and totara *(Podocarpus totara)* and matai *(Podocarpus spicatus)* for canoes. These are the main wood species we receive in the laboratory.

Typical artifact assemblages from excavated wet sites include house timbers and carvings, palisade posts from fences around villages or *pa,* domestic items such as bowls and fern root beaters, horticultural implements such as digging sticks and their footrests, weeding implements, fishing gear, eel traps, toys, ornaments, bundles of vine, and canoes.

At Kauri Point swamp in the Bay of Plenty, (Shawcross 1964) an unusual artifact assemblage of over one hundred combs was found during excavations. Using these combs Shawcross was able to demonstrate a change in style over a period of 150 years, from plain square-topped combs at the beginning of the site's occupation to round-topped ornamented ones at later dates.

Fiber-work is not often found in New Zealand wet sites but there were significant finds at Kohika (Irwin 1975) and Raupa (Prickett 1990). At Raupa several pieces of Maori fiber-work were found including rain capes and plaited pieces dating from the early nineteenth century. These pieces illustrate Maori fiber-working techniques during an important period of change, some fifty years after European contact (Lander 1992). Of particular interest was the eight-stranded rain-cape neck edge (Figure 21.4), a technique no longer used by cloak makers. Replication of the neck edge by a well-known teacher of cloak making skills, Diggeress Te Kanawa , has allowed a technique which had largely

Figure 21.4. Rain cape neck edge with eight-stranded square braid from Raupa excavations. Ngati Tamatera. Auckland Museum.

been forgotten to survive through archaeology and conservation.

Development of Maori art traced through carvings from wetsites

As previously mentioned, the draining of swampland on farms has revealed several nationally-significant artworks, including the Kaitaia carving (Figure 21.3).

While these styles fit into the early end of a sequence of art development (Mead 1995,1997; Simmons 1994), from rectilinear East Polynesian to curvilinear Maori art (Figure 21.2 and 21.5 are typical examples of curvilinear art), as chance finds they do not provide an established stratigraphic or chronological context.

The Kaitaia carving is a product of a carving style different to that documented at the time of European arrival in New Zealand. It has been argued that it is similar to the art of some East Polynesian islands (Barrow 1972, Kaeppler 1978) the generally agreed homeland of the Aotearoa Maori. However, the examples which Barrow and Kaeppler cite as similar to New Zealand "old" carvings are pieces collected in East Polynesia during the 1700s. The apparent time gap between the early New Zealand carvings and their later Polynesian parallels has not been fully explained. Neich (1996) suggested that there may have been similar artistic expression from widespread ancestral East Polynesian culture, or that the parallels may have been merely coincidental. Neich also discussed Sutton's (1987) work which indicated that there may have been multiple arrivals from Polynesia. If this did happen, each new arrival could have brought fresh East Polynesian art influences to New Zealand, possibly over a long period of time.

If this were the case, the idea that Maori art which resembles East Polynesian looking art is old may be incorrect.

Others (Davidson 1987) have suggested that these stylistic changes in Maori art may have been regional .

A carved palisade post found at North Kaipara head eight years ago and conserved in our laboratory, which is also referred to as old (Simmons pers.comm.) is worthy of mention here. This *taonga* has both rectilinear and curvilinear features present in its relief (Figure 21.6), which could be interpreted as a mixture of styles and may have been carved during a period of change from one style to the next.

The Waitore carved panel (Figure 21.7), from an excavation dated to the fifteenth century illustrates what archaeologists Cassels (1979) and Lawlor (1979) have described as "simple spirals which have been punched and notched into the surface of the board." Lawlor surmised that this design could possibly represent an intermediate stage in the progression from rectilinear style to curvilinear style and the beginnings of contemporary Maori art.

Both theories, (that carvings with rectilinear features may be old, *and* that carving styles are different because they come from different regions), have some merit. However, we need more examples of regional carving to be fully documented together with further confidently-dated carvings in order to gain a thorough understanding of the development of Maori art.

It is hoped that the current wetland projects outlined above will help to resolve some of these questions, at least for the Waikato and Taranaki regions.

Conservation

In 1978, a purpose-built conservation laboratory was set up at the University of Auckland in response to the large quantities of material in need of conservation which were being produced by the excavation of wet-sites.

Ambrose's method of impregnating waterlogged artifacts, using 5% to 10% aqueous solutions of polyethylene glycol (PEG) followed by freeze-drying (Ambrose 1976) was adopted. In 1984, I experimentally compared the PEG method with acrylic monomer impregnation followed by gamma-radiation-induced polymerisation (Johns 1985). Results from the latter technique appeared promising but variable. The available Co^{60} source was suitable for only small artifacts, and the use of large volumes of flammable organic solvents to dewater the objects was discouraged within the University.

The PEG method is still used, with sequential impregnation using two different grades being used for particularly degraded objects (Johns 1998).

Figure 21.5. Detail of the ornately-carved canoe prow found at Mason Bay, Stewart Island. Murihiku. Southland Museum and Art Gallery.

An ongoing program of research into techniques suitable for New Zealand wood species and organic remains is in place.

Wet organic archaeological materials are brought to the laboratory from throughout New Zealand for treatment and then returned either to their *iwi* or to museums. A set of protocols has been established with Maori to ensure that cultural values are respected during the conservation process. During the last year several important *taonga* have been received in the laboratory for treatment including a canoe prow roughout from the Ngati Ranganui *iwi* and a *waka kereu* from the Ngai Tahu people in the South Island.

The costs of conservation treatment for chance finds of Maori origin, more than 100 years old, are met by the Department of Internal Affairs. These artefacts are brought to the laboratory by *iwi*, museum personnel, and archaeologists.

Conclusion

New Zealand wetland sites have provided us with valuable archaeological data over the last thirty years and they require current and long-term integrated approaches for management if they are to survive .

This year has seen the beginning of two new wetlands projects being co-ordinated by two government-affiliated institutions, the University of Auckland and the Department of Conservation. These projects herald the beginning of renewed systematic investigations of wetland archaeological sites, which will enable us to understand and protect this rich heritage resource.

Finally, wetsites may be coming of age in New Zealand and promise to be an exciting aspect of our archaeology for years to come.

Glossary

Aotearoa	– "land of the long white cloud", New Zealand
iwi	– tribe
pa	– typically a fortified village, later applied to any village
paepae	– beam, threshold, doorsill, usually to a raised food storehouse
pare	– door lintel
tangata whenua	– people of the land, Maori
taonga	– property, anything highly prized

Acknowledgments

I would like to thank Hamish MacDonald, Maureen Lander, the Te Awamutu and District Museum, and the Photographic Archive, Department of Anthropology, The University of Auckland, for the photographs, and Harry Allen and Ritchie Sims for comments on the script.

References Cited

Allen, H. 1998. Protecting Historical Places in New Zealand, Research in Anthropology and Linguistics, Vol. 1, 1998, Department of Anthropology, The University of Auckland, Auckland, New Zealand.

Ambrose, W.R. 1976. Sublimation Drying of Degraded Wet Wood, in Grosso, Pacific Northwest Wet Site Wood Conservation Conference, Neah Bay, WA, Vol. 1, 1976, pp. 7–15

Barrow, T. 1972. Art and Life in Polynesia Reed Publishing Ltd. Wellington, New Zealand.

Barr, C.E. 1989. An Ecological Approach to the Management of Wet Archaeological Sites in New Zealand, Unpublished M.A. thesis, Anthropology Department, University of Auckland, Auckland, New Zealand.

Bellwood, P.S. 1969. Pa Excavations at Otakanini, South Kaipara, and Lake Mangakaware, New Zealand Archaeological Newsletter, Vol. 12, No. 1, 1969, pp. 38–49.

Bellwood, P.S. 1971. Archaeological Research at Lake

Figure 21.6. Carved palisade post found in the swampy sand dunes of Kaipara Head. Both rectilinear and curvilinear features are present on this piece. Ngati Whatua. Northern Wairoa Museum.

Mangakaware, Waikato: a summary of results, New Zealand Archaeological Newsletter, Vol.14, No. 3, p. 113

Bellwood, P.S. 1978. Archaeological research at Lake Mangakaware, Waikato, 1968–1970, New Zealand Archaeological Association Monograph 9, Wellington, New Zealand Archaeological Association.

Cassels, R. 1972. Human ecology in the prehistoric Waikato, Journal of the Polynesian Society, Vol. 81, No. 2, 1972, pp. 196–248.

Cassels, R.J. 1979. Early Prehistoric Wooden Artefacts from the Waitore Site (N136/16), near Patea, Taranaki, New Zealand Journal of Archaeology, Vol. 1, 1979, pp. 85–108.

Corfield, M. 1998. The Role of Monitoring in the Assessment and Management of Archaeological Sites, in Bernick, Hidden Dimensions, The Cultural Significance of Wetland Archaeology, Vancouver, UBC Press, 1998, pp. 302–316.

Davidson, J. 1987. The Prehistory of New Zealand. Longman Paul Ltd. Auckland, New Zealand.

Edson, S. 1979. Historical Archaeology in the Waikato: An interim report on the Te Miro project, New Zealand Archaeological Association Newsletter, Vol. 22, No. 2, pp. 65–76.

Furey, L. 1996. Oruarangi: The Archaeology and Material Culture of a Hauraki Pa, Bulletin of the Auckland Institute and Museum 17, Auckland, 1996.

Houston, J. 1954. Notes and Queries: The Taranaki Lintel, Journal of the Polynesian Society, Vol. 68, No. 3, 1954, pp. 239–240.

Irwin, G. 1975. The Kohika Site, Bay of Plenty, Whakatane and District Historical Society Newsletter, Vol. 23, No. 2, 1975, pp. 101–104.

Johns, D.A. 1985. Waterlogged Wood Conservation: an investigation of radiation-induced polymerisation of monomers, Unpublished M.A. thesis, Anthropology Department, University of Auckland, Auckland, New Zealand.

Johns, D.A. 1998. Observations Resulting from the Treatment of Waterlogged Wood Bowls in Aotearoa (New Zealand), in Bernick, Hidden Dimensions, The Cultural Significance of Wetland Archaeology, Vancouver, UBC Press, 1998, pp. 317–328.

Kaeppler, A.L. 1978. Artificial Curiosities. Bishop Museum Press. Bernice P. Bishop Museum Special Publication 65, Honolulu, U.S.A.

Lander, M. 1992. Fibre Fragments from the Raupa Site, Hauraki Plains, Records of the Auckland Institute and Museum, Vol. 29, 1992, pp. 7–23.

Lawlor, I. 1979. Stylistic affinities of the Waitore Site (N136/16), near Patea, Taranaki. New Zealand Journal of Archaeology. Vol. 1 p.109.

Mead, H.M. 1995. Te Toi Whakairo. The Art of Maori Carving Reed Publishing Ltd. Auckland, New Zealand.

Mead, S.M. 1997. Maori Art on the World Scene Ahua Design and Illustration Ltd Matua Associates Ltd, Wellington, New Zealand.

Prickett, N. 1990. Archaeological Excavations at Raupa: The 1987 Season, Records of the Auckland Institute and Museum, Vol. 27, 1990, pp. 73–153.

Scott, D.A. 1996. A Directory of Wetlands in New Zealand, Wellington, The Department of Conservation.

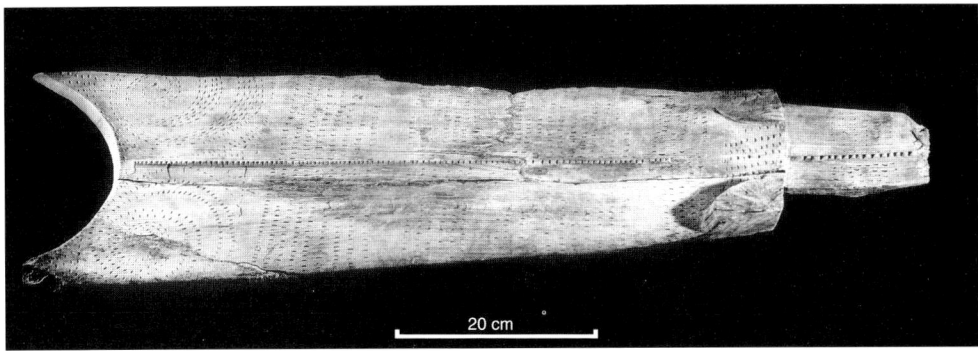

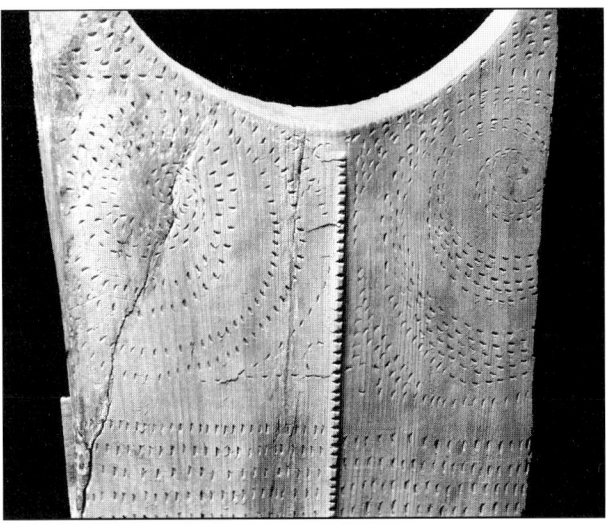

Figure 21.7. Decorated board excavated at the Waitore site near Patea, Taranaki. The Waitore site has been dated to the 1400s. The surface is decorated with punch marks in a series of spirals. Nga Rauru. Tararak Museum.

Shawcross, W. 1963. Kauri Point Swamp, New Zealand Archaeological Association Newsletter, Vol. 6, No. 1, 1963, pp. 50–56.

Shawcross, W. 1968. The Ngaroto Site, New Zealand Archaeological Association Newsletter, Vol. 11, No. 1, 1968, pp. 2–29.

Simmons, D.R. 1994. Whakairo: Maori Tribal Art. Oxford University Press Auckland, New Zealand.

Simmons, D.R. pers. comm. 25 November 1999. Simmons suggests that the Kaipara Palisade post dates from the 1500s.

Sutton, D.G.S. 1987. A Paradigmatic Shift in Polynesian Prehistory: Implications for New Zealand. New Zealand Journal of Archaeology. Vol.9 p. 135.

Wallace, R.T. 1983. Studies on the Conservation of Waterlogged Wood in New Zealand, Unpublished Ph.D. thesis, Chemistry Department, University of Otago, Dunedin, New Zealand.

Wilson, D. 1921. Ancient Drains Kaitaia Swamp, Journal of the Polynesian Society, Vol. 30, No. 2, 1921, pp. 185–188.

22. Wetlands and Archaeology: The Role of Ecosystem Structure and Function

Thomas L. Crisman, Ulrike A. M. Crisman and Joseph Prenger

Introduction

From the earliest suggestion that conifer trunks and stumps buried deep within bog deposits of Denmark were evidence of past forests (Dau 1829) to the application of von Post's techniques of pollen analysis for reconstruction of forest histories from sediments of Danish bogs by Jessen in the 1920's and 1930's (Iversen 1973), wetlands have been regarded as major repositories of historical evidence for climate change. Armed with new paleoecological analytical techniques, Danish scientists began to link pollen and macrofossil evidence for forest change with the stratigraphically contemporaneous archaeological record (Jessen 1935). Shortly thereafter, scientists began separating the relative importance of human interactions and climate change as contributing factors to the paleoecological record of local and regional forest change in Denmark (Iversen 1973), Ireland (Mitchell 1976) and England (Pennington 1974). The integrative science of wetland archaeology had begun.

Wetland paleoecological/archaeological reconstructions over the past four decades have studied both direct human interactions with the wetland proper and the more diffuse activities of humans within the watershed. Most paleoecologists seeking to link the archaeological and paleolimnological records have focused on sediment cores from lakes because of the greater likelihood of obtaining temporally continuous, undisturbed sediment records (Whitmore *et al.* 1996, Curtis *et al.* 1998, Leyden *et al.* 1998).

The interdisciplinary examination of the Huleh wetland of the Jordan River valley in Israel clearly demonstrated the value of wetlands for paleoecological research. Sediment chemistry (Hutchinson and Cowgill 1973, Cowgill 1994), diatoms (Sherman and Patrick 1981), non-siliceous plant and invertebrate remains (Ohlhorst *et al.* 1982) and molluscs (Ohlhorst *et al.* 1977) from a 54-meter core have provided an excellent environmental baseline against which the detailed archaeological record for the region can be linked.

One of the greatest challenges for integrating archaeology and aquatic ecology is the development of common terminology and definitions. Wetland archaeology appears to encompass loosely sites variously designated as underwater, wet or water-logged and includes archaeological records from wetlands, stream riparian zones, lakes and estuaries, thus spanning an array of conditions from salt to fresh, humically colored to clear, acid to alkaline, oligotrophic to eutrophic, and open water to littoral/ecotonal. Such boundless inclusion of terminology and ecosystem types both precludes development of a single set of archaeological methods applicable to all (Cole 1984) and often obscures the ecosystem structural and functional aspects responsible for preserving the archeological record contained within. Recognizing such constraints, Coles (1984) and Nicholas (1991,1992) stressed the need for archaeologists to have detailed understanding of wetland ecology, especially how preservational environments are likely to differ among wetland types.

The first terminalogical problem to overcome is a basic definition for wetland. Mitsch and Gosselink (1993) presented a detailed history of the definition of wetlands, including both scientific and political ramifications of each. Wetlands either can be thought of as ecotones separating upland and pelagic (lakes, oceans, rivers) aquatic ecosystems (Cowardin *et al.* 1979) or can be more broadly defined as any aquatic ecosystem less than six meters deep as recognized by the international Ramsar Convention (Navid 1989). Regardless of whether considered ecotones or ecosystems, wetland extent is usually delineated by vegetation, soils and hydrology.

There is equal confusion over definitions of individual wetland types. Although broadly reflecting either a European or North American bias, there is a general lack of intercontinental and often intracontinental agreement on terminology. Mitsch and Gosselink (1993) discussed sixteen of the most common wetland types recognized today and their intercontinental equivalents, but most, if not all, fall within two broadly defined wetland types. Swamps are wetlands dominated by trees and/or shrubs, and marshes are dominated by herbaceous vegetation. Reedswamps (Europe) and papyrus swamps (Africa) are recognized exceptions of the former, while bogs (usually dominated by *Sphagnum*) are often considered separate from the latter. The remainder of this chapter will follow this broad definitional convention.

Ecosystem Aspects Influencing the Archaeological Record

Hydrology

Hydrology is the most important factor affecting preservation of the archaeological record at wet sites. In most cases, water depth at the site is of relatively little concern given that most wetlands, regardless of depth, remain hypoxic almost continuously. Only in well oxygenated lakes and rivers is depth an important control for decomposition rates of those carbon components of the record protruding from the sediment matrix.

Wet site hydrology is characterized by its uncertainty. Regional patterns in the duration and seasonality of inundation have long been recognized (Mitsch and Gosselink 1993), but the archaeological record, especially within the terrestrial-aquatic ecotone, appears to be controlled mostly by unpredictable and rare extreme hydrological events. Rapidly rising water levels associated with short-term global climatic oscillations or local storm events can bury beached canoes and shoreline structures quickly via wave-generated sediment movement, but similarly rapid lowering of water levels can extend the terrestrial-aquatic ecotone horizontally into the aquatic ecosystem and expose the archaeological record to desiccation and oxidation of carbon artifacts.

One of the most poignant examples of the latter is the ongoing effort to catalogue and preserve over 85 wooden canoes exposed in Gumroot Swamp near Gainesville, Florida as adjacent Newnans Lake declined to historically low levels associated with a pronounced drought during the spring and summer of 2000. Not only is this the largest single find of canoes to date, but the site has yielded abundant, but rarely found, wooden push poles that were used to propel the canoes. Canoes were radiocarbon dated as between 500 and 5,000 years old, with a majority being

between 3,000 and 5,000 years (Miller 2000). The failure to find older canoes does not necessarily mean the absence of a human population, but corresponds with reflooding of lake basins in Florida associated with a regional water table rise approximately 5,000 years ago (Watts and Hansen 1988). The herculean task presented at Newnans Lake has clearly demonstrated that governmental agencies and museums lack the infrastructure and funding to respond adequately to archaeological opportunities associated with rare and extreme hydrological events.

Sediment and Soils

A clear distinction must be made between sediments and soils because of inherent differences in processes affecting the survival of buried archaeological records. Sediment is comprised of organic and inorganic material being deposited, whereas soils are formed through biochemical transformations of sediments as a function of hydrological conditions and time. Preservation of the archaeological record is strongly influenced by the progression of aquatic soil development at the site.

Aquatic (hydric) soils, whether mineral or organic, are defined as "a soil that is saturated, flooded, or ponded long enough during the growing season to develop anaerobic conditions in the upper part" (U.S. Soil Conservation Service 1987). Water percolation into such soils is controlled by both sediment grain size and overall organic content. Vertical percolation rates of water are positively related to sediment grain size and negatively with the thickness of surficial organic deposits. Reduced water exchange promotes increasing anoxia with depth in the soil. Biochemical transformations are strongly influenced by development of anoxia, in the sequence of oxygen reduction, nitrate reduction, iron reduction, sulfate reduction and methanogenesis controlled by time-mediated reduction in redox potential (Mitsch and Gosselink 1993). It is important to note, however, that soil biogeochemical transformations are also influenced by the depth of the root zone for aquatic plants and their species-specific ability to pump oxygen to their root zones to create oxygenated soil micozones.

Water Column Parameters

Three interrelated characteristics of water columns in aquatic ecosystems set the stage for long-term preservation of the archaeological record prior to burial: trophic state, dissolved oxygen and color. Trophic state is defined as the extent of primary production (photosynthesis) by algae and/or macrophytes in aquatic ecosystems and ranges from low (oligotrophic), through intermediate (mesotrophic), to highly (eutrophic) productive systems. Oligotrophic systems tend to be well oxygenated throughout their water columns continuously with slow deposition rates of low to moderately rich organic sediments. In marked contrast, eutrophic lakes display marked vertical reduction in oxygen in their water columns, with deepest waters remaining anoxic or strongly hypoxic with the exception of brief seasonal mixing periods (spring and fall in the temperate zone) of the entire water column. Such ecosystems are characterized by rapid sedimentation of highly organic sediments, which, coupled with reduced bioturbation of sediments from oxygen-stressed benthic invertebrates and fish, present ideal conditions for preservation of archaeological records in eutrophic ecosystems. Most wetlands, regardless of trophic state, can display extended periods of hypoxia or even anoxia throughout the year (Chapman *et al.* 1998).

Dissolved oxygen concentrations in aquatic ecosystems are also influenced by the level of organic color (humic and fulvic acids) in the water column. The principal source of the organic acids producing such stained waters is decomposition of plant material of select species including pine and bald cypress in the southeastern United States, and there can be a marked seasonality in the concentrations of color associated with the timing of leaf fall. Colored lakes are characteristic of regions displaying small fluctuations of water table, thus favoring development of cypress fringes and swamps around lake basins (Crisman 1992). Multiple regression models have been developed to examine the relative importance of lake size, trophic state and color as determinants of interlake variance in percent oxygen saturation for humically colored lakes of central Florida (Crisman *et al.* 1998). On a seasonal basis, percent oxygen saturation displayed a significant negative correlation with color levels for winter, spring and summer, and no relationship was seen with lake trophic state for any season. Lake size was a significant factor during fall and winter, the period of cool water and wind. Colored Florida lakes are naturally oxygen depleted, and recent cultural eutrophication of many lakes has had little pronounced effect on profundal oxygen levels of such lakes.

Linkages of Wetland and Cultural Histories

Both cultural and wetland ecosystems are dynamic temporally, and their juxtaposition in time geographically is often random. When humans and aquatic ecosystems are contemporaneous, unless water is a scarce commodity and there are no or limited choices of ecosystem types in the area, human selection of aquatic ecosystems with which to interact is likely not random but dictated by specific ecosystem attributes necessary to fulfill the culture's religious/ceremonial, dietary, and sustainable resource (water supply, building materials, grazing) needs. The

association between cultures and wetlands has likely never been a passive process, and it is not always evident if humans influenced the aquatic ecosystem more than the reverse and whether such interactions remained consistently positive or negative through time.

Wetlands are often ephemeral features of landscapes and may not provide conditions conducive for preservation of cultural records throughout the entire period of human occupation of the surrounding area. A prime example of this is provided by the history of hydrological variation and the associated presence of swamps in Florida and adjacent southern Georgia (Figure 22.1). The palynological history developed by W.A. Watts and associates for six lakes, Annie (Watts 1975) and Tulane (Watts and Hansen 1994) in south-central Florida, Mud (Watts 1969) and Sheelar (Watts and Stuiver 1980) in north-central Florida, Camel (Watts *et al.* 1992) in the northwest Florida panhandle and Louise (Watts 1971) in southern Georgia provide a detailed history of *Taxodium* (baldcypress and pond cypress), a dominant element of both isolated and fringing swamps of the region. At no site has the representation of *Taxodium*, and presumably hardwood swamps, been stable over the past 30,000 years. The peak of sustained presence of *Taxodium* was greater than 30,000 years ago in Mud and Camel lakes, 11,000–20,000 in Tulane and Sheelar and for approximately the past 3,000 years in Louise and Annie. Athough poorly developed in Tulane and Sheelar, all basins displayed increased representation of *Taxodium* pollen in the past 5,000 years relative to the previous 5,000 years, reflecting relative hydrological stability at higher water levels. Although humans likely occupied Florida for most of the period coverd by the pollen record of these lakes, it is readily apparent that there was great interbasin variability in the hydrological history of individual lakes and wetlands of the region reflecting basin morphology and groundwater proximity. This has affected both interactions of local cultures with water resources and the likelihood of preservation of evidence of this interaction. The latter is strongly influenced by temporal oscillations in the shoreward extent of the zone of most pronounced human interaction, the ecotone or littoral zone. Although humans likely interacted intensely with the water resource at the land-littoral zone interface, interactions lakeward at the littoral zone-open water interface, a zone of concentrated fish biomass, have not been delineated.

In landscapes possessing an array of aquatic ecosystems, the relative importance of individual wetland and lake ecosystem types to local populations for sacred versus secular purposes is poorly known. Partitioning of type and intensity of human interaction according to ecosystem type has likely affected the archaeological record and its interpretation in a profound way due to interecosystem differences in the potential for preservation of cultural records.

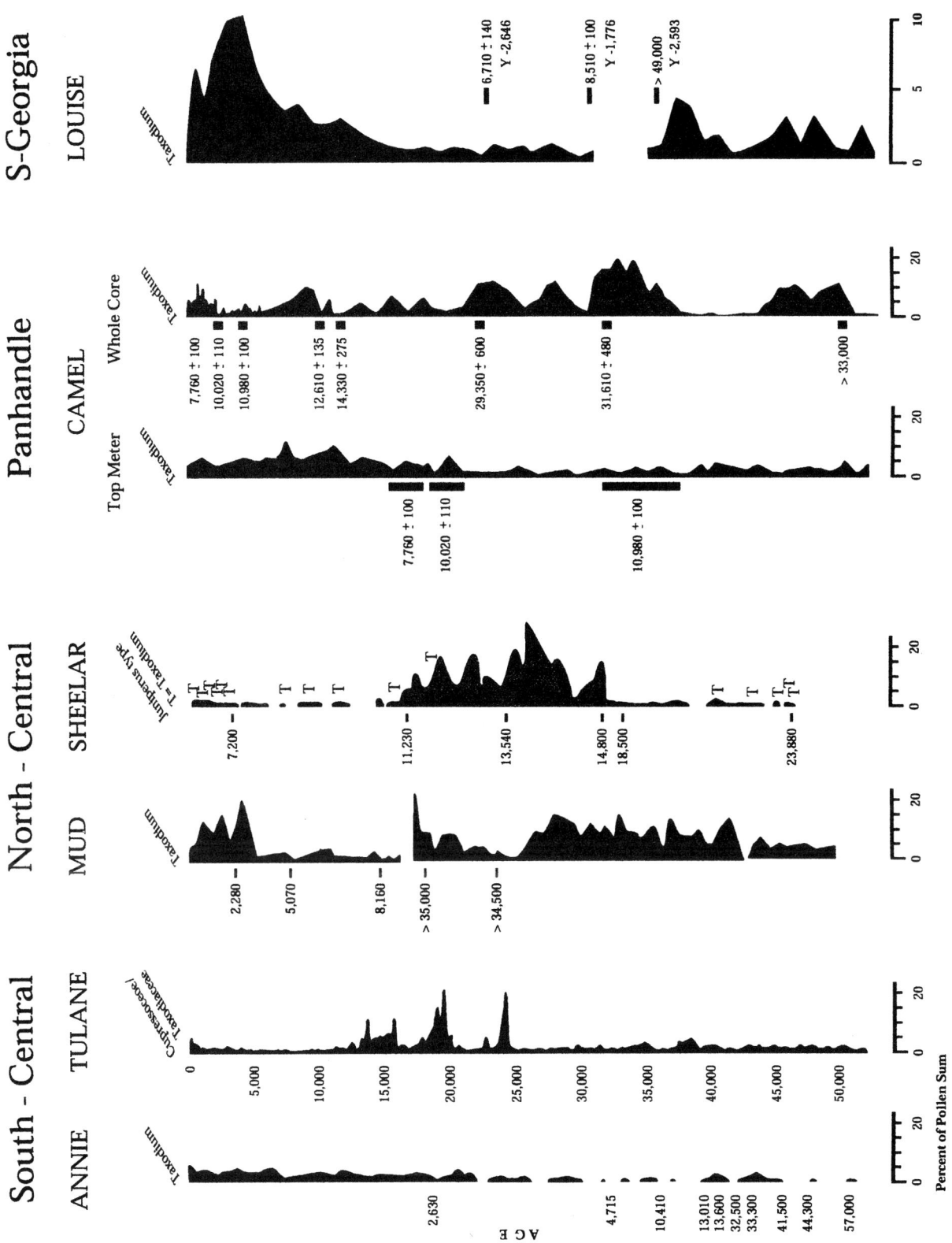

Figure 22.1. History of Taxodium in Florida and southern Georgia based on palynological evidence. Data for Annie, Tulane, Mud, Sheelar, Camel and Louise were adapted from Watts (1975), Watts (1969), Watts and Stuiver (1980), Watts et al. (1992) and Watts (1971), respectively.

Eutrophic lakes are characterized by higher fish biomass and species composition than oligotrophic lakes, and marshes generally possess greater biotic diversity and productivity than swamps. Lakes and swamps with highly humically-stained waters (dystrophic) are generally low in dissolved oxygen and often biotically impoverished (Crisman 1992, Crisman *et al.* 1998). These latter ecosystems often are associated with mystery and forboding, as implied by the aptly named darkly stained River Styx near Gainesville, Florida. The differential importance of dystrophic ecosystems as ceremonial sites in past cultures is poorly known.

Eutrophic lakes are better preservational environments than oligotrophic lakes due to the presence of profundal anoxia for most of the year and rapid sedimentation of autochthonously derived organic matter. Although both marshes and swamps are often strongly hypoxic for pronounced periods, marshes are faster depositional environments for organic matter. Finally, although oxygen-poor dystrophic lakes and wetlands favor artifact preservation, they may yield more of a sacral than secular history.

Temperate Versus Subtropical/Tropical Ecosystems

There are a number of fundamental differences between temperate and subtropical/tropical aquatic ecosystems that can affect the archaeological record. Temperate ecosystems are predominately glacially derived features of uniform age regionally. Subtropical and tropical ecosystems are formed through tectonic-related vertical movements of the landscape, erosion by wind and surface waters, and crustal collapse in karstic areas associated with vertical supression of water tables during prolonged droughts. Thus, it is common to find wetlands representing a broad span of ages juxtapositioned on subtropical/tropical landscapes. The age of individual wetlands in such a temporally heterogenous landscape usually is not readily apparent without detailed paleolimnological investigations, thus the lack of an archaeological record at a particular wetland may be the result of the absence of ecosystem-cultural contemporaneity rather than cultural avoidance of the ecosystem.

Hydrology and its secondary manifestations are the prime factors accounting for differences in the presence and preservation of archaeological records between climate zones. The width of the land-water ecotone is controlled by seasonal differences in rainfall regimes as well as by the frequency and magnitude of extreme events. In general, ecotonal zones in temperate regions are less dynamic and display fewer and less extreme intraannual flucuations than warm latitudes, which are characterized by either one or two alterations of wet and dry cycles annually. In addition, climate in the subtropics and tropics tends to be related more to convectional storm development than broad regional frontal patterns.

Depending on their placement on the landscape and associated hydrology, wetlands of comparable plant community structure can have vastly different potential as preservation sites for archaeological records. Monodominant papyrus swamps filling valleys of southern and western Uganda and serving as the headwaters of streams likely display decreasing potential as preservational sites along their length due to intraswamp responses to discharge events of water from the landscape during peak hydrological events. While the upper portions of such swamps are likely firmly anchored to the substrate, downstream sections detach and float as an intact mat during such events while underlying stormwater scours the stream bed and removes accumulated sediments. Subtropical and tropical wetlands experiencing pronounced intraannual fluctuations in water level also undergo significant system "memory" loss of organic sediments during desiccation either through soil oxidation and enhanced decomposition or increased fire frequency. Lightning frequency in subtropical Florida is greatest at the end of the dry season (Lee *et al.* 1995), and associated fires are thought to have burned individual landscape units approximately every 1–3 year during precontact times (Frost 1995). Fire frequency appears to be higher in isolated cypress domes than alluvial swamps (Mitsch and Gosselink 1993). Some marsh types appear to be maintained by frequent fires (Lee *et al.* 1995), but the complex relationship among plant species composition, hydrology and fire and the general paucity of detailed field investigations preclude any broad comparisons between marshes and swamps. Most wetland fires appear to be of limited regional extent with burn patterns within individual wetlands being very heterogeneous and dictated by the distribution of fire susceptible species of ground cover plants (personal observation).

Finally, interregional differences in the relative balance of sediment accrual from autochthonous primary production versus decomposition along climate gradients is still not fully known. Wetlands from warm areas can have both enhanced photosynthethic and decompositon rates relative to temperate systems (Mitsch and Gosselink 1993). One consequence of rapid decomposition rates in the subtropics/ tropics is the formation of floating islands of living and partially decomposed wetland vegetation through accumulation and rapid release of decomposition gases within the soil column. Formation of such islands not only mixes the stratigraphy of the wetland, but subsequent lateral movement and eventual sinking of the island can deposit older sediments on top of younger sediments (Clark 2000).

Threats To The Wetland Archaeological Record

Differences in historical patterns of wetland loss between the temperate and tropical zones are related to both scale and effectiveness of drainage. Large sections of the 71,000 ha Pontine Marshes of central Italy were drained by Julius Caesar, several subsequent early emperors, and Pope Sixtus V for malaria control, and the operation was completed by Mussolini in the 1930's to turn the remaining marsh into agricultural production. Many of the major discoveries in wetland archaeology have been associated with large-scale drainage of peat bogs of central and northern Europe over the past several centuries. Similar large-scale drainage of wetlands in the United States during the past 100 years, most notably the near complete drainage of the Great Kankakee Marsh (8,100 km^2) in Indiana and Great Black Swamp (4,000 km^2) in Ohio, and large sections of the Great Dismal Swamp (Virginia) and the Everglades (Florida) for expanded agricultural production, have contributed significantly to the 53% loss in total wetlands of the conterminous United States between 1780 and the mid 1980's (Mitsch and Gosselink 1993).

Most large wetlands of the temperate zone are under some sort of protection, and reclamation projects to restore the structure and function of many, including the current $6.8 billion plan for the Everglades, are planned. Similar protection measures for small wetlands of the temperate zone are slow in coming. Prairie pothole wetlands often are considered landscape imperfections needing removal to increase efficiency of large-scale wheat growing operations in the upper Midwest of the United States. Undoubtedly, the most serious threat to small wetlands is government sanctioned destruction of small ecosystems during large urbanization projects in exchange for developers buying into regional wetland mitigation banks at high multipliers for wetland area destroyed. Such tradeoffs, plus continuous attempts to redefine wetland boundaries, highlight the failure to recognize the importance of small, often ephemeral, wetlands worldwide.

The historical pattern of wetland loss in the tropics is the reverse of that of the temperate zone. Large-scale wetland drainage, principally for development of rice agriculture, is a recent phenomenon in the tropics, while drainage historically was mostly of small wetlands for agricultural expansion at the village or family level. Both scales of drainage have increased alarmingly in the past four decades as nations and development agencies struggle to meet the food needs for rapidly expanding populations, and modern implements to facilitate effective drainage are widely disseminated throughout the tropics.

Human encroachment into tropical wetlands for aquaculture and agriculture in recent years in response to rapidly expanding populations is depleting wetland resources at an alarming rate (Crisman and Streever 1996). Africa and Asia have approximately 52% and 35% of their original wetland area remaining, respectively. Government sanctioned resettlement of Indonesians from densely populated Java and Bali (6% of wetlands remaining) to sparsely populated outer islands including Irian Jaya (98% remaining) and Kalimantan (63% remaining) will likely have disastrous consequences for wetland ecosystems. The traditional land-tenure system of Uganda has resulted in such a progressive diminution in the size of inherited farming plots that the rural population has been forced to develop marginally productive lands to make farming sustainable. Only 6% of Ugandan farmers surveyed in 1966 were cultivating wetlands, but this figure jumped to 84% by 1981 (Hamilton 1984). Such family-level drainage of small wetlands has likely resulted in pronounced soil oxidation and decomposition of any contained archaeological record. Fire frequency in such intensively farmed landscapes has also increased to become an annual event and a major threat to archaeological records in drained wetlands. Finally, although coastal marine archaeology is still in its infancy, large areas of mangrove forests along the coasts of Africa, Southeast Asia and northwest South America have been converted into shrimp aquaculture operations. The archaeological records of other coastal areas throughout the world, including Jamestown, Virginia, are threatened by global sea level rise.

Crisman *et al.* (1996) proposed that wetland conservation in the subtropics and tropics is possible only within a complex economic and social framework integrating local populaitons and recognizing the necessity for multipurpose sustainable use of wetlands. Archaeology, unfortunately, has been totally overlooked as an important conservation element.

Conclusion

There is a critical need to develop dialogue among wetland archaeologists, conservationists, ecologists and managers. Archaeologists often lack the fundamental academic background in aquatic ecology to understand the role played by hydrology, fire, sedimentation rates and long-term transformations of aquatic soils on preservation of the archaeological record. Wetlands are not monotypic, rather distinct structural and functional differences of swamps versus marshes, littoral zones versus isolated wetlands, tidal versus non-tidal and salt versus freshwater systems along gradients of trophic state and latitude present a complex array of conditions differentially affecting the archaeological record. While most archaeologists have only a passing knowledge about the ecological conditions of the wet site at the time of human occupation/contact recorded in the sediments, aquatic conservationists, ecologists and managers are likely to have no formal training in archaeology. Until recently, humans were not regarded as part of landscape historical ecology. Rather, they were viewed solely as the defilers of ecosystem structure and function.

Given current and projected intensification of negative human interactions with wetlands worldwide, aquatic conservationists, ecologists, and managers are increasingly having to deal with archaeological questions with which they are unfamiliar. Archaeology and ecology have been sitting adjacent to one another for years without recognizing it. By superimposing a single letter, the hidden relationship is apparent: Archae^{c}ology.

In order for the relationship to work, it must be two-way. Wetland archaeologists need a fundamental background in limnology and wetland ecology in order to understand ecosystem conditions at the time represented by the archaeological record, and aquatic scientists and managers need training in archaeology sufficient to identify potential sites and how to deal with them in an appropriate manner. The time is truly ripe for such academic linkages.

References Cited

Clark, M.W. 2000. Biophysical characterization of floating wetlands (flotant) and its influence on vegetatie succession of a warm-temperate aquatic ecosystem. Ph.D. dissertation, University of Florida, Gainesville. 413 pp.

Coles, J.M. 1984. The Archaeology of Wetlands. Edinburgh University Press, Edinburgh. 111 pp.

Cowardin, L.M., V. Carter, F.C. Golet and E.T. LaRoe. 1979. Classification of Wetlands and Deepwater Habitats of the United States. U.S. Fish & Wildlife Service Pub. FWS/OBS-79/31. Washington, D.C. 103 pp.

Cowgill, U.M. 1994. The waters of Merom: A study of Lake Huleh. IX. The minor chemical constitutients of a 54 m core. Arch. Hydrobiol. 99:97–153.

Crisman, T.L. 1992. Natural lakes of the southeastern United States: origin, structure, and function. pp. 475–538. In: C.T. Hackney, S.M. Adams and W.A. Martin (eds.). Biodiversity of Southeastern United States/Aquatic Communities. Wiley, New York.

Crisman, T.L. and W.J. Streever. 1996. The legacy and future of tropical limnology. pp. 27–42 In: F. Schiemer and K.T. Boland (eds.). Perspectives in Tropical Limnology. SPB Academic Publishing, Amsterdam. 347 pp.

Crisman, T.L., L.J. Chapman and C.A. Chapman. 1996. Conserving tropical wetlands through sustainable use. Geotimes 41:23–25.

Crisman, T.L., L.J. Chapman and C.A. Chapman. 1998. Predictors of seasonal oxygen levls in small Florida lakes: the importance of color. Hydrobiologia 368:149–155.

Curtis, J.H., M. Brenner, D.A. Hodell, R.A. Balser, G.A. Islebe and H. Hooghiemstra. 1998. A multi-proxy study of Holocene environmental change in the Maya lowlands of Peten, Guatemala. J. Paleolimnol. 19:139–159.

Dau, J.H.C. 1829. Allerunterthanigster Bericht an die Konigliche Danische Rentekammer uber die Torfmoore Seelands. Kopenhagen und Leipzig.

Frost, C.C. 1995. Presettlement fire regimes in southeastern marshes, peatlands and swamps. pp. 39–60. In: S.I. Cerulean and R.T. Engstrom (eds.). Fire in Wetlands: A Management Perspective. Proceedings 19th Tall Timbers Fire Ecology Conference. Tall Timbers Research Station, Tallahassee, FL. 175 pp.

Hamilton, A.C. 1984. Deforestation in Uganda. Oxford University Press, Nairobi. 92 pp.

Hutchinson, G.E. and U.M. Cowgill. 1973. The waters of Merom: A study of Lake Huleh. III. The major chemical constituents of a 54 m core. Arch. Hydrobiol. 72:145–185.

Iversen, I. 1973. The Development of Denmark's Nature Since the Last Glacial. Geological Survey of Denmark. Series V, No. 7-C. 125 pp.

Jessen, K. 1935. Archaeological dating in the history of North Jutland's vegetation. Acta Arch. 5:185–214.

Lee, M.A., K.J. Ponzio and B.G. Ormiston. 1995. Fire Effects and fire management in the upper St. Johns River basin marsh, Florida. pp. 142–150. In: S.I. Cerulean and R.T. Engstrom (eds.). Fire in Wetlands: A Management Perspective. Proceedings 19th Tall Timbers Fire Ecology Conference. Tall Timbers Research Station, Tallahassee, FL. 175 pp.

Leyden, B.W., M.Brenner and B.H. Dahlin. 1998. Cultural and climatic history of Coba, a lowland Maya city in Quintana Roo, Mexico. Quartern. Res. 49:111–122.

Miller, J. 2000. Archeologists confirm age of prehistoric canoes. Gainesville Sun, 19 October 2000.

Mitchell, G.F. 1976. The Irish Landscape. Collins, London. 240 pp.

Mitsch, W.J. and J.G. Gosselink. 1993. Wetlands. Wiley, New York. 722 pp.

Navid, D. 1989. The international law of migratory species: the Ramsar Convention. Natural Resources Journal 29:1001–1016.

Nicholas, G.P. 1991. Putting wetlands into perspective. Man in the Northeast 42:29–38.

Nicholas, G.P. 1992. Directions in wetlands reseach. Man in the Northeast 43:1–9.

Ohlhorst, S., G.E. Hutchinson and J.G.J. Kuiper. 1977. The waters of Merom: V. Temporal changes in the molluscan fauna. Arch. Hydrobiol. 80:1–19.

Ohlhorst, S., A. Shmida, M.M. Poulson and G.E. Hutchinson. 1982. The waters of Merom: A study of Lake Huleh. VIII. Non-siliceous plant remains, with appendices on some animal fossils. Arch Hydrobiol. 94:441–459.

Pennington, W. 1974. The History of British Vegetation. The English Universities Press, London. 152 pp.

Sherman, J.W. and R. Patrick. 1981. The waters of Merom: A study of Lake Huleh. VII. Diatom stratigraphy of the 54-m core. Arch. Hydrobiol. 92:199–221.

U.S. Soil Conservation. 1987. Hydric Soils of the United States. Washington, D.C.

Watts, W.A. 1969. A pollen diagram from Mud lake, Marion County, north-central Florida. Bull. Geol. Soc. Amer. 80:631–642.

Watts, W.A. 1971. Postglacial and interglacial vegetation history of southern Georgia and central Florida. Ecology 52:676–690.

Watts, W.A. 1975. A late Quaternary record of vegetation from Lake Annie, south-central Florida. Geology 3:344–346.

Watts, W.A. and B.C.S. Hansen. 1988. Environments of Florida in the Late Wisconsin and Holocene. pp. 307–323. In: B.A. Purdy (ed.). Wet Site Archaeology. Telford Press, Caldwell, N.J. 338 pp.

Watts, W.A. and B.C.S. Hansen. 1994. Pre-Holocene and Holocene pollen records of vegetation history from the Florida peninsula and their climatic implications. Palaeogeography, Palaeoclimatology, Palaeoecology 109:163–176.

Watts, W.A. and M. Stuiver. 1980. Late Wisconsin climate of northern Florida and the origin of species-rich deciduous forest. Science 210:325–327.

Watts, W.A., B.C.S. Hansen and E.C. Grimm. 1992. Camel Lake: A 40,000-yr record of vegetational and forest history from northwest Florida. Ecology 73:1056–1066.

Whitmore, T.J., M. Brenner, J.H. Curtis, B.H. Dahlin and B.W. Leyden. 1996. Holocene climate and human influences on lakes of the Yucatan Peninsula, Mexico: An interdisciplinary, palaeolimnological approach. The Holocene 6:273–287.

23. Wet Sites, Wetland Sites and Cultural Resource Management Strategies

George P. Nicholas

Archaeologists face many challenges in pursuit of the past: raging rivers, restless natives, hungry bears, and fever-maddened colleagues. Perseverance (and a good research design) often leads to new discoveries and ultimately to new knowledge of the history, challenges, and accomplishments of earlier societies. Of no less importance is the quest for representativeness in the archaeological record, itself an important mandate of the discipline. After all, if our desire is to reconstruct and understand past human lifeways, it is essential that we seek the full range of variation that once existed regarding past social and political organization, settlement and subsistence, technological adaptations, and all of the other important facets of human existence. Representativeness is also crucial in cultural resource management, both for ensuring adequate survey coverage and for defining significance. Much effort has thus been devoted to devising methods to explore adequately and appropriately the range and settings of past human behavior.

However, one major component of past human landscapes that has not received adequate or consistent attention by archaeologists and cultural resource managers alike is wetland environments (Table 23.1). Wetlands are transitional zones between dry land and open water that are defined by land that: a) is periodically water-covered or saturated; b) supports hydrophytic vegetation; and/or c) has hydric soils (NRC 1995; see also Mitsch and Gosselink 1993:23–29). These settings include swamps, bogs, fresh and saltwater marshes, fens, and other "permanent" and temporary forms (see Finlayson and van der Valk 1995; Mitsch and Gosselink 1993:30–40). As importantly, wetlands also represent distinct ecological entities (e.g., Gore 1983; Lugo *et al.* 1990) comprised of a variety of associated floral and faunal communities and the hydrological, terrestrial, and other components with which they interact.

Wetlands have long been viewed as marginal and unattractive places to be avoided (e.g., Miller 1989) or developed (Vileisis 1997); alternatively, they may be seen as peripheral to other more "important" landscape features. The limited attention paid to wetland settings has two important implications regarding archaeology. First, it constrains or channels our knowledge of past human societies to coastal, riverine, or other more popular places to work. Swamps, playas, marshes, and bottomlands have been important in human affairs throughout prehistory (Nicholas 1998a), and have at times been among the most attractive areas on the landscape based on their high values for resource diversity, productivity, and reliability (Nicholas 1988:268–269; Niering 1985:29; also see Forman and Godron 1986). The range of activities associated with them is diverse, and includes resource harvesting, settlement, sacred and spiritual uses, and defense (Coles and Coles 1989, 1996; Nicholas 1998a). Thus, if we don't consider the role that wetlands had in past, our knowledge of past human lifeways remains incomplete. The second implication is that this lack of recognition and understanding limits the protection that wetlands should receive as important and often unique cultural heritage sites (Coles 1990; Coles and Coles 1996).

Of course, wetlands are more than just a setting for prehistoric settlement, or a repository for organic artifacts. They have a central role in local and regional ecologies, and are vital to water purification and storage, climate regulation, waterfowl habitat, and biological diversity (Mitsch and Gosselink 1993). For such reasons, wetlands in North America and elsewhere are today subject to extensive study, management, and conservation (e.g., Dennison and Berry 1993); in fact, they probably receive more attention and funding than virtually any other terrestrial ecozone. Despite this, wetland protection methods are

Table 23.1. Wetland Descriptive Terms (after Mitsch and Gosselink 1993:32).

Bog	Peat-accumulating wetland with no significant inflow or outflow, usually supporting *sphagnum* moss.
Bottomland	Lowland along rivers and stream, generally on alluvial floodplain, that is periodically flooded; usually forested.
Fen	Peat-accumulating wetland that receives some drainage from surroundingmineral soil; usually supports marsh-like vegetation.
Marsh	A frequently or continually inundated, freshwater or saltwater wetlands characterized by emergent herbaceous vegetation adapted to saturated soil conditions.
Mire	In Europe, any peat-accumulating wetland
Moor/Peatland	In Europe, any wetland that accumulates partially decayed plant matter.
Muskeg	A large expanse of peatlands or bogs, especially in subarctic regions.
Playa/Pothole	Shallow, marsh-like ponds.
Reedswamp	In Europe, a marsh dominated by *Phragmites*.
Swamp	In North America, a wetland dominated by trees or shrubs; in Europe, a forested fen or reed grass-dominated wetland.

Examples of Common Wetland Types, North America

Coastal Wetlands
- *Spartina* or *Juncus*-dominated tidal salt marsh
- Mangrove swamp

Inland Wetlands
- *Typha*-dominated freshwater marsh
- Riparian forested swamp
- Spaghnum-sedge peatland

contentious, inconsistent, and incomplete (e.g., Holloway 1994; Malakoff 1998; Alper 1992). In addition, wetland management strategies often ignore cultural values and, in some cases, actually contribute to their loss.

This paper challenges these restrictive ways of thinking about and managing wetlands by looking at three complementary issues. The first concerns the difference between wet sites and wetland sites – a distinction considered very important here. The second reviews the relationship between wetland environments and human affairs, as reflected by different types of sites and scales of interaction, which are used as the basis for a discussion on representativeness. The third issue is the need for more effective wetland conservation strategies that recognize the unique archaeological and ecological character of wetlands, and that subsequently engage *both* cultural and natural resource legislation.

Finally, the emphasis of this paper on wetland ecology and hunter-gatherer archaeology reflects my own research focus (Nicholas 1998a). Although they receive only passing

mention here, the use of wetlands by prehistoric farmers and later folk is no less important (see Coles and Coles 1989, 1996 for overviews).

Wet Sites And Wetland Sites

Something long overdue in the field of wetlands-oriented archaeology is the need to differentiate *wet sites* from *wetland sites*. The two terms have had a long association, and in fact have sometimes been used almost synonymously. Wet sites occur in a variety of coastal, lacustrine, and riverine/estuarine environments, and are exemplified by the remarkable preservation of organic artifacts in water-saturated sediments (e.g., Lake Neuchatel [Egloff 1988]; Biskupin [Piotrowski 1998]; Hoko River [Croes 1988]). In contrast, wetland sites have a more restricted distribution, being located within or adjacent to existing swamps, marshes, and comparable settings, or where such features were once present. These sites are exemplified by a variety of human activities associated with such ecozones, including habitation, resource procurement, and travel (e.g., Flag Fen [Hall and Coles 1994]; Stillwater Marsh [Kelly 1990]; various Australian locales [Meehan 1991]). While it is evident that there may be considerable overlap between wet sites and wetland sites, there are significant differences that must be recognized, as discussed below.

Both types of sites have been well represented at conferences and in their published proceedings. Figure 23.1 provides a breakdown of the major types of studies included in four edited volumes: *Wet Site Archaeology* (Purdy 1988); *The Wetland Revolution in Prehistory* (Coles 1991); *Hidden Dimensions: The Cultural Significance of Wetland Archaeology* (Bernick 1998); and finally, *Bog Bodies, Sacred Sites, and Wetlands* (Coles *et al.* 1999). The major types of reports identified are:

1. Overview Studies that included introductions to the volume and broad reviews of wet site and wetland site research.
2. Wet Site studies; articles on the so-called Bog Bodies.
3. Wetland Site studies.
4. Preservation studies that focused on the stabilization of organic artifacts.
5. Other Studies.

This classification scheme was based on very general criteria, and some papers were difficult to type when they spanned several classes. In addition, the number of papers for any one category varied substantially between the volumes, due at least in part to the structure and venue of the corresponding conference. For example, the relatively high number of papers on bog bodies in 1998 (Coles *et al.* 1999) is clearly linked to the conference being held in Denmark. However, when all four years are combined

Figure 23.1. Types of Reports in Conference Proceedings.

(Figure 23.2), the two categories that clearly stand out are wet sites and wetland sites. That these are almost equally represented tells us something important about the research interests of members.

Wet sites and wetland sites provide valuable information about the past. Each has produced a remarkable array of artifacts, especially in regards to organic materials and composite tools. However, it must be emphasized that wet sites are not wetland sites, although there may be substantial overlap between them. The Sweet Track in Britain (Coles and Coles 1986) and the Windover cemetery in Florida (Doran and Dickel 1988), for example, can be considered both wetland sites and wet sites because each was originally situated within a marshy/swampy area, and subsequently preserved in a water-saturated context. On the other hand, Stillwater Marsh is a wetland site only, and Hoko River is a wet site only. The distinction between the two types is clearly a slippery one because a wetland site may have a

wet component where hydric conditions have persisted to the present, or it may be dry where conditions have changed. Likewise, wet sites frequently have no association at all with wetlands, and indeed some sites (e.g., Lake Neuchatel) would normally be considered "underwater sites."

My concern is with more than making an arbitrary distinction between the two, based on the type, amount, or context of organic artifacts found, or whether what was originally a lake or river-oriented habitation site is preserved by subsequent wetland formation or inundation. What has gone largely unnoticed is that they represent two fundamentally different things. Wetland sites are defined here by a *relationship* between people and the particular types of ecological settings represented by wetlands, and the archaeological record it has produced; the site is present primarily because of past human association with, or exploitation of a swamp, marsh, moor, or other wetland types. Many wetland sites do not have a wet component; in

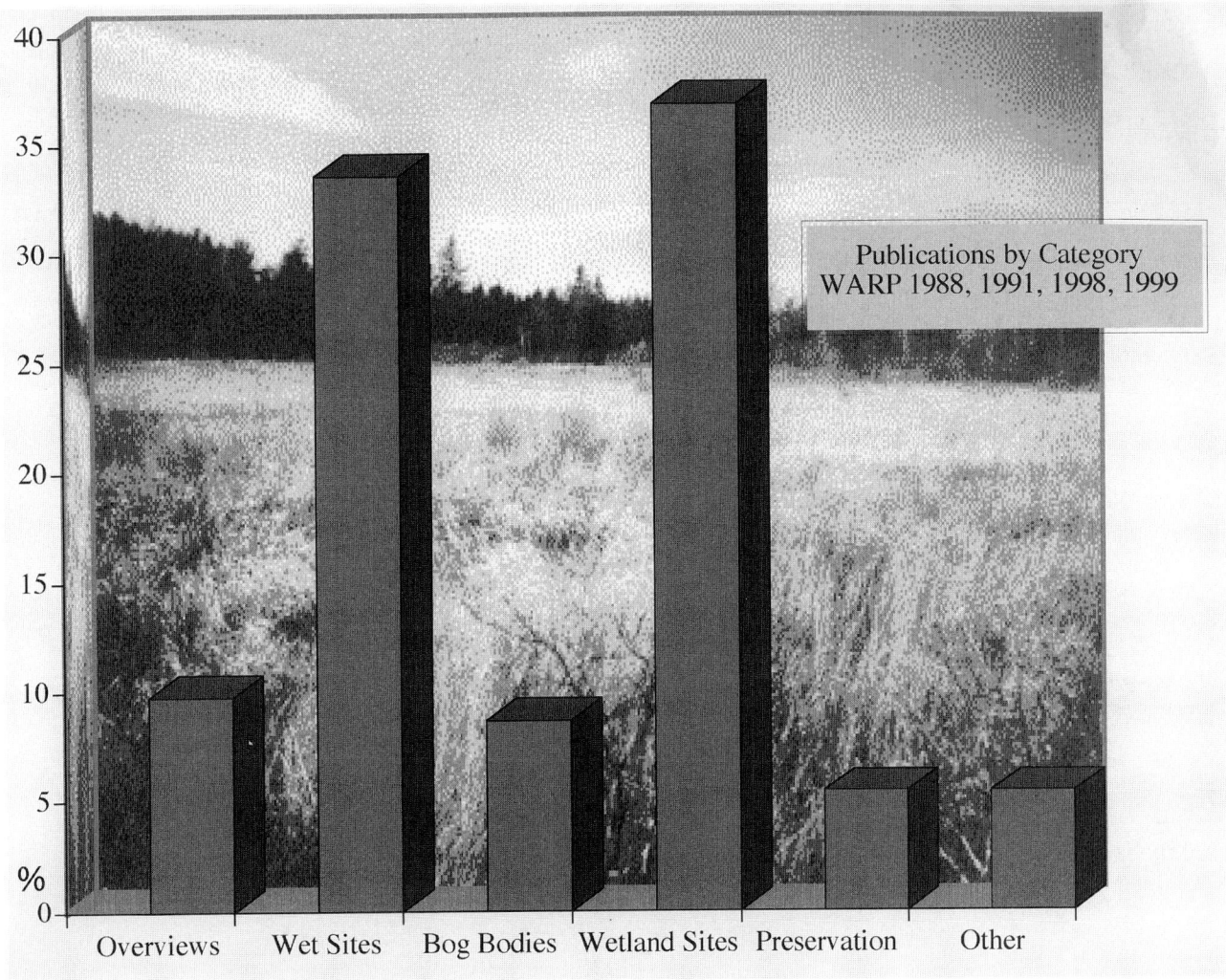

Figure 23.2. Publications by Category: Proceedings 1988, 1991, 1998, 1999.

fact, the wetlands that were associated with some early or middle Holocene sites may have since been transformed into grassland, forest, or desert by changes in hydrology or plant succession.

Wet sites, on the other hand, are defined by the *association* between artifacts and the context of preservation. In this case, the association is not necessarily with wetlands (as distinct ecological features), but rather with wet, often anaerobic conditions. What have been termed wet sites are not only found in swamps, bogs, and so on, but also associated with bays, lakes, estuaries, and other settings. Fish weir sites, by this definition, are wet sites by virtue of the fact that the preservation of stakes and wattling is due to their water-saturated context. As their original context was usually within a riverine or tidal environment, they are not considered wetland sites per se (although fresh or

saltwater marshes may indeed have been nearby, the association is largely secondary). On the other hand, eel traps in southeastern Australia (Lourandos 1987) are considered wetland sites because they were constructed in channels created in swamps.

Scale is also an important factor in the recognition and management of these different types of sites. Wet sites are relatively small, and may consist of anything from isolated or clustered sets of artifacts to entire lake shore communities now underwater. These tend to be relatively visible types of archaeological sites – muck, peat, and water aside – because they are defined by artifacts and features. In fact, the recovery of such normally perishable basketry and fabric, wooden bowls, boats, and paddles, trackways, and even bodies may carry significant weight in public recognition and protection. Wetland sites, on the other hand, have a much greater spatial and temporal extent, ranging from a single site adjacent to a marsh to the site distribution patterns associated with regional ecozones, and from single occupations to thousands of years. In some cases, wetland sites, and the changing ecological relations they reveal, may be more important than the material culture of individual components or sites (see Janetski and Madsen 1990 for Great Basin examples). This emphasis on *relationships*, not *things*, may challenge some conventional cultural resource management strategies. The potential of new approaches to CRM is illustrated by David Sanger's Milford Reservoir Project in Maine for which he obtained $150,000 for wetland history reconstruction (D. Sanger, pers. comm. 2000).

This discussion isn't meant to suggest that wet sites should not be included within the scope of wetlands research in general, for wet and wetland sites are in many regards complimentary and there is considerable overlap between them. They do, however, represent very different types and scales of archaeological phenomena. Thus it is time to utilize these terms with greater precision and awareness. It is also important that we recognize that the distinction between the two (and between the types of environmental settings and ecological relations represented), has implications for both research and resource management.

Wetlands And Archaeological Representation

It is the responsibility of archaeologists and cultural resource managers to ensure that the archaeological record reflects the full range of human behavior to the degree possible. The archaeology of wetland areas has much to contribute in this regard. We can identify three important dimensions of the wetland archaeological record that guide our quest for representativeness: culture history and chronology, culture process and explanation, and interpretation and meaning.

Culture History And Culture Chronology

The dimensions of culture history and chronology document the basic association between people and wetlands (see Nicholas 1998a for overview). This relationship has considerable antiquity, as first evidenced at *H. erectus* sites in Africa (Walker and Leakey 1993) and southern Europe (Klein 1987). Wetland-associated sites are present in the Middle Pleistocene, with evidence of *Typha* processing at the Mousterian site of Combe Grenal in France (Binford, cited in Wendorf 1993:355). The association is strongly developed by the late Pleistocene-Holocene transition. In North America, for example, Paleoindian and Archaic sites often cluster around large, wetland mosaics and other

wetland-dominated environments (Almquist-Jacobson and Sanger 1999; Custer 1989; Dillehay 1988; Langemann and Dempsey 1993; Nicholas 1988).

During the Holocene, the diversity and richness of wetlands contributed to the economic diversity and specialization of hunter-gatherers worldwide, as seen in the American Great Basin (Fowler 1992; Janetski and Madsen 1990; Simms 1987). These factors may have also facilitated increased population size and decreased territory requirements of some hunter-gatherers as evidenced in southeastern Australia (Lourandos 1987) and in central California (Moratto 1984) and the eastern United States (Jefferies 1987; Saunders *et al.* 1997).

The archaeology of wetlands has also contributed significantly to documenting aspects of past technology, subsistence, health, and worldview that are infrequently, if ever, found elsewhere. Both fortuitous discoveries and systematic surveys and excavations fill in gaps in our knowledge; the types of organic artifacts frequently found not only contribute to the historical dimension, but may also humanize it. For example, when we gaze at the stubbled face of the Tollund Man (Glob 1969) or think about the person who made this cloak or wore these shoes, our connection to the past is often more immediate and stronger than when we hold a stone blade or ceramic bowl of the same age and provenience.

Culture Process And Explanation

It is one thing to document the emergence during the Holocene of relatively sedentary and complex, pre-horticultural societies within the context of wetland-rich environments; it is another to explain that emergence. Through examining the processes of change at different scales (Nicholas 1998: Fig. 1), we can see that these societies were active participants in the various interaction spheres, not merely responding to stimuli in an environmentally deterministic manner.

Situations where the same wetlands were used in different ways over time provides the opportunity to investigate both the processes of adaptation and the historical dimensions of land use and social organization (e.g., Jefferies 1987). In some locations, the increasing use of wetland resources (e.g., wapato, rice, taro, eels) may have been a response to, as well as facilitator of, growing local populations (Nicholas 1998a:726) as evidenced Asia, Central and North America, and elsewhere. In addition, intensive harvesting of some wetland-associated plants, such as sumpweed (*Iva annua*) in the Illinois and Mississippi River valleys may have contributed to the process of plant domestication.

The search for explanation may require that we consider the archaeological record in new ways. For example, we can look at how and where the formation of core areas of settlement and social and economic interactions relates to the changing degree of regional ecological contrast. An example of this occurred in southern New England between wetland-dominated settings and coastal and riverine ecozones during the early Holocene, which may help to explain the strong association between wetlands and early postglacial land use in places like Robbins Swamp (Nicholas 1988).

Finally, the ecological focus of processual archaeology enables us to see how hunter-gatherers modified their landscape through a variety of intentional and unintentional means. In North America, Britain, and Australia, burning was used to thin forests, rejuvenate swamps, or otherwise manipulate the local resource base. In those areas where substantial settlements were developing, wetlands were also probably very heavily exploited to the degree that their developmental trajectory could have been influenced by plant harvesting or beaver exploitation (Nicholas 1998b). In some cases, intentional swamp management and manipulation, as in southeastern Australia (Lourandos 1987), may have permitted continued intensive resource harvesting.

Interpretation and Meaning

The third dimension of wetland-oriented research concerns interpretation and meaning – goals that require both processual and postprocessual methods. At issue here is not only our interpretation of settlement and subsistence patterns, but the need to identify those aspects of the wetland archaeological record that may reflect the ideology and worldview of past societies.

Perceptions of wetlands are remarkably broad, as defined by host of aesthetic, economic, spiritual, and environmental values (Giblett 1996; Mitsch and Gosselink 1993). Many today view wetlands as marginal places to be altered; others cherish their natural beauty and wildlife habitat. But beyond these extremes, wetlands are given little thought by most people (including archaeologists). It is obvious, however, that wetlands were of great importance to prehistoric peoples in many regions, as we see with the self-named *Toidikadi*, or Cattail-Eaters, of Stillwater Marsh (Fowler 1992; Wheat 1967). Even today, a strong association with wetlands persists among many aboriginal people worldwide where they are afforded the opportunity.

The archaeological record associated with wetlands, whether viewed as core and peripheral space or something else, can reveal much about how people conceptualized the landscape around them. This is evident by the presence of cemeteries in and around wetlands, such as the Windover site (Doran and Dickel 1988), and by isolated and sometimes sacrificed interments in European bogs (see Coles *et al.* 1999). Both cases indicate that wetlands figured promin-

ently in meeting spiritual needs. Cemeteries also served as territorial markers of places to be returned to and possibly even defended. In those areas where wetlands contributed to semi-sedentary or sedentary lifeway, such as southeastern Australia (Pretty 1977), the elaborate burials are atypical of "classic" hunter-gatherers. The special and sometimes spiritual dimension of wetlands is also found in stories and mythology (e.g., Dean 1992).

The very richness of the archaeological record associated with wetlands (and wet sites) provides opportunities to both test assumptions and revise interpretations. Perhaps the best example of this is the Mesolithic site of Star Carr (Clark 1971), a site that has been revisited and reinterpreted by a succession of archaeologists. Regardless of whether we take a culture historical, processual, or other approach to wetlands, each contributes to a more representative archaeological record. The challenge remains to conserve the rich cultural and natural heritage of wetlands.

Conservation Strategies

Of the estimated 86 million acres of wetlands in the United states present in 1700, less than half remain (Tiner 1984:29). Comparable figures are found worldwide where wetlands have been or are being dredged, channel, drained, or filled for farming, peat mining, or other purposes. The loss of these swamps, marshes, estuaries, and such has had its greatest impact on local and regional ecology, but has also had a profound effect on associated cultural resources. Drainage causes preserved *in situ* organic artifacts to deteriorate, while such activities as dredging may destroy a wet site completely. With enough time and funding, wetlands can be restored or even created – archaeological sites cannot be. There is thus a pressing need for more effective wetland preservation strategies that recognize and integrate cultural heritage values into their charter.

Wetland conservation efforts have had a long history in North America (Vileisis 1997), but intensified substantially in the 1970s and 1980s in the United States (WWF 1992). During that period, studies by ecologists and others led to a greater understanding of the nature of wetlands, and of their overall role in wildlife and human habitat. This research, in turn, contributed to such federal legislation as the Clean Water Act: Section 404, Executive Order 11990: Protection of Wetlands, and the Emergency Wetlands Resources Act, and to such state legislation as the Maine Natural Resources Protection Act and the Illinois Interagency Wetlands Policy Act of 1989. A suite of legislation slowed wetland loss appreciably in the 1980s. Unfortunately, as public opinion shifted in the early 1990s against the regulatory efforts of the U.S. Government, wetlands policies have been challenged on many fronts (Tiner 1998:11–12). Although there are a host of private and special interest conservation strategies, ranging from Ducks Unlimited incentives to land trusts, these lack the strength of federal and state regulations.

One notable wetland conservation strategy used by the federal government and adopted by some states is the "No Net Loss" Policy of 1988. In theory, "no net loss" limits further reduction in the national wetland inventory, based on acreage and function, through the restoration and creation of wetlands. Yet from the perspective of cultural resource management, this strategy contains a serious flaw. As already noted, the archaeological record associated with wetlands is substantial. However, while the creation of new wetlands may do much to enhance biological productivity, it cannot create new archaeological sites. In fact, the idea that wetlands located in prime development areas can, in effect, be relocated without affecting local habitat values and hydrology is dangerous because it does not consider critical cultural resources.

Likewise, in a recent volume published by the World Wildlife Fund, *Statewide Wetlands Strategies: a Guide to Protecting and Managing the Resource* (WWF 1992), proposed conservation strategies are based on flood control, water quality, fisheries, waterfowl habitat, biological diversity, groundwater recharge, erosion and land formation, and recreation. Cultural heritage factors are not included, a perspective that is widespread (although historic/archaeological site presence may be included in some types of wetland evaluation (WWF 1992:212–219). Thus, when the different potential values of wetlands are viewed individually, rather than collectively, much more than wetland acreage may be lost.

Cultural and archaeological resource protection legislation provide the greatest protection to wetland archaeological sites. There is a host of federal and state legislation in the United States, including the Archaeological Resources Protection Act of 1979, that offer strong protection for archaeological sites on federal and Indian land, or that take effect when federal or state monies are involved. Such mandated protection is not available to sites located on private lands, however, such as in the case of the Lafotane peat bog in Connecticut (Nicholas 1998c) where cultural material are found not only around the bog, but within it – one quartz biface was even recovered from the peat processor. In Canada, where comparable legislation is absent at the federal level, cultural resources are protected by provincial mandate. The Heritage Act of British Columbia theoretically protects all archaeological sites, whether on private or public lands, but in reality there are many exceptions as when it is over-ridden by the Municipal Act. However, in both countries archaeological sites associated with wetlands that might not otherwise be exempt from development or disturbance may possibly qualify for protection under existing wetlands protection legislation.

The management of wet sites requires protection not only of the land itself, but of the original hydric conditions that preserved the artifacts and structures in the first place (see Corfield 1998). A case in point in Britain is the Somerset Levels and Moors, which have an exceptionally rich record of trackways and other archaeological remains (Coles and Coles 1986). Now designated an Environmentally Sensitive Area, farmers there receive £400 or more per hectare for adjusting their practices to raised water levels (Coles and Coles 1996:128). Similarly, in the United States, under the "Swampbuster" provision of the 1990 Farm Bill, farmers who drain wetlands may be denied crop subsidies and other agricultural benefits. While this policy is designed to protect water quality and related values primarily, it may also benefit wet and wetland sites.

Currently interaction between wetlands scientists and archaeologists remains limited, presumably because each is largely unaware of what the other is doing. However, there are many examples of collaborations that have proved valuable, as exemplified by Harry Godwin's contribution to the interdisciplinary Star Carr project (Clark 1971), and numerous examples from Scandinavia. The exchange of information and the integration of cultural and biological conservation efforts may strengthen appreciably what can be achieved by all parties involved. This is especially the case for the unique and fragile archaeological sites whose protection can be increased if appropriate environmental protection legislation can be invoked. Archaeologists actively involved in wetlands-oriented research should consider membership in such organizations as the Society for Wetland Scientists (http://www.sws.org). Another opportunity is involvement with the Ramsar Convention on Wetlands (http://www.ramsar.org), which is an intergovernmental treaty signed by 117 members that identifies and conserves wetlands of international importance; social and cultural values are included in the evaluation of wetlands being considered for inclusion.

Conclusions

As has been amply demonstrated in recent decades, the archaeological record associated with both wetlands and wet sides is extraordinary in many regards. Whether we are talking the development of the earliest earthen mounds in America in Louisiana during the middle Holocene (Saunders *et al.* 1997) or the remarkable preservation of water-saturated artifacts and bodies of Europe (Coles *et al.* 1999), the archaeology of wetlands is not only allowing us to refine our knowledge of the past, but in some cases to overturn conventional wisdom. This is certainly the case when we consider the hunter-gatherers who lived in some wetlands-rich areas: they assume airs – building mounds, achieving relatively high population densities, settling down – acting in decidedly atypical hunter-gatherer fashion.

It is thus important that we seek such cultural and ecological diversity, as represented at different scales and in different ways, on the past landscape (Nicholas 1994). Some types of behaviors or aspects of material culture may even be associated only with swamps, bog, or marshes. Without recognizing the importance of wetlands, we will not have a representative view of the past, nor a complete understanding of past and present human ecology. On the other hand, by integrating such perspectives, we can achieve both a more complete understanding of the human condition, and more effective legislation to preserve this record for future generations.

Acknowledgments

For their invitation, encouragement, or assistance, I am grateful to Barbara Purdy, John and Bryony Coles, Catherine Carlson, and Thomas Crisman. David Sanger provided helpful comments on a draft of this paper. Travel was supported by a Vice-President's Research Fellowship, Simon Fraser University.

References Cited

Almquist-Jacobson, H., and D. Sanger. 1999. Paleogeographic Changes in Wetland and Upland Environments in the Milford Drainage Basin of Central Maine in Relation to Holocene Human Settlement History. In Current Northeast Paleoethnobotany, edited by J.P. Hart, pp. 177–190. New York State Museum Bulletin 494.

Alper, J. 1992. War Over the Wetlands: Ecologists v. the White House. *Science* 257:1043–1044.

Bernick, K., (ed). 1998. Hidden Dimensions: The Cultural Significance of Wetland Archaeology. Vancouver: University of British Columbia Press.

Clark, J.G.D. 1971. Excavations at Star Carr (2nd ed). Cambridge: Cambridge University Press.

Coles, B. 1990. Wetland Archaeology: A Wealth of Evidence. In Wetlands: A Threatened Landscape, edited by M. Williams, pp. 145–180. Oxford: Basil Blackwell.

Coles, B., (ed). 1991. The Wetland Revolution in Prehistory. WARP. Occasional Paper 6, Department of History and Archaeology. Exeter: University of Exeter.

Coles, B., and J. Coles. 1989. People of the Wetlands: Bogs, Bodies, and Lake-Dwellers. New York: Thames and Hudson.

Coles, J., and B. Coles. 1986. The Sweet Track to Glastonbury. London: Thames and Hudson.

Coles, J., and B. Coles. 1996. Enlarging the Past: The Contribution of Wetland Archaeology. Society of Antiquaries of Scotland, Monograph Series 11, Edinburgh; Wetland Archaeology Research Project Occasional Papers 10.

Coles, B., J. Coles, and M.S. Jorgensen, (eds). 1999. Bog Bodies, Sacred Sites, and Wetlands. WARP Occasional Paper 12. Department of Archaeology. Exeter: University of Exeter.

Corfield, R. 1998. The Role of Monitoring in the Assessment and Management of Archaeological Sites. In Wet Site Archaeology, edited by B. Purdy, pp. 302–316. Caldwell, N.J.: Telford Press.

Custer, J. 1989. Prehistoric Cultures of the Delmarva Peninsula: An Archaeological Study. Newark: University of Delaware Press.

Dean, J. 1992. Wetland Tales: A Collection of Stories for Wetland Education. Publication 92–17. Washington State Department of Ecology (Olympia, WA).

Dennison, M.S., and J.F. Berry. 1993. Wetlands: Guide to Science, Law, and Technology. Park Ridge, N.J.: Noyes Publications.

Dillehay, T. 1988. Monte Verde: A Pleistocene Settlement in Chile, Vol. I: Paleoenvironment and Site Context. Washington, D.C: Smithsonian Institution Press.

Doran, G.H., and D.N. Dickel. 1988. Multidisciplinary Investigations at the Windover Site. In Wet Site Archaeology, edited by B. Purdy, pp. 263–289. Caldwell, N.J.: Telford Press.

Egloff, M.D. 1988. Recent Archaeological Discoveries in Lake Neuchatel, Switzerland: From the Paleolithic to the Middle Ages. In Wet Site Archaeology, edited by B. Purdy, pp. 31–42. Caldwell, N.J.: Telford Press.

Finlayson, C.M., and A.G. van der Valk, (eds). 1995. Classification and Inventory of the World's Wetlands. Boston: Kluwer. Academic Publishers.

Forman, R.T.T., and M. Godron. 1986. Landscape Ecology. New York: John Wiley.

Fowler, C.S. 1992. In the Shadow of Fox Peak: An Ethnography of the Cattail-Eater Northern Paiute People of Stillwater Marsh. Fish and Wildlife Service, U.S. Department of the Interior, Cultural Resource Series 5.

Glob, P.V. 1969. The Bog People: Iron-Age Man Preserved. New York: Ballatine Books.

Gore, A.J.P., (ed). 1983. Ecosystems of the World, Vol. 4A: Mires: Swamp, Bog, Fen, and Moor. Amsterdam: Elsevier.

Giblett, R. 1996. Postmodern Wetlands: Culture, History, Ecology. Cambridge: Cambridge University Press.

Hall, D., and J. Coles. 1994. Fenland Survey: An Essay in Landscape and Persistence. Archaeological Report 1. London: English Heritage.

Holloway, M. 1994. Nurturing Nature. Scientific American April: 99–108.

Janetski, J.C., and D.B. Madsen, (eds). 1990. Wetland Adaptations in the Great Basin. Museum of Peoples and Cultures Occasional Papers 1, Brigham Young University. Provo, Utah.

Jefferies, R.W. 1987. The Archaeology of Carrier Mills: 10,000 Years in the Saline Valley of Illinois. Illinois University Press, Carbondale.

Kelly, R.L. 1990. Marshes and Mobility in the Western Great Basin. In Wetland Adaptations in the Great Basin, edited by J. Janetski and D.B. Madsen, pp. 259–276. Brigham Young University Museum of People and Cultures Occasional Papers 1. Provo, Utah.

Klein, R. 1987. Problems and Prospects in Understanding How Early People Exploited Animals. In The Evolution of Human Hunting, edited by M.H. Nitecki and D.V. Nitecki, pp. 11–45. Plenum Press, New York.

Langemann, G., and H. Dempsey. 1993. The Vermillion Lakes

Wetlands: Interpreting 10,000 Years of Human Occupation in Banff National Park, Alberta. In Culture and Environment: A Fragile Coexistence, edited by R.W. Jamieson, S. Albroni, and N.A. Mirau, pp. 237–245. Archaeological Association of the University of Calgary, Alberta.

Lourandos, H. 1987. Swamp Managers of Southwestern Australia. In Australians to 1788, edited by D.J. Mulvaney and P. White, pp., 292–307. Sydney: Fairfax, Syme and Weldon.

Lugo, A.E., S. Brown, and M. Brinson. 1990. Concepts in Wetland Ecology. In Forested Wetlands: Ecosystems of the World 15, edited by A.E. Lugo, M. Brinson, and S. Brown. New York: Elsevier.

Malakoff, D. 1998. Restored Wetlands Flunk Real-World Test. Science 280:371–372.

Meehan, B. 1991. Wetland Hunters: Some Reflections. In Monsoonal Australia: Landscape, Ecology, and Man in the Northern Lowlands, edited by C.D. Haynes, M.D. Ridpath, and M.A.J. Williams, pp. 197–206. Rotterdam: A.A. Balkema.

Miller, D.C. 1989. Dark Eden: The Swamp in 19th Century American Culture. New York: Cambridge University Press.

Mitsch, W.J., and J.G. Gosselink. Wetlands (2nd edition). New York: Van Nostrand and Reinhold. 1993.

National Research Council (NRC). 1995. Wetlands: Characteristics and Boundaries. Washington, D.C.: National Academy Press.

Nicholas, G.P. 1988. Ecological Leveling: The Archaeology and Environmental Dynamics of Early Postglacial Land Use. In Holocene Human Ecology in Northeastern North America, edited by G.P. Nicholas, pp. 257–296. New York: Plenum Press.

Nicholas, G.P. 1994. Prehistoric Human Ecology as Cultural Resource Management. In Cultural Resource Management: Archaeological Research, Preservation Planning, and Public Education in the Northeastern United States, edited by J.E. Kerber, pp. 17–50. Greenwich, Conn.: Bergin and Garvey.

Nicholas, G.P. 1998a. Hunter-Gatherers and Wetlands: A Global Perspective. Current Anthropology 39:720–731.

Nicholas, G.P. 1998b. A Light But Lasting Footprint: Human Influences on the Northeastern Landscape. In The Archaeological Northeast, edited by M.A. Levine, K.E. Sassaman,

and M.S. Nassaney, pp. 25–38. Greenwich, Conn.: Bergin and Garvey.

Nicholas, G.P. 1998c. Wetlands and the Mid-Holocene Warming. Journal of Middle Atlantic Archaeology 14:147–160.

Niering, W.A. 1985. Wetlands. New York: Knopf.

Piotrowski, W. 1998. The Importance of the Biskupin Wet Site for Twentieth-Century Polish Archaeology. In Hidden Dimensions: The Cultural Significance of Wetland Archaeology, edited by K. Bernick, pp. 89–106. Vancouver: University of British Columbia Press.

Pretty, G.L. 1977. The Cultural Chronology of the Roonka Flat. In Stone Tools as Cultural Markers, edited by R.V.S. Wright. Canberra: Australian Insitute of Aboriginal Studies.

Purdy, B., (ed). 1988. Wet Site Archaeology. Caldwell, NJ: Telford Press.

Saunders, J.W., R.D. Mandel, R.T. Saucier, E.T. Allen, C.T. Hallmark, J.K. Johnson, E.H. Jackson, C.M. Allen, G.L. Stringer, D.S. Frink, J.K. Feathers, S. Williams, K.J. Gremillion, M.F. Vidrine, and R. Jones. 1997. A Mound Complex in Louisiana at 5400–5000 Years Before Present. Science 277:1796–1799.

Simms, S.R. 1987. Behavioral Ecology and Hunter-Gatherer Foraging: An Example from the Great Basin. BAR International Series 381.

Tiner, R.W. 1998. In Search of Swampland: A Wetland Sourcebook and Field Guide. New Brunswick, N.J.: Rutgers University Press.

Vileisis, A. 1997. Discovering the Unknown Landscape: A History of America's Wetlands. Washington, D.C.: Island Press.

Walker, A., and R. Leakey, (eds). 1993. The Nariokotome Homo erectus Skeleton. Cambridge, Mass.: Harvard University Press.

Wendorf, F. 1993. E-87-S: Occupations Dating to the Grey Phases. In Egypt During the Last Interglacial: The Middle Paleolithic of Bir Tarfawi and Bir Sahara East, edited by F. Wendorf, R. Schild, and A.E. Close, pp. 345–355. Plenum Press, New York.

World Wildlife Fund (WWF). 1992. Statewide Wetlands Strategies: A Guide to Protecting and Managing the Resource. Washington, D.C.: Island Press.

24. Repeated Water Table Lowering in the Dutch Delta: A Major Challenge to the Archaeological Heritage Management of Pre- and Protohistoric Wetlands

Robert Van Heeringen and Liesbeth Theunissen

The amazing observation can be made that large parts of the pre- and protohistoric wetlands in the Netherlands with ages up to 5500 BC are now under agrarian and urban pressure. This is due to the fact that the Low Countries are a man-made landscape from 1000 AD onwards. Actually most of the former wetlands with inhabitation in the distant past are now diked areas with a controlled water level (*polders*). The effect is that desiccation and urbanization have taken their tolls on this unique cultural heritage.

Due to the growing awareness of the rapid erosion of the archaeological record the mission of the National Service for Archaeological Heritage (formerly State Service for Archaeological Investigations in the Netherlands (ROB) has changed in recent years. Large-scale rescue excavations in rural areas and towns were – under strict conditions, and hopefully in the near future under a new Monument Act – handed over to the private sector. In close co-operation with provincial and municipal planning departments, other government agencies, private landowners the work now concentrates on an integrated conservation policy. Preservation of sites and monuments in their historical landscapes is one of the main objectives.

Spearhead program

Wetlands all the way down

Recently, the National Service for Archaeological Heritage launched the spearhead programme *Wetlands all the way down*. The program accommodates a chain of projects. What these projects have in common is their objective to improve the quality of the conservation of archaeological monuments in areas of high archaeological value in the low-lying parts of the Netherlands. These areas can be described as a succession of wet delta landscapes. The value

of these pre- and protohistoric wetlands is mainly determined by the good state of preservation of perishable material such as wooden artifacts, bones, seeds and fruits in a stratified context.

This is also the reason why archaeological sites in these areas have to be well managed: the low-lying parts of the Netherlands have not been marshes for over 1,000 years, but form a modern man-made landscape in which dikes keep out the sea, the polder level has to be lowered periodically to ensure optimal farming and new housing estates are added to towns and villages. Ploughing, water level reduction and building are therefore the main threats to archaeological sites.

In the following sections of this paper the project *Conservation of Neolithic wetland sites in West-Friesland and De Kop van Noord-Holland* will be held up as an example of the way in which the projects in the spearhead program will be elaborated.Conservation of Neolithic wetland sites in West-Friesland and De Kop van Noord-Holland

Some 4,600 years ago, in about 2600 BC, an extensive landscape of channels and creeks developed in the region of West-Friesland and De Kop van Noord-Holland under the influence of both the sea and the fresh water brought in by the rivers IJssel and Overijsselse Vecht (Figure 24.1). Archaeological investigation has shown that the higher-lying areas of this changeable saltmarsh landscape provided the Neolithic inhabitants of this region (Single Grave Culture) with all kinds of possibilities for arable farming, stockbreeding, hunting and fishing.

Traces of this early human presence are certainly worth preserving, but are at the same time extremely fragile. If no measures are taken, all the Neolithic sites in the region will have disappeared in a few years' time.

The aim of the project is to safeguard the current quality

Figure 24.1. Map of the Netherlands showing the locations of the study area.

of the prehistoric sites present as they form an important part of the current man-made landscape.

As a first impetus to the permanent conservation of the sites, a report was presented in mid 2000 comprising all the information on the archaeologically valuable Neolithic settlement sites in De Kop van Noord-Holland and West-Friesland. On the basis of this report an effective policy aimed at the permanent conservation of these sites can be developed and realized. This report will serve as a handle both for the designation of a site as a legally protected monument as well as for the practical arrangement of necessary once-only measures for preservation, and the advisable management and monitoring of these sites.

Once the report has been presented, the ROB will take the initiative to develop measures and plans together with all those concerned, and all organizations and authorities involved, which will enable the step from collective thinking to collective action to be taken. How did the earliest inhabitants of West-Friesland and De Kop van Noord-Holland live?

Archaeological research leaves no room for doubt: just like modern man, Neolithic man did not like getting his feet wet. In the coastal area, he chose to settle on natural levees, higher and therefore drier patches of ground in the otherwise wet landscape (Figure 24.2). Here and there on these natural levees grew oak, hazel, ash and pine, and on the wetter edges willow and alder. There were mudflats and salt marshes, ideal for grazing, and also numerous lakes and extensive reed-lands.

Large dwelling places were near open water, smaller ones further inland. The large dwelling-places were settlements which had been inhabited for a long time, whereas the small seasonal camps were used for specialized activities such as hunting.

People lived in rectangular wooden houses 6 to 14 m long and 3.5 to 4.5 m wide. The buildings had a single row of upright posts supporting the roof. The floors were of clay and some houses had a hearth. These people supported themselves by means of arable farming and animal husbandry (Figure 24.3). They ploughed their land and cultivated primitive types of cereal such as barley and emmer wheat. They also hunted duck, wild boar, brown bear, elk and deer. Large quantities of remains of shellfish (mussels and oysters) and fish (including sturgeon, salmon, eel, grey mullet and cod) indicate that this food too was regularly on the menu.

What is left of them and where are the sites?

Although thousands of years old, the traces left by the earliest inhabitants in the area are still fairly close to the present surface. They consist mainly of organic refuse, food remains such as seeds and fruit, animal bones and mussel shells, but pottery sherds, fragments of querns, broken stone axes, flint tools and sometimes even amber ornaments are also found (Figure 24.4). Under this top layer of finds there is a layer with soil traces. These are discolorations in the soil of rotted posts and postholes, pits (wells), and sometimes even a burial pit/grave (Figure 24.5).

By combining the information from these so-called occupation layers, archaeologists are able to reconstruct a picture of the mode of existence of Neolithic people as well as of the landscape in which they lived. This picture has already been sketched above.

Broadly speaking, the areas inhabited by people in the Neolithic or New Stone Age lie scattered over the municipalities of Anna Paulownapolder, Niedorp, Noorder Koggenland, Opmeer, Wervershoof, Wieringermeer, Wognum and Zijpe (Figure 24.6). Most are situated in the land consolidation area of De Gouw and in the Groetpolder, altogether an area of ca 20 × 30 km, which means that the sites were used for agriculture. There are over 30 sites of differing size. Some settlements are situated on the territory of more than one municipality. Due to the scale of the map shown here, the Neolithic settlements are all indicated by the same symbol.

Figure 24.2. Aerial photograph of Aartswoud with channels and levees dating from the Neolithic.

Figure 24.3. Zeewijk (site 8): excavation trench with ploughing traces and hoof prints of cattle dating from around 2600 BC.

Figure 24.4. Aartswoud (site 11): ditchside profile in which a well-preserved layer of settlement refuse several decimetres thick is visible. The humus layer consists of a large number of small layers containing pottery fragments, mussel shells, broken bones of domestic animals, game, birds and fish.

Figure 24.5. Sijbekarspel-De Veken (site 9): excavation of the remains of a woman aged about 30, from the so-called Enkelgrafcultuur, *ca 2600 BC.*

Time is short!

Even though something has been preserved in the soil for thousands of years, it does not mean it will last forever. The situation at the moment is such that there is a good chance of the Neolithic values in De Kop van Noord-Holland and West-Friesland being lost forever in the not too distant future.

As long as the occupation layers are saturated with water and sealed off from the air there is little to worry about, and the archaeological remains will easily resist the ravages of time. However, problems arise when the water level is lowered for agricultural reasons or when the soil is mechanically disturbed and the occupation layers come into contact with oxygen. This is precisely what is happening now. As a result of the lowering of the polder level, the groundwater level becomes deeper and deeper. The effect

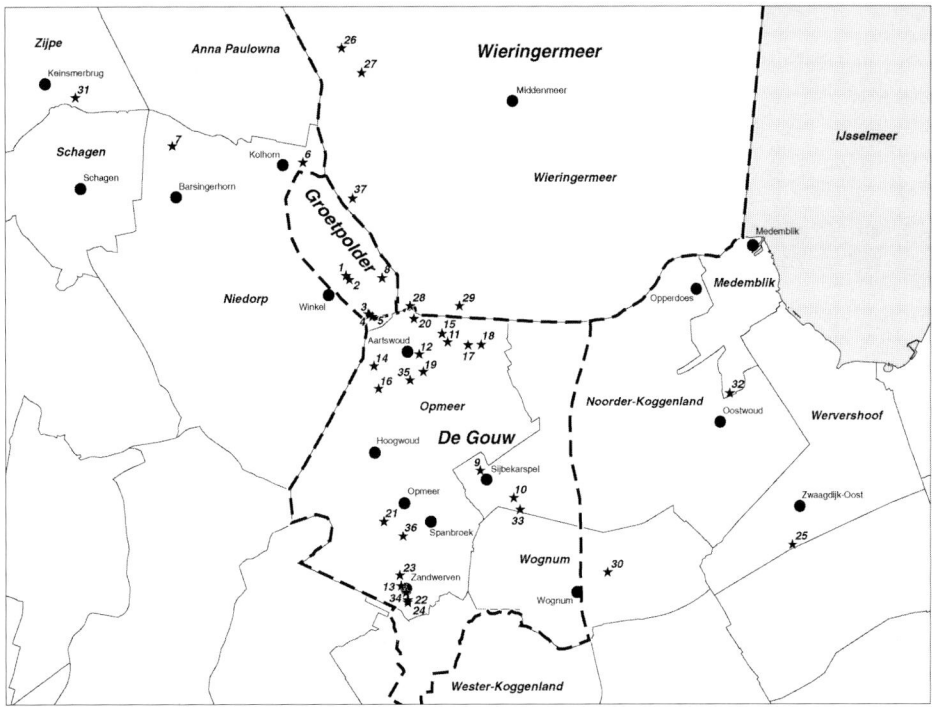

Figure 24.6. *Distribution map of the 37 known Neolithic sites in the project area. The municipality boundaries and the land consolidation area of De Gouw and the Groetpolder are indicated. Legend: 1 De Vrijheid 1; 2 De Vrijheid 2; 3 Flevo 1; 4 Flevo 2a; 5 Flevo 2b; 6 Kolhorn; 7 Poolland; 8 Zeewijk; 9 De Veken; 10 Meester Juffer; 11 Aartswoud; 12 Gouwe; 13 Kogbon; 14 Koningspaadje; 15 Maantjesland; 16 Mienakker; 17 Molenkolk1; 19 Portelwoid; 20 Rhomneyhut; 21 Wijzend; 22 Zandwerven 1; 23 Zandwerven 2; 24 Zandwerven 3; 25 Westfrisiaweg; 26 Bouwlust; 27 Kreukelhof; 28 Land uit Zee; 29 Strijdhamers; 30 Oostenderweg; 31 Keinsmerbrug; 32 Tuithoorn; 33 De Dres; 34 Molentocht; 35 Lange Weid; 36 De Roeper; 37 Tweede beker.*

Figure 24.7. *During ploughing, the upper part of the dark shaded occupation level is disturbed (ploughed up) and then incorporated in the soil.*

of this is that the occupation layer dries out, enabling oxygen to penetrate deeper into the soil.

Mechanical disturbance usually occurs when the occupation layer is relatively high and close to the surface. Due to levelling and erosion a small section of the occupation level disappears every year during ploughing: the organically material ploughed up is incorporated in the soil and quickly disappears through oxidation, erosion or wind. This process repeats itself annually, so the occupation layer is subject to "wear" (Figure 24.7). In the case of the partially excavated site of Zeewijk, it has been estimated that every year at least 1 cm of the 18-cm-thick occupation layer is incorporated in the soil. This could mean that in the course of a single generation the entire occupation level has disappeared and that the site may be regarded as lost for future generations.

Collective action is called for

In view of the ROB's efforts to achieve permanent conservation of archaeological sites, the gradual disappearance of such an important part of the nation's archaeological heritage is an undesirable development. By starting up the project *Conservation of Neolithic sites in West-Friesland and De Kop van Noord-Holland*, the ROB is taking the initiative to check this trend. Responsibility for Dutch archaeological heritage is, after all, a shared responsibility. It is important to all of us that the generations to come know and experience what the landscape is telling us now about the earliest inhabitants of the area. Practical organizational measures such as the local raising of the water level, the covering of sites with soil, the transformation of arable land into pastureland, withdrawal of land from agriculture or purchase of land containing sites for nature conservation are measures which can only be taken and borne collectively by all the parties involved, such as municipalities, provinces, the government, water boards, agricultural organisations, land management authorities and land owners.

References Cited

Borger, G.J. 1992. Draining – digging – dredging: the creation of a new landscape in the peat areas of the Low Countries, in: J.T.A. Verhoeven (ed.), Fens and bogs in the Netherlands: vegetation, history, nutrient dynamics and conservation, Dordrecht. pp. 131–171. (Kluwer).

Geel, B. van, D.P. Hallewas and J.P. Pals. 1983. A Late Holocene deposit under the Westfriese Zeedijk near Enkhuizen, (Prov. of N-Holland, the Netherlands): palaeoecological and archaeological aspects. Review of Palaeobotany and Palynology 38: 269–335.

Ginkel, E. van and W.J. Hogestijn. 1997. Bekermensen aan zee. Vissers en boeren in Noord-Holland 4500 jaar geleden, Abcoude/Amersfoort. (Beaker people by the sea. Fishermen and farmers in North Holland, 4500 years ago).

Hagers, J.K., A. Hagers and J.W.H. Hogestijn. 1999. The Dutch symbol of the battle against water: 'Schokland and surroundings', in: Droste, B. von, M. Rössler and S. Titchen (eds.), Linking Nature and Culture. Report of the global Strategy Natural and Cultural Heritage Expert Meeting 25 to 29 March 1998, Amsterdam, The Netherlands, Paris. pp. 81–90.

Hogestijn, J.W.H. 1992. Functional differences between some settlements of the Single Grave culture in the northwestern coastal area of the Netherlands, in: M. Buchvaldek and C. Strahm (ed), Die kontinentaleuropäischen Gruppen der Kultur mit Schnurkeramik (Praehistoria 19), Prague, pp. 199–205 (Schnurkeramik-Symposium 1990, Prague.

Pons, L.J. 1992. Holocene peat formation in the lower parts of the Netherlands, in: J.T.A. Verhoeven (ed), Fens and bogs in the Netherlands: vegetation, history, nutrient dynamics and conservation, Dordrecht. pp. 7–79. (Kluwer).

Ven, G.P. (ed). 1993. Man-made lowlands. History of water management and land reclamation in the Netherlands, Utrecht.

Willems, W.J.H., H. Kars and D.P. Hallewas (eds). 1997. Archaeological heritage management in the Netherlands, Amersfoort/Assen (ISBN 90 232 3304 2).

25. Science-based Conservation and Management in Wetland Archaeology: The Example of Sutton Common, UK

Robert Van de Noort, Henry Chapman and James Cheetham

Introduction: Sutton Common

Sutton Common comprises the remains of two Iron Age enclosures, which straddle the palaeochannel of the Hampole Beck, which is now completely drained (Figure 25.1). Both enclosures are situated on "islands" of sands and clay of the 25-foot drift/Lake Humber clays within the Humberhead Levels (Van de Noort & Ellis 1997). This is an extensive area of lowlands in eastern England that prior to its drainage in the early seventeenth century was one of the world's great wetlands. Enclosure A is situated on the east side of the former Hampole Beck, enclosure B on the west side. Enclosure A includes two major phases of occupation - the earlier phase is characterized by a timber palisade demarcating the site, the later phase includes multivallete ditch and bank arrangements. Evidence for occupation within enclosure B is limited to the later phase only (Parker Pearson & Sydes 1997). The two enclosures are linked by means of a causeway of sands deposited over the peat of the Hampole Beck palaeochannel and flanked by discontinuous post alignments (Van de Noort & Chapman 1999). Both phases remain poorly dated, but both phases of activity on Sutton Common can be dated after 550 cal BC and before 200 cal BC (Parker Pearson & Sydes 1997).

Until its enclosure in *c.* 1850, the area was wet, with peat forming the main soil on the Common, which was predominantly used as rough pasture. The first drainage ditches were possibly dug as part of the enclosure of the Common, with one ditch clipping the southern tip of enclosure A. Other ditches on the Common provided more effective run-off of precipitation and soil water. However, the site was more effectively drained in 1983, with the installation of plastic underfield drains placed in coarse gravel ditches across the site. Field drains were not installed within enclosure B, but underfield drains were installed within enclosure A, which had been bulldozed in 1980.

A number of archaeological studies and assessments were undertaken between 1987 and 1993, and suggested that desiccation of organic archaeological and palaeo-environmental remains occurred across the site and with little or no potential for *in situ* preservation (Adams *et al.* 1988, Parker Pearson & Merrony 1993, Sydes 1992, Sydes and Symonds 1987). The failure to protect the site from drainage and desiccation was discussed on several occasions (e.g. Parker Pearson & Sydes 1995).

Nevertheless, after lengthy negotiations, the Carstairs Countryside Trust (CCT) bought Sutton Common in 1997, with support from English Heritage and the Heritage Lottery Fund. CCT's primary objective for the future management of the Common was to enable the long-term preservation of the archaeological remains. In 1997, a high-resolution digital terrain model of the Common was created using a differential global positioning system (dGPS), which has become the basis for all further research (Chapman & Van de Noort forthcoming). In 1998, English Heritage commissioned the detailed assessment of the hydrology and the preservation of organic remains across the prehistoric site. This study identified the existence of extensive waterlogged archaeological remains and the opportunities for their *in situ* preservation.

Background to wetland conservation and management

The destruction of wetlands across the world, and with it the archaeological sites contained within these wetlands, is well recorded (e.g. Coles and Coles 1996, Bernick 1998). Most wetland archaeological research has been focused on the excavation of sites that were threatened either by the physical destruction of wetlands or by the indirect effect of the de-watering of areas. In the last three or four decades of the twentieth century, the need for wetland conservation and management has been highlighted by many national

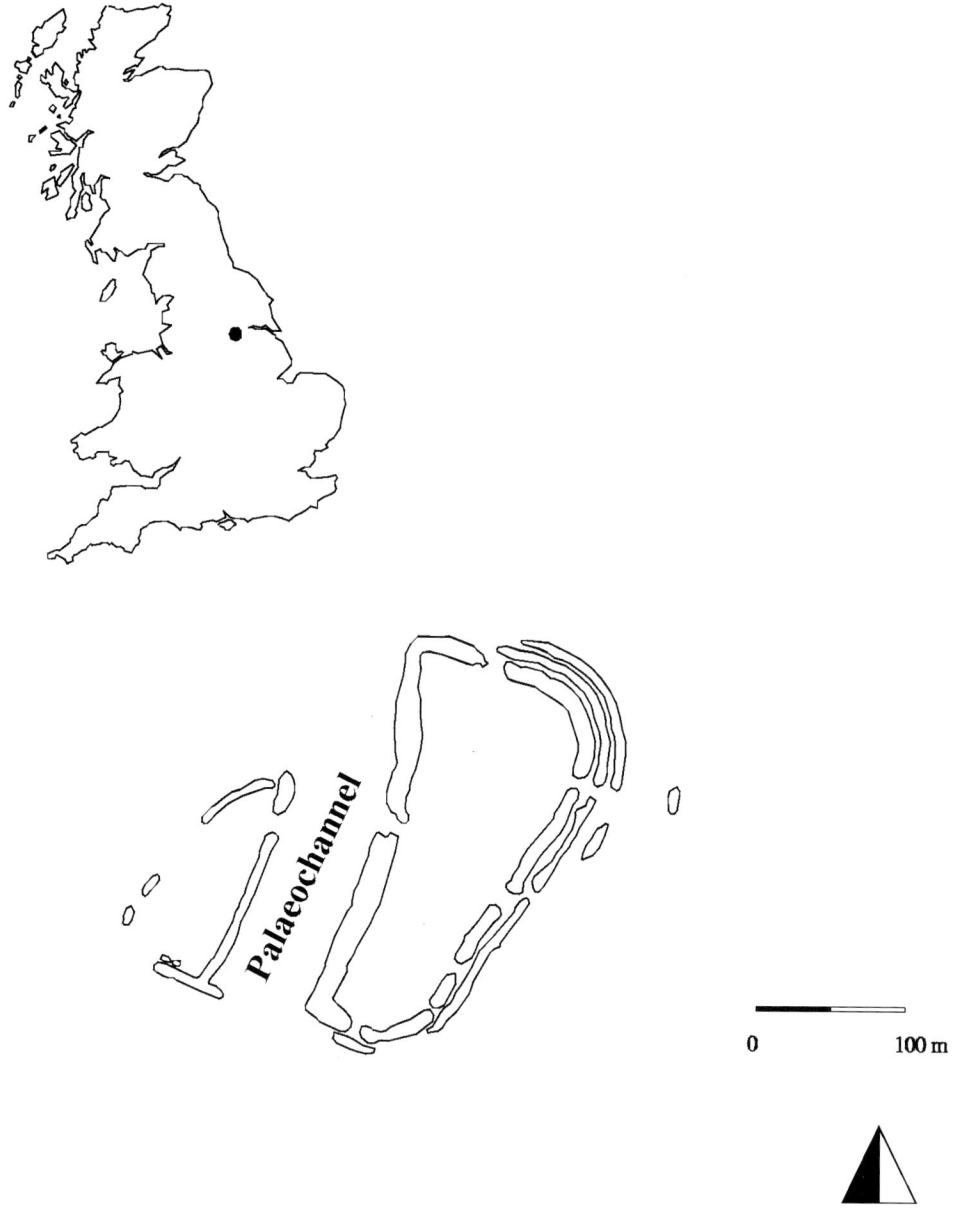

Figure 25.1. Location of the site.

and international bodies, the UN-sponsored 'Ramsar' convention (1971) being the best known (http://www.ramsar.org).

Wetland conservation is now practiced in many countries across the world, by government agencies and voluntary bodies alike (Deny 1995). In all but a few exceptions, the conservation of wetlands is undertaken with the objective of maintaining, creating or recreating wetland habitats that are valued for their contribution to existing faunal and floral communities (e.g., Maltby 1986, Purseglove 1988), for

example as an element in biodiversity strategies. Consequently, wetland conservation as nature conservation is a dynamic process, whereby the management of the wetland is adjusted on a regular basis, for example to accommodate changing priorities or changing trends in the weather. Wetland conservation managers boast a considerable experience in controlling this dynamic process, especially on practical matters that include manipulation of the distribution of water, a crucial element in successful wetland management (e.g., Furniss & Lane 1992).

The conservation of wetlands for the protection of wet-preserved archaeological remains is less common throughout the world. Of course, important archaeological sites may be contained within wetlands that are being managed, but examples of archaeology-led wetland conservation, such as the Sweet Track in the Somerset Levels, England (Brunning 1999), remain rare. The main principle of attempts to preserve wetland archaeological remains *in situ* is to maintain a high water table, and thus to saturate the archaeological site. In the case of the Sweet Track, rehydration or rewetting is achieved by pumping water from the surrounding area to the buffer that surrounds the archaeological monument (Coles 1995). More commonly, for example at the Bronze Age site of Flag Fen in eastern England, or the Corlea (I) trackway in central Ireland, the archaeological remains are surrounded by a "bund," which acts as a water retainer and thus reduces the effects of desiccation (Pryor 1991, O'Donnell 1993).

Important differences within the management of wetlands exist between the archaeology-led and nature conservation instigated projects. Principal among these is the dynamic nature of the wetland management found in nature conservation and the more static approach in archaeological conservation. The scale of operations differs significantly as well, with the archaeologist being essentially concerned with the archaeological monument and possibly a limited buffer area, while nature conservation concerns itself with ever larger areas, sometimes in excess of thousands of hectares (e.g., Ramsar-designated wetlands in the Humber estuary, England). Finally, we note significant differences in the awareness of, and expertise in, wetland management (Coles 1995).

Fundamentally, nature conservation uses the high biomass of wetlands (e.g., Dinnin & Van de Noort 1999) to maximize the variety and quantity of flora and fauna that can be sustained by, or are dependent on, wetlands. The number and abundance of key species, ranging from *Spagnum* to nightjars, can measure the success of this form of wetland management. Archaeological wetland management, on the other hand, aims to create "static" burial environments that minimize further deterioration of the organic archaeological and palaeoenvironmental remains. Its success cannot be expressed in numbers, rather it depends on this absence of change. In certain cases, archaeology-led wetland conservation can successfully adopt nature conservation concerns, and vice versa. Nevertheless, we will concentrate in this paper on wetland conservation work that is instigated with the principal objective of preserving waterlogged archaeological and palaeoenvironmental remains, such as currently practiced at Sutton Common, Yorkshire, UK.

Science-based conservation and management: general concepts

The promotion of the need for science-based wetland management in archaeology rests on several fundamental principles. If our aim is to achieve sustainability of the wet-preserved archaeological resource, or "near-zero" change, and the success of wetland management can only be expressed in those terms, then we must have an approach to wetland management that not only can achieve stability of the burial environment, but that can also demonstrate scientifically the absence of change as an indicator of good management. Furthermore, we cannot progress by "trial and error" but the management must be proactive and informed by empirical findings. After all, the archaeological and palaeoenvironmental resource is limited and non-renewable.

The science-based conservation developed for Sutton Common is, by the Centre for Wetland Archaeology, referred to as the "3M approach", with the 3 M's standing for *m*onitoring, *m*odelling and *m*anagement. The interactive nature of the 3M approach in operation is illustrated in Figure 25.2. Essentially, in the 3M approach the burial environment is monitored at regular intervals. Monitored indicators include the water table, the reduction-oxidation potential of the burial environment (or REDOX), microbial activity in the burial environment, basic chemistry of the soil water and the wet-preserved archaeology itself. Other sites may receive greater benefit from the monitoring of differing sets of parameters. The data from the monitoring are modelled in a Geographical Information System (GIS) environment, providing interpreted information on the monitored parameters. Finally, the models inform the site managers on the need for proactive changes to the management of the wetland in question. In the case of Sutton Common, the instigation of this approach was preceded by a creation of a high-resolution digital-elevation model (DEM) using a differential Global Positioning System (dGPS), which was designed to act as an objective but manipulatable framework.

For the current purpose, three aspects of the 3M approach at Sutton Common will be discussed in detail: the DEM, the water table and the state of preservation of the archaeological wood. These give a flavor, rather than a comprehensive overview, of the kind of work currently being developed at Sutton Common.

The DEM

An objective but manipulatable base-map of Sutton Common was considered essential for the spatial correlation of the various activities that were planned for the site. The survey of the site was undertaken using a *Geotronics©*

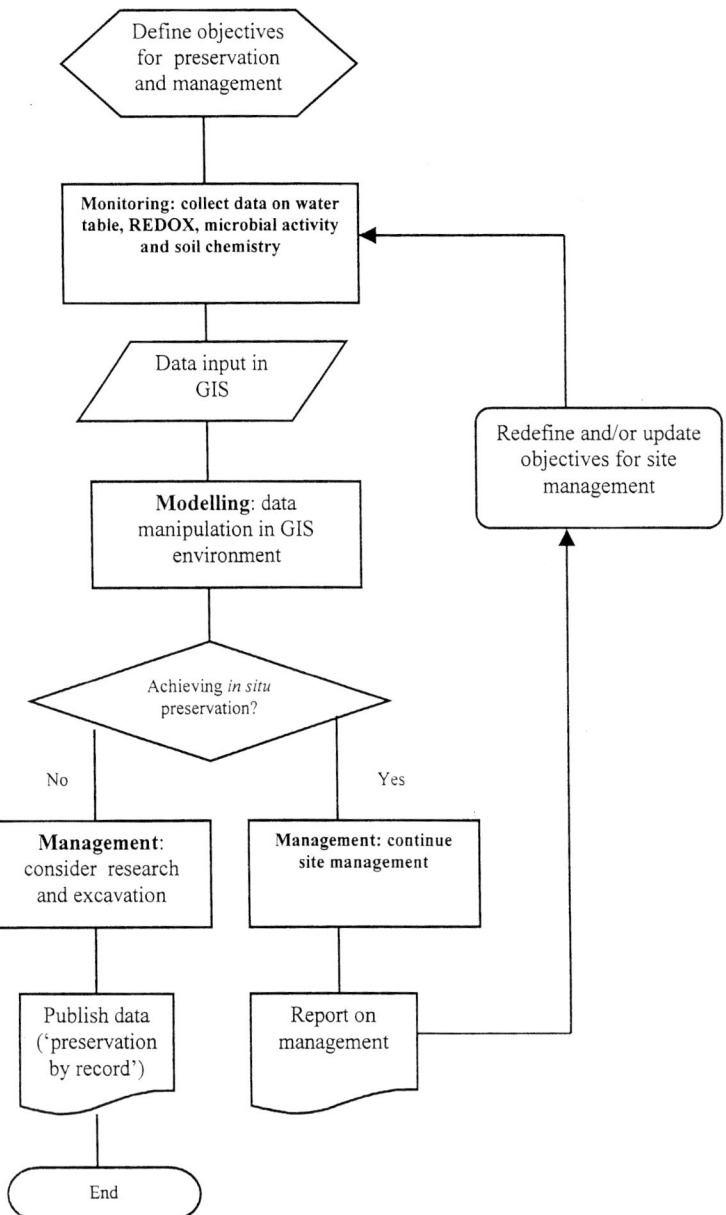

Figure 25.2. 3M approach.

System 2000 L1 – RTK differential GPS. A total of 5,290 points were surveyed, covering an area of approximately 286,754 m², providing a mean density of 0.02 points per m² (184 points per hectare). Points were recorded at between 3 m and 8 m intervals along transects. On areas of high topographic variation, such as across the earthworks within enclosure B, the survey resolution was increased to provide greater detail. The standard deviation error of the GPS was found to be less than 0.054 m for the x-co-ordinate, 0.056 m for the y-co-ordinate and 0.029 m for the z-co-ordinate.

The GPS survey data were recorded in a coded format that was corrected to National Grid values and converted to a comma separated value (CSV) file, consisting of x-co-ordinates, y-co-ordinates and z-co-ordinates, using software developed by Richard Middleton (University of Hull). These data were processed to generate a digital elevation model (DEM) within *ARC/INFO© version 7.2.1* Geographical Information System (GIS) software, run through a UNIX platform.

The variably spaced point data were converted to form

a triangulated irregular surface (TIN) using the CREATETIN command. The accuracy of surfaces inter-polated from a TIN model is dependent upon the function of the triangulation process. ARC/INFO employs a process known as Delaunay Triangulation, which dictates the size and shape of the triangles formed in the generalised surface (Goucher 1997). A cell-based surface was created from the TIN using the TINLATTICE command. This is a vector/raster conversion that interpolates a continuous grid using the TIN as a reference. The function places a grid of cells at a pre-determined density across the area covered by the TIN that are referenced in terms of x- and y-co-ordinates. A height attribute for each cell is then interpolated from the TIN. A number of different ways to convert a TIN to a lattice are allowed by ARC/INFO using this function, and the Quintic interpolation, which applies smoothing to the areas inside the TIN triangles, was used for the DEM of Sutton Common.

Two methods were employed with regard to the representation of the surfaces. The first method involved the basic representation of the surface as contour bands. The second method was to apply a virtual light source to emphasize areas of greater relief. Essentially this technique emphasizes features and slopes, rather than height. A low-positioned light source can highlight the more subtle features. This method can also produce a more realistic-looking surface for comparison with other data sources such as aerial photography.

The resulting DEM of Sutton Common revealed a number of modern and archaeological features despite the intensive agricultural regime of the previous seventeen years. Recent features identified from the survey include the position of the track and drains and, surprisingly, the position of one of the removed drains that was visible on earlier aerial photography of the site. These features were traceable on both models, but were more visible with the application of an assumed light source. However, the hill-shaded model revealed "striping," reflecting the direction of ploughing.

Natural features identified were visible on both models, but the basic contour-banded DEM showed them more clearly (Figure 25.3). The most obvious feature was the relict Hampole Beck (A), which was visible in between the enclosures. Toward the northwestern part of the surveyed area, the braiding channel was visible as identified through previous lithostratigraphical work (Lillie 1997). The two islands occupied by the Iron Age enclosures were high-lighted (B and C), as were three smaller islands to the north (D, E and F). Similarly, in the area to the east the sharp shelving of land towards Shirley Wood was visible (G), which marks one of the faults in the Sutton Common area that includes aquifers. Other areas of lowland were also visible, such as to the southeast of enclosure B (H).

The application of a light source from different positions highlighted a number of buried archaeological features that could not be seen on the ground (Figure 25.4). First, and most obviously, the positions of the undamaged earthworks of enclosure B were clearly visible. A comparison between these and the detailed plan made in the 1930s, before ploughing commenced, by Bennett and Hill (Whiting 1936) demonstrates slight details that were missed, perhaps due to vegetation cover. Of these details the most striking is that its northwestern side was more developed, continuous and cohesive than the early plans suggested (J). Similarly, the break in the eastern side of enclosure B was not as distinctive as the plans had suggested, but rather it had a gradual shape to it (K).

The outline of enclosure A was visible on the models. When compared with the natural topography of the area it was notable that its western edge fell away sharply over the Hampole Beck palaeochannel, but that the width of the enclosure was less than the sandy island, which dropped away gradually to the east up to the shelf near Shirley Wood. The main reasons for the visibility of archaeological features in a ploughed-out landscape is the differential shrinkage of sediments in this actively drained area. Peat-filled structures continue to suffer from desiccation and, consequently, continue to compact at a higher rate than drained minerogenic sediments. A quantitative analysis of the effects of differential shrinkage has been presented previously (Chapman & Van de Noort 2000).

The water table

Essential to the future management of Sutton Common, as an archaeological site that includes many wet-preserved remains, was an accurate understanding of the dynamics of the local hydrology. To provide a model of the hydro-stratigraphy, or shape of the water table, a network of 50 piezometers were laid out in a systematic grid set at 50 m intervals covering the southern and southeastern parts of the Common. Its extension to the east up to the edge of Shirley Wood includes the areas of higher potential for the preservation of palaeoenvironmental material. Piezometer pipes, 2 m long, with a diameter of 19 mm were used with 300 mm long screw-on piezometer tips obtained from MGS Ltd. The tips used were self-contained units consisting of a perforated plastic pipe with an internal permeable mem-brane. The tops of the pipes were sealed using plastic caps to prevent rainwater in-filling and general physical con-tamination.

The grid of piezometers was planned using the DEM. The piezometers were installed by boring 30 mm diameter holes using a spiral augur from Van Walt Ltd. Boreholes were excavated to a depth of 2.3 m and the piezometers with attached tips were placed within the holes. Once

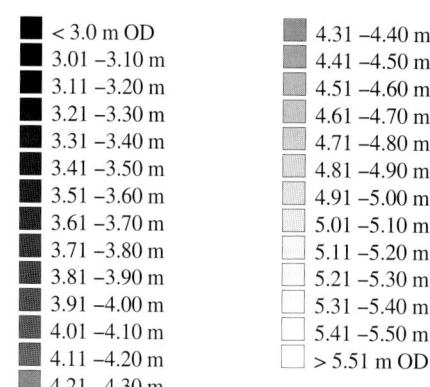

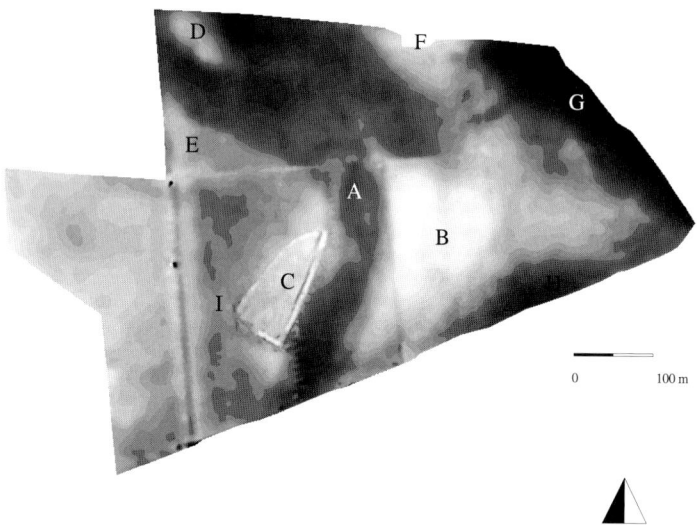

Figure 25.3. Contour band DEM with features mentioned in the text.

installed, the piezometers were left to settle for approximately a week before any readings were taken. After this time the level of water within each piezometer was measured using Van Walt Ltd. sounding apparatus. Measurements were taken on approximately a two-weekly basis following installation and were recorded on pro forma sheets. To provide both relative and absolute levels for the groundwater readings, the positions of the tops of the piezometers were surveyed in three dimensions using dGPS with its accuracy set to a 0.02 m standard deviation. From this the depth of water in each piezometer from the top of the pipe could be subtracted from the absolute height of its top. The series of water table surfaces was generated in the same manner as the contour bands of the DEM, for which the readings taken in September 1998, January 1999 and June 1999 are shown (Figure 25.5).

Readings have demonstrated an overall increasing water

level, which was influenced by the oncoming of winter that was also reflected by the water in the open drains surrounding the site. When modelled the ground water table can be seen to have a dome-shape beneath enclosure A, a wholly natural phenomenon related to precipitation, permeability, through-flow and topography (*e.g.*, Ward and Robinson 1990). The hydromorphic dome became more pronounced in the last months of 1998, when prolonged rainfall had added to the soil and ground water storage. The pronounced character of the dome was enhanced by the size of enclosure A, relative to enclosure B, and the infilling of the ditches surrounding enclosure A in 1980. This rendered these ditches ineffective in drawing down the water table Furthermore, the track to the north of enclosure A acts as a hydrological barrier.

This close relationship between hydrology and topography is not so well defined beneath enclosure B. This may

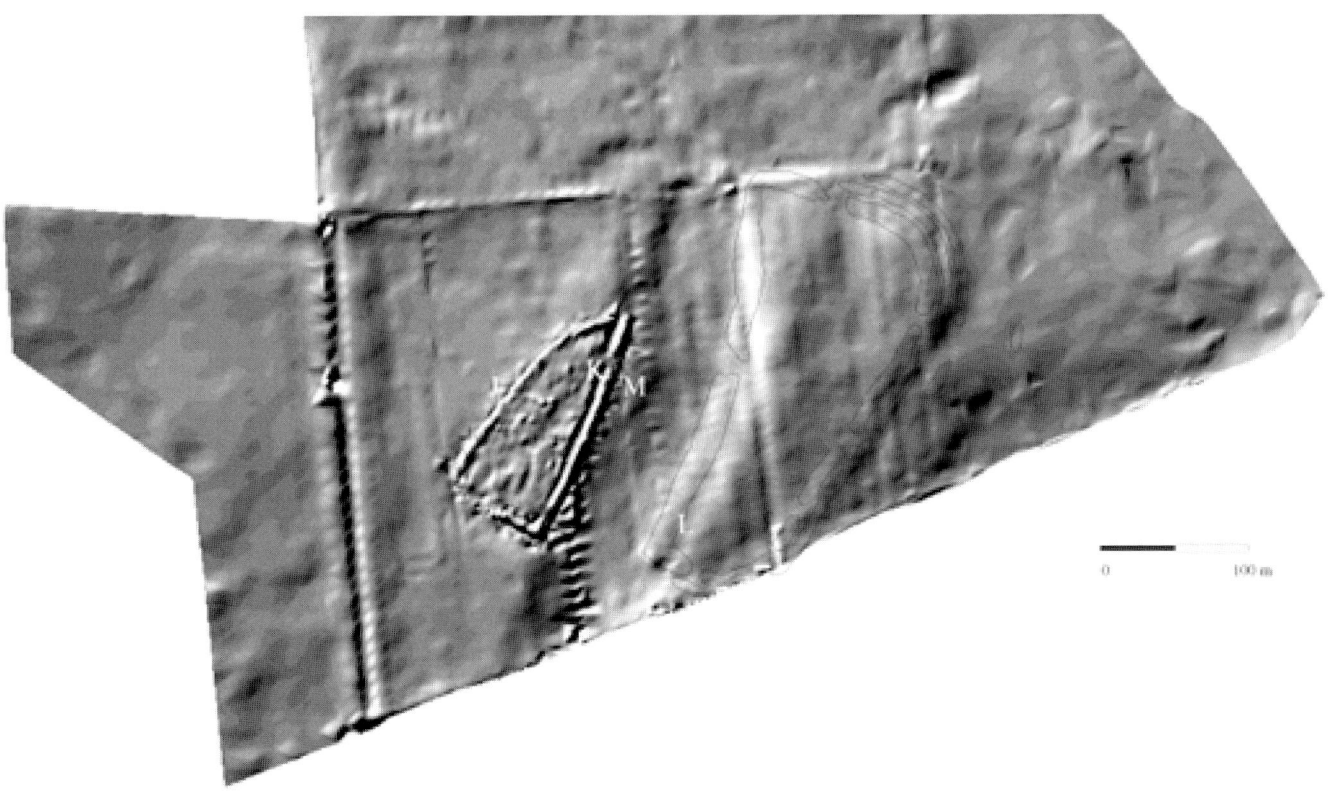

Figure 25.4. Hill-shaded model with features mentioned in the text.

be explained in terms of frustrated soil and ground water storage, which is disturbed through the presence of ditches around enclosure B which draw the water table down, and its relatively small size. Nevertheless, a faint dome-shape was observed in periods of prolonged rainfall, which added to the soil and ground water storage. The soil characteristics of the palaeochannel, which comprises in the upper layers mostly degraded peat with very large pores and therefore high levels of permeability and through flow, are not conducive to hydromorphic domes.

Archaeological wood

Excavations were undertaken in 1998 and 1999 commissioned by English Heritage. In all, sixteen trenches were excavated, ranging in size from 2 × 2 m to 30 × 30 m. Archaeological wood, dated to the Iron Age, was found to survive across the archaeological site. In many places, timber uprights were found to be poor quality and desiccated at the top, but well preserved further down (Figure

25.6). All but a few timbers were found to be *Quercus* (oak).

The preservation of archaeological wood is dependent on a broad range of factors. These factors operate on different scales, ranging from the feature-specific (e.g. wood species, treatment of timber prior to deposition), to the site itself (function, method of deposition) and beyond (water table, drainage). In order to assess the state of preservation of the organic archaeological resource across the site, the highest point of each surviving timber was three-dimensionally recorded by dGPS. On the one hand, this data set represents the hydrology of the last 2500 years, as timbers that would have been exposed to oxygenated environments for any length of time since the Iron Age would not have survived. On the other hand, the data set epitomizes also the effects of the drainage and change in land use of the last two decades of the twentieth century. Nevertheless, the interpolated continuous surface of surviving archaeological wood, which was created in the same manner as the interpolated continuous surface water table,

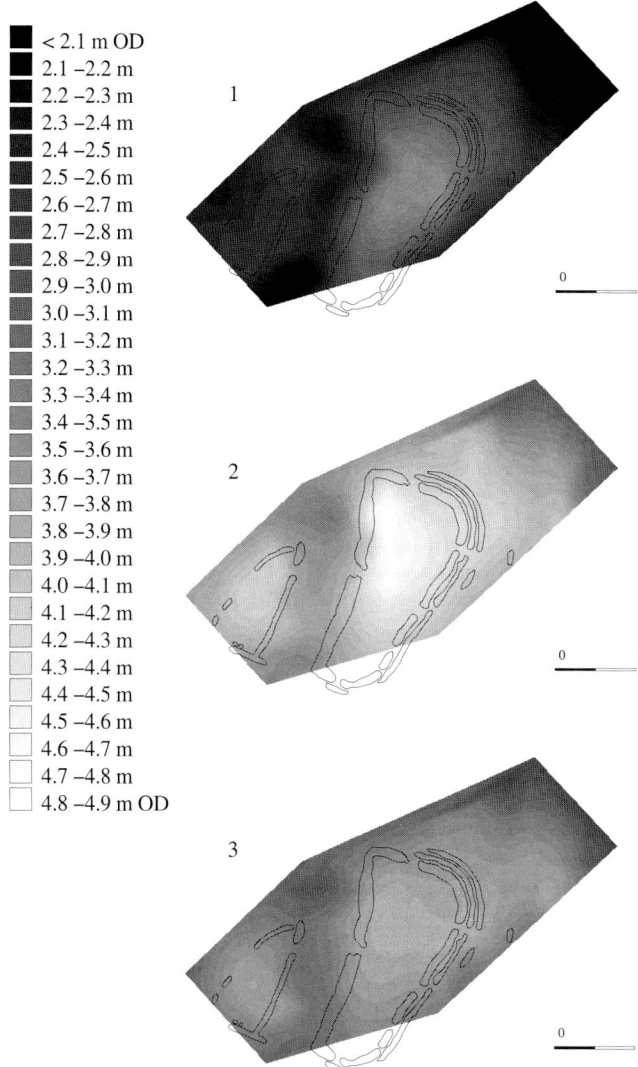

Figure 25.5. Water table in September 1998, January 1999, June 1999.

provides a model of survival of organic archaeological remains that can form the basis for analysis.

Discussion – modelling and management

When integrating the data from Sutton Common, it was evident that the preservation of archaeological wood (and other organic remains) reflects closely the winter water table, as illustrated by the hydrological model. It is reasonable to suggest that the water table observed in the winter of 1998–99 must resemble the more static water table at Sutton Common in the period before its active drainage. Otherwise, archaeological wood preservation would not have mirrored the hydrology. This not only explains the absence of structural archaeological wood in enclosure B,

where the water table is drawn down, but also the surprising presence of waterlogged wood within enclosure A, where the water table is dome-shaped.

The results of the modelling were used to determine the approach to the management of the site. It was decided that:

– First, the dynamic hydrology could be manipulated
– Second, the maximum height of a permanent water table that would not result in the flooding of neighbours' land was 4.1 m Ordnance Datum as measured in one piezometer set within the ditch of encloure B
– Third, that only remains below the 4.1 m Ordnance Datum could be sustainably preserved *in situ* (this

includes the ditches of the Iron Age enclosures and the palaeochannel of the Hampole Beck, the latter containing the causeway linking the two enclosures)

- Fourth, that the interior of enclosure A could not be sustainably preserved as a wetland site
- Fifth, that where *in situ* preservation was not possible, excavation should be considered the most appropriate solution

To assist the management of the site, a model was created of where the water table would breach the surface when the water table was kept permanently at 4.1 m Ordnance Datum, exaggerated by 0.4 m to account for the higher winter tables (Figure 25.7). This shows that much of the palaeochannel of the Hampole Beck would be wet, with areas of standing water to the north and east of enclosure A, and larger areas of standing water to the east of the site, where Sutton Common adjoins Shirley Pool, a protected wetland. On the basis of this model, the varying concerns of landowner, neighboring landowners, archaeologists and nature conservationists were addressed and discussed, and agreements were made on the management of Sutton Common as a wetland.

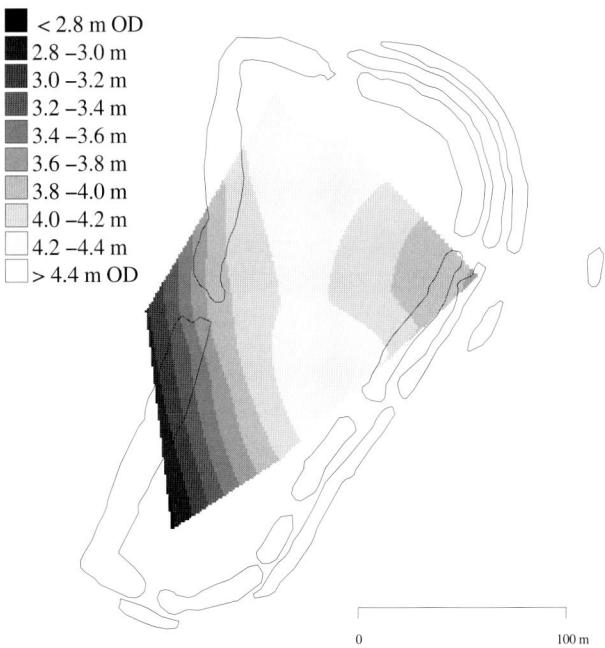

Figure 25.6. Model of surviving archaeological wood.

Conclusion

In many instances of wetland conservation, objectives are to be achieved through reactive management. Where nature conservation leads the management of wetlands, such an approach may well be the most cost-effective and appropriate, but where preservation of waterlogged archaeological remains *in situ* is the main objective, such a trial and error approach is unsuited. Rather, a proactive approach must be developed, that must be based on objective parameters. For the case of Sutton Common, this approach has taken the form of monitoring of key parameters, including the water table, the modelling of information and the prediction of the effects of managing the wet-site archaeology. Elsewhere, the science-based management and conservation of wetlands may involve different or additional parameters.

Postscript, January 2000

Following English Heritage's commissioned evaluation of the interior of enclosure A, undertaken in the late summer of 1999, the recommendations are now published. These include the excavation of the full interior of enclosure A, an area of c. 20,000 m^2, and an integrated education programme that will be developed with local schools. The interior of enclosure A is considered unsuitable for long-term *in situ* preservation, and excavation is the only credible course of action here.

The rest of the archaeological site is believed to be suitable for *in situ* preservation following the rewetting of

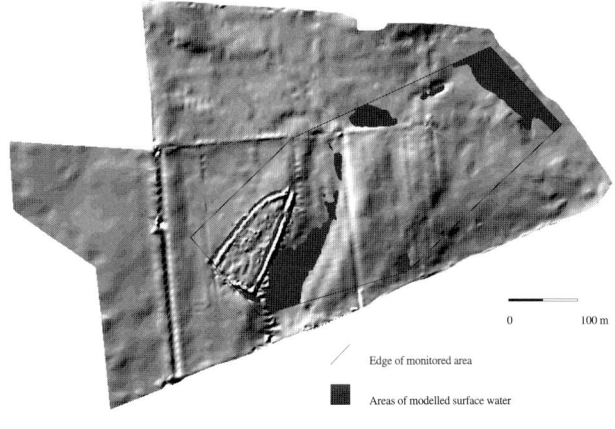

Figure 25.7. Predictions of future situation.

Sutton Common. In the early autumn of 1999, the underfield drain system was modified under supervision of drainage engineers Grantham, Brundell and Farran. Predictions of wetness seem to be justified, with base levels overall higher than in the winter of 1998–99 and the re-creation of a wetland landscape and the *in situ* preservation of much of the wet-preserved archaeology of Sutton Common, are now considered feasible. Monitoring of the burial conditions continues.

Acknowledgments

The "Sutton Common project" is spearheaded by the Trustees of the Carstairs Countryside Trust (CCT), who own the land, in partnership with English Heritage, English Nature, Countryside Agency, the University of Exeter and Hull and Gantham Brundell and Farran.

The project, which includes land acquisitions, wildlife and landscape enhancement, archaeological and paleo-environmental evaluations, research and conservation, and engineering works to raise ground water levels has been made possible through the Sheared Family Trust and financial support from: English Heritage, Heritage Lottery Fund, English Nature, Countryside Agency, Darrington Quarries (Landfill Tax Credits) through WREN – Waste Recycling Environmental, James Goodhart, the Pilgrim Trust, The University of Hull and the Ministry of Agriculture's Countryside Stewardship Scheme. Help has also been given by Doncaster Naturalists Society and Crombie, Wilkinson Solicitors and Carter Jonas, Property Consultants, both of York.

This innovative project forms one of the trial schemes in the Humberhead Levels "Value in Wetness" Land Management Initative. Participation with the Askern Community Partnership over public access and enjoyment of the site seeks to contribute to the environmental and economic regeneration of this Coalfields area in South Yorkshire.

We are especially grateful to Ian Carstairs (CCT), Michael Corfield and Jon Ette (English Heritage), David Patrick (Grantham Brundell and Farran), Allan Hall and Harry Kenward (Environmental Archaeology Unit, University of York) for encouragement and comments on an earlier draft of this paper.

References Cited

Adams, M., C. Merrony and R.E. Sydes. 1988. Excavations at Sutton Common, South Yorkshire 1988. Sheffield: South Yorkshire Archaeology Unit.

Bernick, K. (ed). 1998. Hidden dimensions. The cultural significance of wetland archaeology. Vancouver: University of British Columbia University Press.

Brunning, R. 1999. The *in situ* preservation of the Sweet Track, In B. Coles, J. Coles and M. Schou Jørgensen (eds). Bog bodies, sacred sites and wetland archaeology: 33–8. Exeter: WARP.

Chapman, H. and R. Van de Noort. 2000. Archaeological wetland prospection using GPS and GIS – studies at Sutton Common, South Yorkshire, and Meare Village East, Somerset. Journal of Archaeological Science.

Coles, B. 1995. Wetland management: a survey for English Heritage. Exeter: Wetland Archaeology Research Project.

Coles, J.M. and B. Coles. 1996. Enlarging the past, the contribution of wetland archaeology. Edinburgh: Society of Antiquaries of Scotland, and Exeter: WARP.

Deny, P. 1995. Benefits and priorities for wetland conservation: the case for national wetland conservation strategies, in M. Cox, V. Straker and D. Taylor (eds.) Wetlands: archaeology and nature conservation. London: HMSO: 249–74.

Dinnin, M. and R. Van de Noort. 1999. Wetland habitats, their resource potential and exploitation; a 'case study from the Humber wetlands', in B. Coles, J. Coles and M. Schou Jørgensen (eds) Bog bodies, sacred sites and wetland archaeology: 69–78. Exeter: WARP

Furniss, P and A. Lane. 1992. Practical conservation water and wetlands. London: Hodder & Stoughton.

Goucher, K. 1997. Hill of Tara topographical survey and mapping, in C. Newman, Tara: an archaeological survey (ed). Dublin: Discovery Programme Monograph 2, pp. 245–52.

Maltby, E. 1986. Waterlogged wealth: why waste the world's wet places? London: Earthscan.

O'Donnell, T. 1993. Conservation of Iron Age roadway and associated peatlands at Corlea, Co. Longford. The engineers Journal 46 (3) pp. 44–7.

Parker Pearson, M. and C. Merrony. 1993. Sutton Common desiccation assessment 1993: interim report. Sheffield: Department of Archaeology and Prehistory, University of Sheffield, South Yorkshire Archaeological Services and Doncaster Museum.

Parker Pearson, M. and R.E. Sydes. 1995. Sutton Common: a South Yorkshire wetland, in M. Cox, V. Straker and D. Taylor (eds) Wetlands: nature conservation and archaeology. London: HMSO: pp. 86–96.

Parker Pearson, M. and R.E. Sydes. 1997. The Iron Age enclosures and prehistoric landscape at Sutton Common, South Yorkshire. Proceedings of the Prehistoric Society 63: pp. 221–59.

Pryor, F. 1991. Flag Fen. London: Batsford.

Purseglove, J. 1988. Taming the Flood. Oxford: University Press

Sydes, R.E. 1992. Report on the re-excavation of trench A/C, Sutton Common, South Yorkshire. Unpublished report, Department of Archaeology and Prehistory, University of Sheffield.

Sydes, R.E. and J. Symonds. 1987. Sutton Common 1987 excavation report – investigations of a northern wetland site. Sheffield: South Yorkshire Archaeology Unit.

Van de Noort, R. and H. Chapman. 1999. An archaeological assessment in preparation of a management plan at Sutton Common, South Yorkshire. Hull: Centre for Wetland Archaeology research report.

Van de Noort, R. and S. Ellis. (eds) 1997. Wetland heritage of the Humberhead Levels, an archaeological survey. Hull: Humber Wetlands Project, University of Hull.

Ward, R.C. and M. Robinson. 1990. Principles of hydrology (3rd edition). London: McGraw-Hill.

Whiting, C.E. 1936. Excavations at Sutton Common 1933, 1934 and 1935. Yorkshire Archaeological Journal 33: pp. 57–80.

26. Wetland Archaeology on the Edge: Recent Research into the Nature of Landscape Development and Settlement on the Western Margins of the European Plain

Malcolm Lillie and Henry Chapman
with contributions by Ruth Head (flint) and Benjamin Gearey (palynology)

Introduction

Recent research into the landscape development of western Europe in the Late-glacial to Holocene periods has again highlighted the fact that the modern situation, whereby Britain is isolated from the continental landmass by the North Sea and English Channel, is not analogous to the situation prevailing at the end of the last glaciation (Coles 1998, 1999). The evidence, though occasionally circumstantial, suggests that Britain and Europe were connected by the North Sea Plain, which due to rising sea levels, was effectively inundated by *c.* 5000 cal BC (Coles 1999: 55–6). The landscape of Holderness and the Hull valley has been shown to represent an analogous situation to that inferred by Coles (1999:52) for the Doggerland, with glacial moraines, kettle holes (e.g., The Bog at Roos), gravel banks, numerous lakes (e.g., Hornsea mere), wetlands and minor hills all characterizing the regions under consideration (Dinnin and Lillie 1995 a and b).

The work of the English Heritage funded Humber Wetlands Survey has focussed on the wetlands of the Humber Basin, West Yorkshire, England for the past six years, since 1994 (Figure 26.1). The Humber captures the flow of major rivers such as the Yorkshire Ouse and the river Trent, and numerous tributaries such as the Derwent, Don, Aire, Idle, Torne and Went, which results in a catchment that drains *c.* one-fifth of the landsurface area of England (Pethick 1990). This paper outlines discrete aspects of the results of the first and penultimate regions surveyed; Holderness and the Hull Valley (Figure 26.2). These regions represent the westernmost edges of the 'Doggerland', as defined by Coles (1998, 1999), and are delineated on their western and northern sides by the higher, free-draining uplands of the Wolds. This particular land-block is bordered to the south by the Humber Estuary and to the east by the eroding till coastline of Holderness and

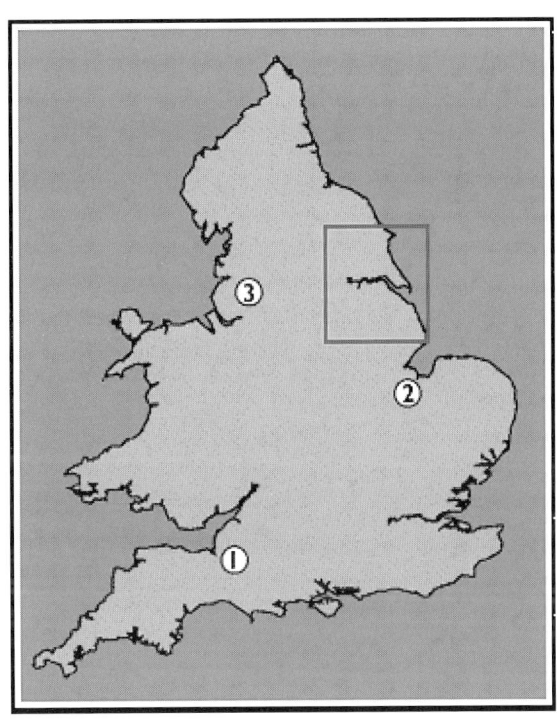

Figure 26.1. Map of Great Britain showing location of Humber Wetlands Survey area (boxed). 1: Somerset Levels survey, 2: Fenland survey, 3: Northwest Wetlands survey.

the North Sea (Van de Noort and Davies 1993, Van de Noort and Ellis 1995). The glacial landscape of Holderness, with its undulating morainic topography, has resulted in the development of a series of lakes and meres which have produced considerable palaeoenvironmental material for the reconstruction of Late-glacial and Holocene landscape developments in the regions considered (e.g., Beckett, 1975,

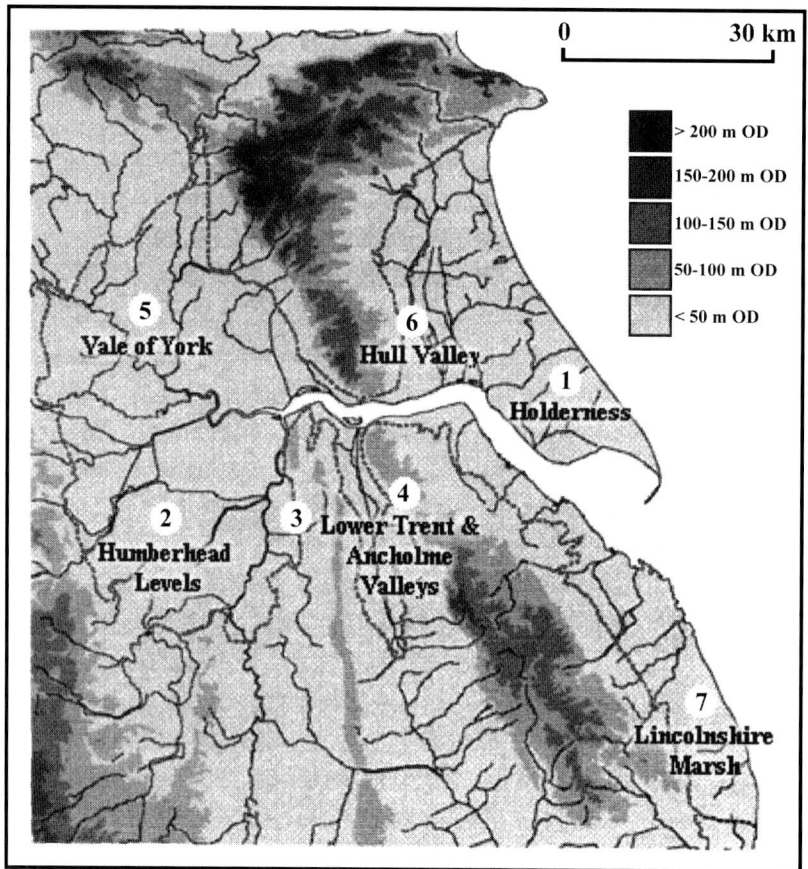

Figure 26.2. Physiographic regions surveyed by the Humber Wetlands Project.

1981, Dinnin and Lillie 1995a and b, Flenley 1987, Gilbertson 1984a, 1984b, 1990, Lowe *et al* 1995, Walker *et al* 1993).

Two sites are considered in this report. At the first, Stone Carr, excavations were carried out in an area where fieldwalking had resulted in the collection of a number of lithics of probable late Mesolithic date. At the second location, an area of fenland developed in the undulating glacial topography at Routh, a gravel extraction site, provides a palaeoenvironmental record for the immediate region.

At Stone Carr (Figure 26.3), preliminary test pit excavations were carried out in April of 1999, in order both to assess the preservation potential of the alluvial deposits and evaluate the state of the buried land surface associated with the lithic assemblage. On the basis of the test pitting at this location, limited excavations were carried out at the end of May and early June of 1999.

The Excavations

Initially, three test-pits were excavated in an east-west direction from a point roughly coinciding with the top of the glaciofluvial "island," westwards into the palaeochannel of the Hull. The easternmost area excavated exhibited a *c.* 0.30 m thick dark gray-brown colored ploughsoil which immediately overlay a thin orange-brown interface comprised of small-medium glaciofluvial gravels and flint in a loose, friable silty-sand matrix with occasional clays which was *c.* 0.02–0.06 m thick on average. This interface grades into the pale brown-brownish yellow glaciofluvial sands and gravels of the 'island'.

A second test-pit was positioned *c.* 1.2 m to the west, at a point where the ploughsoil exhibited evidence for derivation from alluvial sediments. This observation was upheld in the analysis of the ploughsoil, which had more pronounced clay inclusions when compared to the soils overlying the "island." At this location the ploughsoil was again *c.* 0.30 m deep, and exhibited a sharp contact with the underlying alluvial unit. The silt-clay alluvium at this

Figure 26.3. Map of Great Britain showing location of late Mesolithic site of Stone Carr.

location exhibited blue-gray coloration and was oxidized throughout, with columnar peds in evidence. There was a distinct boundary between the alluvium and the gravel "island," which was marked by a transitional horizon. This horizon was shown to be an organic-rich deposit broadly equivalent to the interface noted further east. It was characterized as a very dark gray organic-rich sandy horizon with frequent Fe mottling and gravels throughout and some silts and clays. It probably represents the fluvial reworking of the surface of the "island" along with mixing due to the bioturbation processes accompanying the development of the floodplain flora at this location.

The final test pit in the evaluation produced a similar alluvium derived ploughsoil to that described from the second test-pit, with a distinct contact to the underlying alluvium. Some iron mottling occurs in the upper 0.12 m of the alluvium, which grades into an unoxidized blue-gray silt-clay unit with large collumnar peds in evidence. The water table was encountered at 0.64 m depth in this excavation, necessitating the cessation of the evaluation at this location.

On the basis of the test pit survey a small-scale excavation was undertaken on the raised glaciofluvial area identified on the DEM discussed below (Figure 26.4). The excavation was *c.* 13 m northwestsoutheast and *c.* 16.5 m southeast-southwest, with an area of *c.* 5 × 9 m opened up

in the eastern angle of the L-shape. The rationale behind this strategy was the likelihood that the southern arm of the excavation would produce a sequence from the top of the "island" westwards into the palaeochannel of the Hull, while the 5 × 9 m area would hopefully provide the maximum opportunity to establish the existence of waterlogged archaeological deposits associated with the flint scatter. Finally, the northwest-southeast arm of the "L" would enable evaluation of the degree to which the assemblage had been reworked in the deeper stratified sequences on the northern side of the "island."

The general sequence of deposits identified during the test pit survey was confirmed during the work on the southern part of the excavated area. This sequence revealed ploughsoil over an alluvial unit that was more oxidised towardsthe "island" on the eastern side of the excavation. The glaciofluvial outcrop falls away into the palaeochannel feature from a point *c.* 4 m to the west of the southeastern corner of the excavations, where the base of the sequence is *c.* 0.40 m below the modern ground level. At the westernmost point in the excavation, *c.* 15 m west of the southeastern corner, the surface of the glaciofluvial outcrop was established at 2.20 m below the modern ground surface at a depth of *c.* −1.754 m OD (Figure 26.5). The alluvium that underlies the ploughsoil on the western side of the glaciofluvial outcrop increases in thickness across the floodplain area at this location. The minerogenic-rich organic horizon is overlain by a dark-brown/red-brown silty phragmites reed swamp peat from a point *c.* 8.9 m west of the southeastern corner of the excavation. Upon excavation of the reed peat, large (>1.5 m) branches of oak and alder were extracted from the basal organic horizons and considerable quantities of well preserved macrofossil material was in evidence. These deposits were traced westwards throughout the palaeochannel area, but despite continual pumping, the modern water table prevented the full excavation and recording of these deposits at the westernmost edge of the excavated area.

The main 5 × 9 m area of the excavation encompassed the ploughsoil over a sand and gravel interface onto the glaciofluvial deposits of the outcrop, as identified in the final test-pit discussed above. On the western side of this area the surface of the outcrop follows the same topographic situation as identified in the west-east arm of the "L." A number of plough scrapes cut into the surface of the "island" in the southeastern corner of the excavations, where the ploughzone thins slightly. On the break of slope close to the southern section of the excavation, a shallow fluvial feature or gully was exposed. The dimensions of this feature were proven to be *c.* 1.8 m east to west (running downslope) from the red-brown marginal sands and gravels of the "dry land" part of the "island" into the minerogenic organic-rich sands and gravels. On its eastern side the gully is 0.64 m

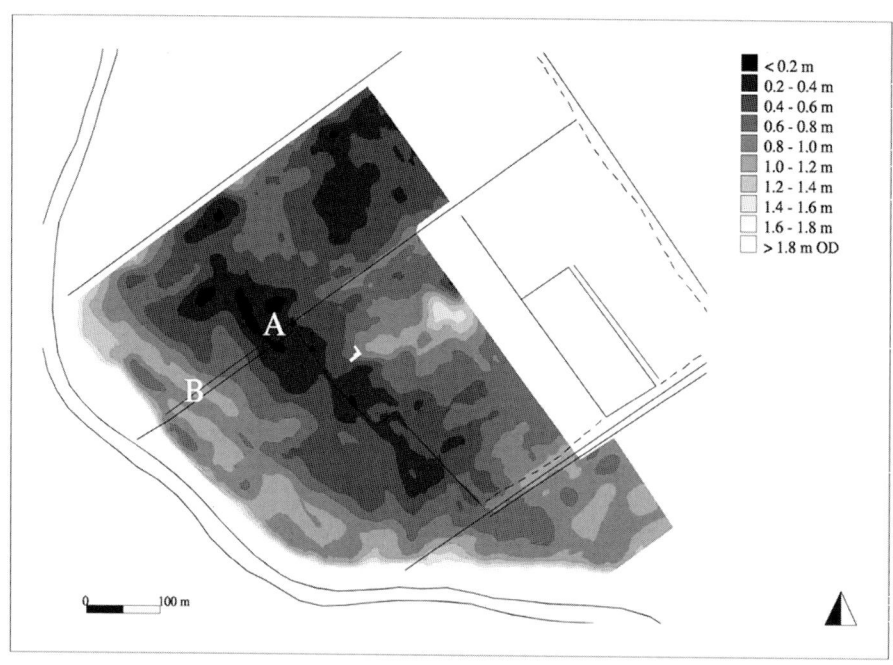

Figure 26.4. DEM of the Stone Carr landscape overlain with detail of the modern landscape features. The River Hull runs around the southwestern side of the model. The position of the old channel (B) is marked by higher ground whereas its floodplain (A) is represented by lower lying topography. This is due to differential shrinage of the clastic and biogenic sediments respectively following de-watering. The position of the excavation trench is marked on the edge of the glacio-fluvial island rising to the northeast.

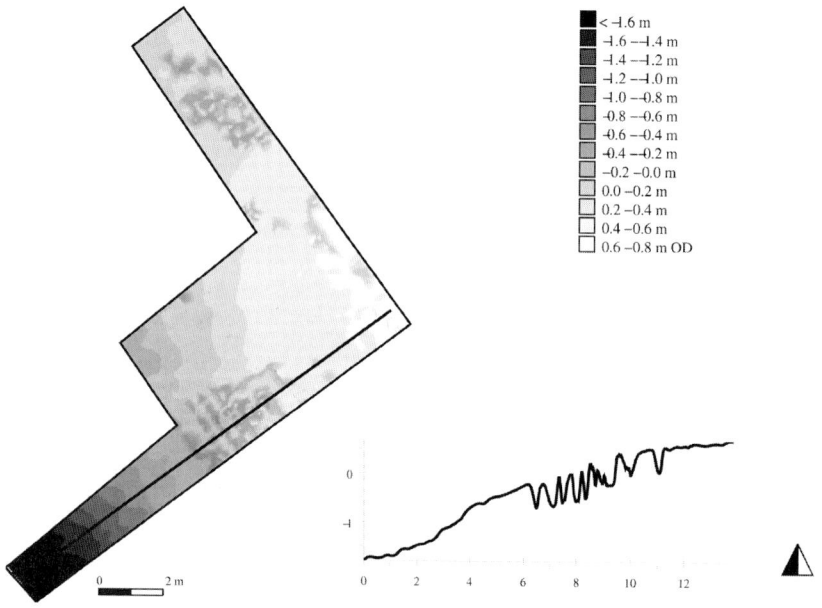

Figure 26.5. DEM of the excavated area showing a profile section along its southwestern arm. The uneven surface of the gully is visible along this line.

wide, with the edges diverging to 0.80 m width at a point *c.* 1.8 m downslope. The maximum proven depth of the feature graded from 0.15–0.28 m east to west. Within this feature fragments of waterlogged deer antler were recovered.

The northsouth "arm" of the "L" has ploughsoil overlying a shallow alluvial unit, which is equivalent to the alluvium identified further south. This alluvial unit has iron mottling throughout and is a silt-clay deposit of dark gray coloration which overlies an organic very dark brown sandy-silt horizon, which upon excavation had the appearance of a buried soil. Further excavation however, showed this deposit to be occupying numerous discrete hollows and dips in the underlying sands that cover the outcrop on its northern side. The appearance of these features suggests a natural (as opposed to anthropogenic) origin, possibly associated with fluvial reworking of the sand surface prior to burial by the alluvium. The precise characteristics of this phenomenon remain to be elucidated by additional investigation at this location. The sands that underlie the alluvium are fine-medium grained with some silts and occasional small gravels (<10 mm). These exhibit a whiteish-gray color, with dark gray concentrations. Lithics were found throughout the brown sandy-silt horizon, occasionally embedded into the surface of the underlying sands (Figure 26.6), but the existence of desiccation cracks penetrating into both horizons necessitates caution as the potential that these lithics are derived remains high.

No discrete structural or hearth features were identified during the excavations at this location, but as the rationale behind the survey was the identification of wetland and waterlogged archaeological materials associated with the activities as indicated by the lithic assemblage, as opposed to the determination of the "dry land" character of the occupation, this is perhaps unsurprising. The excavations cover *c.* 85 m² of the westernmost edge of the glaciofluvial outcrop identified during the coring program, *c.* 36 m² of which are associated with the dry land component of the occupation. The lithic density increases away from the channel, eastwards, suggesting that further insights into the nature of the activities at this site remain to be elucidated. Some indication of the wetland archaeological potential of this location has been identified by the recovery of the antler fragments from the gully feature. However, it must be realized that the full potential, in terms of our understanding of the nature of the activities at this location, remains to be established in the areas to the east, north and south of the locations investigated to date. Some indication of the temporal positioning of this site in relation to Holocene landscape developments was forthcoming from the analysis of the lithic assemblage recovered during the fieldwork. The main conclusions from this area of the investigations are summarized below.

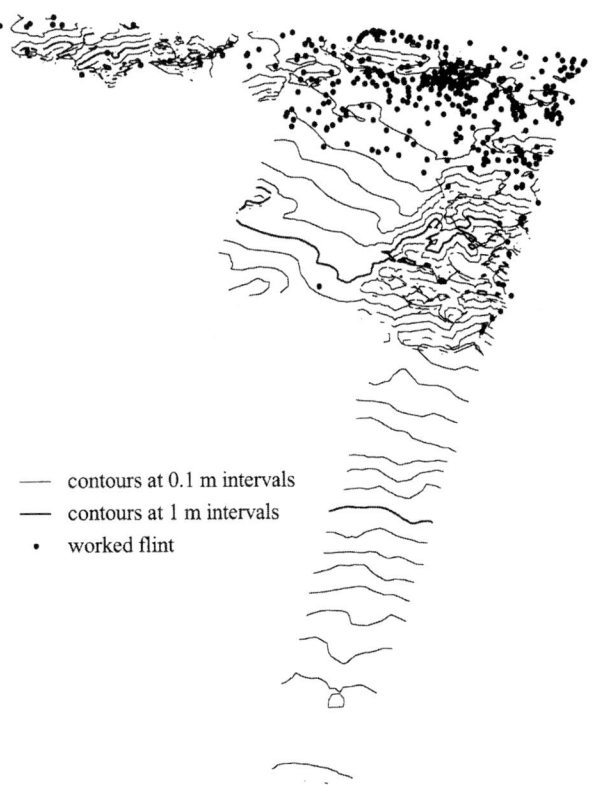

— contours at 0.1 m intervals
— contours at 1 m intervals
• worked flint

Figure 26.6. Perspective view of the excavated area seen from the south, with the distribution of worked lithics draped over the model.

The Lithics

The lithic assemblage comprised 51 pieces from the field walking and a total of 785 lithics that were collected from the excavations. The latter assemblage comprised one scraper, one knife, a microlith, one pick, 16 cores, 659 flakes, 99 chunks and one unmodified piece. This assemblage dates from the late Mesolithic period. The majority of the pieces are knapping debris, with only one bladelet and a handful of retouched flakes present. This proportion of tools to knapping debris, suggests that the assemblage represents a production site, this is also supported by the fact that only one piece shows clear evidence for utilization. A domestic assemblage would be expected to contain a higher proportion of tools and greater evidence for wear and utilization.

The tools are likely to have been removed, to be utilized at a different site. The majority of the flint has recorticated, which is in keeping with other assemblages of this date in the area. While the assemblage predominantly dates from the late Mesolithic period, there are a handful of pieces that suggest a Neolithic date. The majority of the pieces are knapping debris, with very few formal tools present and a few retouched flakes. Several of the flakes, tools and retouched pieces do show signs of abrasion and utilization and it appears likely that this area has produced an assemblage that is part of a larger aggregation, perhaps associated with a nearby domestic complex. On the basis of the borehole survey, higher energy stream and riverine environments clearly occurred on the north and west of the glaciofluvial outcrop, with shallower low energy floodplain conditions prevailing to the south of the outcrop. This would indicate the selection of an optimal vantage-point for hunter-gatherers to prepare tools for food collection and preparation. When considering the assemblage as a whole, it is clear that the artifacts represent a late Mesolithic flint knapping site.

The flint used is generally of good quality for the local raw materials from this area. As such the source is obviously reliable and not in short supply. Few pieces have flaws and faults present within the flint, suggesting that nodule testing is taking place at a different location, near to the raw material source, which in this context may be anywhere in the immediate vicinity of the excavated areas. The assemblage exhibits some good examples of tools, with comparable examples being rarely seen in this region. The blade cores and associated debris show a high level of workmanship, with few hinge terminations and miss-hits being present. In this context it is apparent that the wetland areas are being used not only as a food resource, but also to supply the raw materials needed to produce tools for the exploitation of the contemporary environment. These observations indicate that the human exploitation of the wetland environments of the Hull valley and Holderness was multi-purpose in nature, a scenario that would presumably have prevailed in the adjacent Doggerland areas to the east. An indication of the likely environmental conditions prevailing in the wetlands around Stone Carr has been forthcoming from preliminary palynological analyses at the gravel extraction site of Routh Quarry. The broad late Mesolithic landscape features are summarized below.

Routh Quarry palaeoenvironmental data.

At Routh *Corylus avellana*-type, assumed in this instance to be almost entirely referable to *Corylus avellana*, formed the main component of the early Holocene woodland (Lillie and Gearey 2000). This taxa seems to have remained important locally following the arrival of the other deciduous species. The maintenance of high frequencies *Corylus avellana*-type suggests this taxon was dominant locally. *Quercus* and *Ulmus* percentages indicate that these trees were present in the local vegetation, but were never dominant. A rise in *Tilia* is recorded at Routh at the opening of local pollen assemblage zone RO3a. Elsewhere in the region, at the site of Askham Bog, which is centered on

SE570480 a date of 6310±45BP (OxA-8256) is available for the *Tilia* rise (Gearey & Lillie, 1999). Lime is the last of the main woodland taxa to arrive in the lowland forests as it is slow to colonize new sites and is unable to set seed below a certain temperature. This tree tends to be very poorly represented palynologically and it is probable there is a substantial presence of lime in the Hull valley, as has been suggested for eastern England generally (Gearey & Lillie 1999, Grieg 1982, Girling & Grieg 1977). The palaeoecological record from Routh is notable for the continuing representation of *Betula* at values sufficient to infer a local presence of birch into the mid-Holocene. Birch was probably present on the damper soils, where it had a competitive advantage over oak, elm and lime.

At this time a small increase in Poaceae reflects a more open environment locally, although whether this was derived from on-site wetland grasses such as *Phragmites* or represents an influx of pollen from off-site, dryland habitats is unclear. Grains of Lactuceae undiff., *Pteridium* and Ericales undiff. are present at the level at which *Alnus* levels increases, suggesting that there may have been some opening up of the canopy around the same time that alder became established locally. Although *Pteridium* sometimes occurs in shaded, woodland habitats, the close association between the Poaceae undiff. and *Pteridium* curves may indicate that this species was more prevalent in open locations around the Routh site. Imprecision about the species represented by Lactuceae undiff. prohibits a clear statement about the ecological significance of this pollen type, but the majority of this large group are low growing herbs typically found in open habitats rather than closed, wooded environments.

The pollen data from Routh, which is situated *c.* 5 km to the north of Stone Carr, indicates the presence of open ground within the landscape during the later Mesolithic. Some suggestions have been made regarding the management and exploitation of the landscape of northern England during the Mesolithic at sites such as Star Carr (e.g. Law 1998, Mellars 1998). It is intriguing in this context that we are seeing evidence from the palynology to suggest the presence of open areas within the later Mesolithic regional landscapes. Further analysis will hopefully elucidate the nature of the potential natural versus anthropogenic influences on the early-mid Holocene vegetation communities, and allow us to determine whether these relate to soil maturation or deliberate landscape disturbance by the indigenous hunter-fisher-gatherer populations of the region.

GPS Survey

The excavations and the surrounding landscape were surveyed using Geotronics ©Geotracer System 2000 L1 RTK differential Global Positioning System (GPS) equipment (Figures 26.4–26.6). This equipment tracks the positions of a constellation of orbiting satellites that transmit radio frequency signals containing data on time and velocity. By monitoring the delay it is possible to calculate the position of the receiver relative to the spherical orbit of a number of satellites, and therefore to highlight relative positions accurately. In-built and incidental error is removed by using a differential system that maintains one receiver in a static position while a detail pole records positions. Centimeter accuracy is possible for the detail pole relative to the base station through a calibration of the signals between the two receivers.

The two surveys were carried out on the site at varying resolutions. The first of these examined the overall landscape surrounding the surface lithic scatter (Figure 26.4). The landscape was surveyed in transects at a low resolution providing a surface resolution of 74.04 points per hectare. The second survey was of a much higher resolution over the top of the glaciofluvial outcrop running beneath the alluvium in order to provide a topographic model of the contemporary land surface (Figure 26.5). Points were recorded from the excavated area at a surface resolution of 16 points per square meter. Vertical heights for the two were calibrated to Ordnance Datum from a secondary benchmark within the nearby village of Wawne.

The survey data was modelled using ARC/INFO© Geographical Information System (GIS) software run on a UNIX platform. The GPS data was first transformed into Cartesian coordinates and transformed to British National Grid (1936). This point data was then transformed into a triangulated surface and interpolated into a grid of cells each with an individual attribute relating to its height. For the larger area the cell-size of this grid was 1 m, though on the smaller scale survey cell-size was to the much higher resolution of 0.05 m. Quintic smoothing was also applied to the model to provide a more realistic surface curvature. The data generated by the GIS model allow a number of conclusions to be drawn regarding the nature of the landscape in relation to the fluvioglacial "island" upon which the later Mesolithic activities are located.

1. The surface model of the larger landscape (Figure 26.4) has demonstrated the relative flatness of the modern landscape when compared to the earlier Holocene landscape revealed through boreholes and excavation.
2. An inverted topography was highlighted by the combination of boreholes and surface modeling. The area of the floodplain (Figure 26.4:A) was seen as topographically lower than the area of the main channel (Figure 26.4:B). The sedimentary record indicates that this is a feature of differential sub-surface shrinkage accompanying the de-watering of the site, whereby the fine grained silt-clay and sand sediments representing

the channel have shrunk less than the biogenic and alluvial floodplain deposits.

3. The modeling of the excavated area in relation to the landscape model has highlighted the vast change in topography following Holocene alluviation and the establishment of the modern landscape (Figure 26.5).
4. The relationship between lithic distribution and topography has been demonstrated to extend beneath the surface of the alluvium (Figure 26.6), highlighting the high potential for the preservation of contemporary organic preservation on the wetland margin due to the subsequent rise in base levels.

Conclusions

When considered from the micro- to macro-scales of landscape developments it is clear from the excavated evidence that we are looking at a Late Mesolithic flint knapping site located within a wider alder carr landscape with hazel dominant locally. The palaeoenvironmental evidence for the later Mesolithic period as indicated from the site of Routh, which is located *c*. 5 km to the north, also indicates birch, oak, elm and lime within the local flora, with some evidence for open ground. The topographical situation, on an "island" raised up out of the wetlands would indicate the selection of an optimal vantage-point for hunter-gatherers to prepare tools for food collection and preparation.

The combined GPS and borehole survey has highlighted numerous aspects of the surrounding landscape, with the "island" of glaciofluvial sands and gravel being flanked to the north and south by shallow floodplain alluvial sequences. The western side of the "island" is delineated by the steeply incised palaeochannel of the Hull, with differential sub-surface shrinkage of the fine-grained biogenic alluvial sequences apparently accompanying the de-watering of the site. The evidence indicates that the higher energy fluvial sediments, comprising laminated silt-clays and sands, representing the channel sequences, have shrunk less than the biogenic and alluvial floodplain deposits at this location. In relation to the model generated from the topographic survey, it would appear that we are able to determine those areas representing floodplain, geological or channel units on the basis of the variability in sediment wastage, when we apply the regulating medium of low-resolution borehole survey to the landscapes we study. In effect we are able to predict the nature of the wetlands surrounding the late Mesolithic occupation site with the application of very limited, non-destructive surveying techniques.

The areas considered further highlight the significance of the Holderness and Hull valley regions in terms of our understanding of Late-glacial to Holocene landscape developments and hunter-gatherer exploitation strategies. To date, sites such as Gransmoor, The Bog at Roos and Routh Quarry have produced excellent preservation of sedimentary sequences containing well-preserved palaeo-environmental records for the Late-glacial to Holocene periods. Sites such as Star Carr in North Yorkshire have added considerably to our knowledge of human influences in this respect. It is only with the identification of additional Mesolithic sites such as Stone Carr, with its limited evidence for organic preservation, that a more realistic picture of past human settlement and subsistence activities will be forthcoming from the areas to the west of the Doggerland. Further investigations at Stone Carr will hopefully expand upon the organic evidence obtained to date, with the obtaining of radiocarbon determinations from the floodplain deposits, and the identification of wetland deposits intimately associated with the human activities at this location.

Acknowledgments

The fieldwork element of this paper was funded as part of the English Heritage sponsored Humber Wetlands Survey. The fieldwork team consisted of the main authors, Helen Fenwick, William Fletcher and Gavin Thomas of the CWA, and additional team members who included Jenny Moore and Madeline Solomon (Sheffield University), Doug Joblin who acted as assistant supervisor during the filed work, and Jim Tovey, the site draftsperson and illustrator. Finally, the authors would like to thank the landowner Mr. Malcom Pearson for access to the land.

References Cited

Beckett, S.C. 1975. The Late Quaternary vegetational history of Holderness, Yorkshire. University of Hull PhD thesis.

Beckett, S.C. 1981. Pollen diagrams from Holderness, North Humberside. Journal of Biogeography 8:177–98.

Coles, B.J. 1998. Doggerland: a speculative survey. Proceedings of the Prehistoric Society 64:45–81.

Coles, B.J. 1999. Doggerland's loss and the Neolithic . In Coles, B.J., Coles, J., and M. Schou Jørgensen (eds) Bog Bodies, Sacred Sites and Wetland Archaeology. Exeter: WARP.

Dinnin, M. and M.C. Lillie. 1995a. The palaeoenvironmental survey of the meres of Holderness. In Van de Noort, R., and S. Ellis (eds). Wetland Heritage of Holderness: An archaeological Survey. Hull: Humber Wetlands Project, University of Hull. pp. 49–85.

Dinnin, M. and M.C. Lillie. 1995b. The palaeoenvironmental survey of the meres of Holderness. In Van de Noort, R., and S. Ellis (eds). Wetland Heritage of Holderness: An archaeological Survey. Hull: Humber Wetlands Project, University of Hull. pp. 49–85.

Flenley, J.R. 1987. The meres of Holderness. In S. Ellis (ed.) East Yorkshire field guide. Cambridge: Quaternary Research Association. pp. 73–81.

Gearey. B.R., and M.C. Lillie. 1999. Aspects of Holocene vegetational change in the Vale of York: palaeoenvironmental investigations at Askham Bog. In Van de Noort, R., and S. Ellis (eds). Wetland Heritage of the Vale of York: An archaeological Survey. Hull: Humber Wetlands Project, University of Hull. pp. 109–21

Gilbertson, D.D. 1984a. Late Quaternary environments and man in Holderness. Oxford: British Archaeological reports (British Series) 134.

Gilbertson, D.D. 1984b. Early Neolithic utilisation and management of alder carr at Skipsea Withow Mere, Holderness. Yorkshire Archaeological Journal 56:17–22.

Gilbertson, D.D. 1990. The Holderness meres: stratigraphy, archaeology and environment. In S. Ellis and D.R. Crowther (eds). Humber Perspectives: a region through the ages. Hull: University Press. pp. 89–101.

Girling, M.A. and J.R.A. Greig. 1977. Palaeoecological investigations at Hampstead Heath, London. Nature 268:45–7.

Greig, J.R.A. 1982. Forest clearance and the barrow builders at Butterbump, Lincolnshire. Lincolnshire History and Archaeology 17:11–4.

Law, C. 1999. The Uses and Fire-ecology of Reedswamp Vegetation. In Mellars, P., and P. Dark (eds). Star Carr in Context: new archaeological and palaeoecological investigations at the Early Mesolithic site of Star Carr, North Yorkshire. Cambridge: MacDonald Institute Monographs. pp. 197–206.

Lillie, M.C., and B.R.Gearey. 2000. The palaeoenvironmental survey of the Hull valley. In Van de Noort, R., and S.Ellis (eds). Wetland Heritage of the Hull valley: An archaeological Survey. Hull: Humber Wetlands Project, University of Hull.

Lowe, J.J. G.R. Coope, D.D. Harkness, C. Sheldrick and M.J.C. Walker. 1995. Direct comparison of UK temperatures and Greenland snow accumulation rates, 15000–12000 yr ago. Journal of Quaternary Science 10:175–80.

Mellars, P. 1999. Postscript: Major issues in the interpretation of Star Carr. In Mellars, P., and P. Dark (eds). Star Carr in Context: new archaeological and palaeoecological investigations at the Early Mesolithic site of Star Carr, North Yorkshire. Cambridge: MacDonald Institute Monographs. pp. 215–41.

Pethick, J.S. 1990. The Humber Estuary. In S. Ellis and D.R. Crowther (eds). Humber Perspectives: a region through the ages. Hull: University Press. pp. 54–70.

Van de Noort, R., and P. Davies. 1993. Wetland Heritage: An archaeological assessment of the Humber Wetlands. Hull: Humber Wetlands Project, University of Hull.

Van de Noort, R., and S. Ellis. 1995. Wetland Heritage of Holderness: An archaeological Survey. Hull: Humber Wetlands Project, University of Hull.

Walker, M.J.C., G.R. Coope and J.J. Lowe. 1993. The Devensian (Weichselian) late glacial palaeoenvironmental record from Gransmoor, East Yorkshire, England. Quaternary Science Reviews 12:659–80.

27. On the Conservation of Ships with PEG – The Bremen Cog and Beyond

Per Hoffmann

Introduction

Three magnificent ship finds uncovered within two years, the VASA in Stockholm, the Viking ships in Roskilde Fjord, and the Bremen Cog, started extensive studies into the conservation of waterlogged wood in the grand scale.

For all three ships the conservators chose polyethylene glycol – PEG – as the conservation chemical, although in different applications: The VASA was sprayed for many years with solutions of PEG in water. The Viking ships were dismantled and tank impregnated in a PEG solution. In Bremen the responsible politicians – following the advice of the scientists – decided for a tank treatment of the reassembled ship. This promised to be the safest way to the best possible result.

The medieval Bremen Cog from 1380 is the biggest ship ever that has been sitting in a tank for impregnation. This paper will explain what really is so special about this conservation project, and what impact it has had on the conservation of ship finds and other large wooden constructions, such as houses, wells, and trackways from wetlands.

The conservation of the Bremen Cog

The scientific problem

To identify the wreck of a ship the like of which nobody had seen before, and to salvage it from murky tidal waters in the river Weser was the problem of the historians at the Bremen State Museum (Figure 27.1) (Fliedner 1985, Pohl-Weber 1985). To reassemble the ship of unknown design from 2000 salvaged timbers was the problem of our master shipwright and his small gang at the Deutsches Schiffahrtsmuseum (Figure 27.2) (Lahn 1985). To build a steel tank around the 23-m-ship weighing 30 tons was the problem of our technical director (Figure 27.3). Compared to these three huge problems my problem was rather small. It takes half a million times magnification to make it visible. Figure 27.4 is a scanning electron microscopic image of slightly degraded wood cells in a cross-section. The pits and holes within the cell walls are the result of bacterial attack. I had to get a stabilizing agent into these small voids, and into the still solid cell walls around them.

Archaeological wood destroys itself when it dries, it shrinks, warps, cracks, and collapses (Figure 27.5). When wet wood is excavated, all its pores are filled with water. As water evaporates, the liquid surfaces in the pores and capillaries of the wood structure sink into the wood. These surfaces all have a surface tension pulling at the surrounding wood substance. In degraded wood the weakened cell wall structure cannot always withstand these tension forces, as it does in fresh wood; it shrinks and distorts.

When the degradation process in the wood has proceeded to a state where most of the cell wall substance has been destroyed, digested, and dissolved (Figure 27.6), drying of the wood leads to the total collapse of the structure. This is the worst that can happen, collapse is not reversible. Consequently, the structure of archaeological wood has to be strengthened before drying. And it must dry to sit in a museum or store room.

In the early sixties, when first plans were made for the conservation of the Bremen Cog, PEG was the choice for the chemical to apply for the stabilization of the waterlogged timbers. PEG, however, is produced in various grades, from small molecules, i.e., low molecular weight, to very large molecules, i.e., high molecular weight. The dilemma was: high molecular weight PEG would impact good strength to degraded wood, and it would be only slightly hygroscopic, i.e., only slightly sensitive to high relative humidities of the air. But it could not penetrate into less degraded wood. On

Figure 27.1. The hull of the Bremen Cog appears in the bank of the river Weser, on October 9th, 1962. Photo: Focke-Museum.

Figure 27.2. The fourth strake. Rebuilding the Cog in the Deutsches Schiffahrtsmuseum. Photo: G. Meierdierks/DSM.

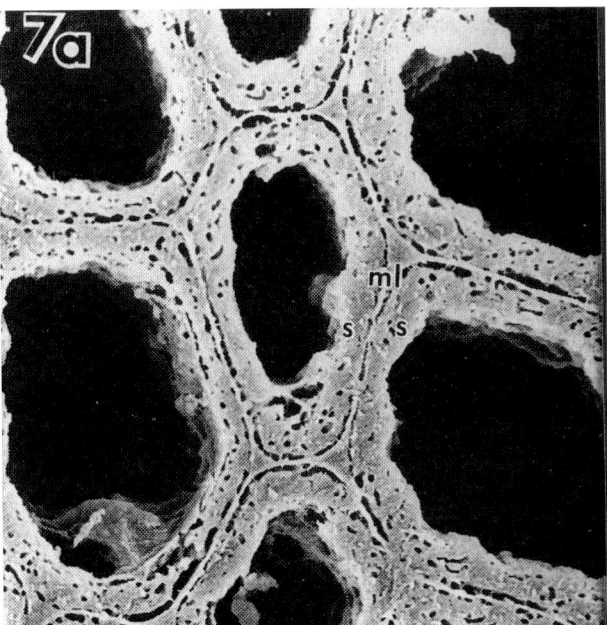

Figure 27.4. Scanning electron micrograph of a cross-section through several wood cells. Minute holes of bacterial degradation can be seen within the cell walls and along the middle lammellae between the cells. Bar = 10 um. Photo: Robert A. Blanchette.

Figure 27.3. Building the conservation tank around the Cog. Photo: Per Hoffmann.

Figure 27.5. A board of wet archaeological wood which has dried without treatment. Photo: Egbert Laska/DSM.

Figure 27.6. Transmission electron micrograph of a cross-section through wood cells. The cell walls are completely destroyed by bacterial degradation, middle lammellae are broken. Photo: Narayan Paramesvaran.

the other hand, low molecular weight PEG would penetrate better into lesser degraded wood, but it would not strengthen highly degraded wood so well, and it would be quite hygroscopic. When I came to our museum to start the conservation of the Cog, I inherited the plan for a compromise: PEG 1000 was to be used, a brand with medium sized molecules and a medium hygroscopicity.

A first pilot scale treatment of some timbers showed it to be a bad compromise: no sufficient stabilization was achieved, and the wood did not dry properly, it stayed damp and sticky. We had to develop a better plan. The situation was further complicated. We could not separate timbers according to their state of degradation, most timbers contained tissue of various degrees of degradation. A layer of highly degraded wood enveloped an inner portion of only slightly degraded wood. Typically in oakwood, the microbial attack had proceeded with a distinct front from the outside towards the inner parts of the timber. An effective conservation treatment would have to stabilize both widely different types of degraded tissue.

Laboratory tests

The first step was to optimize the stabilization of tissues of different degrees of degradation, to find the right PEG for each wood quality. The results derived from dozens of impregnation series were interesting and pleasing: the low molecular weight PEGs 200 and 400 have very good stabilizing properties for only slightly degraded wood with maximum water contents up to 250%; on more degraded wood their effect diminishes. The high molecular weight PEG 3000 is well suited to stabilize heavily degraded wood with maximum water contents higher than 250%. On less degraded wood this PEG has very little to no effect (Hoffmann 1984).

The PEG grades with medium size molecules, PEG 600 and 1500, had the least stabilizing properties, especially PEG 1500 had no convincing effect on any wood quality. This was an astonishing result having in mind the rather widespread use of this PEG in conservation labs at that time. Pleasing, however, was the evidence that obviously only two PEGs, one with small molecules and one with large molecules, would suffice to stabilize the whole range of degraded woods.

The next step was to develop and optimize a two-step treatment, and this is what is special about the conservation of the Bremen Cog (Hoffmann 1986, 1990). PEG 200 is a liquid; it is the PEG with the smallest molecules that will not evaporate (Hoffmann 1988). PEG 3000 is a wax-like solid; it is the PEG with smallest molecules that still has the properties of a typical high molecular PEG: hardness and low hygroscopicity.

The two PEGs are applied separately. The wood is first impregnated with PEG 200: the small molecules rapidly diffuse through heavily degraded tissue and into the cell walls of less degraded wood (Hoffmann 1984, Young and Sims 1987). Here they replace a large part of the swelling water. When the wood dries, the residual water will evaporate; but the PEG stays in the cell wall ultrastructure, and thus keeps it and the whole wood structure in a permanently swollen state. The piece of wood retains its form and dimensions.

In a second bath, in a hot solution of PEG 3000, the large molecules of this PEG diffuse into the heavily degraded tissue. At the same time superfluous PEG 200 diffuses out of this tissue. When the impregnation is ended and the wood cools down, the PEG 3000 sets and crystallizes, forming an inner supporting structure in the frail tissue. The PEG 3000 strengthens the tissue and stabilizes it against shrinkage and collapse when the small amounts of residual water evaporate after treatment. Because the low molecular PEG 200 has left the outer layers of the wood, the wood is not more hygroscopic than the PEG 3000; no harm will occur at relative humidities below 80 %. The big molecules of PEG 3000 cannot enter slightly

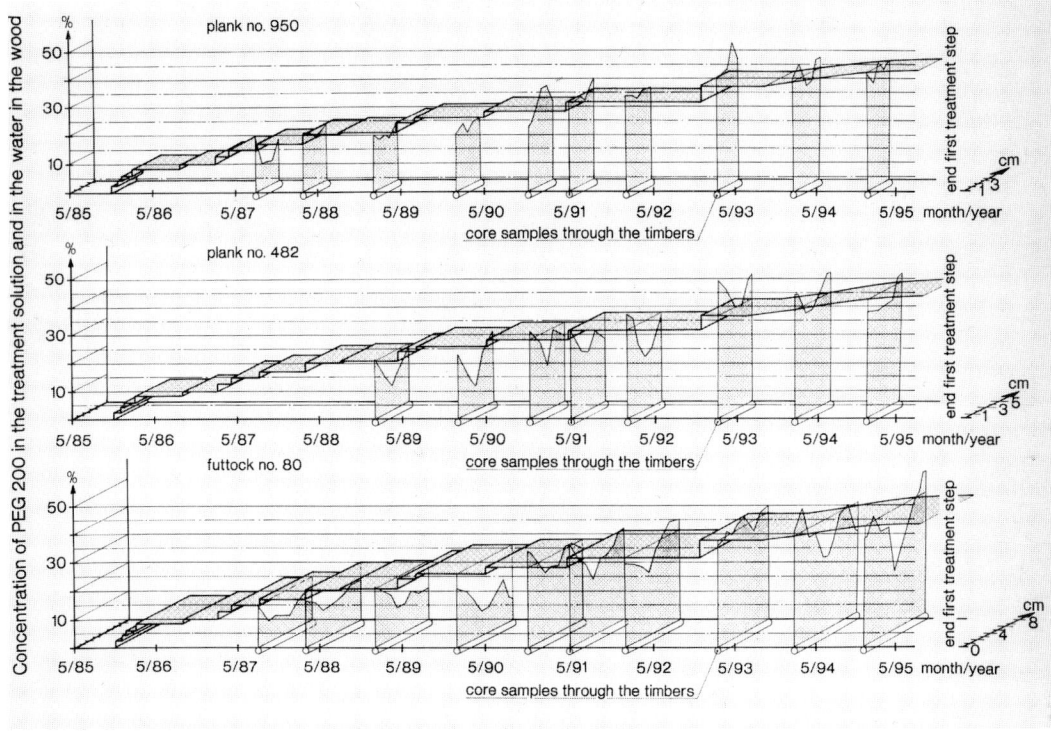

Figure 27.7. Development of the PEG 200 concentration in the treatment solution of the first impregnation step, and in the water in core samples taken at intervals from the same three timbers.

degraded tissue, but it is not necessary to get it there. There the PEG 200 serves the purpose.

In the two-step PEG-treatment two different mechanisms stabilize the two different ranges of wood qualities. The two PEGs are applied in separate baths because a mixed bath would render the outer layer of the wood filled with a soft and sticky PEG-paste that would not become hard and dry.

The exciting and precarious side of the Cog project was the scaling-up from small lab samples to the 23 m ship in a treatment tank of 800,000 litres.

Scaling-up

Before the board of our museum agreed to change the original plan to a two-step PEG-treatment with PEG 200 and PEG 3000, it had the laboratory work evaluated by an international group of leading waterlogged wood conservation scientists. They all recommended to go along with this novel method. Up to today the Cog project is a full-size conservation experiment. But before we have even come to its end, colleagues around the world have already adopted the new two-step method.

The treatment tank built around the Cog, is closed on top. We filled it with water, and over 15 years added PEG 200 to a final concentration of 40%. The impregnation time for the first step could have been much shorter, 2–4 years,

but in this case our budget for buying PEG was the limiting factor. In the basement of our museum a series of huge gravel filters served to remove the constant growth of microbes, which we continuously precipitated with chemical flocculants. The tank had a row of windows. Visitors were supposed to be able to view the ship in its tank, illuminated by underwater floodlights, but the microbial growth threatened to blur the vision.

Two times a year we analysed core samples taken from representative timbers for PEG-penetration, and established PEG-concentration profiles through the timbers (Figure 27.7) (Hoffmann 1987). When the PEG-concentration in the wood had about reached that of the bath, the impregnation was sufficient and could be terminated.

It took a good deal of discussions, experiments, and expert's opinion to convince our local water authorities that PEG is biologically degradable, and that we could be allowed to drain the used PEG solution into the sewage system – very slowly. It took 3 months to pump off the 800 tons solution. Figure 27.8 shows the degradation curves for solutions of PEG 200 and PEG 3000, fed to the active sludge of the biological waste water treatment plant. These curves saved us from spending two million Deutschmarks on having the used solutions of both treatment steps burned as "special waste".

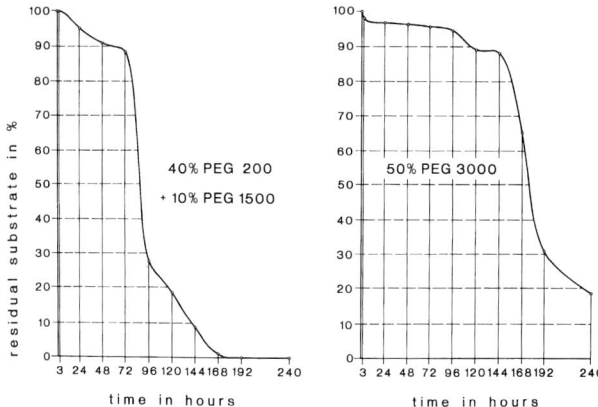

Figure 27.8. Degradation of PEG in conservation solutions containing 40% PEG 200 + 10% PEG 1500 (left), and 50% PEG 3000 (right). The solutions were inoculated with the "bacterial mix" from the biological waste water treatment plant Bremerhaven in a standard test procedure. (Limnological Institute Dr. Nowak, Ottersberg).

Standing inside the Cog for the first time after 15 years we had almost forgotten how beautiful she was.

The second treatment step was more complicated to arrange. PEG 3000 is a solid at room temperature. To avoid having to dissolve 290 tons of PEG in water to a 60% solution – a task we could not imagine how to effect – we ordered the PEG as a hot melt of 90 °C. It came in heated tank lorries, two each week for 7 weeks, from Bavaria, 800 km from Bremerhaven. It was mid winter and freezing heavily, and sometimes the PEG was only 65 °C hot when it arrived, 5 – 8 degrees only above solidification temperature. It threatened to congeal in the pipeline into the museum.

There we had rigged a mixing station: a pump pressed the molten PEG, and water from a fire hydrant through a static mixer and into the tank with the Cog. A static mixer is a tube into which protudes a series of specially arranged lammellae. These produce a heavy turbulence, and two or more components fed into the tube are thoroughly mixed within a very short length of only decimeters. The solution was heated during the whole second treatment step to keep it liquid.

On the days between PEG-deliveries we packed displacement bodies into the void spaces of the tank: under the ship, and into the hull. We had to save money, and tried to reduce the amount of PEG needed. The cheapest means, we found out, were elastomer-coated trevira sacks or balloons, made to our order. We had them in sizes of 1.5, 3.7, and 7 m³ volume, and we filled 160 balloons with 400 m³ of freshly prepared salt water. The density of the salt water was chosen to match that of the PEG solution: the sacks would just sink down without pressing too hard onto the ship. Nor would they swim up and press against the lid

of the tank. We used 80 tons of table salt. The balloons were laid out in the Cog and filled with salt water at the same rates as the level of the PEG solution rose. In the end we succeeded in reducing the active volume of the tank to half its volume.

After one year we removed part of the balloons, and mixed in more molten PEG 3000 to increase the bath concentration to the final value of 72%. The laboratory experiments had shown that this gave the best dimensional stabilization. Higher concentrations led to somewhat increased residual shrinkage on drying, probably due to the contraction of the PEG on solidifying.

After another two years the analysis of the PEG-distribution in the timbers confirmed our expectations: the interior of the timbers still contained PEG 200, while the outer wood layers were filled with PEG 3000. Only small amounts of PEG 200 were left in these outer layers, just enough to slightly reduce the brittleness of PEG 3000 (Figure 27.9).

At this stage we ended the impregnation of the Cog. In accordance with the waste water authority we pumped off the PEG solution at a rate given daily by the actual efficiency of their biological treatment plant, 0–500 l/hour. Again it took three months to drain the tank.

The Cog then sat in the empty tank for 6 months, slowly drying in controlled high relative humidity. The tank is

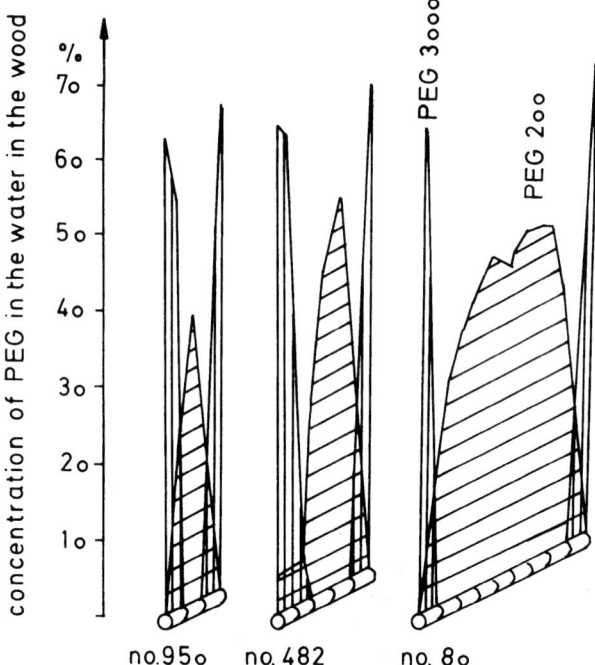

Figure 27.9. PEG-profiles through three timbers at the end of the second impregnation step.

being dismantled just now, and then we will have to clean the ship from adhering PEG.

It is too early to say something about the stabilization result of this full size experiment. The timbers will take some years to dry out, and it is during these years that residual shrinkage and warping might occur. The excitement is yet not over.

Beyond the Cog

The results from the experimental work undertaken to develop a treatment for the Bremen Cog open up a new perspective onto the use of PEG. Taylormade stabilization programs can be fashioned to suit individual cases:

A 20 m cargo ship from the Teufelsmoor near Bremen, with timbers only slightly degraded up to their surfaces was successfully treated with only PEG 200. And as the optimal concentration of 40% PEG 200 has a low viscosity we sprayed the solution onto the ship. Only a few cubic metres of solution were needed in a circulation system composed of simple standard garden sprinklers and hoses connected to a submersible pump, which sent the recollected solution back into the system (Hoffmann 1998). At the moment two 18th century barges from the river Weser, full of sandstone building blocks, are being spray-treated in Lemgo/Germany (Figure 27.10).

A log boat coffin of thoroughly degraded wood from the 4th century grave yard at Wremen near Bremerhaven was best stabilized in a heated bath of PEG 3000. Carrying the concentration to the optimal 70% instead of the traditional 100 % saved money and time.

Even a boat composed of multiquality wood may not always require a two-step treatment as perfect and costly as that for the Bremen Cog. I am investigating the possibilities of compromises: what results does one get when the PEG concentration of the second treatment step is less than the optimal 70%? A 60% PEG 3000 solution stays liquid at room temperature. One would not need to heat the second bath – that would save a lot of money. And without heating no sophisticated conservation tank is necessary; a hole in the ground will do. First results with a 6 m dug-out canoe excavated from the river Leine at Helstorf/Germany, and treated this way are quite satisfying (Figure 27.11). A year after the end of the very cheap back-yard treatment no cross-grain cracks have developed in the surfaces, and only very few along the fibre. The boat has not twisted nor warped.

There are other methods to treat waterlogged ships and timbers. Freeze-drying is one, if you have access to a freeze-dryer large enough to accommodate your ship. Impregnation with sucrose is advocated as cheap, simple, and effective. I have successfully conserved two ships with

Figure 27.10. Two 18th-century river barges from the Weser are being spray-treated with PEG 200 in a temporary greenhouse at Lemgo/Germany. Photo: Gabriele Hoffmann.

Figure 27.11. A log-boat having been treated with a cold two-step PEG-treatment in a plastic-lined hole in the backyard. Photo: Diethardt Hensel.

sucrose, the one a 12 m Renaissance cargo vessel from Friesland, but I would not volunteer to do it again (Hoffmann and Kühn 1998). It is not simple, and the result is somewhat unpredictable. You will need an expert crew of conservators, microbiologists and chemists keeping a close watch on your tank all the time.

For me, modern PEG treatments are first choice when planning the conservation of a large wooden object. The possibilities to design variations for individual cases, and the potentials for technical simplifications make PEG treatments attractive: They are successful and easy to handle, often simple, and even financially bearable.

Figure 27.12. Two cogs engaged in battle. Illustration in an early 14th century English manuscript. From: Asaert et al. 1976.

Today no ship finds, log coffin, or carved idol needs to vanish into hearsay because no one knew what to do with it. Wet wood conservation is standard. And it is worth while to conserve a boat or ship. Boats make extraordinary museums exhibits: they provide many kinds of information from their time, on technical knowledge and skills, on fishing, trade and transport possibilities, on esthetics. They often are a focus for a lost world (Figure 27.12).

References Cited

Asaert, G., J. van Beylen, and H.P.H. Jansen. (eds). 1976. Maritieme geschiedenis der Nederlanden. De Boer Maritiem, Bussum. Vol. 1, p. 256. Illustration from the 'decretes' of Gregorius IX, British Museum London, ms Roy, 10 E IV, fol. 19.

Fliedner, S. 1985. The find of the century in the Weser river. Bremen's Hanse Cog: Discovery and identification. In: The Hanse Cog of 1380, U. Schnall and K.-P. Kiedel (eds). Bremerhaven, pp. 7–14.

Hoffmann, P. 1984. On the Stabilization of Waterlogged Oakwood with PEG – Molecular Size versus degree of degradation. In: Proc. 2nd Waterlogged Wood Working Group Conference, Grenoble, pp. 95–115.

Hoffmann, P. 1986. On the Stabilization of Waterlogged Oakwood with PEG. II. Designing a Two-Step Treatment for Multi-Quality Timbers, Studies in Conservation 31, pp. 103–113.

Hoffmann, P. 1990. On the Stabilization of Waterlogged Softwoods with Polyethylene Glycol (PEG). Four species from China and Korea. Holzforschung 44, pp. 87–93.

Hoffmann, P. 1988. On the Stabilization of Waterlogged Oakwood with Polyethylene Glycol (PEG). III Testing the oligomers. Holzforschung 42, pp. 289–294.

Hoffmann, P. 1987. HPLC for the analysis of PEG in wood. In: Conservation of wet wood and metal, Proc. of the ICOM Groups on Wet Organic Archaelogical Materials and Metal Conference, I.D. MacLeod (ed.), Fremantle, pp. 41–58.

Hoffmann, P. 1998. Das Frachtschiff aus dem Teufelsmoor. Arbeitsblätter für Restauratoren 31, no. 2, pp. 270–276.

Hoffmann, P. and H.J. Kühn. 1998. The candy ship from Friesland. In: Proc. 7th ICOM Group on Wet Organic Archaeological Materials Conference, P. Hoffmann, C. Bonnot-Diconne, X. Hiron, Q.K. Tran (eds), Grenoble, pp. 196–203.

Lahn, W. 1985. A cog is built in the 20th century. Reconstruction of the Cog in the Deutsches Schiffahrtsmuseum. In: The Hanse Cog of 1380, U. Schnall and K.-P. Kiedel (eds), Bremerhaven, pp. 28–41.

Pohl-Weber, R. 1985. Underwater Archaeology on the Weser Bed. Salvaging the Hanse Cog. In: The Hanse Cog of 1380, U. Schnall and K.-P. Kiedel (eds), Bremerhaven, pp. 15–24.

Young, G.S. and Sims, R. 1987. Microscopical determination of PEG in treated wood – the effect of distribution on dimensional stabilization. In: Conservation of wet wood and metal, Proc. of the ICOM Groups on Wet Organic Archaeological Materials and Metals Conference, Fremantle, I.D. MacLeod (ed.), pp. 109–140.